CARBON MONOXIDE TOXICITY

CARBON MONOXIDE TOXICITY

Edited by
David G. Penney

CRC Press
Taylor & Francis Group
Boca Raton London New York

CRC Press is an imprint of the
Taylor & Francis Group, an **informa** business

CRC Press
Taylor & Francis Group
6000 Broken Sound Parkway NW, Suite 300
Boca Raton, FL 33487-2742

First issued in paperback 2019

ISBN-13: 978-0-8493-2065-1 (hbk)
ISBN-13: 978-0-367-39855-2 (pbk)

This book contains information obtained from authentic and highly regarded sources. While all reasonable efforts have been made to publish reliable data and information, neither the author[s] nor the publisher can accept any legal responsibility or liability for any errors or omissions that may be made. The publishers wish to make clear that any views or opinions expressed in this book by individual editors, authors or contributors are personal to them and do not necessarily reflect the views/opinions of the publishers. The information or guidance contained in this book is intended for use by medical, scientific or health-care professionals and is provided strictly as a supplement to the medical or other professional's own judgement, their knowledge of the patient's medical history, relevant manufacturer's instructions and the appropriate best practice guidelines. Because of the rapid advances in medical science, any information or advice on dosages, procedures or diagnoses should be independently verified. The reader is strongly urged to consult the relevant national drug formulary and the drug companies' and device or material manufacturers' printed instructions, and their websites, before administering or utilizing any of the drugs, devices or materials mentioned in this book. This book does not indicate whether a particular treatment is appropriate or suitable for a particular individual. Ultimately it is the sole responsibility of the medical professional to make his or her own professional judgements, so as to advise andtreat patients appropriately. The authors and publishers have also attempted to trace the copyright holders of all material reproduced in this publication and apologize to copyright holders if permission to publish in this form has not been obtained. If any copyright material has not been acknowledged please write and let us know so we may rectify in any future reprint.

Library of Congress Cataloging-in-Publication Data

Carbon monoxide toxicity / edited by David G. Penney.
    p. ; cm.
    Includes bibliographical references and index.
    ISBN-13: 978-0-8493-2065-1 (alk. paper)
    ISBN-10: 0-8493-2065-8 (alk. paper)
    I. Penney, David G.
    [DNLM: 1. Carbon Monoxide Poisoning. 2. Carbon Monoxide--metabolism.
    3. Environmental Pollutants--poisoning. QV 662 C2647 2000]

    RA1247.C2 C396 2000
    615.9'1--dc21   00-022404

Visit the Taylor & Francis Web site at
http://www.taylorandfrancis.com

and the CRC Press Web site at
http://www.crcpress.com

Preface

Carbon Monoxide Toxicity is a new book covering areas of the expansive field involving carbon monoxide (CO) that are not covered in the first book, *Carbon Monoxide*, edited by David G. Penney, Ph.D., and published by CRC Press in 1996. This book is designed to be complementary to the 1996 book, forging into new areas and following new themes.

The scope of this book is broad, encompassing a wide sweep of topics from the history of CO studies, to CO detector technology, to clinical management of CO poisoning around the world, to our emerging understanding of the largely hidden threat of chronic low-level CO exposure to individuals as well as populations. While many chapters present new basic science information, the major thrust of the book is directed toward the clinical (medical, psychiatric, neuropsychological) management of various forms of CO poisoning. A major goal is to provide physicians with a better understanding of the condition and the tools currently available to manage their patients successfully. Aside from the 1996 CRC book, *Carbon Monoxide*, no other book of its type has been published in more than 20 years.

The book contains some unique features:

1. A firsthand look at the setting of health regulations regarding CO in our environment and the remarkable gains made possible by these regulations over the past several decades in the United States.
2. A view of the various approaches to managing CO poisoning around the world.
3. The rapid realization of the threat of chronic low-level and ultralow-level CO exposure to health in the developed and developing countries.
4. A discussion of both high-tech scanning and neuropsychological tools in the diagnosis of damage from CO poisoning.
5. A look at efforts to understand the role of CO in suicide and motor vehicle accidents around the world, and at recent attempts to curtail both.
6. Approaches to understanding the role of CO and other associated gases in the morbidity and mortality of fire situations.

Interest in the impact of CO on human health has grown rapidly during the past decade. This is exemplified by the increased use of CO detectors in public buildings, homes, pleasure boats, and aircraft. The public and the medical community need to obtain quality information about the risks of CO and need the means to manage victims of CO poisoning successfully. As it is clear that this trend will continue into the foreseeable future, it is hoped that this book will in some small way meet these needs.

About the Editor

David G. Penney, Ph.D., is professor of physiology in the School of Medicine and adjunct professor of occupational and environmental health in the School of Allied Health Professions, at Wayne State University in Detroit, Michigan. He is also director of general surgical research at Providence Hospital in Southfield, Michigan, where he directs the scholarly activities of surgical residents and attending staff surgeons.

Dr. Penney obtained his B.Sc. degree from Wayne State University in 1963, and his M.Sc. and Ph.D. degrees from the University of California, Los Angeles, in 1966 and 1969, respectively. Before coming to Wayne State University in 1977, he was a faculty member at the University of Illinois, Chicago.

Dr. Penney's current professional interests are focused on carbon monoxide poisoning, with a special interest in chronic carbon monoxide poisoning, the effects of carbon monoxide on the young and the unborn, and the medical and legal aspects of carbon monoxide toxicology.

During the past 4 years, Dr. Penney has increasingly turned his energy to information technology and the interface of medical and public education and various electronic media. He has participated in major projects in this area including seminal work in the development of the virtual classroom as a World Wide Web-based learning resource for medical students and others. He has also pioneered the development of an expert site on carbon monoxide information, known as Carbon Monoxide Headquarters. It is available at: *http://www.phymac.med.wayne.edu/facultyprofile/penney/ COHQ/CO1.htm.*

Dr. Penney's published work includes more than 100 research papers, 10 or more review articles, various book chapters, and several books in print.

Contributors

Pierre Baume, M.S.
Associate Professor
Central Sydney Area Health Service
Sydney University
Sydney, New South Wales, Australia

Vincent Brannigan, J.D.
Professor
Department of Fire Protection
  Engineering
University of Maryland
College Park, MD

Qing Chen, M.D.
Professor
Division of Environmental Health
School of Public Health
Beijing Medical University
Beijing, People's Republic of China

I.S. Saing Choi, M.D.
Professor
Department of Neurology
Yonsei University College of Medicine
Seoul, Korea

Debbie Davis
Chairman
CO Support (a registered charity)
Leeds, Yorkshire, England

Albert Donnay, M.S.
Graduate Student
Environmental Health Engineering
Johns Hopkins University
Baltimore, MD

Thomas H. Greiner, Ph.D., P.E.
Associate Professor
Agricultural and Biosystems
  Engineering
Iowa State University
Ames, IA

Martin R. Hamilton-Farrell, M.D.
Director
Hyperbaric Unit
Whipps Cross Hospital
London, England

Neil B. Hampson, M.D.
Director
Hyperbaric Medicine
Virginia Mason Medical Center
Seattle, WA

Alistair W.M. Hay, Ph.D.
Reader
Division of Clinical Sciences and
  Molecular Epidemiology Unit
Old Medical School
University of Leeds
Leeds, England

Milan J. Hazucha, Ph.D.
Professor
Pulmonary Division
Department of Medicine
Center for Environmental Medicine and
  Lung Biology
The University of North Carolina
Chapel Hill, NC

Dennis A. Helffenstein, Ph.D.
Neuropsychologist
CRC, Colorado Neuropsychological
  Associates
Colorado Springs, CO

John Henry, M.D.
Professor
Department of Accident and Emergency
St. Mary's Hospital
London, England

Susan Jaffer, Ph.D.
Consultant
CO Support (a registered charity)
Leeds, Yorkshire, England

Richard Kwor, Ph.D.
Professor
Department of Electrical and
  Computer Engineering
University of Colorado
Colorado Springs, CO

Daniel Mathieu, M.D.
Director
Poison Control Centre and
  Critical Care Service
Lille Regional University Hospital
Lille, France

Monique Mathieu-Nolf, M.D., Ph.D.
Director
Poison Control Centre and
  Critical Care Service
Lille Regional University Hospital
Lille, France

James J. McGrath, Ph.D.
Professor
Department of Physiology
School of Medicine
Texas Tech University
  Health Sciences Center
Lubbock, TX

Robert D. Morris, M.D., Ph.D.
Professor
Department of Community Health
Tufts University School of Medicine
Boston, MA

Frederick W. Mowrer, Ph.D., P.E.
Professor
Department of Fire Protection
  Engineering
University of Maryland
College Park, MD

Janusz Pach, M.D., D.Sc.
Professor
Department of Clinical Toxicology
Collegium Medicum
Jagiellonian University
Krakow, Poland

Dieter Pankow, M.D., D.Sc.
Professor
Martin Luther University
Faculty of Medicine
Institute of Environmental
  Toxicology
Halle (Saale), Germany

David G. Penney, Ph.D.
Professor
Dept. of Physiology
Wayne State University
  School of Medicine
Detroit, MI

David A. Purser, Ph.D.
Professor
Centre for Fire Protection Systems
Fire Research Station
Building Research Establishment
Watford, Hertfordshire, England

James A. Raub, Ph.D.
Health Scientist
National Center for Environmental
  Assessment
U.S. Environmental Protection Agency
Research Triangle Park, NC

Charles V. Schwab, Ph.D.
Associate Professor
Agricultural and Biosystems
  Engineering
Iowa State University
Ames, IA

Michaela Skopek, M.D.
Consultant Psychiatrist
Prince of Wales Hospital
Woollahra (Sydney), New South Wales,
  Australia

Jerzy A. Sokal, Ph.D., D.Sc.
Professor
Institute of Occupational Medicine and
  Environmental Health
Sosnowiec, Poland

David K. Stevenson, M.D.
Professor
Department of Pediatrics
  Division of Neonatal and
  Developmental Medicine
Stanford University School of Medicine
Stanford, CA

Hendrik J. Vreman, Ph.D.
Senior Research Scientist
Department of Pediatrics
  (Neonatology Division)
Stanford University School of Medicine
Stanford, CA

Suzanne R. White, M.D.
Assistant Professor
Department of Emergency Medicine
Wayne State University
  School of Medicine
Detroit, MI

Lihua Wang, M.D.
Division of Environmental Health
School of Public Health
Beijing Medical University
Beijing, People's Republic of China

Ronald J. Wong
Research Assistant
Department of Pediatrics
  Division of Neonatal and
  Developmental Medicine
Stanford University School of Medicine
Stanford, CA

Table of Contents

Acknowledgment

I wish to thank my wife, Linda Mae Penney, for her generous help with all aspects of the development of this book. I also wish to thank so many unnamed people who have responded to my Web site with their praise, criticism, and the enthusiasm to volunteer their experiences for inclusion in ongoing CO studies.

David G. Penney

Dedication

I wish to dedicate this book first to my father,
George Donald Penney, a fire captain in Royal Oak, Michigan,
who, when I was a boy, warned me of the dangers of
carbon monoxide. I also dedicate this book to my children,
Loren David, Morgan Donald, Elizabeth Caroline,
and Hannah Vera.

1 History of Carbon Monoxide Toxicology

Dieter Pankow

CONTENTS

1.1 INTRODUCTION

The significance of carbon monoxide (CO) poisonings in this century has not changed, as illustrated by the similarity of remarks by Lewin[1] at the beginning of the century and by Krenzelok et al.[2] at its end:

> Lewin in 1920: "Carbon monoxide causes more health problems and death every year than any other poison or many together" (originally: "Das Kohlenoxyd heischt alljährlich mehr Opfer an Gesundheit und Leben als irgend ein anderes Gift oder viele zusammen").
>
> Krenzelok, et al. in 1996: "Carbon monoxide is responsible for more poisoning fatalities every year than any other toxic agent."

The main clinical manifestations of CO poisoning, such as malfunctions of the cardiovascular system with initial tachycardia and increase in blood pressure, and the effects on the nervous system, with headache, dizziness, myasthenia, paresis, convulsions, and changes in perception, including changes in the visual and auditory system, were reviewed by Lewin[1] (Figure 1.1). Pneumonia and lung edema were

FIGURE 1.1   Louis Lewin (1850–1929), toxicologist and ethnopharmacologist in Berlin, Germany. (Courtesy of Bo Holmstedt, Department of Toxicology, Karolinska Institute, Stockholm, Sweden; Holmstedt[3] characterized the course of Lewin's life.)

observed and the involvement of liver, kidney, and spleen was detected in cases of CO poisoning, but these organs were not regularly affected. Lewin[1] described the important diagnostic procedures such as the determination of CO, lactate, and glucose in the blood, and the importance of excluding alcohol or drug poisoning. Nevertheless, understanding the basic principles of the toxicokinetics and toxicodynamics of CO has advanced considerably since Lewin's work. The diagnosis, treatment, and sequelae of acute CO poisoning are adequately dealt with, and strategies to prevent an impairment of health due to exposure to low CO concentrations have been developed since Lewin's work.

There are several reasons to examine the history of CO toxicology. First, there are a very large number of publications. The review by Lilienthal[4] used over 3000, and the EPA *Air Quality Criteria for Carbon Monoxide* (Table 1.1) refers to a total of 1584 sources. Some important monographs and books on the toxicology of CO are listed in Table 1.1. Second, the classic book by Lewin,[1] the title of which is given in Figure 1.2, contains a chapter on the history of CO poisoning. Lewin showed that the history of CO poisoning is closely associated with the history of civilization. Third, there is a general tendency to neglect the older literature today, and this seems

DIE
KOHLENOXYDVERGIFTUNG

EIN HANDBUCH

FÜR MEDIZINER, TECHNIKER UND UNFALLRICHTER

VON

PROF. DR. L. LEWIN

MIT EINER SPEKTRENTAFEL

BERLIN

VERLAG VON JULIUS SPRINGER

1920

FIGURE 1.2   Title of Lewin's monograph on carbon monoxide toxicology.

not to be justified in every case. Lewin stressed that these times urgently need the benefit of the full, as well as the old, knowledge on all the effects of CO.

1.2   EXPERIENCE OF CARBON MONOXIDE POISONING WITHOUT KNOWLEDGE OF THE POISON

CO is produced by the incomplete combustion of almost any carbon-containing material, such as coal, wood, peat, dry grass, or other fossil fuels. What happened in prehistoric times is not known through written tradition, but excavations have shown that primitive humans discovered the art of making and utilizing fire. In all settlements of the Paleolithic period traces of fires have been detected in the form of ashes, coal, or carbonized wood. Animal bones with singed spots have been found, demonstrating the preference of primitive humans for eating animal meat in a roasted state. Fireplaces have been uncovered and in the cold season tools and weapons as well as ornaments were produced near the fire.[5] At least since that period humankind has experienced the toxic effects of CO. Exposure occurred when humans were exposed to the gas from fires in inadequately ventilated cave dwellings, or when they were confronted with forest fires. This experience should be combined with a lack of understanding, and with mythologizing: the earliest humans could differentiate between harmful and safe substances because animal venoms and toxic plant

TABLE 1.1

Books and Monographs on Carbon Monoxide Toxicology Published in the 20th Century

Year	Author(s)	Title	Publisher
1920	Lewin, L.	*Die Kohlenoxydvergiftung. Ein Handbuch für Mediziner, Techniker und Unfallrichter*	Springer, Berlin
1925	Nicloux, M.	*L'Oxyde de Carbone et l'Intoxication Oxycarbonique*	Masson, Paris
1938	Drinker, C.K.	*Carbon Monoxide Asphyxia*	Oxford University Press, New York
1942	Flandin, C. and Guillemin, J.	*L'Intoxication Oxycarbonee*	Masson, Paris
1944	von Oettingen, W.F.	*Carbon Monoxide: Its Hazards and the Mechanisms of Its Action*	U.S. Government Printing Office, Washington, D.C.
1949	Grut, J.	*Chronic Carbon Monoxide Poisoning*	Munksgaard, Copenhagen
1950	Raymond, V. and Vallaud, A.	*A Study in Occupational Medicine L'Oxyde de Carbon et L'Oxycarbonisme*	Institute National de Securite, Paris
1966	Cooper, A.G.	*Carbon Monoxide. A Bibliography with Abstracts*	U.S. Government Printing Office, Washington, D.C.
1967	Bour, H. and Ledingham, I.M., Eds.	*Carbon Monoxide Poisoning*	Elsevier, Amsterdam
1967	Gras, G.	*L'Intoxication Oxycarbonee Aigue et Ses Manifestations Cardiovasculaires*	Masson, Paris
1970	Bouletreau, P.	*L'Intoxication Aigue par l'Oxyde de Carbone*	Masson, Paris
1977	Coburn, R.F., Allen, E.R., Ayres, S.M., Bartlett, D., Ferrand, E.F., Hill, A.C., Horvath,S.M., Kuller, L.H., Laties, V.G., Longo, L.D., and Radford, E.P.	*Medical and Biological Effects of Environmental Pollutants. Carbon Monoxide*	National Academy of Sciences, Washington, D.C.
1979	Buchwald, H., Cizikov, V.A., Haak, E., Iordanidis, P., Ishikawa, K., Kodat, V., Kurppa, K., Lawther, P.J., McDonald, I.R.C., and Winneke, G.	*Environmental Health Criteria 13: Carbon Monoxide*	WHO, Geneva
1980	Tiunov, L.A. and Kustov, V.V.	*Toksikologia Okisi Ugleroda*	Medizina, Moscow
1981	Pankow, D.	*Toxikologie des Kohlenmonoxids*	Volk & Gesundheit, Berlin
1983	Shepard, R.J.	*Carbon Monoxide: The Silent Killer*	Charles C. Thomas, Springfield, IL
1990	Jain, K.K.	*Carbon Monoxide Poisoning*	Warren H. Green, St. Louis, MO
1991	Raub, J.A. and McMullen, T.B., Eds.	*Air Quality Criteria for Carbon Monoxide*	Environmental Protection Agency, Research Triangle Park, NC
1996	Penney, D.G., Ed.	*Carbon Monoxide*	CRC Press, Boca Raton, FL

extracts were used for hunting and also for warfare. But how could the people of the Stone Age perceive a colorless, odorless, and tasteless gas?

1.3 DESCRIPTION OF CARBON MONOXIDE POISONING WITHOUT KNOWLEDGE OF THE MECHANISMS OF ACTION

The oldest reference to the toxic effects of CO may be from Aristotle (384–322 B.C.).[1] He described how coal fumes led to a heavy head and death. In ancient times coal fumes were used for execution. This was done by locking the victims in bathing rooms with smoldering coals. In another way, tree trunks were set up and the condemned persons were tied up from top to bottom and a fire was lit below to kill them by the flames, the fumes, or anxiety. The Greek physician Galen (A.D. 129–199) supposed even at that time that the changed composition of air containing the fumes of charcoal may be harmful after inhalation.[1]

Recently, research into all the texts of the Byzantine historians from the 4th to the 14th century revealed descriptions of the apparent CO poisonings of the emperors Julian the Apostate and his successor, Jovian, in A.D. 363 and 364, respectively.[6] They suffered as a result of the burning of coal in braziers, a usual method of indoor heating in Byzantium. As a consequence of the increase of exposure to coal fumes in the 16th century, symptoms and signs of poisoning including dizziness, nausea, vomiting, depressed heart action, coma, and bright-red color of the mucous membrane and the skin were described by, for example, the physician Marcellus Donato in 1588.[1]

In 1700, Bernardino Ramazzini (1633–1714) published the book *De Morbis Arteficium Diatriba*, which was translated from the Latin original text into Italian, French, German, English, and Dutch. He stressed that the exposure of craftsmen or artists to the fumes of fires of charcoal or mineral coal was often the cause of disease or death.[7] In 1775, Harmant in Nancy, France[8] described the symptoms, therapy, and pathology of poisonings following exposure to coal fumes. It was clear that the inhalation of fresh air has healing effects. Bloodletting was an accepted method of treatment. Nardius in 1656 mentioned the beneficial effect of acetic acid vapor[1] and, in 1981, Lee et al.[9] presented data that this therapy in CO-poisoned patients reduces the time required for regaining consciousness.

Numerous accounts of tragic events have been related in folklore and mythology. Friedrich Hoffmann[10] (1660–1742), from Halle, Germany delivered a medical expert report on the deadly fumes of charcoal: on Christmas night of 1715 two farmers and a student held a kind of seance. In a hut they conjured up spirits to assist them in a treasure hunt. They used an open charcoal fire. The farmers died, and the student showed signs of severe poisoning. Public opinion and that of theologians was that it was an act of divine providence using the devil against the three men for their blasphemy. Hoffmann explained the event as a severe poisoning. The poison was a component of the charcoal fumes. Some other examples of poisonings due to coal fumes were mentioned.

Regulations for protection against coal gas poisoning already existed in the General Prussian State Laws of 1794.[1] Paragraph 731 states, "The careless use of coal in closed rooms, the fumes of which can be dangerous to persons within the rooms, is, even when no harm has occurred, punishable by a fine of 3 to 10 taler or random imprisonment" (originally: "Der unvorsichtige Gebrauch der Kohlen in verschlossenen Gemächern, wo der Dampf den darin befindlichen Personen gefährlich werden könnte, ist, wenn auch kein Schaden geschehen wäre, mit drey bis zehn Thaler Geld oder willkürlicher Gefängnisstrafe zu ahnden").

1.4 DESCRIPTION OF CARBON MONOXIDE POISONING WITH KNOWLEDGE OF THE MECHANISMS OF ACTION

1.4.1 ASPECTS OF MECHANISMS OF CARBON MONOXIDE EFFECTS AND THERAPY

CO gas was detected by de Lassone in 1776 by annealing zinc oxide with coal.[1] Priestley (1733–1804) described some properties of CO in 1779 and the chemical structure was clarified by Dalton and Henry some years later. As the chemical industry, metallurgy, and the gasification of coal developed during the 19th century, pollution and poisoning from CO under working conditions increased. In 1842, LeBlanc identified CO as the toxic component of coal fumes,[1] introducing a new period of research in the field of CO toxicology. Aspects of the mechanisms of CO effects were first described in 1857 by Bernard[11] and, independently, by Hoppe-Seyler (at that time under the name Hoppe).[12] Bernard pointed out that CO produces hypoxia by its combination with hemoglobin. This tight binding of CO to hemoglobin, which Bernard originally thought to be irreversible, causes a decrease in the oxygen-carrying capacity of the blood. Hoppe-Seyler observed changes in the color of blood after passing CO gas through and concluded a change in the hemoglobin, curtailing its capacity for transporting oxygen to the tissue.

Other targets of the toxicity may involve CO binding to cytochrome oxidase, known as a result of the classic studies by Warburg,[13] as well as Keilin and Hartree,[14] and the CO binding to myoglobin described by Pagniez and Camus,[15] Rossi-Fanelli and Antonini,[16] and later by Coburn et al.[17,18] A binding with other hemoproteins such as cytochrome P-450 is possible, if there is sufficient CO within the cell. This forming of metastable compounds with hemoproteins reduces the availability of oxygen for the cells in a complex manner. An increase in the CO partial pressure in tissues of rabbits, guinea pigs, or rats[19] and a decrease in the oxygen partial pressure in brain tissue of rabbits, cats, or rats[20] were measured.

More recently, it was hypothesized that free oxygen radicals and lipid peroxidation may also play a role in CO poisoning during reoxygenation after the hypoxic phase,[21] and in animal experiments an increase in reactive oxygen species (ROS) was associated with the reoxygenation after CO hypoxia.[22–24] The oxidative stress may explain the development of neuropsychiatric symptoms following an interval during which the patient appears to have recovered and the carboxyhemoglobin (COHb) has been cleared from the circulation. Lewin[1] wrote, "The most important progress in the science of carbon monoxide poisoning was made recently, although only after long battles: namely the knowledge and recognition of the existence of delayed sequelae of the poisoning."

Some newer aspects of histotoxic mechanisms of CO toxicity primarily in the brain, such as apoptosis, excitotoxicity, catecholamine accumulation, are discussed by Piantadosi.[25]

As hypoxia was a recognized response to CO poisoning, treatment with oxygen was the most important therapy. It was first used by Linas and Limousin[26] under normal pressure in 1868. In mice it was shown that the poisonous action of CO diminishes as the oxygen tension increases, and vice versa,[27] and "it is now necessary to discuss more in detail the probability of the hypothesis advanced above that the abolition of the poisonous action of carbonic oxide when the oxygen tension is raised to two atmospheres is due to the fact that the animal can live on the oxygen simply dissolved in the blood." This knowledge led to the treatment of CO poisoning by hyperbaric oxygenation (HBO). Smith and Sharp[28] were the first to treat human patients with HBO; in 68 of 70 patients with CO poisoning the therapy with oxygen under pressure was successful. Some past therapies for CO poisoning which have not proved to be effective are discussed by Jain.[29]

1.4.2 TOXICOKINETICS AND CHEMICAL ANALYSIS

The introduction of illuminating gas, a mixture of hydrogen, CO, methane, and other hydrocarbons, in the last century and its use also for domestic heating purposes increased the risk of CO poisoning as already described by Kirchhoffer,[30] in Switzerland. The oldest case report he mentioned was an event in a storehouse in 1830 with four poisoned men, one of whom died. By around 1900 the intensity of industrialization accelerated and also affected such branches that produce or utilize CO. At this time several great scientists discovered much of what is now known about CO. Douglas and the Haldanes, father and son, derived "laws" that described the equilibrium conditions obtaining when hemoglobin was exposed to a gas mixture containing CO and oxygen[27,31,32]:

- "When a solution containing haemoglobin is saturated with a gas mixture containing O_2 and CO the relative proportions of the haemoglobin which enter into combination with the two gases are proportional to the relative partial pressures of the two gases, allowing for the fact that the affinity of CO for haemoglobin is about 300 times greater than that of O_2." ($[COHb]/[O_2Hb] = M(pCO/pO_2)$ with M = relative affinity constant or Haldane constant; newer determination: M = 240 to 250).[33]
- "Where the pressures of O_2 and CO together is insufficient to saturate the haemoglobin, the dissociation curve, so far as reduced haemoglobin is concerned, will be the same as when oxyhaemoglobin alone, or CO-haemoglobin alone, is present at a pressure equivalent in saturating power to that of the O_2 and CO together; and the O_2 and CO will divide their combined share of haemoglobin in just the same proportions as if they together combined with the whole of the available haemoglobin."
- "Oxygen is given off from oxyhaemoglobin ... in a totally abnormal manner when the blood is highly saturated with carbon monoxide. ...

The dissociation of the oxygen is alterated in such a way that the oxygen comes off less readily, or at a lower pressure than in normal blood" (Haldane effect).

With these "laws" the authors proposed several basic concepts regarding the absorption and elimination of CO. First, attempts to express the toxicokinetics mathematically were basically empiric and reflected the fact that over small ranges the increase in COHb concentration appeared to be proportional to CO concentration and exposure duration.[34-36] Later, theoretical equations were derived containing such variables as duration of exposure, partial pressure of CO in the inhaled air, alveolar ventilation, blood volume, rate of the endogenous CO production, barometric pressure, diffusivity of the lung for CO, and the partial pressure of the oxygen in the lung and capillaries.[37-42] Benignus[43] has developed a computer model, consisting of 35 input and 60 output variables, for predicting COHb and the pulmonary variables involved during exposure to CO, oxygen, and carbon dioxide.

An important prerequisite for the advance of the knowledge in CO toxicology was the improvement in the analysis of CO or COHb in blood. Papers by Vierodt[44] in 1876, by Soret in 1878,[45] by Hüfner in 1894,[45] by van Slyke and Salvesen[46] in 1919, by Sendroy and Liu[47] in 1930, by Hartmann[48] in 1937, and by Havemann[49] in 1940 are examples of pioneer work in this field.

As early as 1898 the existence of a small amount of CO in human blood was demonstrated by Nicloux.[50] Sjöstrand[51] showed CO to be a product of the catabolism of hemoglobin. This was possible with the advances of the analytical methods. In 1879 by Grehant,[1] in 1880 by Le Bon,[52] and in 1899 by Wahl[53] the first studies dealing with the CO concentration in tobacco smoke were published. It is well known that smoking produces elevated COHb levels in the blood.

Two main research groups have investigated the CO uptake in the lungs: Haldane, together with his son and Douglas, as mentioned above, and Roughton and his collaborators Root, Forbes, and Sargent. They carefully studied the fate of CO in human subjects and refined knowledge of the CO kinetics.[34,54-56] In 1970, Roughton[33] summarized the results.

The history of the oxidation of CO to CO_2 by cytochrome c oxidase was presented in 1970 by Fenn,[57] who had started the corresponding investigations about 40 years earlier.[58,59] The extent of this oxidative pathway of CO is too small in human beings to have any real practical importance.

1.4.3   TOXICODYNAMICS

The sensitivity of organs with high oxygen consumption was recognized early. In 1896 Klebs[60] described the pathology of the heart in CO poisoning. He found diffuse punctiform hemorrhages and necrotic foci throughout the heart, and also vasodilatation and relaxation of the smooth muscle. Electrocardiographic changes following severe CO poisoning were demonstrated by Colvin[61] in 1928. Campbell,[62] working with mice, was the first to report that repeated CO exposure induces cardiac enlargement. Since the 1970s, Penney[63,64] has pioneered the use of repeated CO exposure to rats to increase cardiac workload and stimulate additional growth of the myocardium, including the hematological, physiological, biochemical, and morphological changes in CO-induced cardiomegaly. The results are helpful in understanding symptoms of chronic CO poisoning, a term often criticized by pharmacologists and toxicologists. Several research groups have reported data on clinical manifestations of chronic CO poisonings.[65-68]

Investigations by Kolisko[69] in 1893 showed lesions of the brain, especially of the globus pallidus. Cerebral edema and hemorrhages following acute and necrotic lesions in basal ganglia and demyelination in chronic CO poisoning were observed.[1,70-73] CO-induced disorders of mental function were mentioned by Sibelius,[74] Bourguignon and Desoille,[75] and Cohen[76] and damage to the cerebral vasculature by Stewart.[77] As demonstrated by Krause[78] and Schmitt,[79] the function of the peripheral nerve is susceptible to CO.

The first systematic investigation of the electroencephalogram (EEG) following CO poisoning was done by Lennox and Petersen.[80] They studied the EEG of 33 patients and reported that 18 patients had abnormal records. Later, it was demonstrated that the known CO effects in the brain correlate with the findings of a computed tomographic brain scan[81] and of magnetic resonance imaging.[82] The brain syndrome associated with delayed neuropsychiatric sequelae following acute CO poisoning, described principally by Böhm in 1880,[45] has recently been of increasing concern because of its variety of symptoms, because of the high possibility of misdiagnosis, and because of speculations on the possible contributions of ROS, apoptosis, catecholamine release, and nitric oxide (NO) production as mechanisms of CO effects. Effects of CO on behavior where COHb exceeds 20% have been unambiguously demonstrated, but below this level results were less consistent. Recently, behavioral effects below 20% COHb have been demonstrated.[83]

Ventilatory stimulation during the initial phase of CO exposure was first reported by Pokrowksy in 1866.[45] Haldane[84] pointed out hyperpnea 4 min after exposure of a mouse to 2200 ppm CO, but later a dominant decrease in the respiratory frequency was observed. He failed to find a significant increase in the pulmonary ventilation in humans if one third or more of the hemoglobin is held by CO.[27] Chiodi et al.[85] were not able to detect hyperpnea when humans were exposed to 1500 to 3500 ppm CO, producing up to 50% COHb.

Pulmonary edema in acute CO poisoning was described by Lewin.[1] The link to up-to-date experiments by Thom et al.[86] showing lung damage during CO exposure mediated by NO-derived oxidants was forged by the paper published by Finck.[87] Finck reported that congestion, edema, or both were the most frequently noted gross pathological observations, apart from the cherry red color of the tissues. The lungs, tracheae, and bronchi were affected in more than 60% and pulmonary hemorrhages in 10% of cases at autopsy.

The placenta is no barrier for CO. In 1859, Breslau[88] documented CO poisoning caused by illuminating gas. Three cases occurred in pregnant women in the Zürich maternity hospital. Fetal death following CO exposure of the mother was described in a case report in the same year.[89] Fehling[90] in 1877 detected CO in fetal blood following exposure of pregnant rabbits to CO. In 1833, Grehant and Quinquaud found increased CO in the blood of the fetus of a pregnant dog, and in 1890 Lesser described similar results in the human fetal blood following a lethal CO poisoning of the mother.[1] Nicloux[91] exposed pregnant guinea pigs to various concentrations of CO for 90 min and showed that both maternal and fetal COHb levels increased as a function of the inhaled CO concentration. Abortions, resorptions, and abnormal growth of survivors were seen in rats after CO poisoning.[92] Further research in this field was done and reviewed by several research groups,[93–96] pointing out various reasons for the sensitivity of the fetus to CO hypoxia.

Lewin[1] cited a paper by Faure (1856), who observed that it is difficult if not impossible to kill dogs via CO exposure if the animals are repeatedly exposed to sublethal CO concentrations. Several published data showed that, after long-term exposure to moderate CO concentrations, animals can tolerate acute exposure to high CO concentrations which cause collapse or death in "naive" animals.[97–99] With regard to the phenomenon of adaptation, there is unequivocal evidence that, as with other causes of hypoxia, the increased COHb level evoked responses that tend to reduce the hypoxic effects. Responses of short-term compensation to CO hypoxia include increased coronary blood flow,[100] increased cerebral blood flow,[101] and increased glycolysis.[102,103] As early as 1906, Nasmith and Graham[97] showed that prolonged exposure of guinea pigs to CO stimulated hemoglobin synthesis and emphasized for the first time the similarity of this response to that seen in adaptation at altitude. Campbell[104] showed that the increase of red cells and hemoglobin concentration in the blood may be combined with pathological responses such as congestion of various organs, tissue edema, and atrophy of liver cells. Later, Litzner[105] demonstrated an increase in red blood cells in CO-exposed workers and Killick[106–108] studied the adaptation to CO as she experienced it. After exposure to 100 to 500 ppm CO for 6- to 7-h periods at weekly intervals for 23 weeks, she observed that with repeated exposure the subjective symptoms decreased. Killick found no changes in the hemoglobin concentration in her blood and suggested that the lungs, as a result of adaptation, either actively secrete CO from the gas phase against a gradient or prevent the achievement of diffusion equilibrium. These assumptions have been neither confirmed nor refuted by other investigators.

The significance of changes in glucose metabolism which results from CO exposure, first reported by Senff[109] in 1869, Eckard[45] in 1872, Araki[110] in 1891, Ottow[111] in 1893, Straub[112] in 1897, and Rosenstein[113] in 1898, was recently demonstrated by Sokal and Kralkowska[114] and Penney et al.[115–117]

The role of individuality in CO poisoning was stressed by Lewin[1] in a separate chapter of his book. He wrote, "It is effective, but in detail it is a mystery."

1.4.4 RECOMMENDATIONS FOR THRESHOLD LEVELS

The pioneer of environmental hygiene and epidemiology, Max von Pettenkofer (1818–1901), stimulated research on the analysis and toxicity of CO and its indoor occurrence and, as a result, Gruber[118] in Vienna was one of the first to study a threshold for CO effects. He established a threshold of 200 ppm CO using experiments performed on himself and with rabbits. However, he conceded "that one of the most difficult tasks of hygienists confronted with a poison often is to establish the harmfulness or the harmlessness of a long-term exposure to low concentrations of a poison" (originally: "Und allerdings muss man zugestehen, dass es eine der schwierigsten Aufgaben des Hygienikers, die ihm doch so häufig gestellt wird, ist, die Schädlichkeit oder Unschädlichkeit lang andauernder Einwirkung minimaler Dosen eines giftigen Stoffes festzustellen."). It is known that this statement is true today. Sayers et al.[119] reported that exposure of humans to 200 ppm CO caused slight symptoms. Recommendations for an occupational exposure limit were in place in the United States as early as 1946.[120] The maximum permissible CO concentration has been set at 100 ppm of air for exposures not exceeding a total of 8 h daily. But people exposed for 9 h to 100 ppm CO complained of headaches and nausea.[121] In the 1960s, this TLV (threshold limit value) as a TWA (time-weighted average airborne concentration, calculated over an 8-h working day, for a 5-day working week) was reduced to 50 ppm. In the 1990s it was reduced to 25 ppm. The TLV of 25 ppm CO was recommended to keep blood COHb levels below 3.5%, to minimize the potential for adverse neurobehavioral changes, and to maintain cardiovascular exercise capacity. To protect the general population, including not only healthy workers, but also hypersusceptibles, the sick, the embryo and fetus, the neonate and mother, the very young and the elderly, it was determined that a COHb level of 2.5% should not be exceeded.

At present, it is becoming evident that CO is formed physiologically in different tissues and may function as a neurotransmitter. Verma et al.[122] were the first to describe this property of CO.

1.5 SUMMARY AND PROSPECTS

In summary, the history of CO toxicology can be roughly divided into three eras: the era of experience of CO poisoning without knowledge of the poison, the era of describing CO poisoning without knowledge of the mechanisms of action (until 1857), and the era of characterizing CO toxicology involving the mechanisms of action. As stated by Byron, "The best prophet of the future is the past." Some important fields of the research in the future will be the study of the effects of long-term CO exposure on the organism involving the characterization of adaptive responses, the effects of CO exposure on pregnant women, the fetus, and newborn infants, and the combined effects of exposures to CO and other chemicals such as combustion products, drugs, and environmental factors. The existence of hemoxygenase activity in different tissues means that CO is formed there. It may function as a transmitter. Studies of the interactions of the endogenously formed CO and exogenously inhaled CO should also constitute a task for the future. The first sentence of the Lilienthal[4] review, "Carbon monoxide has ever fascinated biologists," has not lost its significance.

REFERENCES

1. Lewin, L., *Die Kohlenoxyvergiftung. Ein Handbuch fur Mediziner, Techniker und Unfallrichter*, Springer, Berlin, 1920.
2. Krenzelok, P., Roth, R., and Full, R., Carbon monoxide ... the silent killer with an audible solution, *Am. J. Emerg. Med.*, 14, 484, 1996.

3. Holmstedt, B. and Louis Lewin, L., Toxikologe und Ethnopharmakologe, in *Der Toxikologe Louis Lewin (1850–1929)*, Müller, R.K., Holmstedt, B., and Lohs, K., Eds., S. N. Leipzig, 1982.

4. Lilienthal, J.L., Carbon monoxide, *Pharmacol. Rev.*, 2, 324, 1950.

5. Hauser, O., *Ins Paradies der Urmenschen*, Hoffmann & Campe Verlag, Hamburg, 1922.

6. Lascaratos, J.G. and Marketos, S.G., The carbon monoxide poisoning of two Byzantine Emperors, *Clin. Toxicol.*, 36, 103, 1998.

7. Weichardt, H., Gewerbetoxikologie und Toxikologie der Arbeitsstoffe, in *Gifte. Geschichte der Toxikologie*, Amberger-Lahrmann, M. and Schmähl, D., Eds., Springer, Berlin, 1988, chap. 6.

8. Larcan, A., Description de l'intoxication oxycarbonee par D. B. Harmant, de Nancy, en 1775, *Ann. Med. Nancy*, 7, 169, 1968.

9. Lee, P.H., Kwon, S.P., Kang, B.S., and Lee, K.S., The beneficial effect of acetic acid vapor on management of carbon monoxide poisoning: clinical and experimental data, *Clin. Pharmacol. Ther. Toxicol.*, 19, 527, 1981.

10. Hoffmann, F., Eines berühmten Medici gründliches Bedenken und physikalische Anmerkungen von dem tödlichen Dampf der Holzkohlen. Auf Veranlassung der in Jena beim Ausgang des 1715, Jahres vorgefallenen traurigen Begebenheiten aufgesetzt und nun zum gemeinen Nutzen dem Druck überlassen, Halle, 1716.

11. Bernard, C., *Leçons sur les Effets des Substances Toxiques et Medicamenteuses*, Bailliere, Paris, 1857.

12. Hoppe, F., Über die Einwirkung des Kohlenoxydgases auf das Hämatoglobulin, *Virchows Arch. Pathol. Anat. Physiol. Klin. Med.*, 11, 288, 1857.

13. Warburg, O., Über Kohlenoxydwirkung ohne Hämoglobin und einige Eigenschaften des Atmungsfermentes, *Naturwissenschaften*, 15, 546, 1927.

14. Keilin, D. and Hartree, E.F., Cytochrome and cytochrome oxidase, *Proc. R. Soc. London Ser. B* (1), 27, 167, 1939.

15. Pagniez, J. and Camus, J., Fixation de l'oxyde de carbone sur la myoglobine, *Bull. Mem. Soc. Med. Hop. Paris*, 339, 1941.

16. Rossi-Fanelli, A. and Antonini, E., Studies on the oxygen and carbon monoxide equilibria of human myoglobin, *Arch. Biochem. Biophys.*, 77, 478, 1958.

17. Coburn, R.F. and Mayers, L.B., Myoglobin O_2 tension determined from measurements of carboxymyoglobin in skeletal muscle, *Am. J. Physiol.*, 220, 66, 1971.

18. Coburn, R.F., Ploegmakers, F., Gondrie, P., and Abboud, R., Myocardial myoglobin oxygen tension, *Am. J. Physiol.*, 224, 870, 1973.

19. Göthert, M., Lutz, F., and Malorny, G., Kohlenoxidpartialdruck im Pneumo-peritoneum von Kaninchen, Meerschweinchen und Ratten in Beziehung zum Kohlenoxidhämoglobin-Gehalt, *Naunyn Schmiedebergs Arch. Pharmakol.*, 260, 122, 1968.

20. Zorn, H., Der Sauerstoff-Partialdruck im Hirngewebe und in der Leber bei subtoxischen Kohlenmonoxid-Konzentrationen, *Staub Reinhalt. Luft*, 32, 161, 1972.

21. Werner, B., Bäck, W., Akerblom, H., and Barr, P.O., Two cases of acute carbon monoxide poisoning with delayed neurological sequelae after a "free" interval, *Clin. Toxicol.*, 23, 249, 1985.

22. Thom, S.R., Experimental carbon monoxide-mediated brain lipid peroxidation and the effect of oxygen therapy, *Ann. Emerg. Med.*, 17, 403, 1988.

23. Thom, S.R., Carbon monoxide mediated brain lipid peroxidation in the rat. *J. Appl. Physiol.*, 68, 997, 1990.

24. Zhang, J. and Piantadosi, C.A., Mitochondrial oxidative stress after carbon monoxide hypoxia in the rat brain. *J. Clin. Invest.*, 90, 1193, 1992.

25. Piantadosi, C.A., Toxicity of carbon monoxide: hemoglobin vs. histotoxic mechanisms, in *Carbon Monoxide*, Penney, D.G., Ed., CRC Press, Boca Raton, FL, 1996, chap. 8.

26. Linas, A.J. and Limousin S., Asphyxie lente et graduelle par l'oxyde de carbone, traitement et guerison par les inspirations d'oxygene, *Bull. Mem. Soc. Ther.*, 2, 32, 1868.

27. Haldane, J., The action of carbonic oxide on man, *J. Physiol.*, 18, 430, 1895.

28. Smith, G. and Sharp, G.R., Treatment of carbon monoxide poisoning with oxygen under pressure, *Lancet*, 905, 1960.

29. Jain, K.K., *Carbon Monoxide Poisoning*, W.H. Green, St. Louis, 1990, chap. 9.

30. Kirchhoffer, J.F., *Ueber die Vergiftung durch Leuchtgas. Dargelegt durch Experimente mit besonderer Berücksichtigung der Mikroskopie und Spectralanalyse*, C.J. Meiselís, Herisau, 1868.

31. Douglas, C.G., Haldane, J.S., and Haldane, J.B.S., The laws of combination of haemoglobin with carbon monoxide and oxygen, *J. Physiol.*, 44, 275, 1912.

32. Haldane, J.B.S., The dissociation of oxyhaemoglobin in human blood during partial CO poisoning, *J. Physiol.*, 45, 22, 1912/1913.

33. Roughton, F.J.W., The equilibrium of carbon monoxide with human hemoglobin in whole blood, *Ann. N.Y. Acad. Sci.*, 174(I), 177, 1970.

34. Forbes, W.H., Sargent, F., and Roughton, F.J.W., The rate of carbon monoxide uptake by normal men, *Am. J. Physiol.*, 143, 594, 1945.

35. Lilienthal, J.L. and Pine, M.B., The effect of oxygen pressure on the uptake of carbon monoxide by man at sea level and at altitude, *Am. J. Physiol.*, 145, 346, 1946.

36. Pace, N., Consolazio, W.V., White, W.A., and Behnke, A.R., Formulation of the principal factors affecting the rate of uptake of carbon monoxide by man, *Am. J. Physiol.*, 147, 352, 1946.

37. Forster, R.E., Fowler, W.S., and Bates, D.V., Considerations on the uptake of carbon monoxide by the lungs, *J. Clin. Invest.*, 33, 1128, 1954.

38. Coburn, R.F., Forster, R.E., and Kane, P.B., Considerations of the physiological variables that determine the carboxyhemoglobin concentration in man, *J. Clin. Invest.*, 44, 1899, 1965.

39. Peterson, J.E. and Stewart, R.D., Predicting the carboxyhemoglobin levels resulting from carbon monoxide exposures, *J. Appl. Physiol.*, 39, 633, 1975.

40. Smith, M.V., Comparing solutions to the linear and nonlinear CFK equations for predicting COHb formation, *Math. Biosci.*, 99, 251, 1990.

41. Tikuisis, P., Kane, D.M., McMellan, T.M., Buick, F., and Fairburn, S.M., Rate of formation of carboxyhemoglobin in exercising humans exposed to carbon monoxide, *J. Appl. Physiol.*, 72, 1311, 1992.

42. Benignus, V.A., Hazucha, M.J., Smith, M.V., and Bromberg, P.A., Prediction of carboxyhemoglobin formation due to transient exposure to carbon monoxide, *J. Appl. Physiol.*, 76, 1739, 1994.

43. Benignus, V.A., A computer model for predicting carboxyhemoglobin and pulmonary parameters associated with exposures to carbon monoxide, oxygen and carbon dioxide, *Aviat. Space Environ. Med.*, 66, 369, 1995.

44. Vierodt, K., *Die quantitative Spektralanalyse*, H. Laupp, Tübingen, 1876.

45. Bock, J., Das Kohlenoxyd, *Heffters Handbuch der Pharmakologie*, Springer-Verlag, Berlin, Vol. 1, 1923, 1.

46. van Slyke, D.D. and Salvesen, H.H., The determination of carbon monoxide in blood, *J. Biol. Chem.*, 40, 103, 1919.

47. Sendroy, J. and Liu, S.H., Gasometric determination of oxygen and carbon monoxide in blood, *J. Biol. Chem.*, 89, 133, 1930.

48. Hartmann, H., Über die Grundlagen der Differenzphotometrie und ihre Anwendung zur Bestimmung geringer Kohlenoxydmengen im Blut, *Ergeb. Physiol.*, 39, 413, 1937.
49. Havemann, R., Die Bestimmung von Kohlenoxyd-Hämoglobin im Blut mit dem lichtelektrischen Colorimeter, *Klin. Wochenschr.*, 19, 1183, 1940.
50. Nicloux, M., Sur lfoxyde de carbone contenu normalement dans le sang, *C. R. Acad. Sci.*, p. 126, 1898.
51. Sjöstrand, T., Endogenous formation of carbon monoxide in man under normal and pathological conditions, *Scand. J. Clin. Lab. Invest.*, 1, 201, 1949.
52. Le Bon, G., Recherches experimentales sur l'influence de l'oxyde de carbone contenu dans le fumee du tabac, *Bull. Soc. Med. Prat.*, p. 82, 1880.
53. Wahl, F., Ueber den Gehalt des Tabakrauches an Kohlenoxyd, *Arch. Ges. Physiol.*, 78, 262, 1899.
54. Roughton, F.J.W., The kinetics of the reaction $CO + O_2Hb = O_2 + COHb$ in human blood at body temperature, *Am. J. Physiol.*, 143, 609, 1945.
55. Roughton, F.J.W., The average time spent by the blood in the human lung capillary and its relation to the rates of CO uptake and elimination in man, *Am. J. Physiol.*, 143, 621, 1945.
56. Roughton, F.J.W. and Root, W.S., The fate of CO in the body during recovery from mild carbon monoxide poisoning in man, *Am. J. Physiol.*, 145, 239, 1945.
57. Fenn, W.O., The burning of CO in tissues, *Ann. N.Y. Acad. Sci.*, 174(I), 64, 1970.
58. Fenn, W.O. and Cobb, D.M., The stimulation of muscle respiration by carbon monoxide, *Am. J. Physiol.*, 102, 379, 1932.
59. Fenn, W.O. and Cobb, D.M., The burning of carbon monoxide by heart and skeletal muscle, *Am. J. Physiol.*, 102, 393, 1932.
60. Klebs, D., Ueber die Wirkung des Kohlenoxyds auf den thierischen Organismus, *Virchows Arch. Pathol. Anat. Physiol. Klin. Med.*, 32, 450, 1865.
61. Colvin, L.T., Electrocardiographic changes in case of severe carbon monoxide poisoning, *Am. Heart J.*, 3, 484, 1928.
62. Campbell, J.A., Hypertrophy of the heart in acclimatization to chronic carbon monoxide poisoning, *J. Physiol.*, 77, 8P, 1932.
63. Penney, D.G., Carbon monoxide induced cardiac hypertrophy, in *Growth of the Heart in Health and Disease*, Zak, R., Ed., Raven Press, New York, 1984, 337.
64. Penney, D.G., Postnatal modification of cardiac development: a review, *J. Appl. Cardiol.*, 5, 325, 1990.
65. Beck, H.G., The clinical manifestations of chronic carbon monoxide poisoning, *Ann. Clin. Med.*, 5, 1088, 1927.
66. Sievers, R.F., Edwards, T.I., Murray, A.L., Russell, A.E., Schrenk, H.H., and Fairhall, T.T., Medical study of men exposed to measured amounts of carbon monoxide in the Holland tunnel for thirteen years, *U.S. Public Health Bull.* 278, 1942.
67. Almgren, S., 12 Jahre Erfahrungen auf dem Gebiete der chronischen Kohlenoxydvergiftung in Schweden, *Arch. Gewerbepathol. Gewerbehyg.*, 13, 97, 1954.
68. Zorn, H., Die chronische Kohlenmonoxidvergiftung, *Med. Klin.*, 70, 441, 1975.
69. Kolisko, A., Beiträge zur Kenntnis der Blutversorgung der Grosshirnganglien, *Wien. Klin. Wochenschr.*, 6, 191, 1893.
70. Hill, E. and Semerack, C.B., Changes in the brain in gas (carbon monoxide) poisoning, *J. Am. Med. Assoc.*, 81, 664, 1918.
71. Grinker, R.R., Über einen Fall von Leuchtgasvergiftung mit doppelseitiger Pallidumerweichung und schwerer Degeneration des tieferen Grosshirnmarklagers, *Z. Gesamte Neurol. Psychiatr.*, 98, 433, 1925.

72. Meyer, A., Über die Wirkung der Kohlenoxidvergiftung auf das Zentralnervensystem, *Z. Gesamte Neurol. Psychiatr.*, 100, 201, 1926.

73. Meyer, A., Über das Verhalten des Hemisphärenmarks bei der menschlichen Kohlenoxidvergiftung, *Z. Gesamte Neurol. Psychiatr.*, 112, 187, 1928.

74. Sibelius, C., Die psychischen Störungen nach akuter Kohlenoxydvergiftung, *Monatsschr. Psychiatr.*, 18, 39, 1906.

75. Bourguignon, G. and Desoille, H., Claudication intermittente et syndrome vasculaire d'une cote et signe de Babinski de l'autre cote, consecutive a une intoxication par l'oxyde de carbone, *Rev. Neurol.* (Paris), 34, 360, 1927.

76. Cohen, L.H., Speech perseveration and astasia-abasia following carbon monoxide poisoning, *J. Neurol. Psychopathol.*, 17, 41, 1936.

77. Stewart, R.M., A contribution to the histopathology of carbon monoxide poisoning, *J. Neurol. Psychopathol.*, 1, 105, 1920.

78. Krause, E., Über den Einfluss des Kohlenmonoxydes auf periphere Nerven, *Acta Psychiat. Neurol. Scand.*, 5, 473, 1930.

79. Schmitt, F.O., On the nature of the nerve impulse. I. The effect of carbon monoxide on medullated nerve, *Am. J. Physiol.*, 95, 650, 1930.

80. Lennox, M.A. and Petersen, P.B., Electroencephalographic findings in acute carbon monoxide poisoning, *Electroencephalogr. Clin. Neurophysiol.*, 10, 63, 1958.

81. Nardizzi, L.R., Computerized tomographic correlate of carbon monoxide poisoning, *Arch. Neurol.*, 36, 38, 1979.

82. Horowitz, A.L., Kaplan, R., and Sarpel, G., Carbon monoxide toxicity: MR imaging in the brain, *Radiology*, 162, 787, 1987.

83. Benignus, V.A., Behavioral effects of carbon monoxide exposure: results and mechanisms, in *Carbon Monoxide*, Penney, D.G., Ed., CRC Press, Boca Raton, FL, 1996, chap. 10.

84. Haldane, J., The relation of action of carbonic oxide to oxygen tension, *J. Physiol.*, 18, 201, 1895.

85. Chiodi, H., Dill, D.B., Consolazio, F., and Horvath, S.M., Respiratory and circulatory responses to acute carbon monoxide poisoning, *Am. J. Physiol.*, 134, 683, 1941.

86. Thom, S.R., Ohnishi, S.T., Fisher, D., Xu, Y.A., and Ischiropoulos H., Pulmonary vascular stress from carbon monoxide, *Toxicol. Appl. Pharmacol.*, 154, 12, 1999.

87. Finck, P.A., Exposure to carbon monoxide: review of the literature and 567 autopsies, *Mil. Med.*, 131, 1513, 1966.

88. Breslau, F., Intoxication zweier Schwangeren mit Holzleuchtgas. Tod und vorzeitige Geburt eines Kindes, *Monatsschr. Geburtskd. Frauenkr.*, 13, 435, 1859.

89. Freund, M.B., Ein Fall von Absterben der Frucht im siebenten Schwangerschaftsmonat infolge von nur mässiger Intoxikation der Mutter durch Kohlenoxydgas, *Monatsschr. Geburtskd. Frauenkr.*, 14, 31, 1859.

90. Fehling, H., Beiträge zur Physiologie des placentaren Stoffverkehrs, *Arch. Gynäkol.*, 11, 523, 1877.

91. Nicloux, M., Passage de l'oxyde de carbone de la mere au foetus, *C. R. Acad. Sci.*, 133, 67, 1901.

92. Wells, L.L., The prenatal effect of carbon monoxide on albino rats and the resulting neuropathology, *Biologist*, 15, 80, 1933.

93. Fechter, L.D. and Annau, Z., Toxicity of mild prenatal carbon monoxide exposure, *Science*, 197, 680, 1977.

94. Longo, L.D., The biological effects of carbon monoxide on the pregnant woman, fetus, and newborn infant, *Am. J. Obstet. Gynecol.*, 129, 69, 1977.

95. Singh, J. and Scott, L.H., Threshold for carbon monoxide induced fetotoxicity, *Teratology*, 30, 253, 1984.

96. Penney, D.G., Effects of carbon monoxide on developing animals and humans, in *Carbon Monoxide*, Penney, D.G., Ed., CRC Press, Boca Raton, FL, 1996, chap. 6.

97. Nasmith, G.C. and Graham, D.A.L., The haematology of carbon monoxide poisoning, *J. Physiol.*, 25, 32, 1906.

98. Campbell, J.A., Tissue oxygen tension and carbon monoxide poisoning, *J. Physiol.*, 68, 81, 1929.

99. Killick, E.M., The acclimatization of mice to atmospheres containing low concentrations of carbon monoxide, *J. Physiol.*, 91, 279, 1937.

100. Ayres, S.M., Giannelli, S., and Mueller, H., Myocardial and systemic responses to carboxyhemoglobin, *Ann. N.Y. Acad. Sci.*, 174, I, 268, 1970.

101. Paulson, O.B., Parving, H.-H., Olesen, J., and Skinhoj, E., Influence of carbon monoxide and of hemodilution on cerebral blood flow and blood gases in man, *J. Appl. Physiol.*, 35, 111, 1973.

102. Thiel, K., Experimentelle Untersuchungen über die akute Kohlenoxydvergiftung und ihre Behandlung, *Z. Gesamte Exp. Med.*, 88, 207, 1933.

103. Swann, H.G. and Brucer, G.M, The cardiorespiratory and biochemical events during rapid anoxic death, IV. Carbon monoxide poisoning, *Tex. Rep. Biol. Med.*, 7, 569, 1949.

104. Campbell, J.A., Comparison of the pathological effects of prolonged exposure to carbon monoxide with those produced by very low oxygen pressure, *Br. J. Exp. Pathol.*, 10, 304, 1929.

105. Litzner, S., Kohlenoxydvergiftung und Polycythämie, *Arch. Gewerbepathol. Gewerbehyg.*, 1, 749, 1931.

106. Killick, E.M., Development of acclimatization to carbon monoxide in the human subject, *J. Physiol.*, 83, 35P, 1935.

107. Killick, E.M., The acclimatization of the human subject to atmospheres containing low concentrations of carbon monoxide, *J. Physiol.*, 87, 41,1936.

108. Killick, E.M., The nature of acclimatization occurring during repeated exposure of the human subject to atmospheres containing low concentrations of carbon monoxide, *J. Physiol.*, 107, 27, 1948.

109. Senff, L., Über den Diabetes nach der Kohlenoxydatmung, dissertation, Univ. Dorpat, 1869.

110. Araki, T., Über die Bildung von Milchsäure und Glucose im Organismus bei Sauerstoffmangel, *Hoppe-Seylers Z. Physiol. Chem.*, 15, 335, 1891.

111. Ottow, H., Über den Glykogengehalt der Leber nach Kohlenoxydvergiftung, dissertation, Univ. Würzburg, 1893.

112. Straub, W., Über die Bedingungen des Auftretens der Glykosurie nach der Kohlenoxydvergiftung, *Arch. Exp. Pathol. Pharmakol.*, 38, 141, 1897.

113. Rosenstein, W., Über den Einfluss der Nahrung auf die Zuckerausscheidung bei der Kohlenoxydvergiftung, *Arch. Exp. Pathol. Pharmakol.*, 40, 363, 1898.

114. Sokal, J.A. and Kralkowska, E., Relationship between exposure duration, carboxyhemoglobin, blood glucose, pyruvate, and lactate, and the severity of intoxication in 39 cases of acute carbon monoxide poisoning in man, *Arch. Toxicol.*, 57, 196, 1985.

115. Penney, D.G., Hyperglycemia exacerbates brain damage in acute severe carbon monoxide poisoning, *Med. Hypotheses*, 27, 241, 1988.

116. Penney, D.G., Helfman, C.C., Dunbar, J.C., and McCoy, L.E., Acute severe carbon monoxide poisoning in the rat: effects of hyperglycemia and hypoglycemia on mortality, recovery and neurologic deficit, *Can. J. Physiol. Pharmacol.*, 69, 1168, 1991.

117. Penney, D.G., Acute carbon monoxide poisoning in an animal model: the effects of
     ·   altered glucose on morbidity and mortality, *Toxicology*, 80, 85, 1993.
118. Gruber, M., Ueber den Nachweis und die Giftigkeit des Kohlenoxyds und sein
     Vorkommen in Wohnräumen, *Arch. Hyg.*, 1, 143, 1883.
119. Sayers, R.R., Yant, W.P., Levy, E., and Fulton, W.B., Effects of repeated daily exposure
     of several hours to small amounts of automobile exhaust gas, Public Health Bull.
     186, U.S. Government Printing Office, Washington, D.C., 1929.
120. American Conference of Governmental Industrial Hygienists (ACGIH), *Documenta-
     tion of the Threshold Limit Values and Biological Exposure Indices*, 6th ed., ACGIH,
     Inc., Cincinnati, OH, 1991.
121. Henderson, Y. and Haggard H.W., *Noxious Gases*, 2nd ed., Reinhold, New York, 1943.
122. Verma, A., Hirsch, D.J., Glatt, C.E., Ronnett, G.V., and Snyder, S.H., Carbon monoxide:
     a putative neural messenger, *Science*, 259, 381, 1993.

2 Carbon Monoxide in Breath, Blood, and Other Tissues

Hendrik J. Vreman, Ronald J. Wong, and David K. Stevenson

CONTENTS

0-8493-2065-8/00/$0.00+$.50
© 2000 by CRC Press LLC

2.1 INTRODUCTION — HISTORICAL PERSPECTIVE

The toxicity of carbon monoxide (CO) has been recorded as early as the third century B.C. by Aristotle. Today, CO still commands much attention not only because people continue to die from overexposure, but also because of the interesting ways that small amounts of CO appear to affect animal and human performance.

A little more than a half century ago in 1946, Roughton and Root[1] were the first to demonstrate conclusively that human blood contains a small, but measurable amount of CO. Subsequent benchmark discoveries and other advances continue to stimulate the imagination of physicians and researchers to learn more about CO. A new era was ushered in by the revealing studies by Sjöstrand, who demonstrated that CO is formed endogenously from the degradation of erythrocytic heme and that hemolytic disease increases CO levels in blood and breath.[2,3] Coburn and co-workers then refined the CO-measuring technology, quantitated CO production rates, and measured total body (and compartment) CO stores.[4-7] In addition, Collison et al.[8] developed a sensitive gas chromatographic (GC) method for the determination of CO in as little as 0.1 ml of blood. Concurrently, Tenhunen and co-workers[9] described the characteristics of heme oxygenase (HO), the rate-limiting enzyme that catalyzes the degradation of hemin derived from the turnover of hemoglobin, myoglobin, and a number of other hemoproteins. The introduction of metalloporphyrins (MPs), other than heme, as *in vitro*[10,11] and *in vivo*[12] competitive inhibitors of HO, as well as the

FIGURE 2.1   "All substances are poisonous. Only the dose differentiates a poison from a remedy." Paracelsus (1493–1541). (Courtesy of PhotoDisc®.)

purification of HO isozymes, have elevated knowledge of the origins of CO to a higher and more-sophisticated plane. Moreover, the availability of pure isozymes, together with the development of molecular and immunochemical tools, opened the door to the identification of amino acid sequences, the characterization of messenger-ribonucleic acids (mRNAs), and the localization of genomic loci, have rapidly advanced knowledge.

Perhaps the most titillating recent advance began with the realization that CO is not only an unavoidable, inert, waste product, but that it may also play a biological role (Figure 2.1, the happy face of CO). In 1991, Marks et al.[13] brought this concept to the fore in his review paper entitled, "Does Carbon Monoxide Have a Physiologic Function?" From the many reports and reviews that have been published since, it appears that this metabolic trace gas is certainly much more than just a waste product.[14–18]

The purpose of this chapter is to review recent CO physiology with emphasis on several aspects of CO-measuring methodology. The chapter will highlight its endogenous production, regulatory mechanisms, tissue distribution, and ultimate elimination from the body and will also discuss the CO quantitation methodologies that form the basis of the authors' research and clinical interests. And finally, the chapter will describe various clinical, as well as *in vivo* and *in vitro* research, applications of CO quantitation in the animal and human body, including their compartments. The authors have written an earlier review[19] on the present subject and have now emphasized some newer and different aspects, in particular, those practiced in their research work. Because it is not possible to cover comprehensively all areas of CO knowledge, interested readers are directed to the many informative reviews,[6,7,14,17–24] monographs,[25,26] books,[27–29] government publications,[30,31] and numerous original research articles, some of which are cited in this chapter.

The hope is to convey the message that this ubiquitous, small, volatile, diatomic molecule, which has a long history of toxicity, now also has become a useful tool in the study of some important physiological processes. Furthermore, the authors hope to emphasize that the heme degradation pathway has a far greater impact on animal biochemistry and physiology than only that of maintaining iron homeostasis.

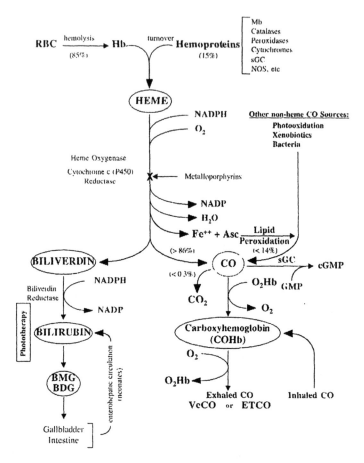

FIGURE 2.2   The origin and fate of endogenous CO and related processes.

2.2   SOURCES OF CARBON MONOXIDE

2.2.1   ENDOGENOUS SOURCES

2.2.1.1   Carbon Monoxide Formation from the Degradation of Heme

Under normal physiological conditions, the degradation of heme by HO is the predominant source (>86%) of endogenous CO[32] (Figure 2.2). The remainder (<14%) is derived from as yet unidentified, non-heme involving processes, such as lipid peroxidation.[33,34] The rate of CO production for adult males is approximately 0.4 ml or 18 µmol/h.[35] Term neonates, whose red blood cells (RBCs) have a shorter life span than those of adults (approximately 70 to 90 days vs. 110 to 120 days, respectively), have a two- threefold higher CO production rate than adults.[35-37] As calculated from bilirubin (BR) production rates, approximately 70% of the CO produced in adult males is derived from the heme of senescent RBCs sequestered by the spleen and the reticuloendothelial system, and up to 9% arises from ineffective erythropoiesis in the bone marrow. The remaining CO (up to 21%) is derived from the turnover of other hemoproteins such as myoglobin, catalases, cytochromes, peroxidases, nitric oxide synthase (NOS), soluble guanylate cyclase (sGC), etc.[38] Pathologic states, such as premature RBC destruction (hemolysis),[4,39] increased ineffective erythropoiesis, and increased hemoprotein turnover, are often affected by oxidative or other cellular stress factors.[40] Under these conditions, the rate of CO generation is increased severalfold to up to 3.6 ml or 160 µmol/h.[41]

Heme oxygenase is the rate-limiting enzyme in the heme catabolic pathway. It cleaves and oxidizes the α-methene bridge of the heme molecule, yielding equimolar amounts of biliverdin, iron, and CO.[42,43] The biliverdin is rapidly reduced by biliverdin reductase to form BR.[44] The CO is bound to the heme in circulating RBCs to form carboxyhemoglobin (COHb), which subsequently dissociates in the lungs through exchange with inhaled O_2, and, in turn, the CO is exhaled. Under steady-state conditions, the rate of CO production, and hence its excretion, parallels the rate of BR production.[32] Thus, total body heme degradation, as well as BR production, can be quantitated through measurements of COHb and total-body CO excretion (VeCO). It can also be qualitatively estimated via measurements of end-tidal breath CO (ETCO).

2.2.1.1.1   Heme oxygenase

The HO enzymes are located in the endoplasmic reticulum in a ternary complex with NADPH cytochrome c reductase and the cytosolic enzyme, biliverdin reductase. Although HO is fairly ubiquitous throughout the body, its activity as measured *in vitro* varies among tissues.[21,29,45] The anucleated RBCs are the only cells that appear to be devoid of HO enzyme activity.[45] On the basis of heme availability, organ capacity, and specific HO activity, the potential for heme degradation may be greatest in the liver, spleen, and erythropoietic tissue.[23] Considering its organ mass and the availability of dietary or biliary heme, the intestine may also be an important, but often overlooked, site of heme degradation.[46–48] Finally, because of the role of HO in the formation of BR (and its association with kernicterus) and CO (and its effect on neuronal function), HO regulation and activity in the brain are also of great interest.[14,16]

Three isozymes of HO have been identified to date: HO-1, the inducible isoform[49]; HO-2, the constitutively expressed isozyme, which is thought to function as the "housekeeping" form, to be upregulated only upon exposure to glucocorticoids[50]; and HO-3, a recently described isozyme with little relative activity and an as yet unknown function.[51] The relative levels of the HO isoforms are, however, not consistent across tissues. For example, HO-1 is the predominant isoform in adult rat spleen, whereas both HO-1 and HO-2 (1:2 ratio) are present in adult rat liver.[21] Although the HO isozymes have not yet been immunologically characterized in the intestine, the epithelial cells display certain biochemical characteristics are consistent with the presence of HO-1.[46,52] Enzyme activity of both isoforms is inhibited by several synthetic and naturally occurring MPs (Figure 2.3),[21] although not necessarily equally.[53]

Much attention is being given to HO-1 because it has been identified as a stress response protein that is induced by a wide variety of conditions and agents, including oxidant stress, heat shock, MPs, heavy metals, xenobiotics, and hormones.[21,29] Induction of HO-1, such as that observed after heme or cadmium ion administration, is due principally to increased gene transcription rates.[54] But no evidence has yet been

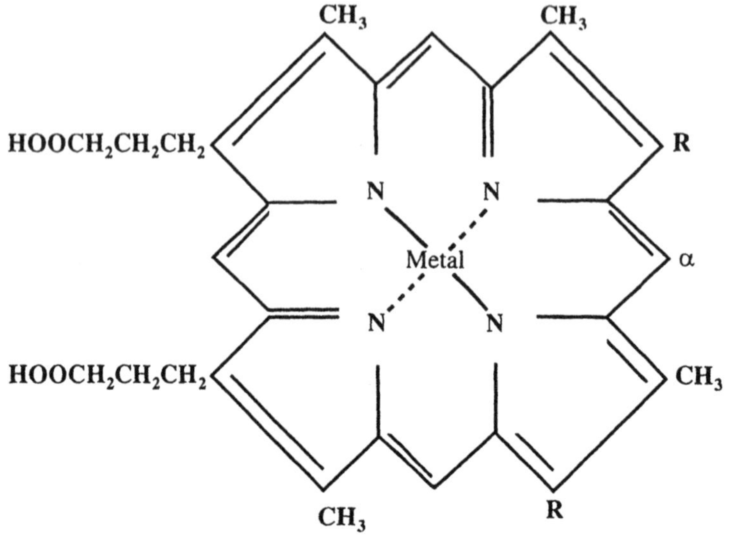

Porphyrin Type Based on Ring Substituent

Metal	Deuteroprophyrin (R = –H)	Mesoporphyrin (R = –CH$_2$–CH$_3$)	Protoporphyrin (R = –CH=CH$_2$)	Bis Glycol Porphyrin (R = –CH$_2$OH–CH$_2$OH)
Iron (Fe^{2+})	FeDP	FeMP	**FePP (Hemin)**	FeBG
Zinc (Zn^{2+})	ZnDP	ZnMP	ZnPP	ZnGB
Tin (Sn^{4+})	SnDP	SnMP	SnPP	SnBG
Chromium (Cr^{2+})	CrDP	CrMP	CrPP	CrBG

FIGURE 2.3 Metalloporphyrin basic chemical structure with central metal and ring substituents and abbreviated nomenclature.

presented to demonstrate conclusively that the *in vivo* upregulation of HO-1 mRNA alone automatically translates into a quantitative increase in HO-1 protein and enzyme activity. Even if true, the source and availability of the heme substrate, i.e., from RBC hemolysis or from hemoprotein turnover, must be accounted for when speculating about the rates of CO formation on the basis of mRNA quantitation. It has been the authors' experience that heme is the limiting factor in the HO reaction in rat tissues.[55] This appears to be consistent with observations by others that only additional endogenous or exogenous heme availability upregulates HO-1 mRNA, protein, and total enzyme activity.[16] However, the minimal concentration of heme required to exceed HO activity reserve capacity needs to be established.

To study the inhibitory effects of MP inhibitors on HO, the authors administer a heme load sufficient to exceed HO capacity (i.e., all reserve HO activity is being used). Thus, when one irreversibly inhibits only a relatively small fraction of HO activity with MPs, the inhibition will be noted directly through a decrease in CO formation.[55] It must be stressed that even an observed increased HO activity, as measured *in vitro* in excised or homogenized tissues, does not necessarily quantitatively relate to the level of *in vivo* CO generation that reflects the regulation of physiological processes (see below). Much work is still required to untangle the interrelated mechanisms affecting the generation and biological activity of CO and the other products of the heme degradation pathway. Nonetheless, it is the combined activities of both isoforms that regulate tissue heme catabolism that lead to subsequent CO, BR, and iron release.

2.2.1.1.2 Bilirubin

Bilirubin, when formed in extrahepatic cells, is excreted into the circulation and transported to the liver bound to albumin. Depending on the plasma BR and albumin concentrations, only a very small portion (approximately 0.3 to 2.0 µg/dl or 0.005 to 0.034 µM) of BR remains unbound (i.e., toxic) in the circulation.[56] In the mature liver, BR is glucuronidated by glucuronyltransferase to BR mono- and diglucuronides, which are subsequently excreted with the bile into the intestine for elimination from the body. In human newborns, BR glucuronidation is virtually absent at birth and develops during the first 10 postnatal days. Thus, in the absence of a fully functioning excretion mechanism, BR progressively accumulates in the circulation and tissues (including skin) leading to the often-observed physiological jaundice in up to two thirds of all well, term infants.[57] Because it has been shown to be an antioxidant, moderate levels of albumin-bound BR may be beneficial to the newborn during the early transitional period after birth.[58,59] During birth, the O_2 environment of the neonate abruptly changes from 5% *in utero* to 21% *ex utero*.[60] The newborn body has several defense mechanisms with which to overcome this oxidative insult and hyperbilirubinemia is hypothesized to be a protective mechanism of last resort.[61–63]

When BR (and, hence, CO) production is elevated during episodes of hemolysis,[64–66] excessive amounts of BR accumulate, especially in the absence of a mature BR excretion mechanism, leading sometimes to neuronal BR toxicity, kernicterus, and even death.[67,68] When excessive hyperbilirubinemia (BR > 20 mg/dl or 340 µM) is diagnosed in neonates, two widely used treatments, phototherapy[69,70] and exchange transfusion,[71,72] are employed to remove injurious BR.

However, the suppression of BR formation is a more logical and preventive strategy, particularly when targeted to those individuals at risk.[73,74] By preventing the increase in BR (and CO) production associated with hemolytic disease, it should be possible to ameliorate the concomitant severe hyperbilirubinemia and ablate the risk of kernicterus.[73,75,76] This preventive strategy is based first on the early, rapid, and accurate identification of those infants at risk for severe hyperbilirubinemia due to elevated BR production. Because CO is produced in equimolar amounts with BR, the rate of BR production can be determined by measuring VeCO,[19,65,66,77] COHb,[78–80] or indexed by measuring ETCO.[19,81–83] All indices should be corrected for inhaled CO to reflect the most accurate values for endogenously produced CO and thus BR. New refinements in CO-monitoring technology have improved the diagnostic utility of these indices (see below).[19,78,83,84]

The second aspect of the preventive strategy is the administration of an effective and safe chemopreventive agent. HO activity, CO formation, and BR production can all be decreased *in vitro* and *in vivo* by treatment with MP inhibitors, which are structural analogues of heme (i.e., iron porphyrin)[12,85–87] (Figure 2.3). These compounds have been proposed as putative chemopreventive agents for the treatment of severe hemolytic neonatal jaundice. Several of these compounds, particularly the naturally occurring zinc protoporphyrin (ZnPP), are also being used in research systems using purified HO to intact animals, for the study of a biological role for CO.[88–90] To date, the identification of effective inhibitors has included several tin, zinc, chromium, copper, and manganese porphyrin complexes.[87,91–94] The tin MPs have been administered to human neonates and were found effective in suppressing excessive jaundice and obviate the need for phototherapy.[76,95] However, the physiological impact of modulating HO activity, iron release, CO formation, and BR production in the developing neonate is not fully known.

Bilirubin is not only of relevance to O_2-stressed neonates, but also to scientists studying the physiological role of CO. Because CO may also be formed through lipid peroxidative and photo-oxidative processes (see below), the presence and metabolic effects of a potent antioxidant that can scavenge singlet O_2 and/or free radical species, are important experimental considerations.

2.2.1.1.3   Iron
Among the products of heme catabolism, iron is perhaps the most essential. Unlike BR and CO, it is not eliminated from the body but enters the iron pool, where it is sequestered by transferrin for return to iron stores as ferritin, the major iron storage protein.[96] The mechanisms and regulation of heme iron reutilization are still poorly understood.[97] Both the Fe^{2+} and Fe^{3+} oxidation states of this metal have been shown to potently catalyze lipid peroxidative reactions[34,98] and have been associated with increased oxidative injury.[99]

With respect to the observations with *in vitro* iron-mediated CO production from cellular membranes[34,100] and the production of antioxidant BR, iron and BR, the products of the heme catabolic pathway, may contribute to total-body CO production and influence homeostasis far beyond that of heme degradation alone.

2.2.1.2   Nonheme Sources

Under normal physiological conditions, only small amounts (perhaps up to 14%) of CO are thought to be derived from other biochemical processes such as lipid peroxidation, photo-oxidation, and, possibly, non-HO-related heme degradation.[24,40] However, pathological processes may increase the rate of CO production to levels that contribute significantly to total CO production, even eclipsing that produced by HO.

2.2.1.2.1   Lipid Peroxidation
There is limited evidence that, under normal physiological conditions, a minor portion of the total-body CO production may be derived from the process of lipid peroxidation. In 1968, Nishibayashi et al.[101] observed the formation of small amounts of CO during NADPH-dependent peroxidation of microsomal lipids. Later, Wolff and Bidlack[100] reported CO production from Fe^{3+}-ascorbate catalyzed peroxidation of microsomes and isolated phospholipids. Exposure to carbon tetrachloride was also found to cause elevated CO excretion, which was attributed in part to the degradation of membrane lipids.[102]

Recently, the authors reported on several parameters that affect *in vitro* CO generation during iron ascorbate–mediated lipid peroxidation in rat tissue homogenate fractions, with particular attention to brain tissue.[34] Much effort is being directed to determine the role of CO in the regulation of this organ[16,88,103]; the process of lipid peroxidation must also be taken into consideration.

Further work is needed not only to demonstrate that lipid peroxidation occurs *in vivo*, but also to determine what conditions affect this process. It appears from recent studies, that this source of CO may also be increased during periods of oxidative stress such as in obstructive pulmonary disease.[104–106] These reports seem promising, but the CO measurements have all been made with handheld electrochemical instruments. These devices (such as the Bedfont EC50 Smokerlyzer) are primarily used in tobacco-smoking cessation programs for the measurement of significantly elevated CO concentrations (>5 ppm) in breath. However, electrochemical sensors have been shown to cross-react with a number of potential breath gases such as H_2, ethanol, H_2O_2, NO, etc.[107] For instance, manufacturer literature lists possible false elevation of 1 ppm CO/4 ppm H_2 for the EC50 instrument. Thus, it is important that clinical or diagnostic measurements be performed with instrumentation proved to be accurate yet insensitive to breath gases other than CO.

Years ago, the authors reported one incident, then unrecognized, of possible lipid peroxidation-mediated increase in CO production.[108] A premature infant with a VeCO of 11.5 μl CO/h/kg, as measured by gas chromatography, at 26 days postnatal was administered 15 mg iron orally on Day 28 and by Day 33, the VeCO had increased to 18.3 μl CO/h/kg. After treatment with the antioxidant vitamin E (300 mg, intramuscular), followed by 25 mg orally/day, the VeCO decreased to 13.6 μl CO/h/kg on Day 40, and to 9.4 μl CO/h/kg on Day 47. Experiments such as these need to be repeated under rigorously controlled conditions.

2.2.1.2.2 Photo-oxidation

Photo-oxidative processes are by nature only superficial because of the strong screening effect of the skin toward light. Studies with rat skin have demonstrated that potently active, short-wavelength blue light reaches into skin tissues below the dermis to the vascular bed.[109] However, even these superficial epidermal effects of light can be fatal when photosensitizers are present.[110,111] The authors have found significant CO production during *in vitro* photodegradation of organic compounds in the presence of MPs and natural (riboflavin) photosensitizers.[112] In addition, neonatal rat studies revealed that light exposure, in the presence of tin protoporphyrin (SnPP) but not ZnPP, was associated with significant increases in hepatic levels of conjugated dienes and malonaldehyde and decreases in polyunsaturated fatty acid levels.[113] Unfortunately, CO excretion was not measured, but based on the *in vitro* results one would expect dermal and, possibly, systemic CO production when small subjects are irradiated with light in the presence of photosensitizers. In fact, earlier work demonstrated that BR itself could be a photosensitizing agent.[114] The authors have also observed that *in vitro* incubation of BR, dissolved in 4% human serum albumin, generates CO during exposure to white light, but less than the rate for riboflavin.[115] Also, CO concentrations within incubators occupied by newborns have been found to increase when the newborns were exposed to fluorescent phototherapy lights.[116] A recent article reported that phototherapy significantly increased end-tidal breath CO, corrected for inhaled CO (ETCOc), in preterm infants from 2.1 to 2.6 ppm.[117] The results were attributed to increased light-mediated hemolysis with subsequent heme-related CO production. However, the authors postulate that CO production resulting from direct photo-oxidative and/or lipid peroxidative origins cannot be excluded entirely.

2.2.1.2.3   Bacteria

Bacteria, such as *Streptococcus fecalis*, *Proteus vulgaris*, *Proteus mirabilis*, *Proteus morganii*, and others can produce CO.[24,28,118,119] Postmortem COHb measurements of thoracic cavity fluids from drowning victims have been found to be increased due to the presence of bacterial growth.[120] Increased COHb levels have also been observed in trauma patients presenting with sepsis.[121]

Intestinal bacteria can also produce CO from heme in the presence of O_2 but not under *in vivo* anaerobic conditions.[116,119] In contrast, the authors found that intestinal bacteria produced CO from excreted heme and thereby masked the efficacy of administered SnPP in neonatal rats.[122] Only when animals were treated with antibiotics that eliminated intestinal flora were the expected inhibitory effects of SnPP observed.[55]

In addition, patients with upper respiratory tract viral infections have presented with increased exhaled breath CO concentrations of 5.6 ± 0.4 ppm, which decreased to 1.0 ± 0.1 ppm upon recovery.[123]

2.2.2   EXOGENOUS SOURCES

Since most of the exogenous sources of CO have been well documented (Chapter 4 by James Raub discusses this in greater detail), this chapter will only briefly discuss some of the sources and focus on those relevant to clinical and research studies; i.e., sources that may contribute to a variable extent to ambient air CO and, therefore, contribute to the amount of inhaled CO. Moreover, these sources may become a significant source of measurement error if their contribution is intermittent or unrecognized.

2.2.2.1   Atmospheric Carbon Monoxide

Of the world's annual global CO production, 40% arises from natural geophysical and biological sources originating from hydrocarbon oxidation (e.g., volcanic, marsh, and natural gases, forest fires), seed germination, algae, kelp, marine hydrozoans, and endogenous hemoprotein degradation by land animals.[28,124]

Human activities, such as fossil fuel combustion (internal combustion engines and heaters), refuse disposal, tobacco smoke, vegetation burning, etc., account for 60% of the annual global CO production.[28,124] This continuous production of global atmospheric CO is counterbalanced by its removal via several sinks such as plants and microorganisms,[28] with the most important being the reaction of CO with ambient hydroxyl radicals to form CO_2 occurring in the lower atmosphere.[124] The concentration of atmospheric CO is less than 1 ppm or 1 $\mu l/l$[28] with a half-life of approximately 2 months.[124]

2.2.2.2 Local Carbon Monoxide Levels

Local CO concentrations are affected by regional characteristics, such as seasonal, altitude, and latitude variations, degree of urbanization, and amount of vehicular traffic and populace.[124] Within a region, each microenvironment dictates the degree of CO exposure for each individual and his/her potential health risks. CO exposure is subject to personal behavior, occupation,[125] tastes, lifestyle (such as type and location of home),[126] type of interior appliances (cooking/heating),[127] mode of transportation (engine type),[128] habits, and exposure (e.g., smoking and/or proximity to smoke or smokers).[129] However, on average, the lower limit of ambient CO concentrations in urban areas, such as Palo Alto, CA, is somewhere between 0.4 and 0.5 $\mu l/l$.[66,80,107,130]

2.2.2.3 Carbon Monoxide from Organic Materials

It appears that a wide variety of natural and anthropogenic organic compounds and materials are subject to spontaneous oxidation resulting in the generation of CO. The authors have found that silicone rubber stoppers used in blood collection tubes leached CO, thereby significantly increasing the COHb concentration in the blood sample.[78] Levitt et al.[131] studied a variety of natural and synthetic materials under a number of different conditions and reported varying rates of CO production (up to 4.1 $\mu mol/g/24$ h) that increased with temperature, pH, and the presence of fluorescent light, but decreased in the absence of O_2. Earlier work by Rodkey et al.[132] showed that acrylic and polycarbonate plastics could absorb and then slowly release CO. In this case, there was no evidence that CO was produced through interaction of O_2 with the plastics. The results of these studies and others[116] emphasize the need for the selection of appropriate materials for sample containers, experimental devices, as well as foodstuffs for test subjects, used in the measurement of CO to rule out artifacts resulting from uncontrollable and nonspecific CO-generating sources.

2.2.2.4 Iatrogenic Evolution of Carbon Monoxide

2.2.2.4.1 Laparoscopy

Modern surgical operating room procedures have also contributed to human CO exposure in several, sometimes unexpected, ways. Laser and bipolar electrocautery surgery used during laparoscopic procedures lead to the production of smoke and relatively large amounts (209 ± 19 ppm) of CO in the body cavities. During prolonged laparoscopic procedures, this CO will enter the circulation and lead to increased COHb levels.[133] However, aggressive smoke evacuation and hyperoxic ventilation during abdominal surgery can actually lead to a decrease in COHb levels [mean ± SD (range)] from 0.70 ± 0.15 (0.44 to 1.20) to 0.58 ± 0.20 (0.30 to 1.33) percent total hemoglobin (tHb).[133,134]

2.2.2.4.2 Anesthesia

Anesthesia equipment has also been found to produce CO to which patients have been inadvertently exposed.[135] Several cases of unexpectedly high COHb values (up to 7.6%) were observed in patients undergoing general anesthesia.[136,137] This phenomenon was attributed to significant CO generation in closed-circuit anesthesia equipment resulting from the interaction of completely dry CO_2 absorbents with volatile anesthetics, desflurane, enflurane, or isoflurane, but not with halothane and sevoflurane. The use of barium hydroxide, instead of soda lime, and increased absorbent temperature exacerbated this problem.[138] However, it was determined that anesthesia could be safely performed when measures were taken to avoid the drying of the CO_2 absorbent.[137]

2.2.2.4.3 Pulmonary function testing

The test of single-breath CO diffusing capacity is the most widely used noninvasive test for the determination of the ability of gases to cross the alveolar–capillary membrane. CO (0.3%) is used because of its great affinity (in comparison to O_2) for hemoglobin and its normally low concentration in the body. This test is indicated to distinguish various conditions of airway obstruction and restriction, such as industrial lung disease, emphysema, pulmonary vascular diseases, and anemia. It is reported as the milliliters of CO that diffuse per minute across the alveolar–capillary membrane per mm Hg CO partial pressure. These measurements are corrected for hemoglobin, COHb, and altitude. The test increases the normal endogenous COHb of 0.5% to approximately 1% tHb.

2.3   INTERACTION OF CARBON MONOXIDE WITH OTHER COMPOUNDS

2.3.1   Hemoglobin

In addition to its primary role of transporting O_2, hemoglobin also binds to CO. This leads to the formation of COHb, which in adults has a half-life 2 to 6.5 h[5,28,139] and a normal range from 0 to 1.5% of total hemoglobin. In neonates, the half-life ranges from 11 to 20 min.[140] CO binds to hemoglobin at $1/10$ the rate of O_2 and alters the geometric configuration of the hemoglobin molecule.[141] As a result, the affinity of hemoglobin for O_2 greatly increases, leading to a decrease in the availability of O_2 to the metabolizing tissues. This in turn leads to CO-induced hypoxia.

The combination of CO with hemoglobin is defined by the Haldane equation[142]:

$$M = (COHb)(pO_2) \ / \ (O_2Hb) \ (pCO) \qquad (2.1)$$

where O_2Hb is the percentage of oxyhemoglobin, pO_2 and pCO are the partial pressures of O_2 and CO, and M is the factor expressing the relative affinities of CO and O_2 when hemoglobin is 50% saturated. The value for M of approximately 200 for normal adult human subjects can be considered to remain constant in normal and CO-poisoned humans. However, the value appears to vary significantly for blood in different subjects.[141]

2.3.2   Myoglobin and Other Hemoproteins

CO also binds to other, mostly extravascular, heme-containing molecules in the body such as myoglobin (the most abundant), cytochromes, catalases, peroxidases, sGC,

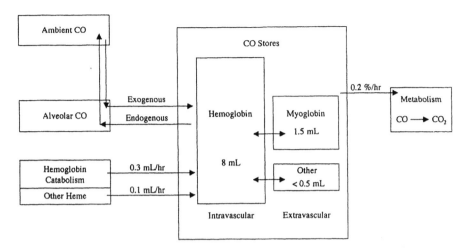

FIGURE 2.4 Schematic representation of CO body stores. (Modified from Coburn, R.F., *Acta Med. Scand. Suppl.*, 472, 278, 1967. With permission.)

NOS, and several others.[7] Although the affinity of CO for these hemoproteins is undoubtedly important, especially for normal oxygenation of heart muscle cells, presumptions should not be made regarding CO pathophysiology for substances other than hemoglobin.[7]

2.3.3 DISSOLVED CARBON MONOXIDE

CO is also slightly soluble in aqueous media such as plasma, which has been found to contain 22.0 ml CO/l H_2O/atm at 24°C.[143] However, this pool of CO is most likely negligible at partial pressures of CO found under normal physiological conditions.

2.3.4 OXIDATION TO CARBON DIOXIDE

In 1932, Fenn[144] described the biochemical removal of CO from the body through its oxidation to CO_2. It was found to be greatest in isolated skeletal muscle and Fenn hypothesized that this conversion is mediated in the mitochondria via cytochrome oxidase. This observation was then confirmed by Stannard[145] in 1940. In 1987, Coburn and Forman[7] concluded that CO metabolism by the tissues is minimal, approximately 0.2%/h of the rate of endogenous CO production. Luomanmäki and Coburn[146] later observed that, as the body stores of CO increase, the rate of CO oxidation also increases. They determined that the rate of tissue CO consumption increased tenfold at COHb levels of 19%.

2.4 CARBON MONOXIDE BODY STORES

CO body stores have been studied extensively by Coburn et al. and the reader is directed to excellent reviews on this subject.[17,30] The human adult body contains approximately 10 ml (448 μmol) CO (see Coburn and Figure 2.4). Approximately 80% of the CO is bound to hemoglobin in circulating erythrocytes. It has been estimated that the remaining 15% is bound to myoglobin and less than 5% to other compounds. Less than 1% is unbound and dissolved in body fluid.[6] The CO body stores are derived from two sources: endogenous production and exogenous pulmonary uptake.

2.4.1 UPTAKE

Inhaled CO readily diffuses across the alveolar–capillary membrane and its uptake is dependent on a number of various physiological variables such as air temperature and humidity, alveolar ventilation, hemoglobin concentration, and the rates of endogenous CO production, CO consumption, and elimination.[28] Because of all these interacting factors, equilibrium between inhaled CO and COHb levels is highly unlikely even in an ideal steady state. However, estimates of CO uptake (and, therefore, COHb formation) and excretion can be fairly accurately predicted using the Coburn–Forster–Kane equation.[147-149]

2.4.2 DISTRIBUTION

Once CO is produced by the reticuloendothelial system, it is removed via the blood as COHb to be excreted by the lungs.[42,150] However, in its course through the body, COHb equilibrates with the extravascular hemoproteins, which are also the recipients of locally produced CO, possibly resulting in different local CO concentrations. During the 1960s Coburn et al.,[6] in a series of elegant and precise studies, considered many of the issues related to CO production and body stores and quantitated the contribution of the most important processes (Figure 2.4).

2.5 ELIMINATION OF CARBON MONOXIDE FROM THE BODY

2.5.1 EXCRETION

Most of the CO inhaled or produced by the body is reversibly bound to hemoglobin and is excreted across the alveolar–capillary membrane during normal respiration. This rate is affected by COHb levels, alveolar ventilation rates and volumes, pulmonary diffusing capacity, and altitude. Because of its high binding coefficient, CO dissociates from hemoglobin at an extremely slow rate. Under standard temperature and pressures, the half-life of COHb in adults is relatively long (3 to 6.5 h).[5,28,139] However, breathing 100% O_2 at 2 atm in a hyperbaric treatment chamber reduces the half-life to 23 min.[28] For neonates, who have higher respiratory rates and smaller blood volumes, the half-life of COHb has been found to range from 11 to 20 min.[140]

At lower atmospheric pressures, the rate of CO excretion is reduced. In addition to affecting hemodynamic changes (such as erythrocyte production rate, hypoxemia), altitude decreases O_2 uptake and prolongs CO excretion. In short, COHb levels tend to increase with increasing altitude.[151] Consequently, CO can have greater toxic effects and sequelae at the high altitudes.[129,152]

Under normal conditions, CO excretion across the skin does not significantly affect total-body CO stores in adults.[41] However, in premature infants who have skin

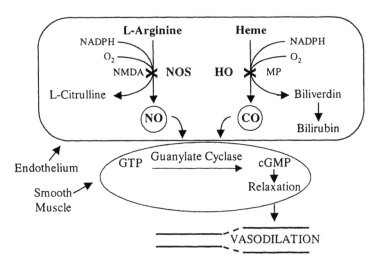

FIGURE 2.5 Schematic representation of the similar affects of NO and CO as biological messengers. NOS is inhibited by *N*-methyl-D-aspartic acid (NMDA) and HO is inhibited by MP derivatives of heme.

that is unusually permeable to H_2O and CO_2,[153] transdermal excretion of CO may be significant. For example, *in vitro* measurements of permeability to salicylate showed that absorption was 102 to 103 times greater in skin from premature (<30 week) infants than that of term infants.[154] However, no values for transdermal CO excretion in newborns have yet been reported.

2.5.2 METABOLISM

As described above, CO is metabolized by oxidation to CO_2. However, in the human body this rate of tissue CO consumption appears to be small. Yet, it would not be too surprising if this process is discovered to be a significant factor in the physiological role of CO (see below).

2.6 PHYSIOLOGICAL ROLE OF CARBON MONOXIDE

"Carbon monoxide — in suitable doses — may improve memory, learning ability, and alertness!?"

— Lewis A. Barness, M.D., 1995[19]

Evidence is accumulating that endogenously produced CO is not just a passive by-product of heme degradation or iron recycling, but that it may also be an important physiologically active gas. Many recent publications suggest that CO may function as a neuronal second messenger involved in heme-dependent signal transduction with properties similar to that of nitric oxide (NO, see Figure 2.5).[13,14,155] NO is produced by NOS from L-arginine.[156] It activates cytosolic sGC through formation of an NO–heme complex to form cyclic guanosyl phosphate (cGMP) from guanosyl monophosphate. In 1987, Brüne and Ullrich[157] found that CO inhibited human platelet aggregation and was mediated by a fourfold activation of sGC.[157] Vedernikov et al.[33] hypothesized that if CO could activate sGC in smooth muscle as well, CO-mediated relaxation in blood vessels, similar to that with NO, should be observed. Their studies with intact and endothelium-denuded canine femoral artery rings confirmed this hypothesis. They further speculated that CO, produced during lipid peroxidation,[100] might also modulate blood vessel tone. They also suggested that the direct effect of CO-mediated relaxation on smooth muscle is involved in its intoxication process. Marks et al.,[13] in a creative paper, compiled the available evidence on NO and CO and concluded that a good case could be made that CO has a physiological function. They suggested what types of experimental investigations could be conducted to test the hypothesis. Most of these suggestions have now been realized. As a result, it is now postulated that CO may regulate the production of cGMP through an activation of sGC.[14,15,17,18]

The observation that HO activity in brain (HO-2) microsomes was similar to that of microsomes isolated from the spleen (HO-1), an organ specifically designed for the degradation of heme[15,29] gave birth to the idea that HO in the brain may have functions other than the turnover of heme. It was subsequently shown that HO activity and inhibition by ZnPP correlated with the generation of cGMP.[88] Zhuo et al.[89] showed that both NO and CO produced presynaptic enhancement of long-term potentiation, a mechanism thought to be involved in learning. HO inhibition by MPs blocked this effect. An increasing number of physiological studies on CO have documented the importance of the HO–CO–cGMP pathway in the central nervous system.[15,17]

CO has also been implicated in vascular smooth muscle relaxation,[158] blood pressure regulation in rats,[159] and inhibition of myometrial contractility in humans,[160] all of which suggest that CO may have a possible role in smooth muscle contractile physiology.[16,18,161,162]

CO has also been associated with cat and rat carotid body sensory activity. Carotid bodies are the sensory organs that regulate ventilation by responding to alterations in blood pO_2, pCO_2, and pH.[163] Olfactory neuroreceptors are also potently activated by low ($K1/2 = 2.9$ μM) concentrations of exogenous CO, again via activation of sGC.[164]

However, it is not only CO that affects physiological processes. Other components of the heme catabolic pathway can also play a role. For instance, HO-1 is induced during periods of oxidative stress.[16] In addition, previous studies have indicated that intracellular heme levels can affect protein phosphorylation,[165] protein synthesis,[166] and cellular differentiation.[167] Furthermore, BR was also found to inhibit protein phosphorylation.[168] Therefore, cellular adaptation to a changing environment (e.g., development and differentiation) may involve the coordinated regulation of intracellular heme, CO, and BR levels through regulation of HO activity.[16]

Many of the studies on the physiological role of CO employ MP inhibitors of HO to suppress the production of HO and, thus, so it is believed, specifically to suppress CO production. However, Ignarro et al.[169] found that ZnPP and manganese protoporphyrin competitively inhibit sGC enzyme activation and activity. Meffert and co-workers[155] have determined that zinc bis-glycol porphyrin, ZnPP, chromium mesoporphyrin (CrMP), and SnPP (Figure 2.3) also inhibit the activity of the hemoprotein NOS. Furthermore, a recent study demonstrated that several MPs are capable of inhibiting all three enzymes, but for some MPs, like CrMP, there exists a concentration at which only HO is inhibited.[170] Like HO and sGC, NOS is also widely distributed in the body. Thus, because these enzymes have essential roles in metabolism, MP administration may have far-reaching consequences. Because of structural similarities and photoreactivity, MPs may affect, directly or indirectly, other heme-containing enzymes by displacing heme from the active site of enzymes or through alterations in tissue heme availability.[171]

In addition, preliminary experiments by the authors[34] and published work by Imai et al.[172] have indicated that MPs can inhibit oxidation of membrane lipids and, therefore, lead to changes in intracellular or pericellular membrane integrity. Although this effect appears to be beneficial, this potential side effect of MPs needs to be studied further.

In contrast to the research describing CO as a toxic compound, the study of a physiological role for CO is only in its infancy. Much work is still needed not only to untangle the interrelationships between CO- and NO-producing processes and their inevitable metabolites, but also to gauge their impact on cellular homeostasis.

2.7 CARBON MONOXIDE QUANTITATION TECHNOLOGIES

2.7.1 SAMPLE COLLECTION AND STORAGE

Before CO can be quantitated, the matrix in which it is contained needs to be collected in suitable vessels for direct analysis or storage for transport. Gaseous samples, such as ambient or breath gas, have been collected for these purposes in glass and plastic syringes,[107,173] latex, polyethylene and Mylar bags,[37,83,174] Tygon tubing,[131] and metal containers (bivalves),[68] etc. Each of these devices and accessories have particular advantages for each sampling application with regard to ease of use, sample stability over time, and transportability.[175] As was discussed above in Section 2.2.2.3 the sample container material may contribute or remove CO or interfering gases from the sample under shipping and handling conditions. The magnitude of these processes, which is very much affected by the volume of the sample, needs to be assessed carefully before sampling or storage is attempted.

Liquid samples, such as blood and tissues, are usually collected into syringes and/or glass tubes of different volumes and closure types.[175-177] Again, in this instance, the appropriateness of a container and its storage conditions need to be carefully examined from the time of sampling until analysis time.[78,178] Storage temperature and the presence of bacteria and enzymes are additional factors that affect the initial level of CO in these samples.[116] CO concentrations in or those generated by solid samples may be less subject to change, but exposure to O_2, light, and elevated temperatures needs to be avoided to maintain sample integrity.

Once samples for CO determination have been collected, they can be analyzed by a variety of techniques. Some of the more common methods used today in the clinical or medical research environments are discussed below. Although the principles of these technologies may be equally applicable to environmental and air quality studies, this type of research usually requires repeated measurements of CO over time in often physically and chemically aggressive environments. These requirements mandate instrument design and characteristics different from that used in a laboratory or clinic.

2.7.2 CARBON MONOXIDE QUANTITATION

CO produced in living cells by any process is removed from tissues by binding to the hemoglobin of circulating RBCs to form COHb. When the erythrocytes reach the lungs, the hemoglobin-bound CO is exchanged for O_2, and the CO is exhaled with the breath. When the test subject is at physiological equilibrium, COHb, VeCO, and ETCO are correlated positively to the COHb concentration.[66,83] When corrected for inhaled CO, COHbc, VeCOc, and the ETCOc, measurements represent estimates of the rate of endogenous CO production.

The COHb level, the VeCO rate, and the ETCO concentration have been reported to correlate well in preterm and term neonates without pulmonary dysfunction over a wide range of heme catabolic rates.[66,81,179] The three CO quantitation methods have been used to estimate the rate of heme degradation to differentiate between physiological and pathological (e.g., hemolysis) rates of BR production.

2.7.2.1 Carbon Monoxide in Blood

The COHb content of blood is measured to assess the effects of CO exposure from environmental sources.[180,181] In addition, this determination is also used to diagnose hemolysis. A wide variety of methods have been employed to determine COHb levels.[182,183] These techniques are based upon two different principles: (1) those that measure the COHb fraction in blood directly via nondestructive (multicomponent) spectrophotometry and (2) destructive methods that measure gaseous CO with different types of detection systems after CO has been liberated from hemoglobin. A major drawback of any type of COHb measurement is that it is an invasive procedure. The drawing of blood carries the risk for infection, is uncomfortable, and often depletes already compromised hospitalized patients of much needed blood.

2.7.2.1.1 CO-oximetry

CO-oximetry, used mostly in clinical laboratories, is based on the observation that COHb has a unique visible light absorption spectrum. Thus, it can be spectrophotometrically distinguished from other hemoglobin derivatives such as O_2Hb, methemoglobin (MetHb), deoxyhemoglobin, and sulfhemoglobin, when the light absorption at several wavelengths is measured.[184]

Method

Using instrument-specific algorithms, the concentration of each hemoglobin derivative, including COHb, can be quantitated within a sensitivity of 0.1%. Up to 100 µl of blood sample is required, although approximately 1 to 2 ml are usually drawn from a subject.[177] For accuracy of O_2Hb measurements, it is important that blood samples are drawn into special syringes and analyzed within 30 min to avoid MetHb formation, even though the COHb concentration remains stable for 2 to 4 h.[178]

Several CO-oximeters are being marketed worldwide. The authors[177,185,186] and others[187] have evaluated the most widely used instruments in the 1990s and compared the results with those obtained with nonspectrophotometric methods such as GC and the cyanmethemoglobin reference methods. The authors found that differences between CO-oximeter tHb measurements were not clinically different for any of the models studied. For COHb measurements, however, the direction of the bias relative to the GC was dependent upon the COHb concentration. It was concluded that five CO-oximeters compared favorably with the reference method for higher concentrations of COHb. But, in general, the CO-oximeters overestimated COHb concentrations at <2.5% tHb. This inaccuracy of the CO-oximeter makes this methodology unsuitable for some applications.

Several studies have reported that CO-oximeter COHb measurements were spuriously elevated in direct proportion to the concentration of fetal hemoglobin in samples from umbilical cords and neonates.[72,188,189] The authors demonstrated that at least one instrument, the Corning 270 (Ciba Corning Diagnostics, Medfield, MA) is not affected by fetal hemoglobin, irrespective of the level of O_2Hb in the sample.[186] These studies also showed that the use of CO-oximeter COHb measurements is excellent for clinical measurements; however, for research studies, particularly when values close to the reference range (<2.5%) are to be expected, these instruments are inaccurate and may be subject to significant error.

GC is the method of choice for determining very low proportions of COHb (<1.5% tHb); (see below). Furthermore, samples to be assayed by GC are more stable than those used for spectrophotometry, and the required sample volume is much smaller, 1 μl vs. 100 μl, respectively.[78,190] Thus, GC is also the method of choice for analyzing blood samples from infants and in situations where immediate, on-site analysis is not possible.[191]

2.7.2.1.2 Gas chromatography

Although CO-oximeters are clinically accurate and convenient instruments for measuring COHb for diagnostic purposes, they are generally not suited for use in research where accuracy and sensitivity are paramount over speed and ease of use. GC is considered to be an excellent reference method for the quantitation of CO because it not only can be accurately calibrated with the National Institute for Standards and Technology or NIST-traceable standard gas, but it also can accurately measure low CO concentrations.[25,26,143,192] The technique is very sensitive, highly precise, and not susceptible to interfering substances. However, the use of this technique is primarily limited to research laboratories because of the bulk of the instrumentation and the need for skilled operators.

Method

Collison et al.[8] described a method by which COHb was dissociated by oxidation with $K_3Fe(CN)_6$ to MetHb. This reaction releases the CO into the reactor headspace from which it was injected into a GC for separation from potential interferents. The CO was quantitated with a flame ionization detector after its catalytic reduction to methane in a methanizer. The method was rapid, specific, and required 100 μl of blood per analysis. This method was subsequently modified by the authors, to increase sample throughput time, accuracy, and sensitivity significantly, and to decrease sample volume.[66,78] Subsequent developments led to the preparation of specialized sample tubes.[190] The tubes, still in use today, measure 76 × 3.0 mm outside diameter × 1.7 mm inside diameter, and contain 1 mg saponin and 0.5 U sodium heparin. They can be filled with up to 160 μl of blood and, after inserting a stainless steel mixing bar, both ends of the tube are sealed with silicone stoppers. The blood in the tubes is stable for up to 2 months at 4°C, but can be stored frozen at −20°C, as well, for an as-yet-undetermined time.

For COHb analysis, 1 μl of blood is injected through a septum cap into a CO-purged 2-ml GC vial into which 20 μl of $K_3Fe(CN)^6$ reagent has been pipetted.[78,190] The vials are placed on ice for 30 min. The CO released into the headspace is injected into the GC via a double-needle assembly attached to the sampling valve. The CO is separated from the other volatile and reducing gases (such as H_2) on a column packed with 13× molecular sieve, operated at 140°C, with an air carrier flow rate of 30 ml/min. CO is detected directly with a reduction gas detector and quantitated via peak area integration. The CO retention time is 0.65 min. The GC is calibrated each analysis day with known volumes of CO in air. The tHb in the blood sample is measured spectrophotometrically with the cyanmethemoglobin method. The COHb concentration is expressed as the percent tHb that is bound to CO. The method is linear for COHb concentration less than 4.5%. Blood dilutions can extend this range to 100%. Sample preparation and analysis time is approximately 2.5 min/sample. Samples are analyzed in duplicate; 1 μl of blood of 1.1% COHb (14 g tHb/100 ml) contains approximately 2.5 nl (120 pmol) of CO.

The concentration of COHb (% tHb) in blood is calculated as follows:

$$COHb\ (\%\ tHb) = [ml\ CO/100\ ml\ blood]\ /\ [(g\ tHb/100\ ml\ blood)(1.39)]\quad (2.2)$$

where 1.39 is the CO/Hb binding constant (Hüfner's factor).[193]

To relate COHb measurements to endogenous CO production, a correction for inhaled CO needs to be performed. For preterm and term neonates, the following equation has been derived:

$$COHbc\ (\%\ tHb) = COHb\ (\%\ tHb) - 0.17\ Room\ Air\ CO\ (ppm)\quad (2.3)$$

For continuous exposures to <100 ppm CO, the California State Department of Public Health derived a correction factor of 0.16 Room Air CO.[180] However, most of the COHb values published to date have not been corrected for inhaled CO. Table 2.1 lists representative values for COHb measured in different subject categories.

2.7.2.2 Carbon Monoxide in Extravascular Tissues

A search of the literature has yielded very little prior work on measurements of CO in extravascular tissue. Previous studies have measured total tissue CO representing vascular as well as extravascular CO in postmortem human tissues.[194] Others have measured CO in rat tissue blood after CO exposure to estimate organ blood volume.[195–197] Only Sokal et al.[198,199] have reported values for extravascular tissue

TABLE 2.1

Representative Values for Total-Body CO Measurements

Subject		COHb, % tHb	VeCO, µL/kg/h	ETCO, µl/l (ppm)
Fetus	Normal	1.11 ± 0.14[80]		
	Hemolytic	1.59 ± 0.72[80]		
Newborn	Premature	0.86 ± 0.28[a39]	17.2 ± 4.4[66]	1.7 ± 0.5[39]
	Ventilated premature	1.06 ± 0.40[226]	22.8 ± 5.5[226]	
	LGA-IDM	0.63 ± 0.13[66]	19.9 ± 7.0[66]	
	Pulmonary dysfunction	0.49 ± 0.15[66]	19.7 ± 6.8[66]	
	Term	0.45 ± 0.13[226]	13.4 ± 3.2[226]	1.6 ± 0.4[a82]
	Rh	2.13 ± 1.11[37]	36.0 ± 22.2[37]	5.7 ± 0.4[173]
	ABO	1.20 ± 0.42[37]	17.4 ± 8.4[37]	2.0 ± 0.7[173]
	Glucose-6-phosphate deficient	0.74 ± 0.14[a208]	2.0 ± 0.5[a207]	
	Down's syndrome	0.92 ± 0.17[a191]		
Adult	Normal	0.56 ± 0.11[107]	6.6 ± 1.9[227]	2.6 ± 0.7[107]
	Smoker	Up to 14[181]		Up to 65[181]
	Asthmatic			5.6 ± 0.6[104]
	Bronchiectasis			6.0[105]
Animal	Rat ($n = 29$)		17.2 ± 4.7[228]	
	Mouse ($n = 70$)	0.34 ± 0.11[228]	44.3 ± 10.7[228]	

Note: The values listed have been selected, for illustrative purposes only, from a wide range of values reported in the literature. The magnitude of each value depends on a number of factors, including the sensitivity and accuracy of the methodology used, the possibility of inhaled CO > 0.5 ppm, usually found in unpolluted air, sampling efficiency, etc.

[a] Corrected for inhaled CO.

CO in rat heart and skeletal muscle after 4 min to 12 h exposure of rat to levels of 0.12 to 1.0% CO. These exposure conditions led to high, near-fatal blood COHb levels of 48 to 76%, or 111 to 179 µl CO/ml tissue. Extravascular CO in heart muscle, representing 49 to 65% COHb, measured 2.8 to 3.4 µl/ml tissue, about a 50-fold concentration difference. CO values for skeletal muscle ranged from 0.73 to 1.26 µl CO/g tissue (~150×). Brain total CO was 30 to 50 times lower than that of blood. However, the methods used were not sensitive enough to measure extravascular CO under normal and pathological conditions. The relatively high COHb content of blood may overwhelm the amounts of extravascular CO produced for physiological functions.

The recent interest in a possible physiological role for CO prompted the authors to determine if the sensitive GC method for measuring COHb could be adapted to measure CO levels in the extravascular space of other tissues. *In vitro* measurements of HO (or lipid peroxidative) activity are not likely to be informative concerning the physiological role of CO because these measurements indicate the potential for

TABLE 2.2
Preliminary Tissue CO Values (pmol CO/mg fresh weight)

Species	Blood	Spleen	Heart	Kidney	Muscle	Liver	Brain	Lung	Intestine	Testes
Rat	47 ± 10	11 ± 3	6 ± 3	5 ± 2	4 ± 4	4 ± 1	2 ± 1	2 ± 1	2 ± 1	1 ± 1
Mouse	45 ± 5	6 ± 1	6 ± 1	7 ± 2	10 ± 1	5 ± 1	2 ± 0	3 ± 1	4 ± 2	2 ± 1

in vivo CO generation.[34,45] They do not reflect the amounts of CO that will actually be formed under various *in vivo* conditions. The speculation is that if CO has a role in tissue or organ regulation, one may be able to measure different tissue CO levels under different physiological conditions. However, to test this hypothesis, the tissues need to be carefully stripped of intravascular COHb and intracellular CO should not be removed too quickly through binding to hemoglobin in the circulation.

Method

The authors have begun to develop a sensitive method for the determination of extracellular CO in organ tissue.[200] Tissues, harvested from adult rats and mice, are blanched or devascularized, diced, and then sonicated with three volumes of cold phosphate buffer. First, 40 μl of the sonicate (representing 10 mg tissue, 2.5 mg for blood) is incubated in ice with 10 μl of 50% (w/v) sulfosalicylic acid in CO-free vials. After 30-min incubation, the CO liberated into the vial headspace is quantitated by GC. The preliminary results for analysis of basal tissue CO concentrations are listed in Table 2.2. Between the rat and mouse tissues, only the spleen and muscle were found to have significantly different ($p < 0.05$) CO concentrations. It was concluded tentatively that while blood has the highest CO concentrations, tissues such as spleen, heart, kidney, skeletal muscle, and liver, which are relatively rich in myoglobin, cytochromes, and other hemoproteins, also contain substantial amounts of CO. In contrast, brain, lung, intestine, and testes were found to contain the lowest concentrations of CO. The authors intend to increase the sensitivity of the assay to study total and extravascular tissue CO concentrations in animals exposed to moderately increased levels of inhaled CO and under various conditions that elevate or inhibit endogenous CO production. These tissue CO measurements will also be correlated with VeCO levels.

2.7.2.3   Carbon Monoxide Air and Breath

2.7.2.3.1   *Ambient air carbon monoxide concentrations*

Ambient air CO concentration has been discussed above. The importance of its measurement relates to the correction of COHb and breath CO measurements to calculate the rate of CO produced endogenously. Inhaled air CO is approximated through measurements of CO in room air by the same method as with breath measurements. For neonates, a straight subtraction of the concentration (μl/l) of CO in room air from that in breath is performed to estimate the rate of endogenous CO formation. For adults, who have substantially greater COHb half-lives (approximately

5 h vs. 40 min) and who are generally exposed to more variable environmental CO pools, a more elaborate correction procedure has been proposed.[201]

2.7.2.3.2   Breath carbon monoxide measurements

Hyperbilirubinèmia continues to be one of the more common serious threats to the well-being of the human infant during the neonatal period. One of the major contributory factors is the production of BR, which is significantly increased in the human neonate compared with adult male subjects on a per unit body weight basis.[57] The ability to estimate the rate of BR production through measurements of CO is contingent upon the assumption that the heme catabolic pathway is by far the major source of CO and that the degradation of heme yields CO and BR in equimolar quantities.[43] Based on the results of simultaneous studies of daily plasma BR turnover and CO production in 37 adults, Berk and co-workers[32] concluded that the quantitative contribution to total CO production from processes other than heme catabolism or from alternate pathways of heme catabolism that may not result in the formation of BR must be small (<14%). During periods of stress, these processes, such as lipid peroxidation (see the appropriate section) and photo-oxidation (see the appropriate section), may produce significant amounts of CO to affect total CO production. However, in a number of separate studies with exogenously administered heme preparations, recoveries of BR and CO were observed to be close to 100%.[4,32,39,202]

Method

Several methods have been devised to measure indirectly the rate of BR production, which equals total-body CO production rate (VCO). The VCO can be estimated through the measurement of COHb, VeCO, and ETCO. All three quantities need to be corrected for inhaled (ambient) CO. It should be noted that for the pediatric population, these tests are performed only to assess the formation of BR. However, the tests have potentially broader applications in that they may also be used to monitor the effects of drug therapy such as the administration of MPs, iron, heme arginate, etc. Furthermore, the tests may also be used to monitor processes other than heme metabolism that generate CO, such as photo-oxidation.

2.7.2.3.3   Total-body carbon monoxide production measurements

Total-body CO production measurements are the most accurate approximations of VCO. However, the measurement is fairly cumbersome for human subjects because it requires that the subject be placed in a totally sealed chamber with purified, reconstituted, and recirculated air (closed system) or with a sealed chamber continuously supplied with a fresh supply of CO-free air (flow-through system). The closed system used in earlier studies relied on measurements of COHb that increased during the study period and from which the rate of CO accumulation (production) could be mathematically derived.[6,64] The system requires CO_2 scrubbers and a continuous monitoring and resupplying of O_2. The flow-through system with its constant supply of air is simpler, but requires a fairly lengthy subject equilibration time when the air CO concentration of the system is less than ambient air.[65,66,87] Furthermore, this method requires more-sensitive CO quantitating equipment because of the dilution of exhaled gas CO with chamber airflow. For human subjects, a sealed head hood

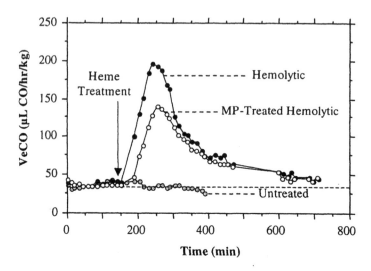

FIGURE 2.6   Representative sequential VeCO (μl CO produced/h/kg) measurements in untreated adult mice, after intraperitoneal injection of 30 μmol heme/kg body weight (hemolytic), and after intraperitoneal injection of 30 μmol heme/kg body weight 20 h after intraperitoneal injection of 20 μmol ZnPP/kg body weight (MP-treated hemolytic).

can be used to differentiate between exhaled pulmonary CO and CO eliminated from the remainder of the body.[81,201]

These measurements have been applied to differentiate physiological from pathological (i.e., hemolysis) heme degradation. In 1979, Bartoletti et al.[65] determined VeCO rates in newborns with various ethnicities, gestational ages, and clinical conditions. They found that the rates decreased as postnatal age increased and that the mean VeCO rates in Asian newborns were higher in comparison with those in Caucasians. In addition, Rh-negative infants displayed higher VeCO rates than ABO hemolytic infants (39.2 ± 20.2 and 32.1 ± 11.4 μl CO/h/kg, respectively), and these infants had increased rates as compared with normal infants as well.

Method

The flow-through system for VeCO measurements in humans has been described[203] and was used for several years.[65,66,108] However, these measurements are no longer performed on human neonates because it is too cumbersome for today's clinical setting. Furthermore, ETCO measurements, which are much more convenient and less time-consuming (see below), have proved to be a good index of COHb and VeCO levels.[81,83,173]

Animals, however, are still most conveniently studied by this method for factors that affect endogenous CO production.[87,204,205] The results of a typical VeCO experiment demonstrating the effect of ZnPP administration to a hemolytic mouse model are given in Figure 2.6. Because of their relatively small size, these animals require only small chambers, low airflows, and short equilibration times. Through development of specialized equipment, including the GC with reduction gas detector, one can study up to 12 neonatal or adult rats and mice simultaneously and reproducibly.

Animals are weighed and placed in airtight Plexiglas chambers supplied with CO-free air at a flow rate of 100 ± 10 (adult rats) or 30 ± 3 ml/min (neonatal rats or adult mice). After 1 h (30 min for small animals) of equilibration, gas exiting the chamber is analyzed for CO concentration by GC.[45] Each chamber is sampled at least once every 30 min. GC is the most sensitive, accurate, and specific method for quantitating CO at levels as low as 0.01 ppm CO. VeCO measurements are calculated as follows:

$$VeCO = W(CO_{out} - CO_{in}) / wt \qquad (2.4)$$

where W is the chamber flow rate in l/h; CO_{in} is the concentration of CO entering the chamber in $\mu l/l$ (ppm); CO_{out} is the concentration in air leaving the chamber in $\mu l/l$; and wt is in weight of the subject in kg. Approximate values for normal and pathologic conditions are presented in Table 2.1 for both human and animal subjects. The VeCO method is especially suited for the study of animals at various developmental stages after treatments with MPs, heme, or other agents, in order to study their effects on CO production, not only via heme degradation pathway, but also via other processes, like photooxidation or lipid peroxidation.

2.7.2.3.4   End-tidal carbon monoxide measurements

Although easily and quickly performed, noninvasive ETCOc measurements are, at best, only estimations of the rate of BR production. However, the collection of uncontaminated (diluted) end-tidal breath samples from spontaneously breathing, but uncooperative subjects, such as neonates, children, and incapacitated adults, posed a problem.[206] This has been resolved recently with the introduction of the CO-Stat™ End Tidal Breath Analyzer (Natus Medical, Inc., San Carlos, CA).[83] This integrated sampling (analysis) instrument measures the respiratory rate, end-tidal CO_2 concentration ($ETCO_2$), and room air CO, in addition to ETCOc. It has been determined to have a sampling efficiency of >97%.[83] The ETCOc test is primarily used to estimate blood COHb levels, which in turn reflect hemolytic conditions, such as Rh[81,206] and ABO isoimmune diseases,[206] glucose-6-phosphate deficiency,[68,207,208] thalassemia,[84,209] and sickle cell disease.[84]

Recently, several publications have reported significantly elevated ETCO levels in patients with asthma and bronchiectasis.[104,105] These CO production rates were tentatively attributed to lipid peroxidative processes isolated to the lungs, because the correlation of ETCO measurements with those of COHb was weak. Although the measurements were made with relatively insensitive electrochemical instruments, which are usually used in smoking clinics, the reports are provocative. Further studies using more sensitive and discriminating instrumentation with no cross-reactivity to other known breath gases, such as H_2, NO, H_2O_2, etc., need to be performed to rule out possible artifactual results. After all, it is intriguing to contemplate that pulmonary CO production alone might nearly double the rate of total-body CO production.

End-tidal CO measurements have been performed for many years with a variety of instruments to monitor exposure to environmental CO[210] and tobacco smoking assessment.[181,211,212] In many cases like those cited above, cross-reactivity with other breath or environmental gases may not be of great import, but this potential confounder needs to be considered for each application to limit the effects of interfering

substances.[107] With the development of accurate ETCO instruments, hemolytic neonates may be identified and treated in a timely manner or observed more closely
during antihemolytic therapy.[213] The dosing and efficacy of the MP or other drug
therapies can also be easily monitored with this instrumentation.

Method

End-tidal breath CO methodology for uncooperative neonates has evolved from
manual sampling with <77% (estimated) efficiency and CO quantitation by GC[81,173]
via automated sampling (60% measured efficiency) and electrochemical CO
quantitation[206] to a fully integrated sampling/quantitation procedure with an instrument that quantitates alveolar air with 97% efficiency based on end-tidal CO_2
measurements.[83] The CO-Stat instrument is designed for use at the bedside of
uncooperative neonates. It samples breath via a 5-French (0.8 mm i.d.) catheter,
passes it through disposable moisture and organic solvent traps, measures the breath
CO_2 concentration with a miniature infrared transducer, and finally measures the
CO concentration with an electrochemical sensor. A second sensor measures breath
H_2, compensates the CO sensor for this concentration, and, if the breath H_2 concentration exceeds 50 ppm, the instrument refuses to measure breath CO. Bench
tests have shown that the CO measurements of model gas mixtures are linear,
accurate, and precise when compared with GC. *In vivo* tests performed on adults
($n = 30$) showed excellent correlation between ETCO and GC. When corrected for
inhaled CO, ETCOc values correlated strongly with COHbc (COHbc = 0.25 ETCOc
− 0.01 µl/l CO, $r^2 = 0.97$) with a coefficient of variation for triplicate determinations
of 11%. Measurements on healthy and hemolytic term neonates indicated that
ETCOc > 3 µl/l correlated with hemolytic condition. Representative ETCO values
are given in Table 2.1.

2.7.2.3.5   In vitro *carbon monoxide assays*

The *in vitro* quantitation of CO by GC, as described for COHb quantitation, is an
important procedure in the authors' research program. It not only is an accurate,
rapid, and versatile method, but also serves as the model for other CO-based assays
described below.[45]

Tissue collection and preparation

Animals are usually sacrificed by decapitation. The organs are removed, blanched,
and rinsed with ice-cold buffer (0.1 M KPO_4, pH 7.4), and homogenized with four
volumes of buffer. Originally, the homogenates were centrifuged at 4°C for 15 min
at 13,000 × g and the supernatant analyzed within 2 h. At present, the authors prefer
to use sonicates so that measured activity can be directly related to the weight of
tissue used as well as to total tissue protein.

Heme oxygenase assay

HO activity of tissue preparations is determined with a GC assay that measures the
amount of CO produced from methemalbumin (50 µM/4.7 µM) in the presence of
NADPH (1.5 mM) as previously described.[45] CO production in the presence of
NADPH (total) and buffer (blank) is determined in duplicate. The protein concentration of the tissue preparation is determined by the method of Lowry et al.,[214] using

bovine serum albumin as a standard. HO activity, defined as the difference in CO production between the total and blank reactions, is expressed as nmol CO produced/h/mg protein. Lately, the authors have been expressing HO activity on the basis of the pmol CO produced/h/mg fresh weight tissue. This normalization procedure more accurately reflects HO activity per weight or volume of tissue.[162]

Lipid peroxidation assay

Supernatants, from 20% tissue homogenates in potassium phosphate buffer, centrifuged for 1 min at 13,000 × g, were incubated for 30 min at 37°C in septum-sealed vials in the dark with ascorbate (100 μM) and Fe^{2+} (6 μM) and/or Fe^{3+} (60 μM). Butylated hydroxytoluene (100 μM) was added for the blank reaction. CO produced into the headspace was quantitated by GC. Thiobarbituric acid reactive substances (TBARS), conjugated dienes, and lipid hydroperoxides in the reaction medium were quantitated by spectrophotometry. Of the tissues studied, the rate of CO (and TBARS) formation was greatest for brain and spinal cord, followed by kidney, lung, spleen, and blood, but no CO (or TBARS) formation was detected from testes, intestine, liver, and heart. Cell fractionation studies indicated that the differences appear to be due to the presence of endogenous soluble antioxidants in the latter tissues. Furthermore, these studies demonstrated that CO was generated exclusively by subcellular fractions that contained membrane materials. To assess the potential for lipid peroxidation–mediated CO production relative to that of heme degradation, the rates of CO formation by each of the two processes for rat brain homogenates were calculated, using *in vitro* data. HO activity was shown to produce approximately 70 pmol CO/30 min/4 mg tissue, whereas lipid peroxidation reactions yielded 250 pmol/30 min/4 mg tissue.[34] Although extrapolation of *in vitro* results to *in vivo* significance is risky, these results show that lipid peroxidative reactions have the potential to contribute significantly to total-body CO stores under conditions of oxidative stress and low antioxidant reserves. Thus, if the peroxidative process could produce CO for the regulation of physiological processes, its potential could be on par with HO regarding CO production capacity.

Other in vitro *assays involving carbon monoxide*

The *in vitro* HO assay procedure can serve as a model for the *in vitro* evaluation of any other CO-producing process. The authors have adapted the HO activity method for the quantitation of the following processes — photosensitizer potency[112] and MP catabolism[87] — as discussed above. The method has also been adapted to measure HO activity in tissue slices[155] and cell cultures.[215]

2.8 APPLICATIONS OF CARBON MONOXIDE-MEASURING TECHNOLOGY

In conclusion, the authors are using *in vitro* and *in vivo* CO-measuring technology in their research efforts to assess parameters that affect endogenous CO production.

In vitro measurements of CO are being made primarily for the study of endogenous CO production as catalyzed by HO.[45] The basal and upregulated HO levels in a number of tissues that were obtained from newborn and adult rats,[216,217] neonatal monkeys,[205,218,219] and mice[204] at various developmental stages[220] and administration routes[217,221] have been determined. Furthermore, the inhibitory effects of a large number of MP inhibitors of HO supplied *in vitro* or administered *in vivo* have been studied.[87,94,170] Because MP administration to living subjects may lead to photosensitization, CO measurements are being used to assess photoreactivity of compounds *in vitro*[70,87,112] and *in vivo*.[111] Lipid peroxidative processes are also being studied *in vitro*[34] as a source for endogenous CO, which in turn may lead to an *in vivo* test that may indicate oxidative tissue damage.

In vivo CO measurements are also, for the time being, primarily performed to assess heme degradation particularly as it relates to BR formation. To this end, the authors use mainly measurements of COHb and ETCOc in human neonates and in several animal models. Hemolytic diseases such as glucose-6-phosphate dehydrogenase deficiency,[68,207,208,222,223] ABO,[65,79] and Rh isoimmune diseases,[64,65] sickle cell anemia,[84] and thalassemia[209] all lead to increased production and excretion of CO (and BR). However, different populations of premature and newborn babies on the basis of sex, ethnicity,[65,224] and health factors, such as infant of diabetic mothers,[225] polycythemia,[65] and sequestered blood[65] have also been studied.

Finally, it is possible that under some conditions other processes, such as lipid peroxidation, may also be found to contribute significantly to the rate of endogenously produced CO. This possibility could complicate measurements of CO due to erythrocyte degradation, but non-CO-based tests may be used to differentiate between the possible origins of CO.

2.9   SUMMARY AND FUTURE DIRECTIONS

The heme catabolic pathway comprises only two reactions. Nonetheless, it and its major enzyme (HO) exert influences not only through the removal of excess heme from various sources but also through the production of metabolically active compounds — the importance of which is just beginning to be appreciated. In addition, the interaction of this pathway with that of the NOS system may lead to even more exciting avenues of research. It may be shown that the integrity of the heme catabolic pathway, which is ever present and plays a role in every tissue, is central to the existence of most complex organisms. Through the study of perturbations on the HO systems, it will be determined whether CO is beneficial or detrimental to life.

Measurements of CO could be applied to expand understanding of (1) the role of CO in biochemical and physiological processes in all tissues and organs; (2) the magnitude and significance of the *in vivo* oxidation of CO (to CO_2) process as a potential regulatory mechanism of physiologically active intracellular CO; (3) the efficacy and safety of the administration of HO inhibitors, iron, and porphyrins; (4) the association between phototherapy and the transdermal excretion of CO in newborns; (5) the noninvasive assessment of RBC life span; (6) the response of hemolytic anemia patients to therapy; and (7) the importance of lipid peroxidative processes as sources of endogenous CO. These are but a few examples of how the application of measurements of CO can expand scientific and public health horizons.

ABBREVIATIONS

BR	bilirubin
cGMP	cyclic guanosine monophosphate
CO	carbon monoxide
COHb	carboxyhemoglobin
CrMP	chromium mesoporphyrin
ETCO	end-tidal carbon monoxide in breath
ETCOc	end-tidal carbon monoxide in breath, corrected for inhaled carbon monoxide
$ETCO_2$	end-tidal carbon dioxide in breath
GC	gas chromatography
HO	heme oxygenase
MetHb	methemoglobin
MP	metalloporphyrin
NO	nitric oxide
NOS	nitric oxide synthase
O_2Hb	oxyhemoglobin
RBCs	red blood cells
sGC	soluble guanylate cyclase
SnPP	tin protoporphyrin
TBARS	thiobarbituric acid reactive substances
tHb	total hemoglobin
VCO	total-body carbon monoxide production rate
VeCO	total-body carbon monoxide excretion rate
ZnPP	zinc protoporphyrin

REFERENCES

1. Roughton, F.J.W. and Root, W.S., The fate of CO in the body from mild carbon monoxide poisoning in man, *Am. J. Physiol.*, 145, 238, 1946.
2. Sjöstrand, T., Endogenous formation of carbon monoxide in man under normal and pathological conditions, *Scand. J. Clin. Lab. Invest.*, 1, 201, 1949.
3. Sjöstrand, T., The formation of carbon monoxide by the decomposition of haemoglobin *in vivo*, *Acta Physiol. Scand.*, 26, 338, 1952.
4. Coburn, R.F., Danielson, G.K., Blakemore, W.S., and Forster, R.E., Carbon monoxide in blood: analytical method and sources of error, *J. Appl. Physiol.*, 19, 510, 1964.
5. Coburn, R.F., Williams, W.J., White, P., and Kahn, S.B., The production of carbon monoxide from hemoglobin *in vivo*, *J. Clin. Invest.*, 46, 346, 1967.
6. Coburn, R.F., Endogenous carbon monoxide production and body CO stores, *Acta Med. Scand. Suppl.*, 472, 269, 1967.
7. Coburn, R.F. and Forman, H.J., Carbon monoxide toxicity, in *Handbook of Physiology, Section 3: The Respiratory System*, Vol. IV: *Gas Exchange*, Farhi, L.E., Ed., American Physiological Society, Bethesda, MD, 1987, 439.
8. Collison, H.A., Rodkey, F.L., and O'Neal, J.D., Determination of carbon monoxide in blood by gas chromatography, *Clin. Chem.*, 14, 162, 1968.

9. Tenhunen, R., Marver, H.S., and Schmid, R., Microsomal heme oxygenase. Characterization of the enzyme, *J. Biol. Chem.*, 244, 6388, 1969.

10. Maines, M.D. and Kappas, A., Study of the developmental pattern of heme catabolism in liver and the effects of cobalt on cytochrome P-450 and the rate of heme oxidation during the neonatal period, *J. Exp. Med.*, 141, 1400, 1975.

11. Maines, M.D. and Kappas, A., Enzymatic oxidation of cobalt protoporphyrin IX: observations on the mechanism of heme oxygenase action, *Biochemistry*, 16, 419, 1977.

12. Maines, M.D., Zinc-protoporphyrin is a selective inhibitor of heme oxygenase activity in the neonatal rat, *Biochim. Biophys. Acta*, 673, 339, 1981.

13. Marks, G.S., Brien, J.F., Nakatsu, K., and McLaughlin, B.E., Does carbon monoxide have a physiological function? *Trends Pharmacol. Sci.*, 12, 185–188, 1991.

14. Dawson, T.M. and Snyder, S.H., Gases as biological messengers: nitric oxide and carbon monoxide in the brain, *J. Neurosci.*, 14, 5147, 1994.

15. Maines, M., Carbon monoxide and nitric oxide homology: differential modulation of heme oxygenases in brain and detection of protein and activity, *Methods Enzymol.*, 268, 473, 1996.

16. Maines, M.D., The heme oxygenase system: a regulator of second messenger gases, *Annu. Rev. Pharmacol. Toxicol.*, 37, 517, 1997.

17. Wang, R., Resurgence of carbon monoxide: an endogenous gaseous vasorelaxing factor, *Can. J. Physiol. Pharmacol.*, 76, 1, 1998.

18. Johnson, R.A., Kozma, F., and Colombari, E., Carbon monoxide: from toxin to endogenous modulator of cardiovascular functions, *Braz. J. Med. Biol. Res.*, 32, 1, 1999.

19. Vreman, H.J., Mahoney, J.J., and Stevenson, D.K., Carbon monoxide and carboxyhemoglobin, *Adv. Pediatr.*, 42, 303, 1995.

20. Longo, L.D., The biological effects of carbon monoxide on the pregnant woman, fetus, and newborn infant, *Am. J. Obstet. Gynecol.*, 129, 69, 1977.

21. Maines, M.D., Heme oxygenase: function, multiplicity, regulatory mechanisms, and clinical applications, *FASEB J.*, 2, 2557, 1988.

22. Haab, P., The effect of carbon monoxide on respiration, *Experientia*, 46, 1202, 1990.

23. Rodgers, P.A. and Stevenson, D.K., Developmental biology of heme oxygenase, *Clin. Perinatol.*, 17, 275, 1990.

24. Rodgers, P.A., Vreman, H.J., Dennery, P.A., and Stevenson, D.K., Sources of carbon monoxide (CO) in biological systems and applications of CO detection technologies, *Semin. Perinatol.*, 18, 2, 1994.

25. World Health Organization, *Carbon Monoxide*, World Health Organization, Geneva, Switzerland, 1979.

26. Kane, D.M., Investigation of the Method to Determine Carboxyhemoglobin in Blood, DCIEM No. 85-R-32, Defense and Civil Institute of Environmental Medicine, Downsview, Ontario, 1985, 1.

27. Shepherd, R.J., *Carbon Monoxide Poisoning: The Silent Killer*, Charles C Thomas, Springfield, IL, 1983.

28. Jain, K.K., *Carbon Monoxide Poisoning*, Warren H. Green, St. Louis, MO, 1990.

29. Maines, M.D., *Heme Oxygenase: Clinical Applications and Functions*, CRC Press, Boca Raton, FL, 1992.

30. EPA 600/8-90/045F, Air Quality Criteria for Carbon Monoxide, Environmental Criteria and Assessment Office, Office of Health and Environmental Assessment, Office of Research and Development, U.S. Environmental Protection Agency, Research Triangle Park, NC, 1991, 1.1.

31. EPA 600/P-99/001, Air Quality Criteria for Carbon Monoxide, Environmental Criteria and Assessment Office, Office of Health and Environmental Assessment, Office of Research and Development, U.S. Environmental Protection Agency, Research Triangle Park, NC, 1999, in press.

32. Berk, P.D., Rodkey, F.L., Blaschke, T.F., Collison, H.A., and Waggoner, J.G., Comparison of plasma bilirubin turnover and carbon monoxide production in man, *J. Lab. Clin. Med.*, 83, 29, 1974.

33. Vedernikov, Y.P., Gräser, T., and Vanin, A.F., Similar endothelium-independent arterial relaxation by carbon monoxide and nitric oxide, *Biomed. Biochim. Acta*, 48, 601, 1989.

34. Vreman, H.J., Wong, R.J., Sanesi, C.A., Dennery, P.A., and Stevenson, D.K., Simultaneous production of carbon monoxide and thiobarbituric acid reactive substances in rat tissue preparations by an iron-ascorbate system, *Can. J. Physiol. Pharmacol.*, 76, 1057, 1998.

35. Coburn, R.F., Blakemore, W.S., and Forster, R.E., Endogenous carbon monoxide production in man, *J. Clin. Invest.*, 42, 1172, 1963.

36. Wranne, L., Studies on erythro-kinetics in infancy. VII. Quantitative estimation of the haemoglobin catabolism by carbon monoxide technique in young infants, *Acta Paediatr. Scand.*, 56, 381, 1967.

37. Fällström, S.P., Endogenous formation of carbon monoxide in newborn infants. IV. On the relation between the blood carboxyhaemoglobin concentration and the pulmonary elimination of carbon monoxide, *Acta Paediatr. Scand.*, 57, 321, 1968.

38. Berk, P.D., Blaschke, T.F., Scharschmidt, B.F., Waggoner, J.G., and Berlin, N.I., A new approach to quantitation of the various sources of bilirubin in man, *J. Lab. Clin. Med.*, 87, 767, 1976.

39. Fischer, A.F., Inguillo, D., Martin, D.M., Ochikubo, C.G., Vreman, H.J., and Stevenson, D.K., Carboxyhemoglobin concentration as an index of bilirubin production in neonates with birth weights less than 1,500 grams: a randomized double-blind comparison of supplemental oral vitamin E and placebo, *J. Pediatr. Gastroenterol. Nutr.*, 6, 748, 1987.

40. Coburn, R.F., Endogenous carbon monoxide production, *N. Engl. J. Med.*, 282, 207, 1970.

41. Coburn, R.F., Williams, W.J., and Kahn, S.B., Endogenous carbon monoxide production in patients with hemolytic anemia, *J. Clin. Invest.*, 45, 460, 1966.

42. Tenhunen, R., Marver, H.S., and Schmid, R., The enzymatic conversion of heme to bilirubin by microsomal heme oxygenase, *Proc. Natl. Acad. Sci. U.S.A.*, 61, 748, 1968.

43. Vreman, H.J., Stevenson, D.K., Henton, D., and Rosenthal, P., Correlation of carbon monoxide and bilirubin production by tissue homogenates, *J. Chromatogr.*, 427, 315, 1988.

44. Yoshida, T., Noguchi, M., and Kikuchi, G., The step of carbon monoxide liberation in the sequence of heme degradation catalyzed by the reconstituted microsomal heme oxygenase system, *J. Biol. Chem.*, 257, 9345, 1982.

45. Vreman, H.J. and Stevenson, D.K., Heme oxygenase activity as measured by carbon monoxide production, *Anal. Biochem.*, 168, 31, 1988.

46. Rosenberg, D.W. and Kappas, A., Characterization of heme oxygenase in the small intestinal epithelium, *Arch. Biochem. Biophys.*, 274, 471, 1989.

47. Vreman, H.J., Gillman, M.J., and Stevenson, D.K., *In vitro* inhibition of adult rat intestinal heme oxygenase by metalloporphyrins, *Pediatr. Res.*, 26, 362, 1989.

48. Vallier, H.A., Rodgers, P.A., and Stevenson, D.K., Oral administration of zinc deu-
    teroporphyrin IX 2,4 bis glycol inhibits heme oxygenase in neonatal rats, *Dev. Phar-
    macol. Ther.*, 17, 220, 1991.

49. Maines, M.D., Trakshel, G.M., and Kutty, R.K., Characterization of two constitutive
    forms of rat liver microsomal heme oxygenase. Only one molecular species of the
    enzyme is inducible, *J. Biol. Chem.*, 261, 411, 1986.

50. Raju, V.S., McCoubrey, W.K., Jr., and Maines, M.D., Regulation of heme oxygenase-
    2 by glucocorticoids in neonatal rat brain: characterization of a functional glucocor-
    ticoid response element, *Biochim. Biophys. Acta*, 1351, 89, 1997.

51. McCoubrey, W.K., Jr., Huang, T.J., and Maines, M.D., Isolation and characterization
    of a cDNA from the rat brain that encodes hemoprotein heme oxygenase-3, *Eur. J.
    Biochem.*, 247, 725, 1997.

52. Rosenberg, D.W. and Kappas, A., Induction of heme oxygenase in the small intestinal
    epithelium: a response to oral cadmium exposure, *Toxicology*, 67, 199, 1991.

53. Vreman, H.J., Wong, R.J., Williams, S.A., and Stevenson, D.K., *In vitro* heme oxy-
    genase isozyme activity inhibition by metalloporphyrins, *Pediatr. Res.*, 43, 202A, 1998.

54. Alam, J., Shibahara, S., and Smith, A., Transcriptional activation of the heme oxy-
    genase gene by heme and cadmium in mouse hepatoma cells, *J. Biol. Chem.*, 264,
    6371, 1989.

55. Posselt, A.M., Kwong, L.K., Vreman, H.J., and Stevenson, D.K., Suppression of carbon
    monoxide excretion rate by tin protoporphyrin, *Am. J. Dis. Child.*, 140, 147, 1986.

56. Funato, M., Tamai, H., Shimada, S., and Nakamura, H., Vigintiphobia, unbound
    bilirubin, and auditory brainstem responses, *Pediatrics*, 93, 50, 1994.

57. Yao, T.C. and Stevenson, D.K., Advances in the diagnosis and treatment of neonatal
    hyperbilirubinemia, *Clin. Perinatol.*, 22, 741, 1995.

58. Stocker, R., Yamamoto, Y., McDonagh, A.F., Glazer, A.N., and Ames, B.N., Bilirubin
    is an antioxidant of possible physiological importance, *Science*, 235, 1043, 1987.

59. Dennery, P.A., McDonagh, A.F., Spitz, D.R., Rodgers, P.A., and Stevenson, D.K.,
    Hyperbilirubinemia results in reduced oxidative injury in neonatal Gunn rats exposed
    to hyperoxia, *Free Radical Biol. Med.*, 19, 395, 1995.

60. Clerch, L.B. and Massaro, D., Oxidation-reduction-sensitive binding of lung protein
    to rat catalase mRNA, *J. Biol. Chem.*, 267, 2853, 1992.

61. Heyman, E., Ohlsson, A., and Girschek, P., Retinopathy of prematurity and bilirubin
    [letter] [see comments], *N. Engl. J. Med.*, 320, 256, 1989.

62. Benaron, D.A. and Bowen, F.W., Variation of initial serum bilirubin rise in newborn
    infants with type of illness [see comments], *Lancet*, 338, 78, 1991.

63. Dore, S., Takahashi, M., Ferris, C.D., Hester, L.D., Guastella, D., and Snyder, S.H.,
    Bilirubin, formed by activation of heme oxygenase-2, protects neurons against oxi-
    dative stress injury, *Proc. Natl. Acad. Sci. U.S.A.*, 96, 2445, 1999.

64. Maisels, M.J., Pathak, A., Nelson, N.M., Nathan, D.G., and Smith, C.A., Endogenous
    production of carbon monoxide in normal and erythroblastotic newborn infants,
    J. Clin. Invest., 50, 1, 1971.

65. Bartoletti, A.L., Stevenson, D.K., Ostrander, C.R., and Johnson, J.D., Pulmonary
    excretion of carbon monoxide in the human infant as an index of bilirubin production.
    I. Effects of gestational and postnatal age and some common neonatal abnormalities,
    J. Pediatr., 94, 952, 1979.

66. Ostrander, C.R., Cohen, R.S., Hopper, A.O., Cowan, B.E., Stevens, G.B., and Steven-
    son, D.K., Paired determinations of blood carboxyhemoglobin concentration and
    carbon monoxide excretion rate in term and preterm infants, *J. Lab. Clin. Med.*, 100,
    745, 1982.

67. Penn, A.A., Enzmann, D.R., Hahn, J.S., and Stevenson, D.K., Kernicterus in a full term infant [see comments], *Pediatrics*, 93, 1003, 1994.
68. Slusher, T.M., Vreman, H.J., McLaren, D.W., Lewison, L.J., Brown, A.K., and Stevenson, D.K., Glucose-6-phosphate dehydrogenase deficiency and carboxyhemoglobin concentrations associated with bilirubin-related morbidity and death in Nigerian infants, *J. Pediatr.*, 126, 102, 1995.
69. Ennever, J.F., Phototherapy for neonatal jaundice. Yearly review, *Photochem. Photobiol.*, 47, 871, 1988.
70. Vreman, H.J., Wong, R.J., Stevenson, D.K., Route, R.K., Reader, S.D., Fejer, M.M., Gale, R., and Seidman, D.S., Light-emitting diodes: a novel light source for phototherapy, *Pediatr. Res.*, 44, 804, 1998.
71. Lee, K.S. and Gartner, L.M., Management of unconjugated hyperbilirubinemia in the newborn, *Semin. Liver Dis.*, 3, 52, 1983.
72. Mahoney, J.J., Wong, R.J., Vreman, H.J., and Stevenson, D.K., Fetal hemoglobin of transfused neonates and spectrophotometric measurements of oxyhemoglobin and carboxyhemoglobin, *J. Clin. Monit.*, 7, 154, 1991.
73. Stevenson, D.K., Rodgers, P.A., and Vreman, H.J., The use of metalloporphyrins for the chemoprevention of neonatal jaundice, *Am. J. Dis. Child.*, 143, 353, 1989.
74. Dennery, P.A., Rhine, W.D., and Stevenson, D.K., Neonatal jaundice — what now? *Clin. Pediatr.*, 34, 103, 1995.
75. Valaes, T., Drummond, G.S., and Kappas, A., Control of hyperbilirubinemia in glucose-6-phosphate dehydrogenase-deficient newborns using an inhibitor of bilirubin production, Sn-mesoporphyrin, *Pediatrics*, 101, E1, 1998.
76. Martinez, J.C., Garcia, H.O., Otheguy, L.E., Drummond, G.S., and Kappas, A., Control of severe hyperbilirubinemia in full-term newborns with the inhibitor of bilirubin production Sn-mesoporphyrin, *Pediatrics*, 103, 1, 1999.
77. Bucalo, L.R., Cohen, R.S., Ostrander, C.R., Hopper, A.O., Garcia, J.F., Clemons, G.K., Schwartz, H.C., and Stevenson, D.K., Pulmonary excretion of carbon monoxide in the human infant as an index of bilirubin production. IIc. Evidence for the possible association of cord blood erythropoietin levels and postnatal bilirubin production in infants of mothers with abnormalities of gestational glucose metabolism, *Am. J. Perinatol.*, 1, 177, 1984.
78. Vreman, H.J., Kwong, L.K., and Stevenson, D.K., Carbon monoxide in blood: an improved microliter blood-sample collection system, with rapid analysis by gas chromatography, *Clin. Chem.*, 30, 1382, 1984.
79. Uetani, Y., Nakamura, H., Okamoto, O., Yamazaki, T., Vreman, H.J., and Stevenson, D.K., Carboxyhemoglobin measurements in the diagnosis of ABO hemolytic disease, *Acta Paediatr. Jpn.*, 31, 171, 1989.
80. Widness, J.A., Lowe, L.S., Stevenson, D.K., Vreman, H.J., Weiner, C.P., Hayde, M., and Pollak, A., Direct relationship of fetal carboxyhemoglobin with hemolysis in alloimmunized pregnancies, *Pediatr. Res.*, 35, 713, 1994.
81. Smith, D.W., Hopper, A.O., Shahin, S.M., Cohen, R.S., Ostrander, C.R., Ariagno, R.L., and Stevenson, D.K., Neonatal bilirubin production estimated from "end-tidal" carbon monoxide concentration, *J. Pediatr. Gastroenterol. Nutr.*, 3, 77, 1984.
82. Balaraman, V., Pelke, S., DiMauro, S., Cheung, S., Stevenson, D.K., and Easa, D., End-tidal carbon monoxide in newborn infants: observations during the 1st week of life, *Biol. Neonate*, 67, 182, 1995.
83. Vreman, H.J., Baxter, L.M., Stone, R.T., and Stevenson, D.K., Evaluation of a fully automated end-tidal carbon monoxide instrument for breath analysis, *Clin. Chem.*, 42, 50, 1996.

84. Vreman, H.J., Wong, R.J., Harmatz, P., Fanaroff, A.A., Berman, B., and Stevenson, D.K., Validation of the Natus CO-Stat™ End Tidal Breath Analyzer in children and adults, *J. Clin. Monit. Comput.*, accepted, 1999.

85. Drummond, G.S. and Kappas, A., Prevention of neonatal hyperbilirubinemia by tin protoporphyrin IX, a potent competitive inhibitor of heme oxidation, *Proc. Natl. Acad. Sci. U.S.A.*, 78, 6466, 1981.

86. Drummond, G.S., Greenbaum, N.L., and Kappas, A., Tin(Sn^{+4})-diiododeuteroporphyrin: an *in vitro* and *in vivo* inhibitor of heme oxygenase with substantially reduced photoactive properties, *J. Pharmacol. Exp. Ther.*, 257, 1109, 1991.

87. Vreman, H.J., Ekstrand, B.C., and Stevenson, D.K., Selection of metalloporphyrin heme oxygenase inhibitors based on potency and photoreactivity, *Pediatr. Res.*, 33, 195, 1993.

88. Verma, A., Hirsch, D.J., Glatt, C.E., Ronnett, G.V., and Snyder, S.H., Carbon monoxide: a putative neural messenger [see comments] *Science*, 259, 381, 1993 (published erratum appears in *Science*, 263 (5143), 15, 1994).

89. Zhuo, M., Small, S.A., Kandel, E.R., and Hawkins, R.D., Nitric oxide and carbon monoxide produce activity-dependent long-term synaptic enhancement in hippocampus, *Science*, 260, 1946, 1993.

90. Labbe, R.F., Vreman, H.J., and Stevenson, D.K., Zinc protoporphyrin: a metabolite with a mission, *Clin. Chem.*, 45, 2060, 1979.

91. Frydman, R.B., Tomaro, M.L., Buldain, G., Awruch, J., Diaz, L., and Frydman, B., Specificity of heme oxygenase: a study with synthetic hemins, *Biochemistry*, 20, 5177, 1981.

92. Kappas, A. and Drummond, G.S., Control of heme metabolism with synthetic metalloporphyrins, *J. Clin. Invest.*, 77, 335, 1986.

93. Vreman, H.J., Lee, O.K., and Stevenson, D.K., *In vitro* and *in vivo* characteristics of a heme oxygenase inhibitor: ZnBG, *Am. J. Med. Sci.*, 302, 335, 1991.

94. Vreman, H.J., Cipkala, D.A., and Stevenson, D.K., Characterization of porphyrin heme oxygenase inhibitors, *Can. J. Physiol. Pharmacol.*, 74, 278, 1996.

95. Kappas, A., Drummond, G.S., Henschke, C., and Valaes, T., Direct comparison of Sn-mesoporphyrin, an inhibitor of bilirubin production, and phototherapy in controlling hyperbilirubinemia in term and near-term newborns, *Pediatrics*, 95, 468, 1995.

96. Meneghini, R., Iron homeostasis, oxidative stress, and DNA damage, *Free Radical Biol. Med.*, 23, 783, 1997.

97. Poss, K.D. and Tonegawa, S., Heme oxygenase-1 is required for mammalian iron reutilization, *Proc. Natl. Acad. Sci. U.S.A.*, 94, 10919, 1997.

98. Braughler, J.M., Duncan, L.A., and Chase, R.L., The involvement of iron in lipid peroxidation. Importance of ferric to ferrous ratios in initiation, *J. Biol. Chem.*, 261, 10282, 1986.

99. Ryan, T.P. and Aust, S.D., The role of iron in oxygen-mediated toxicities, *Crit. Rev. Toxicol.*, 22, 119, 1992.

100. Wolff, D.G. and Bidlack, W.K., The formation of carbon monoxide during peroxidation of microsomal lipids, *Biochem. Biophys. Res. Commun.*, 73, 850, 1976.

101. Nishibayashi, H., Tomura, T., Soto, R., and Estabrook, R.W., *Structure and Function of Cytochromes*, University Park Press, Baltimore, MD, 1968, 658.

102. Lindstrom, A.D. and Anders, M.W., The effect of phenobarbital and 3-methylcholanthrene treatment on carbon tetrachloride-stimulated heme degradation and carbon monoxide expiration *in vivo*, *Toxicol. Lett.*, 1, 307, 1978.

103. Hawkins, R.D., Zhuo, M., and Arancio, O., Nitric oxide and carbon monoxide as possible retrograde messengers in hippocampal long-term potentiation, *J. Neurobiol.*, 25, 652, 1994.

104. Zayasu, K., Sekizawa, K., Okinaga, S., Yamaya, M., Ohrui, T., and Sasaki, H., Increased carbon monoxide in exhaled air of asthmatic patients, *Am. J. Respir. Crit. Care Med.*, 156, 1140, 1997.

105. Horvath, I., Donnelly, L.E., Kiss, A., Paredi, P., Kharitonov, S.A., and Barnes, P.J., Raised levels of exhaled carbon monoxide are associated with an increased expression of heme oxygenase-1 in airway macrophages in asthma: a new marker of oxidative stress, *Thorax*, 53, 668, 1998.

106. Kozma, F., Johnson, R.A., Zhang, F., Yu, C., Tong, X., and Nasjletti, A., Contribution of endogenous carbon monoxide to regulation of diameter in resistance vessels, *Am. J. Physiol.*, 276, R1087, 1999.

107. Vreman, H., Mahoney, J., and Stevenson, D., Electrochemical measurement of carbon monoxide in breath: interference by hydrogen, *Atmos. Environ.*, 27A, 2193, 1993.

108. Stevenson, D.K., Bartoletti, A.L., Ostrander, C.R., and Johnson, J.D., Pulmonary excretion of carbon monoxide in the human newborn infant as an index of bilirubin production: III. Measurement of pulmonary excretion of carbon monoxide after the first postnatal week in premature infants, *Pediatrics*, 64, 598, 1979.

109. Sisson, T. and Wickler, M., Transmission of light through living tissue, *Pediatr. Res.*, 7, 316, 1973.

110. Lipsitz, P.J., Gartner, L.M., and Bryla, D.A., Neonatal and infant mortality in relation to phototherapy, *Pediatrics*, 75, 422, 1985.

111. Hintz, S.R., Vreman, H.J., and Stevenson, D.K., Mortality of metalloporphyrin-treated neonatal rats after light exposure, *Dev. Pharmacol. Ther.*, 14, 187, 1990.

112. Vreman, H.J., Gillman, M.J., Downum, K.R., and Stevenson, D.K., *In vitro* generation of carbon monoxide from organic molecules and synthetic metalloporphyrins mediated by light, *Dev. Pharmacol. Ther.*, 15, 112, 1990.

113. Dennery, P.A., Vreman, H.J., Rodgers, P.A., and Stevenson, D.K., Role of lipid peroxidation in metalloporphyrin-mediated phototoxic reactions in neonatal rats, *Pediatr. Res.*, 33, 87, 1993.

114. Kopelman, A.E., Brown, R.S., and Odell, G.B., The "bronze" baby syndrome: a complication of phototherapy, *J. Pediatr.*, 81, 466, 1972.

115. Vreman, H.J., Wong, R.J., and Stevenson, D.K., unpublished data, 1997.

116. Engel, R.R., Alternative sources of carbon monoxide, in *Chemistry and Physiology of Bile Pigments*, Berk, P.D. and Berlin, N.I., Eds., U.S. Department of Health, Education, and Welfare, Bethesda, MD, 1977, 148–155.

117. Aouthmany, M.M., Phototherapy increases hemoglobin degradation and bilirubin production in preterm infants, *J. Perinatol.*, 19, 271, 1999.

118. Westlake, D.W.S., Roxburgh, J.M., and Talbot, G., Microbial production of carbon monoxide from flavonoids, *Nature*, 189, 510, 1961.

119. Engel, R.R., Matsen, J.M., Chapman, S.S., and Schwartz, S., Carbon monoxide production from heme compounds by bacteria, *J. Bacteriol.*, 112, 1310, 1972.

120. Shigezane, J., Postmortem formation of carbon monoxide by bacteria, *Nippon Hoigaku Zasshi*, 40, 111, 1986.

121. Moncure, M., Brathwaite, C.E., Samaha, E., Marburger, R., and Ross, S.E., Carboxyhemoglobin elevation in trauma victims, *J. Trauma*, 46, 424, 1999.

122. Posselt, A.M., Cowan, B.E., Kwong, L.K., Vreman, H.J., and Stevenson, D.K., Effect of tin protoporphyrin on the excretion rate of carbon monoxide in newborn rats after hematoma formation, *J. Pediatr. Gastroenterol. Nutr.*, 4, 650, 1985.

123. Yamaya, M., Sekizawa, K., Ishizuka, S., Monma, M., Mizuta, K., and Sasaki, H., Increased carbon monoxide in exhaled air of subjects with upper respiratory tract infections, *Am. J. Respir. Crit. Care Med.*, 158, 311, 1998.

124. Khalil, A.K., The global cycle of carbon monoxide: trends and mass balance, in EPA 600/8-90/045F, Air Quality Criteria for Carbon Monoxide, Environmental Criteria and Assessment Office, Office of Health and Environmental Assessment, Office of Research and Development, U.S. Environmental Protection Agency, Research Triangle Park, NC, 1991, 4.1.

125. Akland, G.G., Organizational components and structural features of EPA's new Human Exposure Research Program, *J. Exposure Anal. Environ. Epidemiol.*, 1, 129, 1991.

126. Akland, G.G., Colome, S.D., and Dahms, T.E., Population exposure to carbon monoxide, in EPA 600/8-90/045F, Air Quality Criteria for Carbon Monoxide, Environmental Criteria and Assessment Office, Office of Health and Environmental Assessment, Office of Research and Development, U.S. Environmental Protection Agency, Research Triangle Park, NC, 1991, 8.1.8.

127. Leaderer, B.P., Indoor carbon monoxide, in EPA 600/8-90/045F, Air Quality Criteria for Carbon Monoxide, Environmental Criteria and Assessment Office, Office of Health and Environmental Assessment, Office of Research and Development, U.S. Environmental Protection Agency, Research Triangle Park, NC, 1991, 7.1.

128. Braddock, J.N., Braverman, T.N., and Peterson, W.B., Ambient carbon monoxide, in EPA 600/8-90/045F, Air Quality Criteria for Carbon Monoxide, Environmental Criteria and Assessment Office, Office of Health and Environmental Assessment, Office of Research and Development, U.S. Environmental Protection Agency, Research Triangle Park, NC, 1991, 6.1.

129. Balster, R.L., Horvath, S.M., and McGrath, J.J., Combined exposure of carbon monoxide with other pollutants, drugs, and environmental factors, in EPA 600/8-90/045F, Air Quality Criteria for Carbon Monoxide, Environmental Criteria and Assessment Office, Office of Health and Environmental Assessment, Office of Research and Development, U.S. Environmental Protection Agency, Research Triangle Park, NC, 1991, 11.1.

130. Stevenson, D.K., Vreman, H.J., Oh, W., Fanaroff, A.A., Wright, L.L., Lemons, J.A., Verter, J., Shankaran, S., Tyson, J.E., Korones, S.B., Bauer, C.R., Stoll, B.J., Papile, L.A., Donovan, E.F., Okah, F., and Ehrenkranz, R.A., Bilirubin production in healthy term infants as measured by carbon monoxide in breath, *Clin. Chem.*, 40, 1934, 1994.

131. Levitt, M.D., Ellis, C., Springfield, J., and Engel, R.R., Carbon monoxide generation from hydrocarbons at ambient and physiological temperature: a sensitive indicator of oxidant damage? *J. Chromatogr. A*, 695, 324, 1995.

132. Rodkey, F.L., Collison, H.A., and Engel, R.R., Release of carbon monoxide from acrylic and polycarbonate plastics, *J. Appl. Physiol.*, 27, 554, 1969.

133. Wu, J.S., Monk, T., Luttmann, D.R., Meininger, T.A., and Soper, N.J., Production and systemic absorption of toxic byproducts of tissue combustion during laparoscopic cholecystectomy, *J. Gastrointest. Surg.*, 2, 399, 1998.

134. Nezhat, C., Seidman, D.S., Vreman, H.J., Stevenson, D.K., Nezhat, F., and Nezhat, C., The risk of carbon monoxide poisoning after prolonged laparoscopic surgery [see comments], *Obstet. Gynecol.*, 88, 771, 1996.

135. Moon, R.E., Sparacino, C., and Meyer, A.F., Pathogenesis of carbon monoxide production in anesthesia circuits, *Anesthesiology*, 77, A1061, 1992.

136. Elevated intra-operative blood carboxyhemoglobin levels in surgical patients — Georgia, Illinois, and North Carolina, *Morb. Mortal. Wkly. Rep.*, 40, 248, 1991.

137. Baum, J., Sachs, G., v.d. Driesch, C., and Stanke, H.G., Carbon monoxide generation in carbon dioxide absorbents, *Anesth. Analg.*, 81, 144, 1995.

138. Fang, Z.X., Eger II, E.I., Laster, M.J., Chortkoff, B.S., Kandel, L., and Ionescu, P., Carbon monoxide production from degradation of desflurane, enflurane, isoflurane, halothane, and sevoflurane by soda lime and Baralyme [see comments], *Anesth. Analg.*, 80, 1187, 1995.

139. Landaw, S.A., The effects of cigarette smoking on total body burden and excretion rates of carbon monoxide, *J. Occup. Med.*, 15, 231, 1973.

140. Stevenson, D.K., Estimation of bilirubin production, in *Report of the Eighty-Fifth Ross Conference on Pediatric Research*, Levine, R.L. and Maisels, M.J., Eds., Ross Laboratories, Columbus, OH, 1983, 64.

141. Rodkey, F.L., O'Neal, J.D., Collison, H.A., and Uddin, D.E., Relative affinity of hemoglobin S and hemoglobin A for carbon monoxide and oxygen, *Clin. Chem.*, 20, 83, 1974.

142. Douglas, C., Haldane, J.S., and Haldane, J.B.S., The laws of combination of haemoglobin with carbon monoxide and oxygen, *J. Physiol.* (London), 44, 275, 1912.

143. Dahms, T.E. and Horvath, S.M., Rapid, accurate technique for determination of carbon monoxide in blood, *Clin. Chem.*, 20, 533, 1974.

144. Fenn, W.O., The burning of CO in tissues, *Ann. N.Y. Acad. Sci.*, 174, 64, 1970.

145. Stannard, J.N., An analysis of the effect of carbon monoxide on the respiration of frog skeletal muscle, *Am. J. Physiol.*, 129, 195, 1940.

146. Luomanmäki, K. and Coburn, R.F., Effects of metabolism and distribution of carbon monoxide on blood and body stores, *Am. J. Physiol.*, 217, 354, 1969.

147. Coburn, R.F., Forster, R.E., and Kane, P.B., Considerations of the physiological variables that determine the blood carboxyhemoglobin concentration in man, *J. Clin. Invest.*, 44, 1899, 1965.

148. Stewart, R.D., Baretta, E.D., Platte, L.R., Stewart, E.B., Kalbfleisch, J.H., Van Yserloo, B., and Rimm, A.A., Carboxyhemoglobin levels in American blood donors, *J. Am. Med. Assoc.*, 229, 1187, 1974.

149. Tikuisis, P., Madill, H.D., Gill, B.J., Lewis, W.F., Cox, K.M., and Kane, D.M., A critical analysis of the use of the CFK equation in predicting COHb formation, *Am. Ind. Hyg. Assoc. J.*, 48, 208, 1987.

150. Raffin, S.B., Woo, C.H., Roost, K.T., Price, D.C., and Schmid, R., Intestinal absorption of hemoglobin iron-heme cleavage by mucosal heme oxygenase, *J. Clin. Invest.*, 54, 1344, 1974.

151. Leibson, C., Brown, M., Thibodeau, S., Stevenson, D., Vreman, H., Cohen, R., Clemons, G., Callen, W., and Moore, L.G., Neonatal hyperbilirubinemia at high altitude, *Am. J. Dis. Child.*, 143, 983, 1989.

152. McGrath, J.J., Effects of altitude on endogenous carboxyhemoglobin levels, *J. Toxicol. Environ. Health*, 35, 127, 1992.

153. Evans, N.J. and Rutter, N., Percutaneous respiration in the newborn infant, *J. Pediatr.*, 108, 282, 1986.

154. Barker, N., Hadgraft, J., and Rutter, N., Skin permeability in the newborn, *J. Invest. Dermatol.*, 88, 409, 1987.

155. Meffert, M.K., Haley, J.E., Schuman, E.M., Schulman, H., and Madison, D.V., Inhibition of hippocampal heme oxygenase, nitric oxide synthase, and long-term potentiation by metalloporphyrins, *Neuron*, 13, 1225, 1994.

156. Moncada, S., Palmer, R.M., and Higgs, E.A., Nitric oxide: physiology, pathophysiology, and pharmacology, *Pharmacol. Rev.*, 43, 109, 1991.

157. Brüne, B. and Ullrich, V., Inhibition of platelet aggregation by carbon monoxide is mediated by activation of guanylate cyclase, *Mol. Pharmacol.*, 32, 497, 1987.

158. Morita, T., Perrella, M.A., Lee, M.E., and Kourembanas, S., Smooth muscle cell-derived carbon monoxide is a regulator of vascular cGMP, *Proc. Natl. Acad. Sci. U.S.A.*, 92, 1475, 1995.

159. Johnson, R.A., Colombari, E., Colombari, D.S., Lavesa, M., Talman, W.T., and Nasjletti, A., Role of endogenous carbon monoxide in central regulation of arterial pressure, *Hypertension*, 30, 962, 1997.

160. Acevedo, C.H. and Ahmed, A., Hemeoxygenase-1 inhibits human myometrial contractility via carbon monoxide and is upregulated by progesterone during pregnancy, *J. Clin. Invest.*, 101, 949, 1998.

161. Coceani, F., Kelsey, L., Seidlitz, E., Marks, G.S., McLaughlin, B.E., Vreman, H.J., Stevenson, D.K., Rabinovitch, M., and Ackerley, C., Carbon monoxide formation in the ductus arteriosus in the lamb: implications for the regulation of muscle tone, *Br. J. Pharmacol.*, 120, 599, 1997.

162. Vreman, H.J., Wong, R.J., Nabseth, D., Kim, E.C., Marks, G.S., and Stevenson, D.K., Heme oxygenase activity in human umbilical cord and rat vascular tissues, *Placenta*, submitted, 1999.

163. Prabhakar, N.R., Dinerman, J.L., Agani, F.H., and Snyder, S.H., Carbon monoxide: a role in carotid body chemoreception, *Proc. Natl. Acad. Sci. U.S.A.*, 92, 1994, 1995.

164. Leinders-Zufall, T., Shepherd, G.M., and Zufall, F., Regulation of cyclic nucleotide-gated channels and membrane excitability in olfactory receptor cells by carbon monoxide, *J. Neurophysiol.*, 74, 1498, 1995.

165. Hronis, T.S. and Traugh, J.A., Structural requirements for porphyrin inhibition of the hemin-controlled protein kinase and maintenance of protein synthesis in reticulocytes, *J. Biol. Chem.*, 261, 6234, 1986.

166. Scott, C.D., Kemp, B.E., and Edwards, A.M., Effects of hemin on rat liver cyclic AMP-dependent protein kinases in cell extracts and intact hepatocytes, *Biochim. Biophys. Acta*, 847, 301, 1985.

167. Sassa, S., Heme stimulation of cellular growth and differentiation, *Semin. Hematol.*, 25, 312, 1988.

168. Hansen, T.W., Mathiesen, S.B., and Walaas, S.I., Bilirubin has widespread inhibitory effects on protein phosphorylation, *Pediatr. Res.*, 39, 1072, 1996.

169. Ignarro, L.J., Ballot, B., and Wood, K.S., Regulation of soluble guanylate cyclase activity by porphyrins and metalloporphyrins, *J. Biol. Chem.*, 259, 6201, 1984.

170. Appleton, S.D., Chretien, M.L., McLaughlin, B.E., Vreman, H.J., Stevenson, D.K., Brien, J.F., Nakatsu, K., Maurice, D.H., and Marks, G.S., Selective inhibition of heme oxygenase, without inhibition of nitric oxide synthase or soluble guanylyl cyclase, by metalloporphyrins at low concentrations, *Drug Metab. Dispos.*, 27, 1214, 1999.

171. Odrcich, M.J., Graham, C.H., Kimura, K.A., McLaughlin, B.E., Marks, G.S., Nakatsu, K., and Brien, J.F., Heme oxygenase and nitric-oxide synthase in the placenta of the guinea-pig during gestation, *Placenta*, 19, 509, 1998.

172. Imai, K., Aimoto, T., Sato, M., and Kimura, R., Antioxidative effect of several porphyrins on lipid peroxidation in rat liver homogenates, *Chem. Pharm. Bull.* (Tokyo), 38, 258, 1990.

173. Smith, D.W., Cohen, R.S., Vreman, H.J., Yeh, A., Sharron, S., and Stevenson, D.K., Bilirubin production after supplemental oral vitamin E therapy in preterm infants, *J. Pediatr. Gastroenterol. Nutr.*, 4, 38, 1985.

174. Wranne, L., Studies on erythro-kinetics in infancy. XIV. The relation between anaemia and haemoglobin catabolism in Rh-haemolytic disease of the newborn, *Acta Paediatr. Scand.*, 58, 49, 1969.

175. Ostrander, C.R., Cohen, R.S., Hopper, A.O., Shahin, S.M., Kerner, J.A., Jr., Johnson, J.D., and Stevenson, D.K., Breath hydrogen analysis: a review of the methodologies and clinical applications, *J. Pediatr. Gastroenterol. Nutr.*, 2, 525, 1983.

176. Smith, J.S. and Brandon, S., Morbidity from acute carbon monoxide poisoning at three-year follow-up, *Br. Med. J.*, 1, 318, 1973.

177. Mahoney, J.J., Vreman, H.J., Stevenson, D.K., and Van Kessel, A.L., Measurement of carboxyhemoglobin and total hemoglobin by five specialized spectrophotometers (CO-oximeters) in comparison with reference methods, *Clin. Chem.*, 39, 1693, 1993.

178. Mahoney, J.J., Harvey, J.A., Wong, R.J., and Van Kessel, A.L., Changes in oxygen measurements when whole blood is stored in iced plastic or glass syringes [see comments], *Clin. Chem.*, 37, 1244, 1991.

179. Maisels, M.J. and Kring, E., End-tidal carbon monoxide concentration (ETCO) in the normal and isoimmunized newborn, *Pediatr. Res.*, 35, 239, 1994.

180. Goldsmith, J.R. and Landaw, S.A., Carbon monoxide and human health, *Science*, 162, 1352, 1968.

181. Wald, N.J., Idle, M., Boreham, J., and Bailey, A., Carbon monoxide in breath in relation to smoking and carboxyhaemoglobin levels, *Thorax*, 36, 366, 1981.

182. Akland, G.G., Measurement methods for carbon monoxide, in EPA 600/8-90/045F, Air Quality Criteria for Carbon Monoxide, Environmental Criteria and Assessment Office, Office of Health and Environmental Assessment, Office of Research and Development, U.S. Environmental Protection Agency, Research Triangle Park, NC, 1991, 5.1.

183. EPA 600/8-90/045F, Air Quality Criteria for Carbon Monoxide, Environmental Criteria and Assessment Office, Office of Health and Environmental Assessment, Office of Research and Development, U.S. Environmental Protection Agency, Research Triangle Park, NC, 1991, Table 8.11.

184. Zijlstra, W.G., Buursma, A., and Meeuwsen-van der Roest, W.P., Absorption spectra of human fetal and adult oxyhemoglobin, de-oxyhemoglobin, carboxyhemoglobin, and methemoglobin, *Clin. Chem.*, 37, 1633, 1991.

185. Vreman, H.J., Mahoney, J.J., Van Kessel, A.L., and Stevenson, D.K., Carboxyhemoglobin as measured by gas chromatography and with the IL 282 and 482 CO-oximeters, *Clin. Chem.*, 34, 2562, 1988.

186. Vreman, H.J. and Stevenson, D.K., Carboxyhemoglobin determined in neonatal blood with a CO-oximeter unaffected by fetal oxyhemoglobin, *Clin. Chem.*, 40, 1522, 1994.

187. Zijlstra, W.G., Buursma, A., and Zwart, A., Performance of an automated six-wavelength photometer (Radiometer OSM3) for routine measurement of hemoglobin derivatives, *Clin. Chem.*, 34, 149, 1988.

188. Zwart, A., Buursma, A., Oeseburg, B., and Zijlstra, W.G., Determination of hemoglobin derivatives with IL 282 CO-oximeter as compared with a manual spectrophotometric five-wavelength method, *Clin. Chem.*, 27, 1903, 1981.

189. Wimberly, P.D., Siggaard-Anderson, O., and Fogh-Anderson, N., Accurate measurements of hemoglobin oxygen saturation, and fraction of carboxyhemoglobin and methemoglobin in fetal blood using Radiometer OSM3: corrections for fetal hemoglobin fraction and pH, *Scand. J. Clin. Lab. Invest.*, 203, 235, 1990.

190. Vreman, H.J., Stevenson, D.K., and Zwart, A., Analysis for carboxyhemoglobin by gas chromatography and multicomponent spectrophotometry compared, *Clin. Chem.*, 33, 694, 1987.

191. Kaplan, M., Vreman, H.J., Hammerman, C., and Stevenson, D.K., Neonatal bilirubin production, reflected by carboxyhaemoglobin concentrations, in Down's syndrome, *Arch. Dis. Child. Fetal Neonatal Ed.*, 81, F56, 1999.

192. Stevenson, D.K., Ostrander, C.R., Cohen, R.S., and Johnson, J.D., Trace gas analysis in bilirubin metabolism: a technical review and current state of the art, *Adv. Pediatr.*, 29, 129, 1982.

193. Zwart, A., Spectrophotometry of hemoglobin: various perspectives [editorial], *Clin. Chem.*, 39, 1570, 1993.

194. Blackmore, D.J., The determination of carbon monoxide in blood and tissue, *Analyst*, 95, 439, 1970.

195. Shinomiya, T. and Shinomiya, K., The variation in carbon monoxide release in the blood stain and in visceral tissues, *Acta Med. Leg. Soc.* (Liege), 39, 131, 1989.

196. Shinomiya, K., Orimoto, C., and Shinomiya, T., Experimental exposure to carbon monoxide in rats (II) — blood volume of organs obtained by calculations from amounts of carbon monoxide in organ tissues and in blood, *Nippon Hoigaku Zasshi*, 48, 79, 1994.

197. Shinomiya, K., Orimoto, C., and Shinomiya, T., Experimental exposure to carbon monoxide in rats (I) —- relation between the degree of carboxyhemoglobin saturation and the amount of carbon monoxide in the organ tissues of rats, *Nippon Hoigaku Zasshi*, 48, 19, 1994.

198. Sokal, J.A., Majka, J., and Palus, J., The content of carbon monoxide in the tissues of rats intoxicated with carbon monoxide in various conditions of acute exposure, *Arch. Toxicol.*, 56, 106, 1984.

199. Sokal, J., Majka, J., and Palus, J., Effect of work load on the content of carboxymyoglobin in the heart and skeletal muscles of rats exposed to carbon monoxide, *J. Hyg. Epidemiol. Microbiol. Immunol.*, 30, 57, 1986.

200. Kadotani, T., Vreman, H.J., Wong, R.J., and Stevenson, D.K., Concentration of carbon monoxide (CO) in tissue, *Pediatr. Res.*, 45, 67A, 1999.

201. Strocchi, A., Schwartz, S., Ellefson, M., Engel, R. R., Medina, A., and Levitt, M.D., A simple carbon monoxide breath test to estimate erythrocyte turnover, *J. Lab. Clin. Med.*, 120, 392, 1992.

202. Hintz, S.R., Kwong, L.K., Vreman, H.J., and Stevenson, D.K., Recovery of exogenous heme as carbon monoxide and biliary heme in adult rats after tin protoporphyrin treatment, *J. Pediatr. Gastroenterol. Nutr.*, 6, 302, 1987.

203. Ostrander, C.R., Johnson, J.D., and Bartoletti, A.L., Determining the pulmonary excretion rate of carbon monoxide in newborn infants, *J. Appl. Physiol.*, 40, 844, 1976.

204. Stevenson, D.K., Watson, E.M., Hintz, S.R., Kim, C.B., and Vreman, H.J., Tin protoporphyrin inhibits carbon monoxide production in suckling mice, *Biol. Neonate*, 51, 40, 1987.

205. Vreman, H.J., Rodgers, P.A., Gale, R., and Stevenson, D.K., Carbon monoxide excretion as an index of bilirubin production in rhesus monkeys, *J. Med. Primatol.*, 18, 449, 1989.

206. Vreman, H.J., Stevenson, D.K., Oh, W., Fanaroff, A.A., Wright, L.L., Lemons, J.A., Wright, E., Shankaran, S., Tyson, J.E., Korones, S.B., Bauer, C.R., Stoll, B.J., Papile, L.A., Donovan, E.F., and Ehrenkranz, R.A., Semiportable electrochemical instrument for determining carbon monoxide in breath, *Clin. Chem.*, 40, 1927, 1994.

207. Seidman, D.S., Shiloh, M., Stevenson, D.K., Vreman, H.J., and Gale, R., Role of hemolysis in neonatal jaundice associated with glucose-6 phosphate dehydrogenase deficiency [see comments], *J. Pediatr.*, 127, 804, 1995.

208. Kaplan, M., Beutler, E., Vreman, H.J., Hammerman, C., Levy-Lahad, E., Renbaum, P., and Stevenson, D.K., Neonatal hyperbilirubinemia in glucose-6-phosphate dehydrogenase-deficient heterozygotes, *Pediatrics*, 104, 68, 1999.

209. Chan, G.C., Lau, Y.L., and Yeung, C.Y., End tidal carbon monoxide concentration in childhood haemolytic disorders, *J. Paediatr. Child. Health*, 34, 447, 1998.

210. Wallace, L., Thomas, J., Mage, D., and Ott, W., Comparison of breath CO, CO exposure, and Coburn model predictions in the U.S. EPA, Washington–Denver (CO) Study, *Atmos Environ*, 22, 2183, 1988.

211. King, A.C., Scott, R.R., and Prue, D.M., The reactive effects of assessing reported rates and alveolar carbon monoxide levels on smoking behavior, *Addict. Behav.*, 8, 323, 1983.

212. EPA 600/8-90/045F, Air Quality Criteria for Carbon Monoxide, Environmental Criteria and Assessment Office, Office of Health and Environmental Assessment, Office of Research and Development, U.S. Environmental Protection Agency, Research Triangle Park, NC, 1991, 8.97.

213. Berlin, N.I., Carbon monoxide production: a tool for assessing antihemolytic therapy [editorial], *J. Lab. Clin. Med.*, 120, 361, 1992.

214. Lowry, O.H., Rosebrough, H.J., Farr, A.L., and Randall, R., Protein measurement with the Folin phenol reagent, *J. Biol. Chem.*, 193, 265, 1951.

215. Murphy, B.J., Laderoute, K.R., Vreman, H.J., Grant, T.D., Gill, N.S., Stevenson, D.K., and Sutherland, R.M., Enhancement of heme oxygenase expression and activity in A431 squamous carcinoma multicellular tumor spheroids, *Cancer Res.*, 53, 2700, 1993.

216. Vreman, H.J., Hintz, S.R., Kim, C.B., Castillo, R.O., and Stevenson, D.K., Effects of oral administration of tin and zinc protoporphyrin on neonatal and adult rat tissue heme oxygenase activity, *J. Pediatr. Gastroenterol. Nutr.*, 7, 902, 1988.

217. Vallier, H.A., Rodgers, P.A., and Stevenson, D.K., Inhibition of heme oxygenase after oral vs. intraperitoneal administration of chromium porphyrins, *Life Sci.*, 52, L79, 1993.

218. Rodgers, P.A., Vreman, H.J., and Stevenson, D.K., Heme catabolism in rhesus neonates inhibited by zinc protoporphyrin, *Dev. Pharmacol. Ther.*, 14, 216, 1990.

219. Vreman, H.J., Rodgers, P.A., and Stevenson, D.K., Zinc protoporphyrin administration for suppression of increased bilirubin production by iatrogenic hemolysis in rhesus neonates, *J. Pediatr.*, 117, 292, 1990.

220. Rodgers, P.A., Seidman, D.S., Wei, P.L., Dennery, P.A., and Stevenson, D.K., Duration of action and tissue distribution of zinc protoporphyrin in neonatal rats, *Pediatr. Res.*, 39, 1041, 1996.

221. Vallier, H.A., Rodgers, P.A., Castillo, R.O., and Stevenson, D.K., Absorption of zinc deuteroporphyrin IX 2,4-bis-glycol by the neonatal rat small intestine *in vivo*, *Dev. Pharmacol. Ther.*, 17, 109, 1991.

222. Kaplan, M., Vreman, H.J., Hammerman, C., Leiter, C., Rudensky, B., MacDonald, M.G., and Stevenson, D.K., Combination of ABO blood group incompatibility and glucose-6-phosphate dehydrogenase deficiency: effect on hemolysis and neonatal hyperbilirubinemia, *Acta Paediatr.*, 87, 455, 1998.

223. Kaplan, M., Vreman, H.J., Hammerman, C., Schimmel, M.S., Abrahamov, A., and Stevenson, D.K., Favism by proxy in nursing glucose-6-phosphate dehydrogenase-deficient neonates, *J. Perinatol.*, 18, 477, 1998.

224. Cohen, R.S., Hopper, A.O., Ostrander, C.R., and Stevenson, D.K., Total bilirubin production in infants of Chinese, Japanese, and Korean ancestry, *Taiwan I Hsueh Hui Tsa Chih*, 81, 1524, 1982.

225. Stevenson, D.K., Bartoletti, A.L., Ostrander, C.R., and Johnson, J.D., Pulmonary excretion of carbon monoxide in the human infant as an index of bilirubin production. II. Infants of diabetic mothers, *J. Pediatr.*, 94, 956, 1979.

226. Fischer, A.F., Ochikubo, C.G., Vreman, H.J., and Stevenson, D.K., Carbon monoxide production in ventilated premature infants weighing less than 1500 g, *Arch. Dis. Child.*, 62, 1070, 1987.

227. Lynch, S.R. and Moede, A.L., Variation in the rate of endogenous carbon monoxide production in normal human beings, *J. Lab. Clin. Med.*, 79, 85, 1972.

228. Vreman, H.J., Wong, R.J., Zentner, A.R., and Stevenson, D.K., unpublished data, 1998.

3 Carbon Monoxide Detectors

Richard Kwor

CONTENTS

3.1 INTRODUCTION

3.1.1 DETECTION OF CARBON MONOXIDE

Carbon monoxide (CO) is generated as a result of incomplete combustion of carbon-based fuel. It is formed in houses or buildings when the heating appliances are not working properly. Automobiles can also generate significant amounts of CO. Industrial types of CO detectors for warning and personal monitoring have been manufactured for many years. They are used in mines, garages, factories, and laboratories. First responders from the utility companies and fire departments use them to check houses and buildings for the presence of CO. However, residential CO detectors are a relatively new entry. They became more common only in the last 10 or so years.

3.1.2 EVOLUTION OF RESIDENTIAL CARBON MONOXIDE DETECTORS

Many technologies are available for carbon monoxide (CO) detection. However, only a few technologies are suitable for residential CO detection and monitoring.

They include the technologies using colorimetric, metal oxide, electrochemical, and infrared gas sensors. Color spots on badges, coated paper tapes, and chemical-filled devices that change color upon exposure to CO were available over the years for personal and workplace monitoring. They were the early forms of personal CO detectors but were generally for single-use detection and did not contain any electronics. Modern residential CO detectors can trace their beginning back to early 1960s when the metal oxide gas sensor was invented.

In 1962, N. Taguchi received the Japanese patent[1] on his metal oxide gas sensor and, in the same year, he formed Figaro Engineering to manufacture TGS109, a combustible gas sensor. In 1969, the company became public and started developing CO sensors code-named TGS105 and TGS209.[2] Later, the TGS711 and TGS712 sensors were developed for industrial use (circa 1975) and automatic garage door opener (circa 1980). These early CO sensors suffered low selectivity and frequent false alarms. The TGS711 and TGS712 incorporated a built-in heater to improve selectivity, but still had some long-term drift problems. Then, in 1983, a new CO-specific sensor (TGS203) with improved selectivity and superior long-term drift characteristics was introduced into the Japanese market. The same sensor was introduced to the North American market in 1985 and was adopted by Asahi Electronics (later becoming American Sensors[3]) for residential CO detector application. About the same time, the sensor was also used by Aquameter for monitoring CO in its boats.

In 1993, BRK Brands, Inc., a subsidiary of First Alert, Inc.,[4] started to market a battery-operated CO detector using a colorimetric sensor developed by the Quantum Group of San Diego.[5] The term *biomimetic* was used to describe the sensor because it was considered to mimic the uptake of CO by hemoglobin. Because of the increased public and media awareness, as well as the promotion and advocacy by the U.S. Consumer Products Safety Commission (CPSC) and the American Lung Association, the popularity of residential CO detectors grew rapidly. First Alert CO detectors became an instant success. About the same time, several other U.S. and Canadian companies entered the residential CO detector market, including Nighthawk Systems,[6] S-Tech,[7] Aim Safety,[8] CCI Controls,[9] Atwood,[10] Macurco,[11] and several others. Along with First Alert and American Sensors, they responded to the growing demand for residential CO detectors.

In April 1992, Underwriters Laboratories (UL) released the first edition of the UL-2034 standard for single- and multiple-station CO detectors.[12] Soon, several cities (including Chicago) made it mandatory to install CO detectors in all new houses. The growth of the residential CO detector market took off. However, this growth was not without problems. On December 22, 1994, there were over 1800 CO alarm emergency calls from residents in Chicago.[13] Only one resulted in hospitalization; the rest were termed *false alarms* that were later attributed to a weather-related atmospheric inversion.

This incident triggered the development of the second edition of UL-2034 that went into effect on October 29, 1996.[14] As detector manufacturers responded to the change in the UL standard, which required a reduction in false alarms, the quality of their products continued to improve. The existing CO sensor manufacturers (including Figaro and Quantum Group) as well as new sensor manufacturers began the development of improved versions of the colorimetric and metal oxide sensors. Prototypes of advanced sensors that were smaller, more sensitive, and consumed less power began to appear.

In 1996, International Approval Services* (IAS) established its CO requirements IAS6-96[15] to supplement the existing UL-2034 (second edition) standard. This new requirement came about because the gas industries were not happy with increasing nuisance calls as a result of "false alarms" by UL-approved CO detectors. Because of pressure from the gas industries, fire departments, and CPSC, UL began the process of revising the second edition of UL-2034 in April 1997.

This prompted detector manufacturers to look for sensors that deliver better stability and consume less power than the prevalent colorimetric and metal oxide sensors. The electrochemical sensor manufacturers, on the other hand, also heard the message and accelerated development of their sensors. The result was the arrival of small electrochemical sensors with prices attractive to the detector manufacturers. Soon, several major CO detector manufacturers developed detectors using electrochemical sensors. Meanwhile, the colorimetric and metal oxide sensor manufacturers also improved the performance of their sensors and readied them for the new UL revision.

On June 1, 1998, IAS published the second edition of IAS6-96, and on October 1, 1998, the revision to the second edition of UL-2034 went into effect. By late fall of 1998, many manufacturers had their CO detectors approved to the new standards, and in time for the heating season. At present, a great majority of the CO alarms sold in the United States have either UL or IAS approval, or both. All three major types of CO sensor technology (colorimetric, metal oxide, and electrochemical) are represented and have received UL approval.

As will be discussed in the Section 3.3, CO detectors based on nondispersive infrared (NDIR) technology are superior to the detectors using any of the other three types of sensors; however, but because of the high cost, no residential CO detectors use NDIR sensors at present. This situation will certainly change, but whether or not NDIR detectors will take any significant share in the residential CO detector market will depend on how much cost reduction can be achieved.

The development of residential CO detectors is no different from the development of other consumer electronics. Consumer awareness stimulates demand and demand encourages development. Development brings better products and lower cost, which in turn creates more demand. Along the way, standards/regulations help to improve quality and uniformity that eventually benefit consumers. Today, the CO detectors on the market are significantly better than the ones sold when the first edition of UL-2034 was in effect. However, the development of CO detectors never pauses. The quality will improve and the cost will drop. Some day, reliable and low-cost CO detectors may be found in every household, just as with smoke detectors today.

---

* IAS was a joint venture of the American Gas Association (AGA) and the Canadian Gas Association (CGA). It was acquired by Canadian Standards Association (CSA) in July 1997 and is now a standards and product testing organization in the United States and Canada for appliance and accessories fueled by natural and liquified petroleum gases. On January 27, 1999 the CSA became known as CSA International.

3.1.3 STANDARDS AND REQUIREMENTS

In the early 1990s, as more residential CO detectors were manufactured and sold, it became evident that a common standard was needed to ensure uniformity and quality. In 1992, UL took the initiative and developed the Standard No. 2034 for residential CO detectors. Since then, there have been several editions and one revision to UL-2034. While it may still not be perfect, the most recent revision is certainly a significant improvement over the original version in 1992.

At present, there are several standards established around the world for domestic CO detectors. The most comprehensive ones are from UL and IAS in the United States, the CSA, the British Standards Institute (BSI),[16] and the Japan Gas Appliances Inspection Association (JGAIA). These standards specify not only the performance of electrically operated CO detection apparatus, but also the general requirements for their construction and testing. The UL-2034, CGA6.19, IAS6-96, BS7860, and JGAIA standards all cover devices designed for continuous use in domestic premises that give a warning alarm in the event of a hazardous accumulation of CO. The UL-2034 even includes recreational vehicles and mobile homes. In North America, dialogues and communications among industrial groups and agency task forces have brought the requirements of UL, CSA, and IAS remarkably close. The UL-2034 (second edition), IAS6-96 (second edition), and CGA6.19 (second edition) have identical requirements for CO detector response time (time to reach alarm when exposed to a given CO level), while those for BS7860 and JGAIA are very different, as illustrated in Table 3.1.

3.1.4 "ALARM" VS. "DETECTOR"

On October 1, 1998, the revision of the second edition of UL-2034 went into effect. In this revision, residential CO detectors are referred to as CO "alarms," because the products covered in the standard not only sense CO, but also provide an audible alarm signal when CO reaches a certain level. UL wants to emphasize that these are meant to be warning devices for the protection of the public. The adoption of the term *alarm* also signifies the belief that residential CO detectors should not be considered as an instrument for the precise detection of CO concentration in the location. Instead, they are designed to alert people in the event that dangerous levels of CO exist. This is a sensible consideration because the UL standard does not require a capability for precision measurement of CO for each detector approved. A reasonable tolerance is allowed for deviation of measurement due to environmental effects such as humidity and temperature. Without this provision, low-cost residential CO detectors would not be available, at least not at present. Current industrial CO detectors are capable of measuring CO levels with an accuracy of ±2 ppm, but they require periodic recalibration and cost hundreds or thousands of dollars. Only time will tell when low-cost residential CO detectors will be available with that level of accuracy.

3.2 COMMON FEATURES OF RESIDENTIAL CARBON MONOXIDE DETECTORS

Residential CO detectors have come a long way from the days of color patches. Today consumers can find sophisticated CO alarms with many useful features at a

TABLE 3.1
Comparison of Detector Response Time for Various Standards

Standards	Response Time in CO	
	Conc. (ppm)	Response Time (min)
UL-2034 (10/01/98)	70 ± 5	60–189
CGA 6.19 (06/01/98)	150 ± 5	10–50
IAS6-96 (06/01/98)	400 ± 10	4–15
BS7860 (1996)	45	>60
	150	10–30
	350	<6
JGAIA (08/86)		
Bathroom type	250	<5
Kitchen and dining room type	200	<15
	550	<5
	False Alarm — CO Conc. Resistance Specifications	
	Conc. (ppm)	Response Time (min)
UL-2034	30 ± 3	No alarm within 30 days
IAS6-96	70 ± 5	No alarm within 60 min
CGA6.19		
BS7860	No specification	No specification
JGAIA		
Bathroom type	<100	No alarm
Kitchen & dining room type	<50	No alarm

Note: The Occupational Safety and Health Administration (OSHA) has specified a permissible exposure limit of 50 ppm over an 8-h period.

very low cost. Residential CO detectors/alarms are used in ordinary homes and also in recreational vehicles (RVs) and mobile homes. Although the detectors for different applications may be exposed to different environments (e.g., RV CO detectors have to survive more extreme temperatures), their basic features are similar. They can be broadly divided into two major categories: main powered (AC) and battery powered (DC). The AC detectors usually are operated on the house current and have power cords that are plugged into a wall socket. Many of these AC detectors have backup batteries to keep the unit running during a power outage. Depending on the type of CO sensor used, the backup battery can last anywhere from hours to weeks. The DC detectors can operate over a year on a single 9-V alkaline battery, or a set of AA batteries. Both AC and DC detectors with battery backup feature low-battery warnings. The DC detectors invariably use either the colorimetric or the electrochemical sensors because of their low power requirement. There is no battery-operated CO detector at this point that uses a metal oxide CO sensor, although there has been significant advance in the development of low-power metal oxide sensors.

The major features of a residential CO detector include:

Metal oxide, electrochemical, or colorimetric sensors;
Permanent or replaceable sensor;
85-dB audible alarm when CO accumulation reaches alarm level;
Red light-emitting diode (LED) for visual alarm warning and/or detector trouble
    signal;
Green LED for indication of normal operation;
Test/reset button for testing the detector operation, and resetting during alarm;
Approval by UL, ULC, CSA, IAS, or BSI.

It should be noted that ULC is the acronym of Underwriters Laboratories of Canada.
ULC does not have its own standard, it uses the same standard as CSA.

Some detectors feature a digital display that shows the instant CO concentration.
The digital display also makes it easier to indicate the status or any possible trouble
with the detector, such as low battery and failed sensor. It can provide more information on the CO levels, including an instant reading and rate of change. There can
also be a digital peak-level feature that shows the highest CO level the detector has
experienced since the button was last pressed. However, the inclusion of digital
display in a CO detector was at one time controversial. There were objections to the
use of a digital display because the CO levels displayed might not be accurate. To
eliminate unwarranted user concern and confusion, UL required that CO levels below
30 ppm not be displayed. Lately, the advantages of digital display may have convinced manufacturers otherwise; more of them have added digital display models
to their product lines.

As CO detectors become more reliable, they also come with more features.
Recently, CO detectors have been incorporated into tabletop alarm clocks. There are
also units with dual-detection capability for CO and smoke. Some detectors even
feature a voice alarm/warning system. CO detectors for special populations such as
the hearing impaired and people with low CO tolerance/high susceptibility may also
be available in the not too distant future.

3.3 SURVEY OF CARBON MONOXIDE SENSOR
TECHNOLOGIES

A good consumer CO detector should have the following characteristics:

CO sensitivity down to parts per million (ppm) range;
Superior stability;
No false alarms;
Fast response;
Good accuracy and resolution;
Long life;
Easy operation;
Low maintenance;
No requirement for recalibration;

Mechanical ruggedness;

Electronical ruggedness (immunity to interference);

Low cost.

Although superior mechanical and electronic design/construction helps to achieve a good detector, the determining factor still lies in the choice of the CO sensor. Many different technologies are available for the sensing of CO. However, only four sensor technologies are suitable for use in residential CO detectors. These four are colorimetric, metal oxide, electrochemical, and infrared sensors. There is another type, called catalytic or pellistor sensors, that is used in some industrial applications, but it is not discussed here because of its overall inferior performance. Here, only a brief survey of the four sensor technologies will be attempted. Readers who are interested in a more-detailed technical account of these technologies are referred to Reference 17, which contains the most comprehensive and detailed discussion of residential CO detectors now available.

3.3.1 COLORIMETRIC SENSORS

Colorimetric sensors contain chemicals that have two major functions: (1) to catalyze the reaction of CO with a chemical or chemicals and (2) to change color. They are widely used in passive CO detectors (color spots on badges, continuous spools of paper, and in tubes) for personnel and workplace monitoring, and for single-use detection. In the late 1980s and early 1990s, the Quantum Group in San Diego spearheaded the development of a colorimetric sensor for residential CO alarms. The term biomimetic was adopted to describe this sensor, as the manufacturer claimed that the sensor mimics the human response to CO, changing its spectral response in the presence of CO in a manner similar to hemoglobin in human blood.[5] The Quantum sensor is very simple in construction. A clear plastic housing contains two porous transparent sensing disks holding a chemical complex. Upon exposure to CO, one or both of the sensing disks change their spectral characteristics. An infrared LED and photodiode pair detect the darkening (reduction of optical transmittance) of the sensor. Their output is monitored electronically to indicate the level of CO. Figure 3.1 shows the front and back views of a colorimetric sensor, and a module pack incorporating a 9-V battery.

Chemical reagents that produce color change upon exposure to CO have been known since the early 1900s.[17] Progressive innovations since then have resulted in a formulation that is used in today's commercial CO detectors. A palladium salt is usually used as the catalyst for CO oxidation, and molybdenum or tungsten compounds are used as color indicators. In addition, there are three other essential ingredients in the formulation: a regeneration catalyst that promotes the reoxidation of the palladium and molybdenum catalysts, a hygroscopic agent that maintains the desired moisture in the mixed salts, and an encapsulant that stabilizes the reactions.[18,19] Upon exposure to CO, a series of complex reactions take place involving the various components in the formulation. In the forward reaction, CO is oxidized, while the palladium salt is reduced to its metal. At the same time, the molybdenum salt is reduced to a species that has a darker color. Water concentration plays a critical

FIGURE 3.1 The front and back views of a colorimetric CO sensor, and a module pack incorporating a 9-V battery.

role in this forward reaction. In the reverse reaction, the metallic palladium and the molybdenum salt are reoxidized, giving up water. Here copper ions are usually used as the catalyst. The net kinetics of the reaction depends on the relative magnitudes of the forward and reverse reaction rates and the overall reaction is simply

$$CO + 1/2\ O_2 \rightarrow CO_2$$

In human blood, the rate of COHb formation is dependent upon the ambient air CO concentration. Given a constant air CO level, COHb will reach a steady-state saturation that is a function of the air CO concentration. For a colorimetric sensor to behave biomimetically, the rates of the forward and reverse reactions must be well controlled and balanced. This, plus the fact that the reactions are also strongly dependent on temperature and humidity, as well as traces of environmental gases or vapors, has generated a unique challenge to sensor manufacturers. Early colorimetric sensors in domestic CO detectors (circa 1994) suffered many premature failures and false alarms. The recent colorimetric sensors, however, have a much improved performance.

Colorimetric sensors have a good sensitivity to CO and a simple construction. They exhibit a unique accumulation effect of CO exposure, require uncomplicated associated electronics, and therefore are suitable for low-cost domestic CO detectors, particularly in battery-operated units. These sensors do, however, have several characteristics that are considered weaknesses compared with other types of sensors. Temperature dependence is normally not considered a serious weakness because it can be compensated for *in situ* or electronically. Compensation of humidity dependence is more difficult, but is still achievable. The sensors respond to some interference gases and vapors, but their selectivity as well as susceptibility to poisoning is at levels comparable with other types of sensors. The major weakness is their slower response reversibility. In general, these sensors take a longer time to recover in the absence of CO. The reversibility can also be exposure dependent, making it harder to reset the detector quickly. Furthermore, because colorimetric sensors cannot measure instantaneous CO levels, a colorimetric detector that sounds an alarm cannot be tested to determine the level at which the alarm was actuated. Colorimetric detectors are thus rarely equipped with digital displays. Despite these weaknesses, newer colorimetric detectors do pass the requirement of resetting as specified by UL-2034. Once a detector enters the alarm mode, it can be reset by pressing the test/reset button. If the ambient CO level is above 100 ppm, the detector must realarm within 6 min. If, on the other hand, the ambient CO level is below 100 ppm, the detector should not realarm. Finally, the lifetime of early-generation colorimetric sensors had been shorter compared with that of most other types of sensor. The reagents in the sensor suffered from decomposition or loss through vaporization, thereby affecting their long-term stability and lifetime. Later sensors have been significantly improved and their lifetime is now guaranteed to be 6 years, although no extensive field data are available to support that yet. As time passes, reliable field data will be available for a more accurate assessment of the average lifetime of these sensors.

3.3.2 Metal Oxide Semiconductor Sensors

Many metal oxides change their electrical resistance upon exposure to CO, and the change depends on the CO concentration. The CO level can thus be found by electronically measuring the resistance change of the material. This type of sensor is called a metal oxide semiconductor sensor because its resistance lies somewhere between that of a good conductor and a good insulator (some metal oxides are good insulators, such as aluminum oxide and tantalum oxide). The use of metal oxides as gas sensors was pioneered by Taguchi and Figaro Engineering, Inc. Today, several companies produce metal oxide CO sensors, but Figaro is still the leading manufacturer of this type of sensor in the world. Figure 3.2 shows different generations of metal oxide sensors.

Metal oxide CO sensors can be divided into two major groups. One group is made of metal oxides that display a so-called n-type semiconductor behavior; n-type means "negative-type," a term used to describe an abundance of electrons that carry a negative charge. The most prominent metal oxide in this group is tin oxide (SnO_2), which is used in the majority of today's metal oxide CO sensors. The n-type metal oxide CO sensor has a very high resistance in fresh air. However, when exposed to CO, the resistance drops rapidly. A few hundreds of ppm of CO can bring about two to three orders of magnitude of resistance change (100 to 1000 times lower than the fresh air value). The second group of metal oxide CO sensors involves the p-type semiconductor; p-type means "positive-type," or an abundance of positive charges that result from missing electrons. The p-type metal oxide CO sensors have a low electrical resistance in fresh air, and their resistance increases when exposed to CO. Capteur Sensors and Analysers[20] is the leading

FIGURE 3.2 Different generations of metal oxide CO sensors.

company manufacturing p-type sensors. Capteur uses chromium titanium oxide rather than the typical tin oxide for its CO sensor.[21]

The first generation of metal oxide CO sensors typically consists of a bead of sintered tin oxide and some binding materials such as glass. Platinum or palladium is usually added as a catalyst. Traces of other materials may also be added to improve the performance of the sensor. Since the sensor must operate at an elevated temperature, one or two small heater coils are included inside the bead. Depending on the design, a separate electrode may be incorporated into the sensor. Figaro TGS203 has a representative structure that has two heater coils and therefore four pins.[22] The two heater coils are not connected; thus a measurement between the two coils gives the sensor resistance. Later designs of metal oxide CO sensors adopt a thick-film structure in which the electrode, the heater, and the sensing material are prepared as pastes and then coated in turn on an alumina substrate.[23] The first layer of paste is dried and fired before the second layer is applied. The newest design utilizes a thin-film approach by which thin layers of the electrode, the heater, and the sensing material are deposited in turn on a silicon substrate using methods borrowed from micromachining and integrated circuit technologies.[24]

CO detection by metal oxide sensors is a result of the interaction between the adsorbed oxygen and CO. Essentially, atmospheric oxygen is adsorbed on the surface of the metal oxide, trapping an electron from the material. The resulting oxygen ions at the surface repel electrons in the bulk, creating a region depleted of electrons and increasing the electrical resistance. When CO is adsorbed on the surface, it reacts with the oxygen there. This oxidation of CO to CO_2 consumes the adsorbed ionic oxygen, thus reducing the resistance of the sensor. The change in resistance is measured electronically and gives an indication of the level of CO. This seemingly simple mechanism is in reality very complex. It is affected by the microstructure of the surface and by the geometry of the sensor. Other factors include defects and vacancies in the material, bulk and surface reaction rates, and the contribution of additives. Typical ingredients in a metal oxide sensor include the following: a metal oxide, semiconductor dopants (impurities added intentionally), a catalyst(s), binders, and fillers. Other materials may be added to help control and stabilize the structure of the sensor. Different ingredients are added depending on the structure and the technology used (e.g., thick film, thin film, or sintered bead). For most metal oxide sensors, an active carbon filter is usually incorporated to improve selectivity.

One characteristic of the metal oxide sensor is that it requires heating for its operation. This contrasts with other types of sensor and is responsible for its higher power consumption. The heating (typically around 300 to 400°C) is needed to drive out the moisture and contaminants, and to stabilize the sensor. On the other hand, the sensor is most sensitive to CO at about 80 to 100°C. As a result, most metal oxide sensors require a pulsing heating pattern — each cycle consisting of a high-temperature period followed by a period of lower temperature. TGS203, for example, has a 2½-min heating cycle (60 s at 350°C and 90 s at 85°C).

Metal oxide sensors have some advantages over competing technologies. They include long life, simple electrical measurement, and the potential for microfabrication. However, their main strength lies in their long history. Because of the difficulty in using accelerated tests to predict sensor life and reliability, actual sensor long-term data are valuable. Metal oxide sensors have had the longest history of field usage and therefore provide the most extensive long-term data. Compared with other types, the metal oxide sensor has several weaknesses, including high input power, lower resolution, and higher temperature and humidity dependence. Of these, humidity dependence is the most significant factor limiting the widespread use of this type of sensor. This is so because the higher power requirement and lower resolution may not be problems in some applications, and temperature dependence can be easily compensated for with electronics. In general, different sensor designs have different humidity characteristics. Sensors from some manufacturers have lower dependence than others, but none has eliminated the humidity dependence completely. Significant research-and-development effort has been invested in overcoming these weaknesses. Better and improved devices are on the horizon.

3.3.3 ELECTROCHEMICAL SENSORS

Electrochemical sensors generate a small amount of electricity upon exposure to the target gas. They have very good resolution (down to 1 ppm) and, with regular recalibration, they also exhibit excellent accuracy. This type of sensor has been extensively used for monitoring industrial atmospheres and gas chambers. They are also used widely in portable and fixed toxic gas monitors. However, residential electrochemical sensors are relatively new because low-cost sensors have been available only in the last couple of years. Figure 3.3 shows several electrochemical CO sensors from different manufacturers.

Electrochemical CO sensors behave in a fashion similar to fuel cells that consume gas, generating electricity. Gases are intentionally fed into the fuel cell to be used

FIGURE 3.3 Electrochemical CO sensors from different manufacturers.

as fuel, while electrochemical sensors detect traces of gas in the environment. As required by the intended applications, fuel cells generate amperes of current, while electrochemical CO sensors generate only milliamperes or less. Thus, this type of sensor is sometimes called a "microfuel cell." Another name that is sometimes used is "amperometric sensor," which describes the current-generation nature of the sensor.[25,26] Electrochemical CO sensors have a simple structure, consisting of two or three electrodes separated by an electrolyte. CO enters the cell through a small diffusion hole on the top. When it reaches the top electrode (the sensing electrode), it reacts with water under the catalytic action of platinum, producing electrons, protons, and carbon dioxide. The electrons are removed by the external circuit while the positively charged protons pass through the electrolyte and reach the back electrode (the counter electrode) where they combine with oxygen and electrons to form water.

At the sensing electrode:

$$CO + H_2O \rightarrow CO_2 + 2H^+ + 2e^-$$

At the counter electrode:

$$1/2O_2 + 2H^+ + 2e^- \rightarrow H_2O$$

The overall reaction is the oxidation of CO to CO_2:

$$CO + 1/2O_2 \rightarrow CO_2$$

For stable sensor operation, the diffusion hole is usually made very small to ensure that the sensor is in a diffusion-limited mode. This means that the amount of CO consumed by oxidation in the sensor is limited by its diffusion through the hole rather than its reaction rate at the sensing electrode. A dust filter is often put on top of the hole to reduce the chance of clogging by dust particles. An active carbon filter is usually put above or below the diffusion hole to improve the resistance of the sensor to interfering gases and to prevent poisoning of the sensor. The electrodes are in contact with the electrolyte that can be in liquid, gel, or solid form. The sensing electrode must allow the diffusion of CO into the cell while maintaining the integrity of the electrolyte. Thus, it is usually made of a porous but hydrophobic membrane coated with a thin porous platinum layer. For efficient catalysis, the platinum particles have diameters on the order of a micrometer so that the total surface area is very large. Sometimes, the membrane can be made by mixing a Teflon emulsion with platinum powder. The counter electrode is also made of a similar material. For some sensors, a third electrode (called the reference electrode) is incorporated to provide a more stable reference. Such an arrangement helps to improve sensor performance and to extend its range. For some electrochemical cells, two additional electrodes are added. These electrodes allow the generation of a minute quantity of hydrogen when a voltage is applied to them. The hydrogen is used to check and monitor sensor sensitivity. The conventional electrochemical CO sensors on the market today use concentrated sulfuric acid (H_2SO_4) or phosphoric acid (H_3PO_4) as the electrolyte. The acid electrolyte may be immobilized in a solid polymer (in this case, an acid electrolyte reservoir is needed),[27] or it can be embedded in a glass wool wick inserted between the electrodes. The acid electrolytes work very well except during a prolonged period when humidity is either very high or very low. Under extreme humidity conditions, the acid concentration can change, affecting sensitivity of the sensor. Increasing the volume of acid can reduce this humidity dependence, but at the expense of the size and cost of the sensor. A more recent attempt to solve this problem is the use of a solid polymer electrolyte.[28] The most common choice to date is a polymer membrane called Nafion,[29] which is a good proton conductor when hydrated, providing an excellent medium for the sensor action. Since the degree of hydration will affect the sensor sensitivity, a water reservoir is incorporated into the sensor to maintain a constant hydration level in the membrane.[30]

Electrochemical sensors have been regarded as the best choice for residential CO detector applications. They combine reasonable cost with high resolution, linearity and selectivity, and low power consumption. However, some of today's electrochemical sensors, particularly the acid type, continue to suffer from humidity and temperature dependence (solid polymer electrolyte sensors have a much lower humidity dependence). While some of the more expensive electrochemical sensors are stable over a large humidity range, the low-cost ones can lose a significant percentage of their sensitivity when exposed to very low humidity (approximately 10%) over a few months. The electrochemical sensors manufactured by the Nighthawk Division of Kidde Safety have a much lower humidity dependence because of the solid polymer/water reservoir combination.[6] Another weakness of the low-cost electrochemical sensors is the lack of long-term sensitivity data since these sensors were only developed in the last few years. The sensitivity of electrochemical sensors is known to drift over time. In the first month or so after assembly, the sensitivity drops at a high rate. The decrease then slows and the sensitivity stabilizes. This change has been attributed to a variety of causes, including catalyst degeneration or fouling, irreversible reactions at the electrodes, and pH changes in the electrolyte.[26] With periodic recalibration, as is done in industrial applications, this weakness can be easily overcome and accuracy can be maintained. However, in residential applications, where consumers do not have any facility or equipment to recalibrate the detector, this can be a problem. Thus, it is of the utmost importance to manufacture the detectors using sensors that have already stabilized. Over the life of an electrochemical sensor, the previously mentioned periodic hydrogen puff can be of some help in checking the sensitivity of the sensor. But this adds to the cost of the detector, shortens battery life, and may still fail to detect some sensor problems, as in the case when the diffusion hole is clogged by dust or grease. In that case, the internal hydrogen puff will still indicate that the sensor is working properly, even though the sensor can no longer detect any CO. Of course, for sensors with a predictable sensitivity change rate, it is not too difficult to use the microcontroller in the detector circuit to compensate for the drop in sensitivity, thereby producing a very stable detector output to a given CO level.

One last concern with electrochemical sensors is that like colorimetric and metal oxide sensors, they also suffer from possible cross-contamination and/or poisoning effects of some gases and vapors. Traces of some gases/vapors may seriously poison the sensor and reduce its sensitivity. Much work has thus been done in testing these sensors against poisoning gases and vapors. Such testing will always be an integral part of the development of new sensors.

3.3.4 INFRARED SENSORS

The infrared CO sensors discussed here refer to units that utilize NDIR technology. In an NDIR sensor, some absorption of infrared light occurs when CO is present. This absorption is characteristic of CO and is different from that by other gases. By detecting changes in the infrared light, the CO level can be measured. NDIR technology is by far the best among all of the four sensor technologies discussed. However, sensor cost is significantly higher than the other three. Currently, only one company (Engelhard)[31] has produced a "low-cost" NDIR CO detector, selling for around $800. This is intended for industrial applications and is thus much more expensive than the residential detectors using the other technologies. Telaire Systems, Inc., developed a residential CO detector using NDIR technology, but the product did not reach the market even after Engelhard purchased Telaire.

A typical low-cost NDIR sensor consists of an optical chamber with mirrors, an infrared source, and two infrared detectors (measuring detector and reference detector). As the infrared radiation from the source travels through (often in multipaths) the chamber to the measuring detector, it is partially absorbed by the gases or vapors in the chamber. The absorption is characteristic of the particular gas or vapor, with different species absorbing at different wavelengths. CO, for example, absorbs infrared radiation at a wavelength of 4.6 μm. This absorption produces a decreased electrical signal at the measuring detector. In such an NDIR design, part of the radiation from the source is used as a reference, which travels to the reference detector without any attenuation. By measuring the electrical signal from the measuring detector, and comparing it with that from the reference detector, the concentration and type of the gas in the chamber can be found. Low-cost residential NDIR sensors are CO specific (measures only CO concentration). To achieve that, the sensor incorporates a filter that determines the wavelength at which CO has the strongest absorption of infrared.

NDIR sensors are stable and linear. They have very fast response and recovery that are limited only by the sampling rate of CO (rate at which CO enters the chamber). All in all, except for the cost, NDIR technology is superior in every sensor characteristic including sensitivity, selectivity, resolution, detection range, temperature and humidity dependence, long-term stability, and lifetime. The limitation for NDIR detectors is not from the NDIR technology itself, but is determined by the components and associated electronics of the detector. Despite their overwhelming superiority, NDIR sensors/detectors are not yet found in residential applications, primarily because of the high cost. This situation may change in the future when low-cost NDIR CO sensors/detectors become available.

3.3.5 Comparison of Sensor Technology

A summary and comparison of CO sensor technologies would be appropriate here. However, an accurate assessment is difficult. The following list highlights the complications of the intended comparison.

1. There is a lack of literature describing the detailed specifications and characteristics of most commercial sensors. The manufacturers' published data do not cover all aspects.
2. Improvement is often made to existing sensors. New sensors and detectors are constantly being developed. Any innovation and/or improvement will change the performance characteristics of these sensors.
3. Within each technology, there is a variation in performance characteristics between the products from different manufacturers. Even for the same manufacturer, the products can have a large variation in specifications.
4. While a certain sensor technology may be inferior in certain characteristics compared with another technology, a detector using an inferior sensor technology may actually perform better than a detector using a superior sensor technology. This can result from use of superior electronics (better design, better components) incorporated into the detector. Better components often translate to higher cost. Thus, to render a fair comparison, the cost factor must also be examined.
5. For new sensors, long-term and contamination performance data are collected. New information can change the original assessment of sensor properties.

TABLE 3.2
Comparison of Sensor Technologies for Domestic Applications

Sensor Property	Sensor Technology			
	Colorimetric	Metal Semiconductor	Electrochemical	Infrared
Short-term stability	Difficult to assess	Fair	Good	Excellent
Lifetime	>5 years Data being collected	5–10 years	>5 years Data being collected	>5 years
Resolution	Fair	Fair	Good	Excellent
Immunity to false alarm	Fair	Good	Good	Excellent
Immunity to false negative	Good	Good	Good	˙Excellent
Immunity to poisoning	Good	Good	Good	Excellent
Humidity dependency	Fair	Fair	Good-to-excellent	Excellent
Temperature dependency	Fair	Fair	Fair	Excellent
Sensitivity drift	Unknown	Moderate	Moderate	Low
Response time	Fair	Fair	Good	Excellent
Selectivity	Good	Good	Good	Excellent
Power consumption	Low	High	Low	Medium
Cost of sensor	Low	Low	Low	High

Note: False alarm = detector alarms even though the CO level is low; false negative = detector fails to alarm when CO level is high; resolution = the figure below which circuit noise is likely to make measurement difficult; response time = the time taken by the sensor to reach 90% of its final output; selectivity = ability to distinguish between CO and other gases.

For the above reasons, only a very general comparison of the sensor technologies is provided here (Table 3.2). No precise specifications are included and words like *poor* and *weak* are avoided. The colorimetric, metal oxide, and electrochemical technologies may not be perfect and are inferior to the NDIR technology, but sensors exist using all three technologies that meet the latest UL-2034 requirements.

3.3.6 MARKET SHARE OF DIFFERENT SENSOR TECHNOLOGIES

As discussed in previous sections, colorimetric CO detectors involve the lowest cost, metal oxide CO detectors have the longest life, and electrochemical CO detectors are relatively new, but exhibit the best overall combination of cost and performance. This is reflected in the CO detector market share change in the last few years. Colorimetric and metal oxide–based detectors dominated the market up until 1997.[32] Since that time, they have lost significant market share to the electrochemical detectors, as illustrated in Figure 3.4.

3.4 OPERATING PRINCIPLES OF CARBON MONOXIDE DETECTORS

Early residential-type CO detectors generally featured a single alarm point; i.e., the alarm would come on when the ambient CO detected exceeded a certain level. This

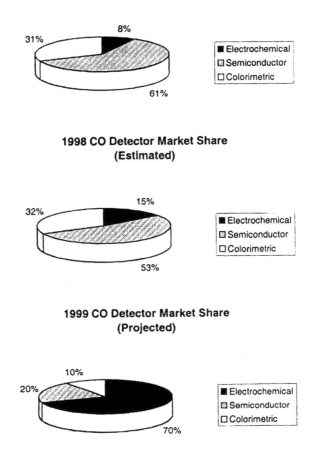

FIGURE 3.4 Residential CO detector market share.

alarm point was usually set at a CO level between 150 and 250 ppm. Sometimes, to ensure a quick response to high concentrations of CO, a second alarm trigger level was set between 350 and 550 ppm. For these detectors, simple and straight-forward electronic circuits were generally adequate. At this time a great majority of the CO sensors were of the metal oxide type, the circuits usually measuring the electrical resistance of the sensor and comparing it with a reference value. If the measured resistance exceeded the reference level, a buzzer would be energized, sounding an alarm. Because of the rapid advance of microcontrollers and micropro-cessors, it was not uncommon at that time for sensor manufacturers to recommend the use of an on-board microcontroller for more complicated circuit functional requirements. As described in its *Gas Sensor Technical Reference*,[22] Figaro Engi-neering in 1991 already had a 4-bit microcontroller included in its TGS203 circuit. This came about because of the more complex heating pattern required by the TGS203 sensor.

The first edition of UL-2034 (April 1992) reflected the general consensus that for the protection of the public, it is better to use the cumulative value rather than the instantaneous level of CO for triggering an alarm. The Coburn, Forster, and Kane model[33] was adopted for the estimation of COHb level in blood after CO exposure. It was agreed that when the COHb level exceeded 10% as predicted by the Coburn model, the alarm must sound. On the other hand, if the COHb level was below 2.5%, no alarm should sound. This was designed to reduce nuisance calls, which puts an unnecessary workload on first responders from fire departments and/or utility companies. The response of the detectors to COHb levels between 2.5 and 10% was left optional for the manufacturers to decide. Some manufacturers opted to add a warning level in the range, while others lowered the alarm level from 10% down to this range. It became obvious that simple electronics were no longer sufficient to provide such an alarm. Microcontrollers were adopted for the residential electronic CO detectors sold in the United States after 1992. Later editions of UL-2034 made some changes to the alarm level and also tightened the tolerances of detector response under different ambient conditions. The most recent revision of the second edition of UL-2034 (October 1998) changed the buzzer alarm pattern so it can be distinguished from that of smoke alarms. With so many new circuit requirements added, a modern CO detector cannot function properly unless a microcontroller is used.

Figure 3.5 shows the functional block diagram of a typical CO detector. A majority of residential CO detectors on the market contain similar functional components, even though the actual electronic circuitry may be different. The operation of the CO detector is rather straightforward. When CO reaches the sensor (1), it causes changes in the characteristics of the sensor, which are then measured by the matching electronics. For example, if the sensor is an electrochemical type, the presence of CO will cause the sensor to generate a minute current that is measured by the measurement electronics (2). The measurement electronics then send the information to the microcontroller (3), which is the control center of the detector. The microcontroller analyzes the data and converts them to digital form if they have not already been converted. It then includes information on the accumulated time and calculates the COHb level based on the Coburn formula and the UL standard (or any other standard the detector is designed to meet). If the COHb level exceeds the alarm limit, it sends a signal to the audible alarm circuit (5) and the visual alarm circuit (6). In modern CO alarms, the audible alarm from a buzzer must have a particular pulsing pattern while the visual alarm can be a flashing red LED. Several manufacturers have incorporated the digital display (7) feature, which shows the instantaneous CO level. Nighthawk Systems pioneered the peak-level indicator that recalls the highest CO level the detector has been exposed to since the last pressing of the reset button. This feature is very useful when first responders arrive at the scene and are unable to find any CO. The signal from the microcontroller is used sometimes to control a relay circuit (8), which can be used to shut off the gas to the furnace or heater, or to energize a remote alarm. For example, in a large building with several CO detectors installed in different locations, one alarming detector can notify the occupants in another room through the alarming of the detector in the vicinity of that room. The same signal can also be used to open the windows/doors and/or turn on the exhaust fan to remove the CO.

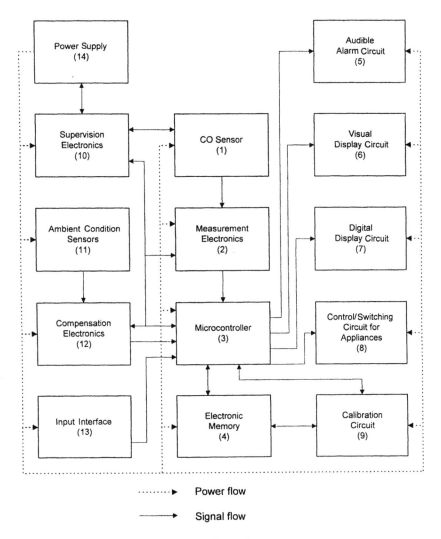

........▶ Power flow

——————▶ Signal flow

FIGURE 3.5 Schematic of CO detector electronics.

Most of the low-cost CO sensors commercially available at this time have some dependence on ambient conditions. Metal oxide, electrochemical, and colorimetric sensors, for example, are all affected to various degree by temperature and humidity. Even though some sensors may have less dependency than others, some form of compensation is always needed. This is accomplished by the ambient condition sensors (11) and compensation electronics (12). All of the residential CO detectors have some form of temperature compensation, while a few have humidity compensation circuitry built in. Humidity compensation is rather difficult mainly because of the mismatch of response time between the CO sensor and the humidity sensor. Although pressure compensation may also improve the performance of CO detectors, few residential CO detectors include this additional compensation.

CO detectors are designed to warn people when the ambient CO reaches dangerous levels. As with all safety devices, they must be made as reliable as possible. These devices all have dedicated circuitry on board to monitor any failure modes resulting from both hardware malfunction and software problems. This important task is performed by the supervision electronics (10), which checks various parts of the detector and alerts the user when it detects some problems. For example, if the microcontroller connection is broken, or if the timer device in the circuit stops, an alarm signal is energized to warn the user. The digital display circuit (7) mentioned earlier is certainly a desirable feature in this case because it can tell the user that the alarm has been caused by a defect in the circuit and not by a high CO level. This same feature can also positively alert the user to low battery power. A good detector usually has a circuit that monitors many failure modes in the circuit. Often, the microcontroller is a major part of the supervision electronics, although it itself may also be monitored.

Most residential CO detectors have simple and user-friendly input interfaces (13). These usually consist of one or two push-buttons on the front face. One of the buttons is the test/reset button. It is used for several purposes: (1) to check if the unit is functioning properly when first installed, (2) to test the detector when it is alarming, and (3) to act as a hush button and to be able to reset the unit temporarily. For example, when the unit is alarming, the user can push the button to stop the alarm. However, if ambient CO concentration exceeds 100 ppm, the detector will alarm again within 6 min. This eliminates any false alarm that may result from a glitch in the power line or a sudden, temporary generation of CO. The second push-button may be used for other purposes, including the recalling of peak CO concentration. More-sophisticated detectors may have some switches or buttons for other purposes, such as data logging and time/date indication.

Modern residential CO detectors all have on-board electronic memories (4). Some detectors may have memories as part of a powerful microcontroller, while others may have a separate memory for storing important electronic information, such as calibration reference, temperature compensation data, CO alarm levels, and COHb calculation formulas. The memory works closely with the microcontroller.

The calibration (9) of the fully assembled detector is usually checked during manufacture by placing it in a chamber containing a known quantity of CO. When the calibration circuit is energized, the sensor output is set at a reference level, corresponding to a given concentration of CO. This reference level is stored in the electronic memory. Later, when the detector is exposed to different CO concentrations, the microcontroller compares the sensor output with the stored reference level and calculates the correct CO concentration. For metal oxide sensors, which may not have a linear response, two-point calibration is usually done (e.g., 100 and 300 ppm). For electrochemical sensors, which have a very linear response, one-point calibration is sufficient.

The power supply (14) supplies electric power to the detector electronics. It can be either from a main AC source (120 V), or from a battery or a set of batteries. The AC power supply usually consists of a step-down transformer and associated rectifying and regulating circuit. Depending on the transformer used, a fuse may or may not be needed. Sometimes a manufacturer may include the optional battery backup feature for the unit. During power failure, the backup battery supplies energy to the detector to keep it running. If the detector is completely battery operated, UL-2034 requires that the battery or set of batteries must be able to run the detector for over 1 year. The detector must perform normally during this period of time. Some detectors use a single 9-V battery, while others use several 1.5-V batteries. At present, only electrochemical or colorimetric sensors are used in battery-only units, because of their low power consumption. Metal oxide sensors (including the newest micropower sensors) at present consume too much power to be used in the battery-only units.

Modern residential CO detectors are sophisticated electronic devices, made possible by advanced sensor and microcontroller technologies. As the sensor technology advances and as UL or other regulatory/standard agencies tighten the standard, detector circuitry will get even more complicated. The results will be better and more reliable CO safety devices for protection of the public.

REFERENCES

1. Taguchi, N., Japan Patent No. 45-38200, 1962.
2. Unno, K., personal communication, Figaro USA, Inc., 3703 West Lake Avenue, Suite 203, Glenview, IL 60025.
3. American Sensors went bankrupt in 1997. Its brand and assets have been acquired by North American Detectors, Inc., 100 Tempo Avenue, Toronto, Ontario, M2H 3S5, Canada.
4. First Alert, Inc., 3901 Liberty Street Road, Aurora, IL 60504.
5. Quantum Group, Inc., 11211 Sorrento Valley Road, San Diego, CA 92121.
6. Nighthawk Systems Division, Kidde Safety, 4980 Centennial Blvd., Colorado Springs, CO 80919.
7. S-Tech (USA), Inc., 2208 Landmeier Road, Elk Grove Village, IL 60007.
8. Aim Safe-Air Products, Ltd., 7-1600 Derwent Way, Delta, BC V3M 6M5, Canada.
9. CCI Controls, Cecelia St., South Gate, CA 90280.
10. Atwood Mobile Products, 4750 Hiawatha Drive, Rockford, IL 61103.
11. Macurco, Inc., 3946 S. Mariposa St., Englewood, CO 80110.
12. Underwriters Laboratories, Inc. (UL), *UL Standard for Safety for Single and Multiple Station Carbon Monoxide Alarms, UL2034*, 1st ed., 1992.
13. Carbon monoxide creates air of unease in Chicago (AP), *San Jose Mercury News*, Dec. 23, 1994.
14. Underwriters Laboratories, Inc. (UL), *UL Standard for Safety for Single and Multiple Station Carbon Monoxide Alarms, UL-2034*, 2nd ed., 1996.
15. International Approval Services – U.S., *IAS U.S. Requirements for Carbon Monoxide Alarms for Residential Use*, IAS No. 6-96, 1996.
16. British Standard Institution, Specification for Carbon Monoxide Detectors (Electrical) for Domestic Use, BS7860, 1996.
17. Clifford, P.K. and Dorman, M.G., Test Protocols for Residential Carbon Monoxide Alarms, Phase I, Final Report, The Gas Research Institute, September, 1996.
18. U.S. Patent No. 4,043,934.
19. U.S. Patent No. 5,063,164.
20. Capteur Sensors and Analysers, 11 Moorbrook Park, Didcot, OX11 7HP, U.K.
21. Capteur Sensors and Analysers, *Product Guide, Solid-State Semiconductor Sensors for Oxygen, Flammable, Toxic and Refrigerant Gases.*

22. Figaro Engineering, Inc., *Gas Sensor Technical Reference*, 1-5-3 Senbanishi, Mino, Osaka 562, Japan.
23. Figaro Engineering, Inc., *Technical Information for TGS2440*.
24. MicroChemical Systems SA, Rue de Porcena 15, CH-2035 Corcelles, Switzerland.
25. Cao, Z., Buttner, W.J., and Stetter, J.R., The properties and applications of amperometric gas sensors, *Electroanalysis*, 4, 253, 1992.
26. Chang, S.C., Stetter, U.R., and Cha, C.S., Amperometric gas sensors, *Talanta*, 40, 461, 1993.
27. *Product Data Handbook*, City Technology Limited, England.
28. Yan, H. and Liu, C.C., A solid polymer electrolyte-based electrochemical carbon monoxide sensor, *Sensors and Actuators B: Chemical*, 17, 165, 1994.
29. Nafion is a registered trademark of E.I. du Pont de Nemours and Company.
30. U.S. Patent No. 5,573,648.
31. Engelhard Sensor Technologies, 6489 Calle Real, Goleta, CA 93117.
32. Data partially derived from information published in *Home Improvement Executive*, 45 West 21st Street, New York, NY 10010.
33. Coburn, R.F., Forster, R.E., and Kane, P.G., Considerations of the physiological variables that determine the blood carboxyhemoglobin concentration in man, *J. Clin. Invest.*, 44, 1899, 1965.

4 The Setting of Health-Based Standards for Ambient Carbon Monoxide and Their Impact on Atmospheric Levels*

James A. Raub

CONTENTS

* The views expressed in this chapter are those of the author and do not necessarily reflect the views or policies of the U.S. Environmental Protection Agency.

0-8493-2065-8/00/$0.00+$.50
© 2000 by CRC Press LLC

4.1 INTRODUCTION

Carbon monoxide (CO) is one of six ubiquitous ambient (outdoor) air pollutants covered by the Federal Clean Air Act (CAA) requiring an assessment of the latest scientific knowledge as a requisite step in the development of standards to protect public health and welfare. The other pollutants are nitrogen dioxide, ozone, sulfur dioxide, particulate matter, and lead. The U.S. Environmental Protection Agency (EPA) is required under the CAA to reevaluate the National Ambient Air Quality Standards (NAAQS) for these "criteria" pollutants every 5 years.

CO is a trace gas in the troposphere produced by both natural processes and human activities. Because plants can both metabolize and produce CO, trace levels are considered a normal constituent of the natural environment. Although ambient concentrations of CO in the vicinity of urban and industrial areas can exceed global background levels, there are no reports of these currently measured levels of CO producing any adverse effects on plants or microorganisms. Ambient concentrations of CO, however, can be detrimental to human health and welfare, depending on the levels that occur in areas where humans live and work and on the susceptibility of exposed individuals to potentially adverse effects. Ambient air quality standards are meant to protect the most sensitive portion of the public from experiencing the effects of CO from controllable outdoor sources (e.g., motor vehicles and industrial activities involving fossil fuel burning). Other sources of CO exist indoors, however, where the regulatory authority is shifted to occupational exposure limits, building codes, and local public health policies (e.g., no-smoking policies).

This chapter presents a brief summary of the legislative and regulatory history of the CO NAAQS, the rationale for the existing ambient standards, and gives a brief overview of the impact that air quality standards have made on ambient concentrations of CO in the United States.

4.2 LEGISLATIVE REQUIREMENTS FOR AIR QUALITY STANDARDS

Two sections of the CAA govern the establishment, review, and revision of NAAQS. Section 108[1] directs the Administrator of EPA to identify and issue air quality criteria for pollutants that may reasonably be anticipated to endanger public health or welfare. These air quality criteria are to reflect the latest scientific information useful in indicating the kind and extent of all identifiable effects on public health or welfare that may be expected from the presence of the pollutant in ambient air.

Section 109(a) of the CAA[1] directs the Administrator of EPA to propose and promulgate primary and secondary NAAQS for pollutants identified under Section 108. Section 109(b)(1) defines a primary standard as one that "the attainment and maintenance of which in the judgment of the Administrator, based on such criteria and allowing an adequate margin of safety, [is] requisite to protect the public health." The secondary standard, as defined in Section 109(b)(2), "shall specify a level of air quality the attainment and maintenance of which in the judgment of the Admin-istrator, based on such criteria, is requisite to protect the public welfare from any

TABLE 4.1
National Ambient Air Quality Standards for
Carbon Monoxide

Date of Promulgation	Primary NAAQS	Averaging Time
August 1, 1994	9 ppm[a] (10 mg/m^3)	8-h[b]
	35 ppm[a] (40 mg/m^3)	1-h[b]

[a] 1 ppm = 1.145 mg/m^3, 1 mg/m^3 = 0.873 ppm at 25°C, 760 mm Hg.
[b] Not to be exceeded more than once per year.

Source: Federal Register.[4]

known or anticipated adverse effects associated with the presence of such air pollutant in the ambient air." Welfare effects generally include, but are not limited to, effects on soils, water, crops, vegetation, anthropogenic materials, animals, wildlife, weather, visibility and climate, damage to and deterioration of property, and hazards to transportation, as well as effects on economic values and on personal comfort and well-being.

Section 109(d) of the CAA[1] requires periodic review and, if appropriate, revision of existing criteria and standards. If, in the Administrator's judgment, the EPA review and revision of criteria make appropriate the proposal of new or revised standards, such standards are to be revised and promulgated in accordance with Section 109(b). Alternatively, the Administrator may find that revision of the standards is inappropriate and conclude the review by leaving the existing standards unchanged.

4.3 REGULATORY BACKGROUND FOR AIR QUALITY STANDARDS

On April 30, 1971, EPA promulgated identical primary and secondary NAAQS for CO at levels of 10 mg/m^3 (9 ppm) for an 8-h average and 40 mg/m^3 (35 ppm) for a 1-h average, not to be exceeded more than once per year. The scientific basis for the primary standard, as described in the first criteria document,[2] was a study suggesting that low levels of CO exposure resulting in carboxyhemoglobin (COHb) concentrations of 2 to 3% were associated with neurobehavioral effects in exposed subjects.[3]

In accordance with Sections 108 and 109 of the CAA, the EPA periodically has reviewed and revised the criteria on which the existing NAAQS for CO (Table 4.1) are based. On August 18, 1980, EPA proposed certain changes in the standards based on scientific evidence reported in the revised criteria document for CO.[5] Such evidence indicated that the Beard and Wertheim[3] study no longer should be considered as a sound scientific basis for the standard. Additional medical evidence accumulated since 1970, however, indicated that aggravation of angina pectoris and other cardiovascular diseases would occur at COHb levels as low as 2.7 to 2.9%. On

August 18, 1980, the EPA proposed changes to the standard[6] based on the findings of the revised criteria. The proposed changes included

1. Retaining the 8-h primary standard level of 9 ppm;
2. Revising the 1-h primary standard level from 35 ppm to 25 ppm;
3. Revoking the existing secondary CO standards (because no adverse welfare effects have been reported at or near ambient CO levels);
4. Changing the form of the primary standards from deterministic (this form of the NAAQS is based on the number of exceedances of an extreme value over a period of multiple years, e.g., the current CO NAAQS is not to be exceeded more than once per year over a 2-year period) to statistical (the NAAQS is based on an average concentration, e.g., annual 2nd, 3rd, or 4th highest daily nonoverlapping maximum 8-h value, over a period of multiple, 2 or 3 years);
5. Adopting a daily interpretation for exceedances of the primary standards, so that exceedances would be determined on the basis of the number of days on which the 8- or 1-h average concentrations are above the standard levels.

The 1980 proposal was based in part on a number of key health studies by Aronow et al. (see Reference 5). In March 1983, the EPA learned that the Food and Drug Administration (FDA) had raised serious questions regarding the technical integrity of several studies conducted on experimental drugs by the same laboratory and clinical investigator, leading the FDA to reject use of its drug study data. Therefore, the EPA convened an expert committee to examine the Aronow et al. CO studies before any final decisions were made on the NAAQS for CO. In its report,[7] the committee concluded that the EPA should not rely solely on these key data because of concerns regarding components of the research that substantially limited the validity and usefulness of the results.

An addendum to the 1979 criteria document for CO[8] reevaluated the scientific data concerning health effects associated with exposure to CO at or near ambient exposure levels in light of the committee recommendations and taking into account findings reported subsequent to those previously reviewed. On September 13, 1985, the EPA issued a final notice[9] announcing retention of the existing primary NAAQS for CO and rescinding the secondary NAAQS for CO.

The criteria review process was initiated again on July 22, 1987, and notice of availability of the revised draft criteria document was published in the *Federal Register*[10] on April 19, 1990. This draft document included discussion of several new studies of the effects of CO on patients with angina that had been initiated in light of the Aronow controversy discussed above. The Clean Air Scientific Advisory Committee (CASAC) reviewed the draft criteria document at a public meeting held on April 30, 1991. The EPA carefully considered comments received from the public and from CASAC in preparing the final criteria document.[11] On July 17, 1991, CASAC sent to the EPA Administrator a "closure letter" outlining key issues and recommendations and indicating that the document provided a scientifically balanced and defensible summary of the available knowledge of the effects of CO. A revised "staff paper" based on the scientific evidence was released for public review in February 1992, followed by two CASAC review meetings held on March 5 and April 28, 1992. The CASAC came to closure on the final staff paper[12] in a letter to the Administrator dated August 11, 1992, indicating that it provided a scientifically adequate basis for the EPA to make a regulatory decision on the appropriate primary NAAQS for CO. On August 1, 1994, the EPA issued a final decision[4] that revisions of the NAAQS for CO were not appropriate at that time.

In keeping with the requirements of the CAA, the EPA has started to review and once again revise the criteria for CO. The current status of the CO NAAQS is as follows:

National Ambient Air Quality Standards for Carbon Monoxide

The U.S. Environmental Protection Agency, which is required under the Clean Air Act to reconsider air quality standards every five years, has recommended ... for CO after an extensive review of the available scientific data.

Information on the recommended standards for CO and current air quality attainment status for CO is available on the Web at: *http://www.epa.gov/oar/oaqps*

4.4 RATIONALE FOR THE EXISTING CARBON MONOXIDE STANDARDS

The following discussion describing the bases for the existing CO NAAQS set in 1994 has been excerpted and adapted from "National Ambient Air Quality Standards for Carbon Monoxide — Final Decision."[4] The discussion includes the rationale for selection of the level and averaging time for the NAAQS that would be protective of adverse effects in the most sensitive subpopulation and the EPA assessment that led to a decision not to revise the existing standards for CO.

4.4.1 CARBOXYHEMOGLOBIN LEVELS OF CONCERN

In selecting the appropriate level and averaging time for the primary NAAQS for CO, the EPA Administrator must first determine the COHb levels of concern, taking into account a large and diverse health effects database. Based on the assessments provided in the criteria document[11] and in a staff paper,[12] judgments were made to identify the most useful studies for establishing a range of COHb levels to be considered for standard setting. In addition, the more uncertain or less quantifiable evidence was reviewed to determine the lower end of the range that would provide an adequate margin of safety from effects of clear concern. The following discussion summarizes the most critical considerations for the Administrator's 1994 decision on the CO NAAQS.

The Administrator of EPA concluded that cardiovascular effects, as measured by decreased time to onset of angina (chest pain) and by decreased time to onset of significant electrocardiogram (ECG) ST-segment depression, were the health effects of greatest concern to be clearly associated with CO exposures at levels observed in the ambient air. These effects were demonstrated in patients with exercise-induced angina at postexposure COHb levels that were elevated to 2.9 to 5.9% (CO-oximetry, or CO-Ox, measurement), representing incremental increases of 1.5 to 4.4% from baseline levels. (See later discussion in Section 4.5.5.) Time to onset of significant ECG ST-segment change, which is indicative of myocardial ischemia in patients with documented coronary artery disease and a more objective indicator of ischemia than angina pain, provided supportive evidence of health effects occurring at exposures as low as 2.9 to 3.0% COHb (CO-Ox). The clinical importance of cardiovascular effects associated with exposures to CO resulting in COHb levels less than 2.9% remains less certain and was considered only in evaluating whether or not the existing CO standards provide an adequate margin of safety.

The Administrator of EPA also considered the following factors in evaluating the adequacy of the existing CO NAAQS.

1. Short-term reductions in maximal work capacity were measured in trained athletes exposed to CO sufficient to produce COHb levels as low as 2.3%.
2. The wide range of human susceptibility to CO exposures and ethical considerations in selecting subjects for experimental purposes, taken together, suggest that the most sensitive individuals may have not been studied.
3. Animal studies of developmental toxicity and human studies of the effects of maternal smoking provide evidence that exposures to high concentrations of CO can be detrimental to fetal development, although little is known about the effects of ambient CO exposures on the developing human fetus.
4. Although little is known about the effects of CO on other potentially sensitive populations besides those with coronary artery disease, there is reason for concern about visitors to high altitudes, individuals with anemia or respiratory disease, and the elderly.
5. Impairment of visual perception, sensorimotor performance, vigilance, and other central nervous system effects have not been demonstrated to be caused by CO concentrations commonly found in ambient air; however, short-term peak CO exposures may be responsible for impairments that could be a matter of concern for complex activities such as automobile driving.
6. Limited evidence suggests concern for individuals exposed to CO concurrently with drug use (e.g., alcohol), heat stress, or coexposure to other pollutants.
7. Large uncertainties remain regarding modeling COHb formation and estimating human exposure to CO that could lead to over- or underestimation of COHb levels associated with attainment of a given CO NAAQS in the population.
8. Measurement of COHb made using the CO-Ox technique may not reflect the COHb levels in patients with angina studied, thereby creating uncertainty in establishing a lowest effects level for CO.

The Administrator concluded that the lowest COHb level at which adverse effects have been demonstrated in persons with angina is around 2.9 to 3.0%, representing an increase of 1.5% COHb above baseline when using the CO-Ox to measure COHb. These data serve to establish the upper end of the range of COHb levels of concern. Taking into account the above data uncertainties, the less significant health end points, and less quantifiable data on other potentially sensitive groups, the lower end of the range of concern was established at 2.0% COHb.

4.4.2 RELATIONSHIP BETWEEN CARBON MONOXIDE EXPOSURE AND CARBOXYHEMOGLOBIN LEVELS

To set ambient CO standards based on an assessment of health effects at various COHb levels, it is necessary to estimate the ambient CO concentrations that are likely to result in the above COHb levels of concern. The best all-around model for predicting COHb levels is the Coburn, Forster, Kane (CFK) differential equation.[11,13,14] Baseline estimates of COHb levels expected to be reached by nonsmokers exposed to various constant concentrations of CO can be determined by the CFK equation.[12] There are, however, two major uncertainties involved in estimating COHb levels resulting from exposure to CO concentrations. First, the large distribution of physiological parameters used in the CFK equation across the population of interest is sufficient to produce noticeable deviations in the COHb levels. Second, predictions based on exposure to constant CO concentrations can under- or overestimate responses of individuals exposed to widely fluctuating CO levels that typically occur in the ambient environment.

4.4.3 ESTIMATING POPULATION EXPOSURE

The EPA review included an analysis of CO exposures expected to be experienced by residents of Denver, CO, under air quality scenarios where the 8-h NAAQS is just attained. Although the exposure analysis included passive smoking and gas stove CO emissions as indoor sources of CO, it did not include other indoor sources that may be of concern to high-risk groups (e.g., woodstoves, fireplaces, faulty furnaces) or outdoor residential sources (e.g., gasoline-powered lawn equipment). The analysis indicated that, at the 8-h standard, fewer than 0.1% of the nonsmoking cardiovascular-disease population would experience a COHb level $\geq$2.1%.[12] A smaller population was estimated to exceed higher COHb percentages.

4.4.4 DECISION ON THE PRIMARY STANDARDS

Based on the exposure analysis results described above, the Administrator of EPA concluded that relatively few people of the cardiovascular-sensitive population group analyzed would experience COHb levels $\geq$2.1% when exposed to CO levels in the absence of indoor sources when the current ambient standards were attained. Although indoor sources of CO may be of concern to high-risk groups, their contribution cannot be effectively mitigated by ambient air quality standards.

The Administrator of EPA also determined that both the 1-h and 8-h averaging times for CO were valid because the 1-h standard provided reasonable protection from health effects that might be encountered from very short duration peak (bolus) exposures in the urban environment, and the 8-h standard provided a good indicator for tracking continuous exposures that occur during any 24-h period. The Administrator concurred with staff recommendations[12] that both averaging times be retained for the primary CO standards.

For these reasons, the EPA Administrator determined under CAA Section 102(d)(1) that revisions to the current 1-h (35 ppm) and 8-h (9 ppm) primary standards for CO were not appropriate at that time.[4]

4.5 SCIENTIFIC ISSUES OF CONCERN FOR THE CURRENT REVIEW OF AIR QUALITY STANDARDS

The following is a brief summary of scientific issues that are addressed in the ongoing review of ambient air quality criteria for CO.[14] These issues are based on available information published in the scientific literature and are intended to help assess the current state of understanding of the sources, atmospheric cycle, and health effects of CO. They will be used to judge the adequacy of the existing standards for the protection of public health and welfare.

4.5.1 GLOBAL TROPOSPHERIC CHEMISTRY

In nonurban areas, tropospheric CO has a significant role in affecting the oxidizing capacity of Earth's atmosphere. Reaction with CO is a principal process by which hydroxyl radicals are removed from the atmosphere. Reaction with hydroxyl radicals is also the primary process for removing many other anthropogenic and natural compounds from the atmosphere and, therefore, is important in determining the concentrations of many environmentally important trace gases (e.g., methane, hydrochlorofluorocarbons). CO is also linked closely to the cycle of tropospheric ozone and may be responsible for 20 to 40% of the ozone formed in nonurban areas. Global CO emissions from fossil fuel and biomass burning and from photochemical sources need to be better estimated to determine future changes in tropospheric chemistry.

4.5.2 REGIONAL AND URBAN AMBIENT AIR QUALITY

CO also plays an important role in atmospheric photochemistry in regional and urban environments. In urban areas, CO either can produce or destroy ozone, depending on the concentrations of nitrogen oxides and hydrocarbons. Between 1988 and 1997, national average CO emissions and concentrations in the United States decreased 25 and 38%, respectively (see later discussion on this trend). On- and off-road mobile sources account for 77% of the 1997 nationwide emissions inventory for CO.[15] Declines in ambient CO levels in the United States follow approximately the decline in motor vehicle emissions of CO. However, it is not known how variations in diurnal and seasonal patterns in CO will change due to the interaction among decreasing motor vehicle emissions, increasing number of vehicles and vehicle miles traveled, changing urban and suburban traffic patterns, and meteorological trends such as global warming.

TABLE 4.2
Estimated Carbon Monoxide Concentrations from Various Sources and Microenvironments

	Average Conc., ppm[a]	Conc. Range, ppm[a]
Source		
Global background	0.05–0.13[b]	0.04–0.22[b]
Urban air, 8-h max.	5	<1–20
Automobile exhaust	7,700[c]	0–40,000
Small engine exhaust	43,000[d]	21,000–72,000
Cigarette smoke	45,000	20,000–60,000
Microenvironment		
Indoor air without combustion sources	1	<1–5
Room with properly operating gas stove	2	<1–9
Smoke-filled room	4	<1–10
In-vehicle automobile compartment	4	<1–12
Room with properly operating unvented space heater	12	<1–50

[a] Concentrations vary widely depending on the conditions and methods of measurement.

[b] Annual average background concentrations in the Southern and Northern Hemispheres, respectively.

[c] Mean model year was 1985.

[d] Average of two- and four-stroke small (3 to 5 hp) gasoline engines.

Source: Modified from Reference 16.

4.5.3 INDOOR EMISSIONS AND CONCENTRATIONS

Personal tobacco smoking and other indoor sources of CO exposure represent a significant portion of the total human exposure to CO. However, more information is needed on the interaction of ambient CO exposures and personal CO exposures occurring in different indoor microenvironments, especially in at-risk subpopulations. CO occurs indoors directly through emissions from various indoor combustion sources or indirectly as a result of infiltration or ventilation from outdoor sources (Table 4.2). Emissions of CO from the use of adequately vented combustion appliances (e.g., gas and oil furnaces, gas water heaters, gas dryers) will not contaminate indoor air unless the units or venting systems are malfunctioning. The major sources of CO for the nonsmoker in residential microenvironments are sidestream tobacco smoke and unvented or partially vented combustion appliances (e.g., gas cooking stoves, space heaters, woodstoves, fireplaces).

4.5.4 POPULATION EXPOSURE ASSESSMENT

The reduction in mobile source emissions brought about by the CAA have reduced in-traffic CO exposures and traffic-related ambient CO concentrations well below those measured in the past decade when large-scale population exposure studies were

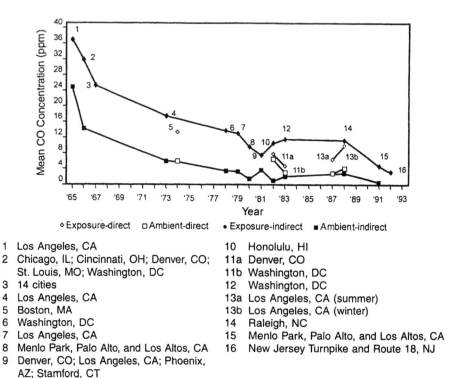

◇ Exposure-direct   □ Ambient-direct   ◆ Exposure-indirect   ■ Ambient-indirect

1   Los Angeles, CA
2   Chicago, IL; Cincinnati, OH; Denver, CO;
    St. Louis, MO; Washington, DC
3   14 cities
4   Los Angeles, CA
5   Boston, MA
6   Washington, DC
7   Los Angeles, CA
8   Menlo Park, Palo Alto, and Los Altos, CA
9   Denver, CO; Los Angeles, CA; Phoenix,
    AZ; Stamford, CT

10   Honolulu, HI
11a   Denver, CO
11b   Washington, DC
12   Washington, DC
13a   Los Angeles, CA (summer)
13b   Los Angeles, CA (winter)
14   Raleigh, NC
15   Menlo Park, Palo Alto, and Los Altos, CA
16   New Jersey Turnpike and Route 18, NJ

FIGURE 4.1   Trends in ambient CO concentrations and in-vehicle CO exposures, 1965 to 1992. (The upper and lower lines are provided to make a clear distinction between exposure and ambient CO data reported for each city; these lines do not imply that results for cities are related.)[14,17] (From Flachsbart, P.G., *J. Exposure Anal. Environ. Epidemiol.*, 5:473–495. With permission.)

conducted in major cities of the United States (Figure 4.1). There currently is not a good estimate of the present CO exposure distribution for the population. Fixed-site monitors often are used in urban areas to estimate ambient CO exposures; however, CO concentrations are not spatially homogeneous throughout urban atmospheres. In addition, only 10% of the sites in the network are designated as "neighborhood"-scale monitors. The remaining are mostly "hot-spot" monitors located near major roadways. As a result, CO measurements from fixed-site monitors tend to overestimate 8-h exposures for people living in areas of lower traffic and underestimate exposure of people living in areas of higher traffic. In other words, nonsmokers exposed to heavy traffic fumes, tobacco smoke, and other indoor sources of CO will have higher body burdens of CO than would be predicted from ambient air quality data alone. Personal exposure studies are needed to determine better what portion of total human exposure to CO is directly attributed to ambient sources controlled by air quality standards.

4.5.5   HEALTH EFFECTS

The most prominent pathophysiological effect of CO is hypoxemia caused by the binding of CO to hemoglobin. The formation of COHb reduces the oxygen-carrying capacity of blood and impairs release of oxygen from red blood cells to tissues. The brain and heart are especially sensitive to CO-induced hypoxia and cytotoxicity because these tissues have the highest resting oxygen requirements. It is not surprising, therefore, that the effects associated with CO are due to physiological limitations in brain and heart function, especially in compromised individuals. The best indicator of potential health risk is the COHb level; unfortunately, the method used for analysis determines the lowest-observed-effect level and the most commonly used method (CO-Ox) is not very accurate in the low levels of concern (<5% COHb) that are predicted to result from typical ambient CO exposures. Gas chromatography (GC) generally is considered a more accurate method of analysis. Controlled human exposure studies[14] resulting in relevant COHb levels indicate that maximal exercise duration and performance in healthy individuals (e.g., athletes) can be reduced at 2.3 and 4.3% COHb (GC), respectively. Decreased exercise tolerance also has been observed in patients with coronary artery disease and reproducible exercise-induced angina (chest pain) at COHb levels as low as 2.4% (GC). In epidemiological studies,[14] daily fluctuations in ambient CO concentrations have been associated with increased hospital admissions for patients with congestive heart failure, who also may have significant underlying coronary disease. Unfortunately, these population-based studies use urban fixed-site monitors to estimate ambient CO exposure. Because of the extremely heterogeneous spatial pattern of ambient CO across a given urban area, and the extremely low levels (≤5 ppm daily 1-h max.) of monitored CO, it is difficult to determine the impact of ambient sources of CO on hospital admissions relative to other, uncontrolled sources especially those sources occurring indoors. In addition, CO may be acting as a marker for other covarying pollutants from fossil fuel combustion (e.g., fine aerosols).

4.6 IMPACT OF AIR QUALITY STANDARDS ON AMBIENT LEVELS OF CARBON MONOXIDE

Ambient air monitoring data for the United States are reported to the EPA Aerometric Information Retrieval System (AIRS) from national, state, and local air monitoring stations. Each year, air quality data summaries from AIRS are published by the EPA in the National Air Quality and Emissions Trends Report. The following information is extracted from the latest available summary report.[15]

Since 1970, when the first air quality and automobile emissions standards were issued (Table 4.3), ambient pollutant concentrations and emissions have decreased 31% despite economic growth and associated increases in the total U.S. population (31%), gross domestic product (114%), and vehicle miles traveled (127%). Air quality trends recorded over the past 20 years for specific ambient air pollutants are the result of the EPA working with states, industry, and other groups to establish and implement clean air laws and regulations. Between 1978 and 1997, for example, monitored concentrations of CO decreased 60%, an impressive reduction. In fact, except for lead (–97%, due to the introduction of unleaded gas), CO had the largest reduction of any of the other major criteria pollutants (nitrogen dioxide, –25%; ozone, –30%; sulfur dioxide, –55%; particulate matter, no data). The reduction of

TABLE 4.3
Milestones in Automobile Emissions Control

1970   New CAA sets auto emissions standards
1971   Charcoal canisters appear to meet evaporative standards
1972   Exhaust gas recirculation (EGR) valves appear to meet NO_x standards
1974   Fuel economy standards are set
1975   The first catalytic converters appear for hydrocarbon, CO; unleaded gas appears for use in
           catalytic equipped cars
1981   Three-way catalysts with on-board computers and oxygen sensors appear
1983   Inspection/maintenance (I/M) programs are established in 64 cities
1989   Fuel volatility limits are set for Reid vapor pressure (RVP)
1990   CAA Amendments set new tailpipe standards
1992   Oxyfuel introduced in cities with high CO levels
1993   Limits set on sulfur content of diesel fuel
1994   Phase-in begins for new vehicle standards and technologies

Source: U.S. EPA.[15]

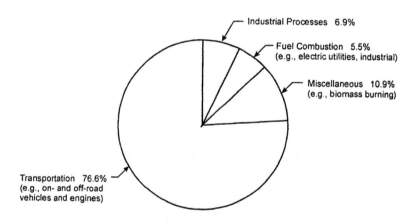

FIGURE 4.2   CO emissions by source category, 1997.[14,15]

ambient CO has been accomplished by implementation of air quality standards and
control strategies for reducing CO emissions from industrial activities and motor
vehicles (Table 4.3). Because the largest source of CO emissions is from transportation sources (Figure 4.2), the greatest impact on CO emissions has resulted
from the introduction of electronic controls, oxygen and pressure sensors, and the
three-way catalytic converter on motor vehicles. (Strict stoichiometric control of the
air–fuel ratio results in lower levels of CO and hydrocarbon, or HC, production
relative to fuel-rich operation, and lower levels of nitrogen oxide production relative
to fuel-lean combustion. Stoichiometric air–fuel mixtures are also required for the
treatment of exhaust gases by three-way catalytic converters, which simultaneously
oxidize HC and CO to carbon dioxide and reduce nitric oxide to nitrogen. New
vehicles equipped with these and other emission controls typically emit less than

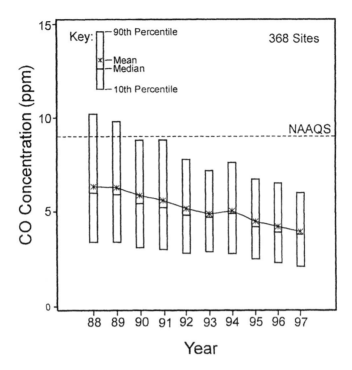

FIGURE 4.3   Variability in the annual second highest 8-h CO concentrations across all sites in the United States, 1988 to 1997.[15]

5% of the pollutants emitted by precontrol vehicles.[18]) The control technologies continue to improve and newer vehicles are replacing older more-polluting cars and light-duty trucks every year. (Results from remote sensing detection of CO levels from vehicles in California[19] indicate that 10% of the vehicles contributed 53% of the CO emissions.) Over the past decade (1988 to 1997), CO concentrations in the United States decreased 38%, as measured by the composite average of the annual second highest 8-h concentration (Figure 4.3). Of the 365 monitoring sites followed for trends, CO concentrations decreased at all three types of locations: 39% at 208 urban sites, 35% at 145 suburban sites, and 46% at 12 rural sites (Figure 4.4). Between 1996 and 1997, national average CO concentrations decreased 7%, showing continued improvement in air quality. In the last year of fully analyzed and published data,[15] only 6 of the 537 monitoring locations reporting CO data failed to meet the CO NAAQS (Table 4.4).

4.7   SUMMARY

Carbon monoxide is a trace constituent of the natural environment and plays an important role in atmospheric photochemistry. Biological organisms have adapted to background atmospheric levels of CO, and produce and utilize CO for normal life processes. Unfortunately, the increasing needs of a growing human population often have negative influences on the ambient (outdoor) and indoor environments,

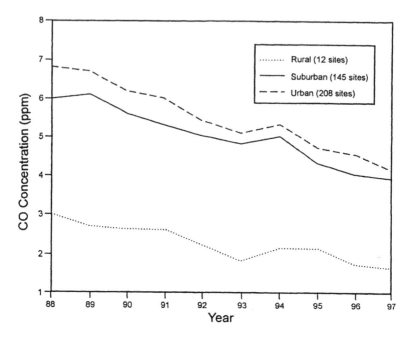

FIGURE 4.4 Composite average of the annual second highest 8-h CO concentrations for rural, suburban, and urban sites, 1988 to 1997.[15]

TABLE 4.4
Air Quality Monitoring Sites Not Meeting the Carbon Monoxide NAAQS in 1997

Location	AIRS Site ID No.	1st max, ppm[a]	2nd max, ppm[b]	No. of Exceedances[c]
Calexico, CA	06 025 0005	17.8	16.7	12
Lynwood, CA	06 037 1301	17.1	15.0	12
Fairbanks, AK	02 090 0020	12.8	10.6	4
	02 090 0002	13.3	12.1	3
	02 090 0013	12.2	10.8	2
Calexico, CA	06 025 0006	16.3	9.6	2

[a] Highest 8-h average CO concentration.
[b] Second highest 8-h average CO concentration.
[c] Number of exceedances of the CO NAAQS.

Source: U.S. EPA.[20]

---

especially in urban areas. Concentrations of CO from the incomplete combustion of biomass and the fossil fuels used for transportation, energy production, industrial processes, and heating and cooking may become a problem if they overwhelm natural background concentrations. This is exactly what happened during the 20th century with increased industrialization and urbanization in developed countries like the United States, and it is occurring now in developing countries.

The mandate of the CAA to regulate ambient air quality in the United States resulted in the promulgation of NAAQS for CO by EPA in the early 1970s. These air quality standards and associated emission control programs have very successfully reduced urban CO concentrations over the past 30 years and brought many urban areas into attainment at levels considered to be safe for the general public. This success story should set an example for other countries with similar ambient air problems.

Unfortunately, the urban air concentrations of CO measured by fixed-site monitors may not represent population exposures to ambient CO because of the spatial variability of CO in a given urban area and the reliance on network monitors placed near major sources. Personal exposures to ambient CO sources associated with motor vehicle emissions, such as inside vehicles on congested roadways or inside parking garages, also have decreased dramatically with use of automobile emission control technologies, making sources in or near residences more important contributors to total human exposure to CO. The most important CO sources for public health concern are most often internal combustion engines or some type of unvented, or malfunctioning, or misused combustion space or water-heating system or cooking device. It may turn out that the predominance of total CO exposures, and virtually all serious CO exposures, occur indoors despite a successful national program for regulating outdoor CO levels.

National Ambient Air Quality Standards for Carbon Monoxide

The U.S. Environmental Protection Agency, which is required under the Clean Air Act to reconsider national ambient air quality standards (NAAQS) every five years, has completed an updated review of the available scientific data for carbon monoxide (CO) and is assessing the public health protection provided by the current CO NAAQS. A proposal detailing the results of both the health and exposure assessment for CO will be published for public review and comment in the fall of 2000, followed by a final decision on the CO NAAQS in the Spring of 2001. Information on the recommended standards for CO and current air quality attainment status for CO is accessible through the Internet at http://www.epa.gov/oar/oaqps. The updated scientific review (U.S. Environmental Protection Agency, 1999) also is available through the Internet at http://www.epa.gov/ncea.

ABBREVIATIONS

AIRS	Aerometric Information Retrieval System
CAA	Clean Air Act
CASAC	Clean Air Scientific Advisory Committee
CFK	Coburn, Forster, Kane
CO	carbon monoxide

COHb   carboxyhemoglobin
CO-Ox   CO-oximetry
EPA   U.S. Environmental Protection Agency
ECG   electrocardiogram
FDA   U.S. Food and Drug Administration
GC   gas chromatography
HC   hydrocarbon
NAAQS   National Ambient Air Quality Standards

ACKNOWLEDGMENTS

The author gratefully acknowledges the contributions to the ongoing review of air quality criteria for carbon monoxide[14] by Drs. Russell R. Dickerson, Peter G. Flachsbart, Milan J. Hazucha, and Stephen R. Thom. The author also wishes to thank Drs. David J. McKee, David T. Mage, James J. McGrath, and Joseph P. Pinto for their advice in preparation of this manuscript.

REFERENCES

1. U.S. Code, Clean Air Act, §108, air quality criteria and control techniques, §109, national ambient air quality standards, U.S.C., 42, §§7408–7409, 1991.
2. National Air Pollution Control Administration, Air Quality Criteria for Carbon Monoxide, NAPCA-PUB-AP-62, U.S. Department of Health, Education, and Welfare, Public Health Service, Washington, D.C., 1970. (Available from NTIS, Springfield, VA, PB-190261.)
3. Beard, R.R. and Wertheim, G.A., Behavioral impairment associated with small doses of carbon monoxide, *Am. J. Public Health*, 57, 2012, 1967.
4. *Federal Register*, National ambient air quality standards for carbon monoxide — final decision, *Fed. Regist.*, (August 1), 59, 38, 906, 1994.
5. U.S. Environmental Protection Agency, Air Quality Criteria for Carbon Monoxide, EPA-600/8-79-022, Office of Health and Environmental Assessment, Environmental Criteria and Assessment Office, Research Triangle Park, NC, 1979. (Available from NTIS, Springfield, VA, PB81-244840.)
6. *Federal Register*, Carbon monoxide; proposed revisions to the national ambient air quality standards: proposed rule, *Fed. Regist.*, (August 18), 45, 55, 066, 1980.
7. Horvath, S.M., Ayres, S.M., Sheps, D.S., and Ware, J., [Letter to Dr. Lester Grant, including the peer-review committee report on Dr. Aronow's studies], Central Docket Section, OAQPS-79-7 IV. H.58, U.S. Environmental Protection Agency, Washington, D.C., 1983.
8. U.S. Environmental Protection Agency, Revised Evaluation of Health Effects Associated with Carbon Monoxide Exposure: An Addendum to the 1979 EPA Air Quality Criteria Document for Carbon Monoxide, EPA-600/8-83-033F, Office of Health and Environmental Assessment, Environmental Criteria and Assessment Office, Research Triangle Park, NC, 1984. (Available from NTIS, Springfield, VA, PB85-103471/HSU.)
9. *Federal Register*, Review of the national ambient air quality standards for carbon monoxide; final rule, *Fed. Regist.*, (September 13), 50, 37,484, 1985.

10. *Federal Register*, Draft criteria document for carbon monoxide; notice of availability of external review draft, *Fed. Regist.*, (April 19), 55, 14,858, 1990.

11. U.S. Environmental Protection Agency, Air Quality Criteria for Carbon Monoxide, EPA/600/8-90/045F, Office of Health and Environmental Assessment, Environmental Criteria and Assessment Office, Research Triangle Park, NC, 1991. (Available from NTIS, Springfield, VA, PB93-167492.)

12. U.S. Environmental Protection Agency, Review of the National Ambient Air Quality Standards for Carbon Monoxide: 1992 Reassessment of Scientific and Technical Information, OAQPS Staff Paper, EPA-452/R-92-004, Office of Air Quality Planning and Standards, Research Triangle Park, NC, 1992. (Available from NTIS, Springfield, VA, PB93-157717.)

13. Coburn, R.F., Forster, R.E., and Kane, P.B. Considerations of the physiological variables that determine the blood carboxyhemoglobin concentration in man, *J. Clin. Invest.*, 44, 1899, 1965.

14. U.S. Environmental Protection Agency, Air Quality Criteria for Carbon Monoxide [External Review Draft], EPA/600/P-99/001, National Center for Environmental Assessment, Research Triangle Park, NC, 1999.

15. U.S. Environmental Protection Agency, National Air Quality and Emissions Trends Report, 1997, EPA/454/R-98-016, Office of Air Quality Planning and Standards, Research Triangle Park, NC, 1998. (Available at: *www.epa.gov/oar/aqtrnd97*)

16. Waffle, C.M., Carbon monoxide poisoning, in *Comprehensive Management of Respiratory Emergencies*, Brenner, B.E., Ed., Aspen Systems Corp., Rockville, MD, 1986, 259.

17. Flachsbart, P.G., Long-term trends in United States highway emissions, ambient concentrations, and in-vehicle exposure to carbon monoxide in traffic, *J. Exposure Anal. Environ. Epidemiol.*, 5, 473, 1995.

18. Singer, B.C., Kirchstetter, T.W., Harley, R.A., Kendall, G.R., and Hesson, J.M., A fuel-based approach to estimating motor vehicle cold-start emissions, *J. Air Waste Manage. Assoc.*, 49, 125, 1999.

19. Bishop, G.A., Stedman, D.H., and Ashbaugh, L., Motor vehicle emissions variability, *J. Air Waste Manage. Assoc.*, 46, 667, 1996.

20. U.S. Environmental Protection Agency, AIRSData: monitor values report. Washington, DC: Office of Air & Radiation, 1999. Available at: *www.epa.gov/airsdata/monvals.htm*

5 Effect of Carbon Monoxide on Work and Exercise Capacity in Humans

Milan J. Hazucha

CONTENTS

0-8493-2065-8/00/$0.00+S.50
© 2000 by CRC Press LLC

5.1   INTRODUCTION

The primary hypoxic action of carbon monoxide (CO) is due to its high affinity for and subsequent strong binding to hemoglobin (Hb), forming carboxyhemoglobin (COHb) and thus affecting the transport and release of O_2 into tissues. At the tissues level, CO binds to extravascular hemoproteins, e.g., myoglobin (Mb), cytochrome oxidases, and interferes with the energy metabolism of cells as well. Although the principal pathophysiological mechanisms of CO hypoxia are well understood, much less is known about the effects on delivery of O_2 to tissues, cellular metabolism, histotoxic effects, and cellular mechanisms of action. When bound to the intracellular Mb of skeletal and cardiac muscle, CO impairs the transport of O_2 to mitochondria and subsequently their respiratory function, leading to muscle dysfunction.[1,2] Recent evidence indicates that reduction in maximum O_2 consumption (VO_2 max) by exercising muscles in the presence of CO may be linked to a decrease in the blood-to-mitochondria O_2 conductance.[3] This is because CO in the blood may affect the off-rate kinetics of oxyhemoglobin and the carrier function of Mb in O_2 transport.

During exercise, the increased demand for O_2 requires adjustment of the cardio-pulmonary system to provide an adequate supply of O_2. Depending on exercise intensity, physiological changes may range from minimal to substantial, involving the cardiovascular, respiratory, and other organ systems to actuate systemic as well as local compensatory changes. Exercise improves the alveolar ventilation to perfusion (VA/Q) ratio in the lung and increases the respiratory exchange ratio (RER), cardiac output, and lung diffusion capacity for CO (DLCO). It also mobilizes red blood cell (RBC) reserves from the spleen. Strenuous exercise leads to hemoconcentration and a decrease in blood volume. Of the many functional responses to exercise, VA and cardiac output both play critical roles in providing adequate energy to working muscle.[4,5] Although some exercise-induced compensatory changes may slow CO loading into the blood (e.g., the decrease in DLCO during heavy exercise), most of these variables facilitate CO transport. Thus, by increasing gas exchange efficiency, exercise promotes CO uptake and COHb formation in proportion to its intensity.

Depending on the severity of CO exposure and the intensity of exercise, the combination of exercise and elevated blood level of COHb may stress the cardiovascular and respiratory systems to the maximum much faster than either CO or exercise would alone. Many of the compensatory physiological responses triggered by CO hypoxia or exercise are identical. At modest to moderate COHb during submaximal exercise, these responses and their effects appear to be additive. During maximal performance and severe CO hypoxia, however, the compensatory mechanisms to sustain O_2 demand under these conditions may diverge or even interfere with each other. Although CO is an omnipresent pollutant and frequently a toxicant, the interaction effects of exercise and CO hypoxia have not been studied extensively in humans. With the exception of one study by Klausen and colleagues[6] who looked at the effects of CO hypoxia on young males periodically exercising over an 8-day period, all of the studies to date have used various incremental exercise stress tests with exercise bouts of 5 to 60 min to determine the effects of CO on exercise performance. But the acute effects and compensatory changes induced by either CO hypoxia or exercise differ from the physiological adjustments that are brought about when these conditions last for an extended period or arise from long-term adaptive changes.

To date, no human study has examined the potential exercise and work performance limits or other health effects of combined physical activity and CO hypoxia beyond the acute condition of exercise stress testing. Even more importantly, it is not known how prolonged CO hypoxia affects the performance of the myocardium under increased demand from prolonged or continual exercise. Although there is some parallelism between the compensatory changes induced by hypoxic hypoxia and CO hypoxia, the magnitude of the effects for the same level of hypoxia and many of the mechanisms by which these responses are effected are not the same.[7] These studies may be helpful in identifying some of the control mechanisms, but they cannot entirely elucidate the effects induced by CO-hypoxia. Similarly, the many studies on the effects of smoking and exercise are also of limited use in examining the interaction of CO hypoxia and exercise because it is difficult, if not impossible, to separate the effects of CO (the principal gaseous component of smoke) from other substances (e.g., nicotine), which have strong physiological effects of their own.

5.2   CARBON MONOXIDE AND THE MUSCLE

5.2.1   CARBOXYMYOGLOBIN

Mb has a CO affinity constant approximately eight times lower than Hb (M = 20 to 40 vs. 218, respectively).[1,8] As with Hb, the combination velocity constant between CO and Mb is only slightly lower than for O_2, but the dissociation velocity constant is much lower than for O_2. The combination of this greater affinity (Mb is 90% saturated at PO_2 of 20 mm Hg) and the lower dissociation velocity constant for CO favors retention of CO in muscular tissue, and thus a considerable amount of CO potentially can be stored in the skeletal muscle. Any of the Mb functions may be affected by the presence of CO. The binding of CO to myoglobin (COMb) has been demonstrated at <1% COHb from the biopsied skeletal muscle and at <1.5% from the biopsied myocardium of dogs.[1,2,9] At arterial PO_2 above 40 torr, the average COMb/COHb ratio for skeletal muscle is close to unity (range 0.4 to 1.5). This ratio does not increase with an increase in COHb (up to 50%) and appears to be independent of the duration of exposure.[1] However, severe hypoxemia (PO_2 < 40 torr) doubles (range 1.4 to 2.5) the COMb/COHb ratio for skeletal muscle[2] and more than doubles (range 1.2 to 2.7) this ratio for cardiac muscle (PO_2 < 35 torr).[1,2] These ratios are generally lower in rats, but the myocardial COMb/COHb ratio was nevertheless twice as high as that measured in skeletal muscle.[10] During exercise, the relative rate of CO binding increases more for Mb than for Hb, and CO diffuses from blood to skeletal muscle.[11,12] Consequently, the ratio of COMb/COHb increases for skeletal and even more so for cardiac muscles where it exceeds unity. Figure 5.1 shows the COHb concentration in blood, skeletal muscle, and myocardium of resting and exercising rats at different concentrations of CO.[12] A similar shift in CO has been observed under hypoxic conditions because a fall in intracellular PO_2 below the critical level increases the relative affinity of Mb to CO.[1] The consequent reduction

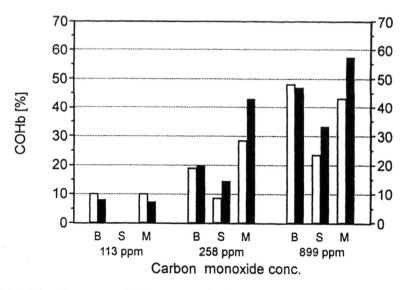

FIGURE 5.1 The average COHb concentration in blood (B), skeletal muscle (S), and myocardial (M) tissue samples obtained from resting (empty bars) and exercising (solid bars) rats (n = 7 to 14) exposed for 40 min to three different concentrations of CO. (Modified from Sokol, J.A. et al., *J. Hyg. Epidemiol. Microbiol. Immunol.*, 30, 57, 1986. With permission.)

in the O_2-carrying capacity of Mb might have a profound effect on the supply of O_2 to organ tissues.

Mb, the respiratory hemoprotein of muscular tissue, undergoes a reversible reaction with CO in a manner similar to O_2. The greater affinity of O_2 for Mb than Hb (hyperbolic vs. S-shaped dissociation curve) is in this instance physiologically beneficial, since a small drop in tissue PO_2 releases a large amount of O_2 from O_2Mb. The lowest intracellular PO_2 that will not result in anoxia or hypoxia has been found to be 1.5 torr in contracting skeletal muscle[13,14] and approximately 2 to 5 torr in the myocardium.[15] Low intracellular PO_2 and other characteristics of Mb (such as uniform spatial distribution of PO_2 gradients) may act as an intracellular buffer when PO_2 < 5 torr, to maintain cell respiration,[16] to promote diffusion of O_2 from Hb to cytochrome via MbO_2, and to assist in the release of O_2 from capillaries.[17,18]

Studies of isolated skeletal muscle preparations have demonstrated that Mb supplies a significant fraction of O_2 to intracellular muscle mitochondria.[17,19] Similarly, cardiac mitochondria are supplied with O_2 by two concurrent pathways: via Mb-bound O_2 to the mitochondrial inner membrane (the site of respiratory chain components and ATP synthesis), and via dissolved O_2 to cytochrome oxidase.[20] Thus, depending on the PO_2 and O_2 demands of working muscle, CO binding to Mb and the subsequent reduction in available O_2Mb may have a profound impact on intracellular respiration. In the presence of COMb, the contractive ability of isolated myocytes from rat hearts was clearly diminished. This is probably related to reduced oxidative-chain phosphorylation and a lower yield of ATP (the energy source for muscle contraction), due to decreased O_2 uptake.[21] Thus, the main function of Mb appears to be as a temporary store of O_2 and as a diffusion facilitator between Hb and the tissues.[22]

5.2.2   Effects of Carbon Monoxide on Muscle Function

5.2.2.1   Skeletal Muscle

In situ studies of dog gastrocnemius muscle have shown that isometric contractions at both 25 and 50% of VO_2 max for a prolonged period of time (36 min) are not substantially affected by moderate CO hypoxia (10 and 25% COHb).[23] Increased muscle demand for O_2 was adequately maintained by a compensatory increase in blood flow and O_2 extraction. Since O_2 extraction was approaching the limit, any increase in the demand for O_2 had to be met by a further decrease in venous PO_2.[23,24] At these intensities of contraction and COHb levels, lactate production was highly variable, and although proportionally increased on average, did not appear to be closely related to O_2 deficit. The fatigue of the muscle was not related to CO but rather to the reduction in VO_2, since both the hypoxic hypoxia and CO hypoxia generated the same fatigue for the same level of VO_2.[25,26]

5.2.2.2   Myocardium

Even at rest, O_2 extraction by the myocardium is almost complete; thus any increase in myocardial O_2 demand must be satisfied by increased coronary blood flow. There is a close relationship among cardiac performance, myocardial O_2 consumption, and coronary blood flow. CO hypoxia puts an extra burden on the myocardium, since a compensatory increase in cardiac output demands a greater supply of O_2 to the heart muscle by the blood, which has a greater Hb affinity for CO and consequently carries a smaller amount of O_2. Since during exercise the myocardial energy requirement is proportionally increased, a combination of CO hypoxia and exercise may severely strain the O_2 supply system to the heart muscle.

Studies in intact anesthetized dogs have shown that moderate hypoxia (26% COHb) results in a nearly twofold increase in coronary blood flow, while severe hypoxia (42% COHb) resulted in an almost fivefold increase. At this blood flow rate, left ventricular perfusion approaches maximal or near maximal level. However, during moderate hypoxia the average ratios of right and left ventricular subendocardial-to-subepicardial blood flow (1.18 and 1.20, respectively) already are reduced from the respective control values (1.25 and 1.33, respectively), and become even more reduced at high COHb levels (1.09 and 1.13, respectively). For the right ventricle, the lower regional subendocardial-to-subepicardial blood flow ratios are more or less uniformly distributed at both COHb levels. But for the left ventricle, the distribution became more heterogenous at both COHb levels. The lowest ratio (1.02) was measured in the posterior papillary muscle region of the left ventricle. According to the authors, even small reductions in subendocardial perfusion may progress from localized to transmural hypoxia, and ultimately to myocardial damage.[27]

Further supporting evidence for a ventricular dysfunction at moderate COHb levels (10 to 30%) comes from observations on resting dogs, which exhibit decreased myocardial contractility during CO exposure, suggesting both myocardial ischemia and depression.[28] Although there is no information on regional changes in blood flow when exercise is performed in the presence of CO hypoxia, the above observations suggest that added cardiac workload due to exertion may accelerate localized

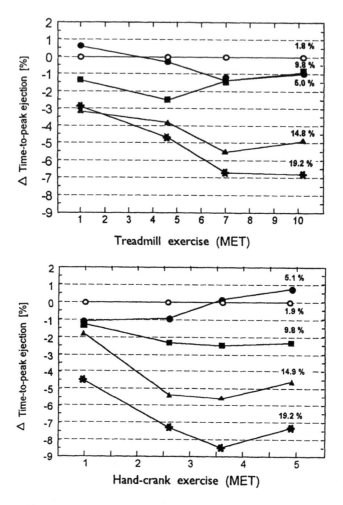

FIGURE 5.2 Changes in a time-to-peak ejection fraction measured at rest (MET = 1), moderate (MET = 4.6, 7), and strenuous (MET = 10.2) treadmill exercise (5 min at each load) top panel, and at rest (MET = 1), light (MET = 2.6, 3.6), and moderate (MET = 4.9) hand-crank exercise (5 min at each load) bottom panel, during four different levels of COHb (5.0/5.1, 9.8/9.8, 14.8/14.9, and 19.2/19.2%, respectively). The average values are expressed in percent difference from respective sham exposure condition (COHb = 1.8, 1.9%) at each respective exercise load. (From Kizakevich et al., unpublished observations.)

ischemia and lead to cardiac impairment even at modest COHb levels. In the only study to date, Kizakevich and colleagues[29] in 1994 showed that in exercising males, 5 to 20% COHb reduces the time-to-peak ejection fraction (measure of contractility) more than the same exercise does under normoxia. The shorter time was not only related to exercise intensity but to COHb levels as well (Figure 5.2). Another measure of cardiac contractility, aortic blood acceleration, progressively increased during moderate-to-heavy exercise loads at COHb below 10%, but peaked at COHb >10% during moderate exercise and decreased during heavy exercise. These observations suggest that the ventricular function of exercising individuals may be compromised at blood COHb >15%.[29]

5.3 PHYSIOLOGICAL ADJUSTMENTS TO CARBON MONOXIDE HYPOXIA AND EXERCISE

The exercising muscle requires an increased supply of O_2, which is achieved by complex cardiovascular and metabolic responses at both systemic and local levels. The primary systemic compensatory mechanisms include increased cardiac output, redistribution of blood flow between organs, a decrease in mean arterial pressure, a decrease in peripheral vascular resistance, and an increase in systemic O_2 debt. Local adjustments in the working muscle involve increased blood flow and increased O_2 extraction. The dominant compensatory mechanisms during CO hypoxia are essentially the same as those recruited during exercise. In addition, at above the lactic acid threshold (LAT)* augmented ventilation above that elicited by heavy exercise in normoxic or hypoxic hypoxia will facilitate O_2 supply. Thus, during exercise in a CO atmosphere or when blood COHb is already elevated by prior exposure to CO, the great majority of physiological adjustments are likely to be additive or even potentiating. Under certain combinations of exercise intensity and CO hypoxia, however, some of these mechanisms may induce less than additive changes or may even be mutually interfering.

5.3.1 SYSTEMIC COMPENSATORY MECHANISMS

Almost all the compensatory mechanisms activated in exercise and CO exposure involve the cardiopulmonary system either directly or indirectly. Maximal exercise performance under both normoxic and hypoxic conditions stress both the cardiovascular and the ventilatory systems, but the cardiovascular component is generally the limiting factor. In the following discussion, the relative contribution and adaptive responses of each system and related key mechanisms separately will be examined; however, one should keep in mind that exercise and CO hypoxia elicit an integrated cardiopulmonary response. Treatment of each factor as a discrete element is thus for didactic purposes only.

5.3.1.1 Cardiac Output

The immediate and fundamental response of the cardiovascular system to exercise is to increase cardiac output, which can be achieved either by an increase in heart rate (HR), an increase in stroke volume (SV), or a combination of the two. The increase in cardiac output, which is directly related to O_2 consumption, is achieved primarily through an increase in HR and a small increase in SV. CO hypoxia, which reduces the O_2 content and potentially the O_2 delivery to working muscle, elicits a response very similar to exercise. The initial adjustment to increased O_2 demand is

---

* The concept of a lactic acid threshold has been challenged recently by a number of investigators, who have observed that production of lactate appears to be a progressive process rather then a threshold-like function.[30]

an increase in cardiac output combined with other compensatory mechanisms to maintain O$_2$ delivery to working tissues. Thus, the combination of exercise and CO hypoxia for the most part will activate the same compensatory mechanisms that primarily load the cardiovascular system.

In dogs at rest, 10% COHb has been found to increase both HR and cardiac output. Moderate COHb levels (20 to 30%) increased the HR by 57% and the cardiac output by 23%.[28] Although HR in the presence of CO as compared with clean air exposure may not increase at rest or at very low work intensity, it becomes higher at a given workload for submaximal exercise but equivalent at maximum aerobic capacity.[31-33] The cardiovascular response of dogs exercising at 32 and 50% VO$_2$ max for 25 to 30 min differed little between normoxia and CO hypoxia (<21% COHb). There was a trend toward higher cardiac output and diminished O$_2$ extraction with increasing COHb and heavier exercise, but the increases were small. The authors noted the general lack of any significant cardiovascular adjustment to CO hypoxia as well as some inconsistencies in their findings, which they explained by the higher physical fitness of their study animals as compared with the lesser fitness of animals in comparable studies.[34]

Young healthy individuals exposed to CO (6.9% COHb) during progressive submaximal exercise testing showed no difference in cardiovascular changes between normoxic and CO hypoxic conditions.[35] In a similar group of subjects, submaximal exercise at low to moderate COHb (6 to 20%) increased the HR by 3 to 15%; however, maximal exercise under similar conditions actually decreased the HR by 1 to 4% over normoxic condition at the same exercise load.[29,36] Kizakevich and colleagues[29] in 1994 compared the effects of CO hypoxia on key cardiopulmonary variables during treadmill and hand-crank exercise. As can be seen from Figures 5.2 and 5.3, at rest (metabolic equivalent, MET = 1) and during light exercise (MET < 3), cardiac output is higher than the output reached during comparable exercise with a sham exposure (COHB 1.8 to 1.9%). As the workload in CO atmosphere becomes more strenuous, the cardiac output generally converges to the level reached during sham exposure.

An HR is determined by a complex interaction of neural and humoral factors. The autonomic nervous system regulates the HR through the sinoatrial node and the frequency of beating is the product of a balance between sympathetic (acceleration) and vagal (retardation) system discharges. The autonomic system integrates signals from other central nervous system (CNS) regions and sensory afferents from a variety of receptors (stretch receptors in a carotid sinus and aortic arch, baroreceptors, etc.). In addition, other physiological adjustments to exercise (increased body temperature, release of catecholamines such as histamine, changes in blood pressure, etc.) modulate HR and the cardiovascular response. Earlier studies have suggested that the increase in HR and cardiac output during CO hypoxia is dominated by an adrenergic system, particularly by beta-2-adrenoceptors.[28,37] Indeed, the selective blockade of beta-2-adrenoceptors during severe CO hypoxia (38% COHb) reduced cardiac output in anesthetized dogs substantially (35%) when compared with the unblocked group. The lower cardiac output resulted primarily from a higher peripheral vascular resistance (up 34%) and only minimally due a direct effect on cardiac function. Thus, vasodilation triggered by activation of beta-2-adrenoceptors

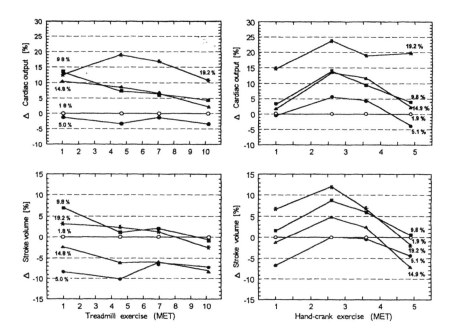

FIGURE 5.3   Changes in cardiac output and SV measured at rest (MET = 1), moderate (MET = 4.6, 7), and strenuous (MET = 10.2) treadmill exercise (5 min at each load) left panels, and at rest (MET = 1), light (MET = 2.6, 3.6), and moderate (MET = 4.9) hand-crank exercise (5 min at each load) right panels, during four different levels of COHb (5.0/5.1, 9.8, 14.8/14.9, and 19.2%). The average values are expressed in percent difference from respective sham exposure condition (COHb = 1.8, 1.9%) at each respective exercise load. (From Kizakevich et al., unpublished observations.)

appears to be an important compensatory response.[38,39] Selective blockade of either alpha-adrenoceptors[40] or beta-1-adrenoceptors[28] had little effect on cardiac output, although in the latter study blockade of beta-1-adrenoceptors lowered the HR but also increased the SV, resulting in no net change in cardiac output. Other selective pharmacological manipulation of adrenoceptors suggests that neither alpha- nor beta-1-adrenoceptors are substantially involved in modulating the cardiovascular response to CO hypoxia[28,39,40].

The SV changes during progressive exercise are not as closely related to increasing CO hypoxia as are the HR changes. This is in part because SV is determined by a number of other factors, primarily by venous return to the heart, ventricular contraction, and systemic vascular resistance. Resting dogs responded to 10 to 30% COHb with increased cardiac output but at a lower SV (10 to 18% of the control).[28] In sedentary individuals, an increase in cardiac output during exercise is almost entirely due to an increase in HR. In trained individuals, SV generally increases more during exercise than in untrained individuals. Young healthy males exposed to CO (up to 20% COHb) respond in a similar fashion. The compensatory changes in SV during both treadmill and hand-crank (Figure 5.3) exercise, although similar in pattern to respective cardiac output adjustments, showed the maximum difference from the sham exercise to be only 12%. Taken together with HR changes, these findings suggest that at COHb < 20% and low exercise load the primary contributor to increased cardiac output is a higher SV. As the intensity of exercise increases, SV contribution begins to diminish while the HR increases progressively, contributing more and more to cardiac output.

5.3.1.2 Redistribution of Blood Flow

The rate of CO uptake is sigmoidal in shape. The brief initial delay in COHb formation upon exposure to CO is followed by a rapid uptake of CO, which slows as COHb approaches equilibrium. Depending on CO concentration and minute ventilation there may be a considerable difference between the venous and the arterial blood COHb level, which during CO uploading is transiently higher.[41,42] Although mixing of air in the lung is responsible at least in part for these attributes, the main determinants of the arterial–venous COHb difference are regional differences in blood volume and circulation time. During a transition period from rest to exercise, diffusing capacity and CO uptake rise faster than O_2 consumption for each respective exercise intensity.[43] Moreover, decreased circulation time during exercise flattens the O_2 saturation curve[41] and redistribution of blood to high energy demand organs minimizes the arterial–venous COHb difference. In anesthetized dogs, severe hypoxia (61 to 67% COHb) causes redistribution of the blood from skeletal muscle to other organs, most likely to boost cardiac output.[44]

During moderate to heavy exercise the blood flow in the skin (measured as peripheral tissue conductance) of young males has been found to be lower in CO hypoxia (2.8 to 33% COHb) than normoxia.[45,46] Milder vasodilation (in CO hypoxia) does not seem to be a result of central thermoregulatory or passive hemodynamic adjustment, but rather an active redistribution of blood flow to boost cardiac output.[45] Indirect supporting evidence for a redistribution of blood flow from the skin to other tissues comes from a recently reported study of resting males exposed to CO. Based on the arteriovenous COHb differences measured at the forearm, the authors hypothesize that the blood is preferentially shunted from a superficial to a deep venous network.[47] Although this finding and the observations of Raven et al.[46] in 1974 suggest that the skin blood flow begins to shift even at modest COHb levels, the importance of this mechanism in the compensatory chain of events is unclear. Perhaps the redistribution of blood between organs promotes venous return and thus enhances cardiac output.

Severe CO hypoxia in sheep elicited approximately the same changes in the blood flow of various organs as did severe hypoxic hypoxia, with the exception of the myocardium, brain, and skeletal muscle. CO hypoxia increased myocardial blood flow fourfold, and brain and muscle blood flow each twofold, while hypoxic hypoxia increased the respective blood flows by factors of three, four, and 1.5 when compared with normoxia. Most of the other organs, except spleen, showed either no or only small increases in blood flow under both hypoxic conditions. The spleen, in contrast, showed a substantial decrease in blood flow under both conditions.[48]

5.3.1.3 Blood Pressure

A considerable number of studies have explored the effects of CO hypoxia on blood pressure. The evidence comes primarily from early studies on both anesthetized and conscious animals employing moderate to very high COHb saturation, but this evidence is equivocal. The inconsistency of blood pressure changes is more likely related to experimental conditions than to COHb levels. Moreover, experiments on conscious animals were done at rest and the potential interaction of exercise and CO was not studied.[7,48–50]

Of the few human studies, Chiodi and colleagues[51] reported an increase in systolic and a decrease in diastolic blood pressure of resting subjects with high COHb (48%) levels. Much lower blood COHb (<15%) did not elicit any systolic or diastolic blood pressure changes in resting males.[50,52] The most comprehensive data on blood pressure and other cardiovascular variables due to the interaction of CO and exercise come from a series of experiments by Klausen and colleagues[6] in 1968. In this study, eight subjects were briefly exposed five times a day (between 7:00 and 23:00 hours) to CO to maintain the blood COHb levels near the target concentration of 15%. The CO exposures repeated over an 8-day period brought the average daytime COHb levels from 11.5 to 13.6%. The COHb varied between 5 and 25% during the 24-h period.[53] Throughout this 8-day period, both systolic and diastolic blood pressure remained about the same when subjects were at rest, but systolic pressure increased by about 10% and diastolic pressure decreased by about 15% of the nominal rest value measured during once-a-day 15-min light exercise on a bicycle ergometer. Unfortunately, no control exercise experiments without CO loading were performed; therefore, it is difficult to separate the effects of CO and exercise. Under normoxic conditions, the mean blood pressure driven by systolic pressure increases almost linearly with increasing intensity of exercise until VO_2 max is nearly reached and the pressure begins to level off. The diastolic pressure remains about the same during submaximal exercise, but begins to decrease at loads above 50% VO_2 max. Based on the ventilatory and VO_2 uptake values in Klausen's study one can surmise that the change in systolic blood pressure was dominated by exercise, but a 15% decline in diastolic pressure at this exercise intensity suggests some CO-induced vasodilatation.[6] In another study on humans, direct measurement of the mean arterial blood pressure (measured in the branchial artery) at moderate COHb (19 to 21%) showed that for the same VO_2 the blood pressure was about the same as observed at normoxic conditions (Figure 5.4, upper panel).[32]

That even very high COHb levels either do not or only minimally alter the blood pressure suggests that the baroreceptors are unaffected by CO and that the changes in blood pressure are only marginally involved in a compensatory response to CO. That part of the circulatory response that is modulated by the carotid chemoreceptors appears to be similarly unaffected by CO, which contrasts with stimulation of the carotid chemoreceptors by hypoxic hypoxia and subsequently aroused cardiovascular reflexes. The blood pressure changes observed during exercise and elevated blood COHb are almost entirely in response to exercise. CO may have a marginal effect on the diastolic blood pressure which may be lower due to

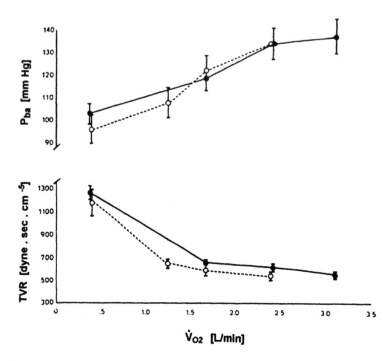

FIGURE 5.4   The average mean brachial artery pressure ± SEM (top panel) and the average total vascular resistance ± SEM (bottom panel) in subjects ($N = 8$) breathing air (solid circles) and CO (18.6 to 20.5% COHb; open circles) plotted against VO_2. (From Vogel, J.A. and Gleser, M.A., *J. Appl. Physiol.*, 32, 234, 1972. With permission.)

a vasodilation. Any blood pressure effects of CO are transient, since circadian variations of blood pressure in patients with sequelae of CO poisoning did not differ from those of healthy normotensive individuals.[54]

5.3.1.4   Peripheral Vascular Resistance

Reduction of peripheral vascular resistance is yet another mechanism by which the body augments cardiac output. Vascular resistance is determined by vasomotor tone, which is under the control of the vasomotor center located in the medulla oblongata. Although sympathetic nerve activity dominates this response, particularly in the skin, there are a number of well-known vasoactive metabolites, substances, and mediators which, acting either locally or generally, may further modulate the vasomotor tone. Moderate to severe COHb level progressively decreased total peripheral vascular resistance in anesthetized dogs, and upon reaching a severe hypoxic level (61 to 67% COHb), the resistance decreased by almost 50% from control conditions.[7,40,44] A carotid body resection even during severe CO hypoxia had only a minimal effect on peripheral resistance.[7] This finding is in agreement with observations from other laboratories that have reported the reduction in vascular resistance induced by CO hypoxia is not due to an increased peripheral chemoreceptor drive, but is controlled by the sympathetic input.[7,44,55]

During exercise, an increase in cardiac output is at least partially achieved by peripheral vasodilation. Incremental exercise by young and elderly persons progressively decreases systemic vascular resistance until the maximal exercise load is reached, when the resistance increases slightly.[56] Repeated direct measurement (up to 60 min) of sympathetic nerve activity to muscle blood vessels, forearm blood flow, and the forearm vascular resistance of resting young healthy subjects showed that ~8.2% COHb had a negligible effect on these variables.[50] Although submaximal and maximal exercise of young males with moderate COHb blood level (19 to 21%) substantially decreased their peripheral vascular resistance, the difference in resistance from that elicited by breathing clean air at the same workloads was (Figure 5.4, lower panel) minimal.[32]

5.3.1.5 O_2 Debt

At the beginning of exercise until the steady-state level is reached, the O_2 supply to the working muscle is insufficient and thus the required energy is supplemented from other sources. At the end of exercise this O_2 debt is recovered by an increased (above the immediate muscle energy needs) O_2 consumption until the non-O_2 sources of energy are replenished. Modest exposure to CO (COHb < 4.3%) of maximally exercising subjects results in the same O_2 debt as that measured under similar conditions in air.[57,58] The higher concentration of blood lactate found in healthy subjects exercising at submaximal and maximal intensity in CO atmosphere (18 to 20% COHb) than that observed under controlled (clean air) conditions suggests a greater O_2 debt. However, the investigators interpret this observation as simply reflecting a higher relative workload performed by subjects when breathing CO.[32]

5.3.1.6 Ventilation

At rest and during submaximal exercise, modest to moderate COHb levels (<16%) have been reported to have no effect on minute ventilation of male subjects when compared with control conditions.[31,50] At levels below the LAT, compensatory mechanisms (primarily increased cardiac output) apparently keep the O_2 supply to muscle at a sufficient level without additional boost of ventilation. Under similar CO hypoxia (<20% COHb) and maximal exercise, minute ventilation of young men increased, but the augmented response was not significantly different from that elicited under normoxic conditions.[31,36,58,59] However, when expressed in terms of ventilatory equivalent (VE/VO$_2$), higher ventilation was required during CO hypoxia for the same amount of O_2 utilized.[31] The increase in ventilation above the LAT during CO hypoxia does not appear to be due to stimulation of the carotid bodies, since breathing 100% O_2 did not affect the response (Figure 5.5).[59] Rather, as the animal studies show, the CO-induced hyperpnea is centrally mediated and most likely triggered by acidosis of the cerebrospinal fluid.[60,61] Collectively, these studies suggest that the LAT is the critical level in muscle energy demand at which ventilatory response to CO increases above the response elicited under normoxic or hypoxic hypoxia conditions. During high-intensity exercise, the rate of lactate production exceeds the rate of lactate removal, leading to lactate accumulation, acidosis, and associated

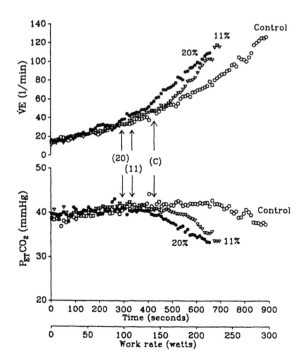

FIGURE 5.5 Ventilation (upper panel) and end-tidal PCO_2 (lower panel) responses to incremental exercise performed at three blood COHb levels (1.5% [control], 11%, and 19.6%) plotted against time and work rate for one subject. Each point is the average of 10 s of data. Arrows indicate the time/work rate of the lactic acidosis threshold for the three levels of COHb. (From Koiker, A. et al., *Respir. Physiol.*, 85, 169, 1991. With permission.)

hyperventilation. The additional contribution to central acidosis by CO hypoxia further augments the ventilatory response.

Long-term exposure to CO does not seem to elicit a ventilatory response that is different from the pattern under normoxic conditions. An 8-day study of males periodically exposed to CO (11.5 to 13.6% COHb) by Klausen et al.[6] showed that at both rest and during light exercise, the minute ventilation and frequency of breathing varied around baseline values. No CO effects on the ventilatory parameters were observed.

5.3.1.7 Thermoregulation

Effective body temperature regulation is an essential requirement of optimal muscle performance during exercise. The cardiovascular system, which is critical in supplying energy to the working muscle, also plays a key role in thermoregulation. Both hypoxic hypoxia and CO-induced hypoxia have been found to lower both the core temperature and the metabolic rate of rats. These effects become more pronounced as ambient temperature is decreased.[62]

The core temperature of young males does not appear to differ between normoxic and modest COHb (<5% COHb) conditions in response to maximal exercise.[46]

However, sustained exercise (60 min) at 25 to 33% COHb increases the core temperature of cycling males by 0.3 to 0.5°C above that found during normoxia or hypoxic hypoxia. The mean of averaged skin temperatures (measured simultaneously at 15 different places for each individual) was lower by up to 1°C during CO hypoxia, but did not differ between normoxia and hypoxic hypoxia during exercise.[45,63] Such a shift in core temperature, as compared with normoxia and hypoxic hypoxia, is not likely to be a central thermoregulatory response, but a consequence of limited skin vasodilation to protect venous return and maintain cardiac output. The maintenance of energy supply thus appears to override the need for systemic cooling. Although an increase in a sweat rate partially compensates for diminished convection and radiation of heat through the skin, this adjustment does not prevent the central temperature from rising to a new level.[45,63]

5.3.2 LOCAL COMPENSATORY MECHANISMS

5.3.2.1 Muscle Blood Flow

Compensatory increases in muscle blood flow during CO hypoxia have been reported in many animal and a few human studies.[23,39,45] Systemic blood flow increased more than twofold and hindlimb blood flow by 50% over control in anesthetized and severely CO hypoxic animals.[39] In electrically stimulated isolated *in situ* dog gastrocnemius muscle, the blood flow increased substantially more in CO hypoxia than with comparable hypoxic hypoxia (Figure 5.6).[24] In a similar experimental setup the increase in muscle blood flow at a respective duration and intensity of stimulation has been found to be proportionally and significantly associated with COHb levels but only at higher frequency of stimulation. This muscle blood flow correlated highly with increased O_2 demand.[23] In a study on humans, Nielsen[45,63] reported finding a greater blood flow in the vastus lateralis muscle of exercising males during moderate CO hypoxia (25 to 33% COHb) than that found either under normoxic or hypoxic hypoxia conditions. Increased muscular blood flow is a result of decreased vascular resistance due to active vasodilation of muscle capillaries and is primarily mediated by beta-2-adrenoceptors.[39] To what extent local hypoxia contributes to an overall change in tissue blood flow, however, is unclear.

5.3.2.2 O_2 Uptake and Extraction

When O_2 delivery is constant, VO_2 max is lower during CO hypoxia and has been shown to decrease approximately 0.9 to 1.2% for every 1% COHb.[36,46] At low to moderate COHb levels (<30%), the primary mechanism of VO_2 max reduction appears to be reduced O_2 conductance from the capillaries to the mitochondria.[26] Although the precise location of an O_2 diffusion impairment along this pathway has not yet been elucidated, O_2 dissociation kinetics appears to have a major role in limiting VO_2 max.[16]

During severe CO hypoxia, to maintain muscle O_2 delivery in anesthetized dogs at rest, in addition to a decreased O_2 content, concomitant decrease in O_2 uptake appears to be inversely related to both systemic and a local (hindlimb) increase in O_2 extraction.[39] In *in situ* electrically stimulated gastrocnemius muscle preparation,

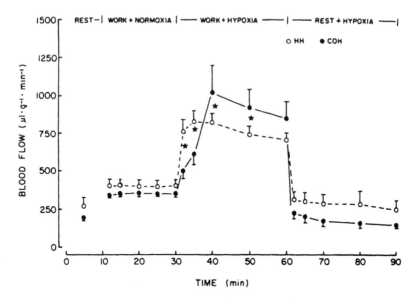

FIGURE 5.6 The average blood flow ± SEM in *in situ* isolated dog gastrocnemius muscle plotted vs. time during hypoxic hypoxia (*n* = 6, open circles, dashed line) and CO hypoxia (~15% COHb; *n* = 6, closed circles, solid line) at rest and work (isometric contractions at 1 twich/s). The experimental conditions with respective duration for two groups of animals are indicated on top of the figure. A star indicates the value is significantly different (*p* < 0.05) between groups. (From King, C.E. et al., *J. Appl. Physiol.*, 63, 726, 1987. With permission.)

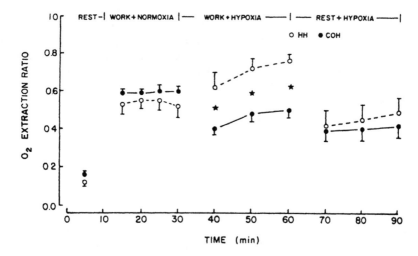

FIGURE 5.7 The average O_2 extraction ratio ± SEM in *in situ* isolated dog gastrocnemius muscle plotted vs. time during hypoxic hypoxia (*n* = 6, open circles, dashed line) and CO hypoxia (~15% COHb; *n* = 6, closed circles, solid line) at rest and work (isometric contractions at 1 twitch/s). The experimental conditions with respective duration for two groups of animals are indicated on top of the figure. A star indicates the value is significantly different (*p* < 0.05) between groups. (From King, C.E. et al., *J. Appl. Physiol.*, 63, 726, 1987. With permission.)

the working muscle O_2 extraction ratio decreased by about 33%, while the ratio increased by 20% during hypoxic hypoxia when compared with stimulation during normoxia (Figure 5.7).[24] Under similar experimental conditions, duration of work (muscular contractions) has been shown to be positively associated with O_2 extraction as well as COHb levels.[23] Several plausible mechanisms have been proposed that could lead to incomplete O_2 extraction by working muscle even though the PaO_2 remains normal.[24,32] One of these mechanisms, inhibition of the cytochrome oxidases system by CO, can indeed affect O_2 transport to tissues and impair cellular respiration, but this mechanism comes into play only at severe CO poisoning when COHb levels are approaching 50% and above.[24,64] A much more likely effect on O_2 extraction may be due to formation of COMb, which would decrease the O_2 content of the muscle and impair the diffusion of O_2. Another mechanism, the leftward shift of the O_2 dissociation curve, could also account for some reduction in O_2 extraction, since it will interfere with release of O_2 to the tissues.[32] At rest, leftward shift of the curve has no effect on VO_2.[65] During exercise the CO-induced shift may limit O_2 delivery to tissues due to the tighter binding of O_2 to Hb; however, the importance of this mechanism is unclear.[26] There are still other mechanisms that can potentially interfere with O_2 extraction, such as the time course of deoxygenation, shunting, perfusion heterogeneity, and others, but the involvement of these mechanisms during CO hypoxia has not yet been studied.

5.3.3 OTHER FACTORS POTENTIALLY MODULATING THE RESPONSE

Such physical characteristics as age, sex, race, and pregnancy do not modify the basic mechanisms of CO uptake, COHb formation, or elimination. However, the relative contribution and the extent of response of many cardiopulmonary variables to exercise and hypoxia depend on physical characteristics, which in turn may influence COHb kinetics.

Older individuals are at a greater risk from a cardiovascular system overload during physical activity under CO hypoxic conditions. Performance of the cardiopulmonary system declines with increasing age; e.g., myocardial contractility decreases, which can lead in turn to a reduction of SV. The same intensity of exercise will generally stress the O_2 supply system of older individuals more than that of younger persons.[56] Any functional reduction of the cardiovascular system beyond age-related changes will further compromise the hemodynamic response to exercise and CO hypoxia. Since the compensatory mechanisms involved in CO hypoxia are the same as those maintaining the energy demands of working muscles, work performance will be limited or even impossible under certain disease conditions. The CO uptake and elimination rates either at rest or during exercise have been shown to decrease with age. During the growing years (2 to 16 years), the COHb elimination half-time increases rapidly with age in both sexes, and is relatively shorter for boys than girls. After the teenage years, the half-time for CO elimination continues to grow longer, but at a slower rate. In contrast to adolescence, during the adult years the COHb half-time has consistently been found to be shorter (~6%) in females than in males. As expected, exercise will shorten the half-time; the more strenuous the exercise, the shorter the half-time. With increasing intensity of exercise,

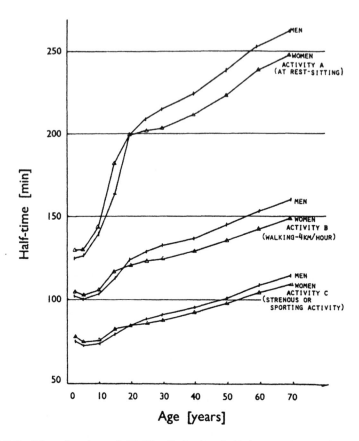

FIGURE 5.8   Plot of estimated COHb elimination half-time at three different physical activity levels for males and females over a span of 2 to 70 years (From Reference 66.)

the difference between half-time for males and females also is progressively reduced (Figure 5.8).[66]

It also is well established that with age the DLCO, one of the determinants of the rate of COHb formation, decreases.[67] The rate of DLCO decline is lower in middle-aged women than in men; however, at older ages the rates even out and are about the same for both sexes.[68] The decrease in DLCO combined with the increase in the VA/Q mismatch means that with increasing age it takes longer to both load and eliminate CO from the blood. During a transition period from rest to exercise while exposed to CO (500 ppm), the DLCO and CO uptake have been shown to rise faster than VO_2 at each new level of exercise intensity.[43] The intensity of exercise also is positively associated with the DLCO, which will plateau near the maximal exercise.[69]

It is not known whether or not the dynamics of COHb formation and elimination or the absolute COHb levels for the same exposure conditions are in any way different among races. Blacks have a lower diffusion capacity than whites,[68] and this should transiently slow both CO loading and unloading. However, as the sensitivity analysis of these physiological variables shows, the influence of the DLCO and other phys-iologic variables on CO kinetics is only transient.[70] Exercise will accelerate these transitional changes.

During pregnancy, the increased requirement for iron may lead to iron deficiency and anemia (see Section 5.4.2.1). Pregnant women who smoke exhibit a more pronounced shift of the O_2 dissociation curve to the left (~5% COHb) than nonpreg-nant women. Increased O_2 affinity when combined with the decreased O_2-carrying capacity of the blood of CO-exposed women thus may promote fetal hypoxia.[71] Studies have shown that protein deficiency in pregnant mice has no modulating effect on maternal COHb, but results in greater concentration of placental COHb.[72-74] Although the effects of exercise and CO hypoxia on pregnancy have not been studied in either women or animals, any exercise would be expected to accelerate as well as deepen fetal hypoxia.

5.4   PHYSICAL PERFORMANCE IN CARBON MONOXIDE HYPOXIA

5.4.1   HEALTHY POPULATION

5.4.1.1   Sea Level

In normal subjects at higher COHb levels, both the time to an onset of anaerobic threshold and the level of O_2 consumption at which anaerobic threshold are reached is reduced.[59,75] At submaximal workloads even as high as 70% of VO_2 max at COHb < 20%, VO_2 was proportionally related to the exercise load but is not different from comparable exercise in air (Figure 5.9).[36] Apparently, at submaximal work intensities the supply and extraction of O_2 is sufficient to satisfy the energy requirements of working muscle. At a maximal workload, however, the VO_2 max at elevated COHb was lower than that measured during control exercise in a sham atmosphere. Even at modest COHb, VO_2 max was lower by 7% when compared with the values measured during normoxic exercise.[76] This decrease in VO_2 max is strongly and positively ($r = 0.85$) associated with blood COHb concentration.[31,36,46]

Moderate CO-induced hypoxia (<20% COHb) accelerates the rate of muscle deoxygenation during exercise. At both below and above the LAT, muscle deoxy-genation during hypoxia was more rapid than during air-breathing. During work performed below the LAT, no additional lactate due to CO hypoxia is produced, and the minute ventilation, VO_2, and $PETCO_2$ are not different from the exercise-related changes under normoxic conditions (Figure 5.5).[31,36,59] The compensatory mecha-nisms (primarily increased cardiac output) apparently keep the O_2 supply to muscle at a sufficiently high level without any additional boost of ventilatory drive.[31] How-ever, at work intensity above the LAT, when working muscle becomes more sensitive to the balance between metabolic rate and O_2 supply, it progressively deoxygenates. This deoxygenation is further accelerated by a reduction in arterial O_2 content. Under these conditions, muscle blood flow probably is insufficient to satisfy the O_2 demand even though lactic acidosis during heavy exercise promotes O_2 extraction[77] and

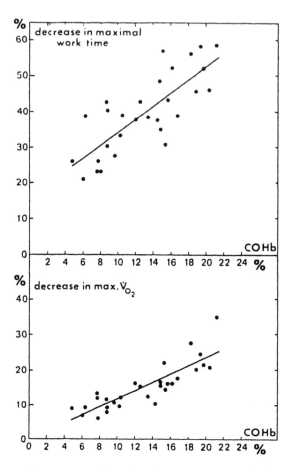

FIGURE 5.9 Individual values for the relationship between a percent decrease in maximal work time and COHb concentration (top panel; $r = 0.79$) and a percent decrease in VO$_2$ max and COHb concentration (bottom panel; $r = 0.85$). Each of 10 individuals performed maximal treadmill exercise at three different levels of COHb which varied between subjects. (From Ekblom, B. and Huot, R., *Acta Physiol. Scand.*, 86, 474, 1972. With permission.)

stimulates ventilation. This compensatory increase in ventilation is positively related to lactic acidosis but inversely related to PETCO$_2$ (Figure 5.5).[59]

The intensity of symptoms (exertion, fatigue, etc.) experienced by subjects exercising at 35 and 60% VO$_2$ max in a thermally neutral environment (25°C, 40% relative humidity or RH) and <11% COHb does not differ from exercise performed without CO exposure (<1% COHb).[78] However, adding heat stress (35°C and 60% RH or 40°C and 30% RH) to the same experimental conditions increases the symptom scores, which appear to be linearly related to the COHb level.[79] Comparison of these scores to those reported in a thermoneutral environment shows that the scores are dominated by exertion-related symptoms (heart rate perception, breathing, body temperature, etc.). The authors surmise that the additional load of CO to heat + exercise makes the symptoms more apparent to the subjects. However, a more likely explanation is that the higher exertion-related and temperature-related symptom scores simply reflect a physiological adjustment to added heat stress.

The duration of maximal exercise is negatively associated with blood COHb concentration.[31] Although the maximal work time is not the best index of work performance since it can be strongly influenced by psychological factors, it still is a convenient measure of work intensity. In young healthy males, 3 to 4.2% COHb shortened the duration of the maximal exercise test by 5%,[80] and 5.26% COHb shortened it by 10%, on average.[76] At 7% COHb, maximal exercise (treadmill or bicycle) was shorter by about a quarter (to 4 min, 4 s), and at 20% COHb the performance time was about half (2 min, 54 s) of the time the exercise lasted during air-breathing (6 min, 32 s). The positive relationship ($r = 0.79$) between the decrease in maximal work time and COHb at concentrations below 22% seems to be linear.[36]

5.4.1.2 Altitude

The combination of altitude, exercise, and CO hypoxia presents a significant challenge to cardiovascular and respiratory systems.[81-83] At a barometric pressure (PB) of 760 torr (sea level), the partial pressure of O_2 in inspired gas (PI O_2) saturated with water vapor at 37°C (BTPS conditions) is 149 torr. At an altitude of 3000 m (9840 ft; PB = 526 torr), the PI O_2 is only 100 torr, resulting in acute hypoxic hypoxia. Direct measurements of blood gases from over 1000 not acclimatized individuals at this altitude have revealed that the partial pressure of O_2 in alveolar air (PA O_2) was only 61 torr.[84] The hypoxic drive triggers a complement of compensatory physiological mechanisms (to maintain O_2 transport and supply), the extent of which will depend on elevation, exercise intensity, and the length of time at that altitude. During the first several days, the pulmonary ventilation at a given O_2 uptake (work level) increases progressively until a new quasi-steady state is achieved.[85,86] The DLCO does not change substantially at elevations below 2200 m (7218 ft), but has been reported to increase above that altitude. Spirometric lung function is reduced as well.[87] Maximal aerobic capacity and total work performance also decrease and the RER increases.[82,88] Due to a decrease in plasma volume (hemoconcentration), the Hb concentration is higher than at sea level.[89] The blood electrolytes and the acid–base equilibrium are readjusted to facilitate transport of O_2. For the same CO concentration as at sea level, these compensatory changes thus favor CO uptake and COHb formation.[85] By the same token, these adaptive changes not only affect CO uptake, but also CO elimination. Young women have been found to be more resistant to altitude hypoxia than men, but the physiological factors for this difference remain unexplored.[82,88] The COHb level at a given altitude has been shown to increase in both animals and men.[90,91] Breathing CO (9 ppm) while resting at 3517 m (11,540 ft) produces a higher COHb concentration than at sea level. No differences between males and females in blood COHb have been found.[91]

Short-term acclimatization (within a week or two) stabilizes these compensatory changes. During prolonged periods at a high altitude (over a few months), most of the early adaptive changes gradually revert to sea level values, and some long-term adaptive changes such as an increase in tissue capillarity and myoglobin content in the skeletal muscle begin to take place. Smokers appear to tolerate short-term hypoxic

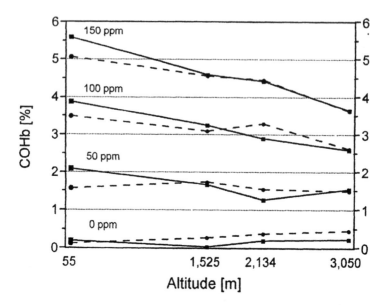

FIGURE 5.10   COHb concentration in blood of males (squares, solid lines) and females (circles, dashed lines) at different altitudes and ambient CO concentration. (Modified from Horvath, S.M. et al.[82,88] With permission.)

hypoxia due to high altitude (7620 m/ 25,000 ft) better than nonsmokers, who experience more severe subjective symptoms and a greater decline in task performance.[92] Perhaps smokers develop partial tolerance to hypoxic hypoxia because of chronic hypoxemia (due to chronically elevated COHb). Although the mechanisms of COHb formation in hypoxic hypoxia and CO hypoxia conditions are different, the resultant decrease in O_2 saturation and activation of compensatory mechanisms (e.g., increased cerebral blood flow) appear to be additive or even synergistic.[93] Psychophysiological studies in particular seem to support the possibility of physiological equivalency of hypoxic effects, whether induced by altitude at equilibrium or by ambient CO concentration. However, it must be kept in mind that although some of the mechanisms of action of hypoxic hypoxia and CO hypoxia are the same, CO elicits additional toxic effects not necessarily related to O_2 transport mechanisms.[94,95]

Horvath and colleagues,[82,88] who studied the effects of CO atmosphere (0, 50, 100, and 150 ppm) at a simulated altitude (0, 55 m/180 ft, 1524 m/5000 ft, 2134 m/7000 ft, and 3048 m/10,000 ft), have reported that under these conditions strenuous exercise appears either to suppress COHb formation partially or to shift CO storage, or both. Regardless of an altitude and the intensity of exercise, as CO concentration increases the blood COHb increases as well. However, as altitude increases, the attained COHb levels are progressively lower for the same CO concentration (Figure 5.10). The highest COHb attained at 3048 m while breathing 150 ppm CO during exercise was 58 and 70% lower on average for males and females, respectively, than at sea level, where the mean COHb values were 4.42 and 3.98%, respectively. This difference can, at least in part, be explained by reduced partial pressure of CO at higher altitudes. Of all the cardiopulmonary variables measured (VO_2 max, HR max, plasma lactate, RER, ventilation, total work) only the VO_2 max decreases consistently with increasing CO level and altitude, and this effect is more marked for men than for women. One other observation of this comprehensive study deserves attention. At every altitude and for every CO concentration, the postexposure (5 min) COHb values were 27 to 43% higher than those observed at maximum exercise. Such an increase in blood COHb level during the early recovery period when subjects were already breathing CO-free air was attributed to a release of CO accumulated in the skeletal muscle during preceding exercise when the tissue PO_2 was low.[82,88] Generally, no synergism between CO and altitude hypoxia was found; however, some effects appear to be additive. A more recent study also supports this conclusion.[96]

5.4.2 POPULATIONS AT RISK

Even for a healthy individual, strenuous exercise presents a significant challenge to the cardiovascular system. When combined with CO-induced hypoxia, the system will become taxed much earlier and performance may decrease. An individual with any pathophysiological conditions that reduce the blood O_2 content will be at a greater risk from CO exposure, since additional reduction in the O_2-carrying capacity of blood due to COHb formation will increase hypoxia. Increased O_2 demand during exercise will further lower the O_2 content, and depending on the initial level of hypoxia, the O_2 delivery to working muscles may become insufficient. Although a detailed review of the effects of CO hypoxia on exercise in patients with blood and cardiopulmonary diseases is beyond the scope of this chapter, the subsequent sections will briefly outline the most salient interactions between exercise and CO in at-risk populations.

5.4.2.1 Blood Disorders and Respiratory Diseases

The anemias encompass a wide range of etiologically varied diseases, all characterized by decreased O_2-carrying capacity of the blood. Anemia is a result of either impaired formation of RBCs or increased loss or destruction of RBCs. (Anemia reported in athletes in training is consequent to an increased plasma volume with RBC volume remaining the same.) The former category includes disorders of altered O_2 affinity, methemoglobinemias, and diseases with functionally abnormal and unstable Hbs. By far the most prevalent disorder in this group is sickle cell anemia. The O_2-carrying capacity of individuals afflicted with sickle cell anemia is not only reduced due to a lower amount of Hb, but the O_2 dissociation curve also is shifted to the right and O_2 affinity is reduced. The black population with a higher incidence of sickle cell anemia, may be at greater risk for CO hypoxia, particularly when combined with exercise.

An incremental exercise test of children with mild to severe anemia showed proportionally and substantially shorter exercise time, lower metabolic equivalent values, and generally diminished compensatory capabilities of the cardiovascular system as severity of the anemia increased.[97] To maintain the O_2 supply to the muscle during submaximal exercise, young males with acute anemia had increased cardiac output and muscle blood flow above the values obtained from control exercising subjects with normal RBC count.[98] Adults with sickle cell anemia showed abnormally decreased work capacity, peak heart rate, and blood pressure, when compared with normal values.[99] The arterial O_2 content appears to be the primary factor affecting the compensatory changes in exercising anemic individuals, with or without the presence of hypoxic hypoxia.[100] Although the cardiopulmonary response of anemic individuals to CO and exercise has not yet been studied, the parallelism of compensatory mechanisms elicited by anemia and CO hypoxia strongly suggests that the effects will be additive when both conditions are present.

Polycythemia (the opposite of anemia) is an increased number of RBCs in the blood. Although in polycythemia the total amount of Hb is generally elevated, under certain conditions the arterial O_2 saturation may be decreased, leading to a higher risk of additional hypoxia when such an individual is exposed to CO.

One of the characteristic symptoms of chronic obstructive pulmonary disease (COPD) is an increased dead space (VD) and an increased VA/Q inequality.[101] Impaired gas mixing due to poorly ventilated lung zones eventually will result in decreased arterial O_2 saturation and hypoxia. These pathophysiological conditions slow both CO uptake and elimination. Any COHb formation further lowers the O_2 content of the blood and increases hypoxia. COPD patients very often operate at the limit of their O_2 transport capability and therefore exposure to CO may severely compromise tissue oxygenation. Since the exercise capacity of patients with mild to moderate disease is already limited even under normoxic conditions, the addition of CO hypoxia may further decrease the exercise performance of these individuals. Depending on the severity of a disease, activation of compensatory mechanisms by CO and exercise (Section 5.3) may put the patient at serious risk and even lead to an overload and ultimate collapse of the cardiovascular system.

5.4.2.2 Cardiovascular Diseases

Because O_2 extraction by the myocardium is high, a greater O_2 demand by the myocardium of healthy individuals is met by increased coronary blood flow. Patients with coronary artery disease (CAD) have a limited ability to increase coronary blood flow in response to increased O_2 demand during physical activity. If this compensatory mechanism is further compromised by decreased O_2 saturation due to CO inhalation and subsequent hypoxia, the physical activity of patients with CAD may be severely restricted because of more rapid development of myocardial ischemia. The issue whether modest levels of CO hypoxia have clinically significant effects on the work or exercise performance of patients with cardiovascular disease has been revisited periodically and reviewed extensively in the literature.[102,103] Nevertheless, the issue still remains unsettled.[104]

Hinderliter and colleagues[105] reported that low levels of COHb (<6%) had no arrhythmogenic effects on patients with ischemic heart disease. Although exercise increased the frequency of premature ventricular beats of patients with diagnosed cardiac arrhythmias, surprisingly CO hypoxia (<5.1% COHb) appeared to have an antiarrhythmic and stabilizing effect on the heart since the frequency of ventricular ectopic beats decreased at the 5.1% COHb level. The frequency and complexity of ectopic beats did not differ between preexposure, during exposure at rest or exercise, during the exercise recovery period, and at 6 and 10 h after CO exposure. This held true even when patients were stratified by left ventricular ejection fraction, exercise ST-segment changes, and arrhythmia rate at a baseline.[106,107] In another dose–response study, 4% COHb had no proarrythmic effect, but 6% COHb increased the frequency of premature ventricular beats of increased complexity. Older individuals were found to be much more susceptible to the development of complex arrhythmias.[108,109]

The already limited compensatory ability and capacity of the cardiovascular system during exercise in patients with CAD is further reduced by exposure to CO. Less than 4% COHb causes some reduction in maximal exercise time, time to the onset of angina, and change in the left ventricular ejection fraction of ischemic patients, but these changes are clinically insignificant. The magnitude and time to ST-segment depression was about the same for both air and CO exposure.[110] These observations have been essentially confirmed by later studies.[106,107] Other laboratories, however, have reported that 2 to 4% COHb significantly reduces time to angina during exercise and time to significant ST-segment depression. In some cases, it increases the ST-segment depression at the end of exercise.[111-114] Higher COHb (6%) significantly reduces the duration of exercise and induces smaller (adaptive) changes in the left ventricular ejection fraction at submaximal exercise, but still had no effect on maximal exercise.[115] The HEI Multicenter CO study evaluated the most comprehensively CO-induced ECG changes and symptoms in patients with stable angina.[111,112] The principal study findings were that at COHb levels of less than 4.1%, decrements in the time to onset of ST-segment changes (Figure 5.11a) and to development of angina pectoris (Figure 5.11b) were negatively and significantly associated with postexercise COHb level. The authors concluded that low levels of CO have significant effects on cardiac function in patients with CAD during exercise. Recently, Kleinman et al.[96] studied the effects of CO administered near sea level and at a simulated altitude of 2100 m to a group patients with stable angina pectoris, who were then exercise stress-tested after the exposure. Both altitude and CO hypoxia (3.9% COHb) significantly reduced the total duration of exercise and the time to an onset of angina. No other cardiopulmonary variables were significantly affected by either CO or altitude. However, although they were small, the effects elicited by the two hypoxic stimuli were additive.

Based on the above findings, blood COHb levels around 4% seem to represent a concentration at which CO-induced ischemic changes in patients with CAD have attained statistical significance in most studies. The differences in observations among these studies at levels below 4% COHb may be due to differences in clinical characteristics and treatment regimens, which may limit the generalization of the findings. Collectively, however, these studies suggest that exposure to modest levels of CO (<6% COHb) are unlikely to predispose the heart to extrasystole, even in patients with diagnosed cardiac arrhythmias. This interpretation is further supported by studies of both anesthetized and conscious dogs with both induced acute and chronic myocardial ischemia. In these canine models, COHb as high as 20% had no effect on the electrical stability of the heart.[116,117] However, other indices of cardiac ischemia (such as ST-segment depression and angina pectoris) may worsen at concentrations below 4% COHb and may further destabilize cardiac function.

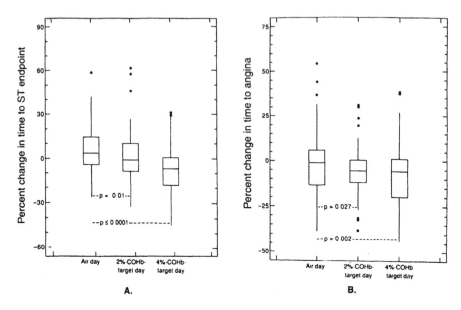

FIGURE 5.11   Box-and-whisker plots of percent change in time to ST end point (A) and percent of change in time to angina (B) between the pre- and postexposure exercise tests on the air day, 2% COHb, and 4% COHb target days. The bar across the box is a median and the ends of the box are the quartiles. The whiskers (lines) extend to within 1.5 times of the interquartile range from the box. The dots are individual values outside that range. (From Allred, E.N. et al., *Res. Rep. Health Effects Inst.*, 1, 1989. With permission.)

Individuals with congestive heart failure, a right-to-left shunt with congenital heart disease, or cerebrovascular disease also may be at greater risk from CO exposure because of already compromised O_2 delivery.

5.5 SUMMARY

Increased skeletal muscle demand for O_2 during work is adequately maintained by a compensatory increase in blood flow and O_2 extraction, and ultimately by a decrease in venous PO_2. Anything less than severe CO hypoxia does not appear to have an appreciable effect on any of these variables with the exception of acceleration of the rate of muscle deoxygenation. The great affinity and low dissociation constant of myoglobin for CO promotes retention of CO in muscle, which can potentially store a large amount of the gas. The main function of myoglobin appears to be as a temporary energy store and a diffusion facilitator of O_2 (and CO as well) between myoglobin and the tissue. Even at rest, cardiac muscle extracts O_2 almost completely; thus any increase in myocardial O_2 demand must be satisfied by increased coronary blood flow. There is a close relationship and interdependence among cardiac performance, myocardial O_2 consumption, and coronary blood flow. CO hypoxia puts an extra burden on the myocardium, since a compensatory increase in cardiac output demands a greater supply of O_2 to the heart muscle by the blood which, because of greater Hb affinity for CO than O_2, carries a smaller amount of O_2. Since during exercise the myocardial energy requirement is proportionally increased, a combination of CO hypoxia and exercise may severely strain the O_2 supply to both the skeletal and heart muscle. The myocardium is very sensitive to hypoxia and even moderate decrease in PO_2 may induce functional instability.

The increased supply of O_2 to the tissues is achieved by complex cardiovascular and metabolic adjustments at both systemic and local levels. The primary systemic compensatory mechanisms include increased cardiac output, redistribution of blood flow between organs, a decrease in mean arterial pressure, a decrease in peripheral vascular resistance, an increase in systemic O_2 debt, and an increase in minute ventilation. The increase in cardiac output is achieved by an increase in HR, and to a lesser degree, of SV as well. Most other cardiovascular compensatory mechanisms are recruited as needed to boost the cardiac output. The autonomic nervous system, particularly beta-2-adrenoceptors, have a dominant role in modulating the cardiovascular response to CO hypoxia. Neither carotid chemoreceptors nor baroreceptors appear to respond to CO; thus any activation of these receptors is in response to exercise alone. The O_2 debt is only negligibly increased by CO. However, VO_2 max was found to be limited by CO. The most plausible mechanism for such a reduction seems to be slower O_2 conductance from the capillaries to the mitochondria. Minute ventilation, which during heavy exercise may exceed the resting value by an order of magnitude, was augmented only when the lactic acid threshold was reached and blood COHb level was moderate or above. Local adjustments in the working muscle involve increased blood flow and increased O_2 extraction. The muscle blood flow is closely and inversely related to vascular resistance and vasomotor tone. Vasomotor tone is regulated by the autonomic nervous system, dominated by the sympathetic nerve activity and modulated by vasoactive substances, including CO. Moderate to severe CO hypoxia progressively decreases peripheral vascular resistance.

In conclusion, the dominant compensatory mechanisms during CO hypoxia are essentially the same as those seen during exercise, although the relative magnitudes of these adaptive changes may differ. Experimental evidence has shown that during muscular exercise with elevated blood COHb level, the great majority of physiological responses to these two conditions are likely to be additive to potentiating. Under certain combinations of work intensity and CO hypoxia, however, some of these mechanisms may induce less than additive changes or even mutually interfering ones.

There is a convincing body of evidence demonstrating that the duration of maximal exercise is negatively associated with blood COHb concentration and this relationship appears to be linear for COHb concentrations below 22%. The combination of an altitude, exercise, and CO hypoxia presents a significant challenge to cardiovascular and respiratory systems, since the compensatory changes triggered by these conditions appear to be additive.

Older and sick individuals are at a greater risk from a cardiovascular system overload during physical activity under CO hypoxic conditions. With advanced age, the performance of the cardiopulmonary system declines; thus the same intensity of exercise when combined with CO hypoxia will generally stress the O_2 supply system of older individuals more than that of younger persons. Similarly, individuals with any pathophysiological conditions that reduce the blood O_2 content (e.g., blood disorders), decrease arterial O_2 saturation (e.g., chronic obstructive lung disease), impair coronary blood flow (e.g., CAD), or impair in any way the energy supply to working organs (e.g., congenital heart disease) will be at a greater risk from CO exposure, since additional reduction in the O_2-carrying capacity of blood due to COHb formation will increase hypoxia and may subsequently lead to cardiac impairment. Under these conditions, additional demand on O_2 by exercise will further lower the O_2 content of blood and accelerate hypoxia. Depending on the initial level of hypoxia, the O_2 delivery to working skeletal and heart muscle may become insufficient, and can potentially lead to ischemia. For at-risk individuals, the severity of the disease is the critical limiting factor on any physical activity (even light exercise) at modest COHb levels.

It should be noted that the findings summarized above are based, with one exception,[6] entirely on acute studies of the interaction between CO hypoxia and exercise. To date, no study has examined the potential exercise and work performance limits or other health effects of combined physical activity and CO hypoxia on humans, beyond the acute condition of exercise stress testing. Therefore, it is not known how prolonged CO hypoxia affects work performance or myocardial function under the increased demands of energy during sustained or prolonged repeated exercise.

REFERENCES

1. Coburn, R.F. and Mayers, L.B., Myoglobin O_2 tension determined from measurement of carboxymyoglobin in skeletal muscle, *Am. J. Physiol.*, 220, 66, 1971.
2. Coburn, R.F., Ploegmakers, F., Gondrie, P., and Abboud, R., Myocardial myoglobin oxygen tension, *Am. J. Physiol.*, 224, 870, 1973.
3. Haab, P., The effect of carbon monoxide on respiration, *Experientia*, 1202, 1990.
4. Irvin, C.G., Exercise physiology, *Allergy Asthma Proc.*, 17, 327, 1996.
5. Plowman, S.A. and Smith, D.L. *Exercise Physiology*, Viacom, Needham Heights, MA, 1997.
6. Klausen, K., Rasmussen, B., Gjellerod, H., Madsen, H., and Petersen, E., Circulation, metabolism and ventilation during prolonged exposure to carbon monoxide and to high altitude, *Scand. J. Clin. Lab. Invest. Suppl.*, 103, 26, 1968.
7. Sylvester, J.T., Scharf, S.M., Gilbert, R.D., Fitzgerald, R.S., and Traystman, R.J., Hypoxic and CO hypoxia in dogs: hemodynamics, carotid reflexes, and catecholamines, *Am. J. Physiol.*, 236, H22, 1979.
8. Haab, P.E., and Durand-Arczynska, W.Y., Carbon monoxide effects on oxygen transport, in *The Lung: Scientific Foundations*, Crystal, R.G. and West, J.B., Eds., Raven Press, New York, 1991, chap. 5.3.2.6.
9. Coburn, R.F., Mean myoglobin oxygen tension in skeletal and cardiac muscle, *Adv. Exp. Med. Biol.*, 37A, 571, 1973.
10. Sokal, J.A., Majka, J., and Palus, J., The content of carbon monoxide in the tissues of rats intoxicated with carbon monoxide in various conditions of acute exposure, *Arch. Toxicol.*, 56, 106, 1984.
11. Werner, B. and Lindahl, J., Endogenous carbon monoxide production after bicycle exercise in healthy subjects and in patients with hereditary spherocytosis, *Scand. J. Clin. Lab. Invest.*, 40, 319, 1980.
12. Sokal, J.A., Majka, J., and Palus, J., Effect of work load on the content of carboxymyoglobin in the heart and skeletal muscles of rats exposed to carbon monoxide, *J. Hyg. Epidemiol. Microbiol. Immunol.*, 30, 57, 1986.

13. Gayeski, T.E., Connett, R.J., and Honig, C.R., Oxygen transport in rest–work transition illustrates new functions for myoglobin, *Am. J. Physiol.*, 248, H914, 1985.
14. Gayeski, T.E. and Honig, C.R., O_2 gradients from sarcolemma to cell interior in red muscle at maximal VO_2, *Am. J. Physiol.*, 251, H789, 1986.
15. Honig, Ċ.R. and Gayeski, T.E., Comparison of intracellular PO_2 and conditions for blood-tissue O_2 transport in heart and working red skeletal muscle, *Adv. Exp. Med. Biol.*, 215, 309, 1987.
16. Honig, C.R., Gayeski, T.E., Federspiel, W., Clark, A., and Clark, P., Muscle O_2 gradients from hemoglobin to cytochrome: new concepts, new complexities, *Adv. Exp. Med. Biol.*, 169, 23, 1984.
17. Cole, R.P., Sukanek, P.C., Wittenberg, J.B., and Wittenberg, B.A., Mitochondrial function in the presence of myoglobin, *J. Appl. Physiol.*, 53, 1116, 1982.
18. Gayeski, T.E. and Honig, C.R., Intracellular PO_2 in individual cardiac myocytes in dogs, cats, rabbits, ferrets, and rats, *Am. J. Physiol.*, 260, H522, 1991.
19. Wittenberg, B.A., Wittenberg, J.B., and Caldwell, P.R., Role of myoglobin in the oxygen supply to red skeletal muscle, *J. Biol. Chem.*, 250, 9038, 1975.
20. Wittenberg, B.A. and Wittenberg, J.B., Myoglobin-mediated oxygen delivery to mitochondria of isolated cardiac myocytes, *Proc. Natl. Acad. Sci. U.S.A.*, 84, 7503, 1987.
21. Wittenberg, B.A. and Wittenberg, J.B., Effects of carbon monoxide on isolated heart muscle cells, *Res. Rep. Health Effects Inst.*, 1, 1993.
22. Peters, T., Jurgens, K.D., Gunther-Jurgens, G., and Gros, G., Determination of myoglobin-diffusivity in intact skeletal muscle fibers. An improved microscope-photometrical approach, *Adv. Exp. Med. Biol.*, 345, 677, 1994.
23. King, C.E., Muscle oxygenation and performance during low level carbon monoxide exposure, *Adv. Exp. Med. Biol.*, 277, 533, 1990.
24. King, C.E., Dodd, S.L., and Cain, S.M., O_2 delivery to contracting muscle during hypoxic or CO hypoxia, *J. Appl. Physiol.*, 63, 726, 1987.
25. Hogan, M.C., Roca, J., Wagner, P.D., and West, J.B., Limitation of maximal O_2 uptake and performance by acute hypoxia in dog muscle in situ, *J. Appl. Physiol.*, 65, 815, 1988.
26. Hogan, M.C., Bebout, D.E., Gray, A.T., Wagner, P.D., West, J.B., and Haab, P.E., Muscle maximal O_2 uptake at constant O_2 delivery with and without CO in the blood, *J. Appl. Physiol.*, 69, 830, 1990.
27. Einzig, S., Nicoloff, D.M., and Lucas, R.V., Myocardial perfusion abnormalities in carbon monoxide poisoned dogs, *Can. J. Physiol. Pharmacol.*, 58, 396, 1980.
28. Cramlet, S.H., Erickson, H.H., and Gorman, H.A., Ventricular function following acute carbon monoxide exposure, *J. Appl. Physiol.*, 39, 482, 1975.
29. Kizakevich, P.N., Hazucha, M.J., Van Hoose, L., Bolick, K., Jochem, W.J., McCartney, M.L., and McMaster, L., Reproducibility of impedance cardiogram waveform analysis in lower-body and upper-body exercise, in *IEEE 7th Symposium on Computer-Based Medical Systems*, IEEE Computer Society Press, Los Alamitos, CA, 182, 1994.
30. Myers, J. and Ashley, E., Dangerous curves. A perspective on exercise, lactate, and the anaerobic threshold, *Chest*, 111, 787, 1997.
31. Pirnay, F., Dujardin, J., Deroanne, R., and Petit, J.M., Muscular exercise during intoxication by carbon monoxide, *J. Appl. Physiol.*, 31, 573, 1971.
32. Vogel, J.A. and Gleser, M.A., Effect of carbon monoxide on oxygen transport during exercise, *J. Appl. Physiol.*, 32, 234, 1972.
33. Vogel, J.A., Gleser, M.A., Wheeler, R.C., and Whitten, B.K., Carbon monoxide and physical work capacity, *Arch. Environ. Health*, 24, 198, 1972.
34. Wagner, J.A., Horvath, S.M., and Dahms, T.E., Cardiovascular adjustments to carbon monoxide exposure during rest and exercise in dogs, *Environ. Res.*, 15, 368, 1978.

35. Turner, J.A. and McNicol, M.W., The effect of nicotine and carbon monoxide on exercise performance in normal subjects, *Respir. Med.*, 87, 427, 1993.

36. Ekblom, B. and Huot, R., Response to submaximal and maximal exercise at different levels of carboxyhemoglobin, *Acta Physiol. Scand.*, 86, 474, 1972.

37. Adams, J.D., Erickson, H.H., and Stone, H.L., Myocardial metabolism during exposure to carbon monoxide in the conscious dog, *J. Appl. Physiol.*, 34, 238, 1973.

38. Chapler, C.K., Melinyshyn, M.J., Villeneuve, S.M., and Cain, S.M., The role of beta-adrenergic receptors in the cardiac output response during carbon monoxide hypoxia, *Adv. Exp. Med. Biol.*, 227, 145, 1988.

39. Melinyshyn, M.J., Cain, S.M., Villeneuve, S.M., and Chapler, C.K., Circulatory and metabolic responses to carbon monoxide hypoxia during beta-adrenergic blockade, *Am. J. Physiol.*, 255, H77, 1988.

40. Villeneuve, S.M., Chapler, C.K., King, C.E., and Cain, S.M., The role of alpha-adrenergic receptors in carbon monoxide hypoxia, *Can. J. Physiol. Pharmacol.*, 64, 1442, 1986.

41. Tikuisis, P., Kane, D.M., McLellan, T.M., Buick, F., and Fairburn, S.M., Rate of formation of carboxyhemoglobin in exercising humans exposed to carbon monoxide, *J. Appl. Physiol.*, 72, 1311, 1992.

42. Benignus, V.A., Hazucha, M.J., Smith, M.V., and Bromberg, P.A., Prediction of carboxyhemoglobin formation due to transient exposure to carbon monoxide, *J. Appl. Physiol.*, 76, 1739, 1994.

43. Kinker, J.R., Haffor, A.S., Stephan, M., and Clanton, T.L., Kinetics of CO uptake and diffusing capacity in transition from rest to steady-state exercise, *J. Appl. Physiol.*, 72, 1764, 1992.

44. King, C.E., Cain, S.M., and Chapler, C.K., Whole body and hindlimb cardiovascular responses of the anesthetized dog during CO hypoxia, *Can. J. Physiol. Pharmacol.*, 62, 769, 1984.

45. Nielson, B., Thermoregulation during work in carbon monoxide poisoning, *Acta Physiol. Scand.*, 82, 98, 1971.

46. Raven, P.B., Drinkwater, B.L., Ruhling, R.O., Bolduan, N., Taguchi S., Gliner, J., and Horvath, S.M., Effect of carbon monoxide and peroxyacetyl nitrate on man's maximal aerobic capacity, *J. Appl. Physiol.*, 36, 288, 1974.

47. Smith, M.V., Hazucha, M.J., Benignus, V.A., and Bromberg, P.A., Effect of regional circulation patterns on observed HbCO levels, *J. Appl. Physiol.*, 77, 1659, 1994.

48. Koehler, R.C., Traystman, R.J., and Jones, M.D.J., Regional blood flow and O_2 transport during hypoxic and CO hypoxia in neonatal and adult sheep, *Am. J. Physiol.*, 248, H118, 1985.

49. Theissen, J.L., Loick, H.M., Traber, L.D., Herndon, D.N., and Traber, D.L., Carbon monoxide and pulmonary circulation in an ovine model, *J. Burn. Care. Rehab.*, 13, 623, 1992.

50. Hausberg, M. and Somers, V.K., Neural circulatory responses to carbon monoxide in healthy humans, *Hypertension*, 29, 1114, 1997.

51. Chiodi, H., Dill, D.B., Consolazio, F., and Horvath, S.M., Respiratory and circulatory responses to acute carbon monoxide poisoning, *Am. J. Physiol.*, 134, 683, 1941.

52. Stewart, R.D., Peterson, J.E., Fisher, T.N., Hosko, M.J., Baretta, E:D., Dodd, H.C., and Herrmann, A.A., Experimental human exposure to high concentrations of carbon monoxide, *Arch. Environ. Health*, 26, 1, 1973.

53. Astrup, P., Pauli, H.G., Kjeldsen, K., and Petersen, C.E., Introduction and general description of the study and of the procedures for prolonged exposure to carbon monoxide and hypoxia, *Scand. J. Clin. Lab. Invest. Suppl.*, 103, 1968.

54. Fukuhara, M., Abe, I., Matsumura, K., Kaseda, S., Yamashita, Y., Shida, K., Kawash-ima, H., and Fujishima, M. Circadian variations of blood pressure in patients with sequelae of carbon monoxide poisoning, *Am. J. Hypertens.*, 9, 300, 1996.

55. Lahiri, S., Penney, D.G., Mokashi, A., and Albertine, K.H., Chronic CO inhalation and carotid body catecholamines: testing of hypotheses, *J. Appl. Physiol.*, 67, 239, 1989 (published erratum appears in *J. Appl. Physiol.*, 67 (4), preceding 1311, 1989).

56. Bogaard, H.J., Woltjer, H.H., Dekker, B.M., van Keimpema, A.R., Postmus, P.E., and de Vries, P. M., Haemodynamic response to exercise in healthy young and elderly subjects, *Eur. J. Appl. Physiol.*, 75, 435, 1997.

57. Chevalier, R.B., Krumholz, R.A., and Ross, J.C., Reaction of non-smokers to carbon monoxide inhalation. Cardiopulmonary responses at rest and during exercise, *J. Am. Med. Assoc.*, 198, 1061, 1966.

58. Horvath, S.M., Raven, P.B., Dahms, T.E., and Gray, D.J., Maximal aerobic capacity at different levels of carboxyhemoglobin, *J. Appl. Physiol.*, 38, 300, 1975.

59. Koike, A., Wasserman, K., Armon, Y., and Weiler Ravell, D., The work-rate-dependent effect of carbon monoxide on ventilatory control during exercise, *Respir. Physiol.*, 85, 169, 1991.

60. Santiago, T.V. and Edelman, N.H., Mechanism of the ventilatory response to carbon monoxide, *J. Clin. Invest.*, 57, 977, 1976.

61. Gautier, H. and Bonora, M., Ventilatory response of intact cats to carbon monoxide hypoxia, *J. Appl. Physiol.*, 55, 1064, 1983.

62. Gautier, H. and Bonora, M., Ventilatory and metabolic responses to cold and CO-induced hypoxia in awake rats, *Respir. Physiol.*, 97, 79, 1994.

63. Nielsen, B., Exercise temperature plateau shifted by a moderate carbon monoxide poisoning, *J. Physiol.* (Paris), 63, 362, 1971.

64. Piantadosi, C.A., Toxicity of carbon monoxide: hemoglobin vs. histotoxic mecha-nisms, in *Carbon Monoxide*, D.G. Penney, Ed., CRC Press, Boca Raton, FL, 1996, 163–186.

65. Ross, B.K. and Hlastala, M.P., Increased hemoglobin-oxygen affinity does not decrease skeletal muscle oxygen consumption, *J. Appl. Physiol.*, 51, 864, 1981.

66. Joumard, R., Chiron, M., Vidon, R., Maurin, M., and Rouzioux, J.M., Mathematical models of the uptake of carbon monoxide on hemoglobin at low carbon monoxide levels, *Environ. Health Perspect.*, 41, 277, 1981.

67. Guenard, H. and Marthan, R., Pulmonary gas exchange in elderly subjects, *Eur. Respir. J.*, 9, 2573, 1996.

68. Neas, L.M. and Schwartz, J., The determinants of pulmonary diffusing capacity in a national sample of U.S. adults, *Am. J. Respir. Crit. Care Med.*, 153, 656, 1996.

69. Turcotte, R., Kiteala, L., Marcotte, J.E., and Perrault, H., Exercise-induced oxyhe-moglobin desaturation and pulmonary diffusing capacity during high-intensity exer-cise, *Eur. J. Appl. Physiol.*, 75, 425, 1997.

70. McCartney, M.L., Sensitivity analysis applied to Coburn-Forster-Kane models of carboxyhemoglobin formation, *Am. Ind. Hyg. Assoc. J.*, 51, 169, 1990.

71. Grote, J., Dall, P., Oltmanns, K., and Stolp, W., The effect of increased blood carbon monoxide levels on the hemoglobin oxygen affinity during pregnancy, *Adv. Exp. Med. Biol.*, 345, 145, 1994.

72. Singh, J., Smith, C. B., and Moore Cheatum, L., Additivity of protein deficiency and carbon monoxide on placental carboxyhemoglobin in mice, *Am. J. Obstet. Gynecol.*, 167, 843, 1992.

73. Singh, J., Aggison, L., and Moore Cheatum, L., Teratogenicity and developmental toxicity of carbon monoxide in protein-deficient mice, *Teratology*, 48, 149, 1993.

74. Singh, J. and Moore Cheatum, L., Gestational protein deficiency enhances fetotoxicity of carbon monoxide, *Ann. N.Y. Acad. Sci.*, 678, 366, 1993.
75. Hirsch, G.L., Sue, D.Y., Wasserman, K., Robinson, T.E., and Hansen, J.E., Immediate effects of cigarette smoking on cardiorespiratory responses to exercise, *J. Appl. Physiol.*, 58, 1975, 1985.
76. Klausen, K., Andersen, C., and Nandrup, S., Acute effects of cigarette smoking and inhalation of carbon monoxide during maximal exercise, *Eur. J. Appl. Physiol.*, 51, 371, 1983.
77. Maehara, K., Riley, M., Galassetti, P., Barstow, T.J., and Wasserman, K., Effect of hypoxia and carbon monoxide on muscle oxygenation during exercise, *Am. J. Respir. Crit. Care Med.*, 155, 229, 1997.
78. Bunnell, D.E. and Horvath, S.M., Interactive effects of physical work and carbon monoxide on cognitive task performance, *Aviat. Space Environ. Med.*, 59, 1133, 1988.
79. Bunnell, D.E. and Horvath, S.M., Interactive effects of heat, physical work, and CO exposure on metabolism and cognitive task performance, *Aviat. Space Environ. Med.*, 60, 428, 1989.
80. Aronow, W.S. and Cassidy, J., Effect of carbon monoxide on maximal treadmill exercise. A study in normal persons, *Ann. Intern. Med.*, 83, 496, 1975.
81. U.S. Environmental Protection Agency, Altitude as a Factor in Air Pollution, EPA Rep. 600/9-78-015, 1-1, 1978.
82. Horvath, S.M., Bedi, J.F., Wagner, J.A., and Agnew, J., Maximal aerobic capacity at several ambient concentrations of CO at several altitudes, *J. Appl. Physiol.*, 65, 2696, 1988.
83. Leaf, D.A. and Kleinman, M.T., Urban ectopy in the mountains: carbon monoxide exposure at high altitude, *Arch. Environ. Health*, 51, 283, 1996.
84. Boothby, W., Lovelace, W.R., Benson, O.O., Jr., and Strehler, A.F., Volume and partial pressures of respiratory gases at altitude, in *Respiratory Physiology in Aviation*, Boothby, W.M., Ed., Air University, USAF School of Aviation Medicine, Randolph Field, TX, 1954, chap. 4.
85. Burki, N.K., Effects of acute exposure to high altitude on ventilatory drive and respiratory pattern, *J. Appl. Physiol.*, 56, 1027, 1984.
86. Bender, P.R., Weil, J.V., Reeves, J.T., and Moore, L.G., Breathing pattern in hypoxic exposures of varying duration, *J. Appl. Physiol.*, 62, 640, 1987.
87. Ge, R.L., Matsuzawa, Y., Takeoka, M., Kubo, K., Sekiguchi, M., and Kobayashi, T., Low pulmonary diffusing capacity in subjects with acute mountain sickness, *Chest*, 111, 58, 1997.
88. Horvath, S.M., Agnew, J.W., Wagner, J.A., and Bedi, J.F., Maximal aerobic capacity at several ambient concentrations of carbon monoxide at several altitudes, *Res. Rep. Health Effects Inst.*, 1, 1988.
89. Messmer, K., Oxygen transport capacity, in *High Altitude Physiology and Medicine*, Brendel, W. and Zink, R.A., Eds., Springer-Verlag, New York, 1994, 16.
90. McGrath, J.J., Effects of altitude on endogenous carboxyhemoglobin levels, *J. Toxicol. Environ. Health*, 35, 127, 1992.
91. McGrath, J.J., Schreck, R.M., and Lee, P.S., Carboxyhemoglobin levels in humans — effects of altitude, *Inhal. Toxicol.*, 5, 241, 1993.
92. Yoneda, I. and Watanabe, Y., Comparisons of altitude tolerance and hypoxia symptoms between nonsmokers and habitual smokers, *Aviat. Space Environ. Med.*, 68, 807, 1997.
93. McGrath, J.J., Carbon monoxide studies at high altitude, *Neurosci. Biobehav. Rev.*, 12, 311, 1988.

94. Ludbrook, G.L., Helps, S.C., Gorman, D.F., Reilly, P.L., North, J.B., and Grant, C., The relative effects of hypoxic hypoxia and carbon monoxide on brain function in rabbits, *Toxicology*, 75, 71, 1992.

95. Zhu, N. and Weiss, H.R., Effect of hypoxic and carbon monoxide-induced hypoxia on regional myocardial segment work and O_2 consumption, *Res. Exp. Med.* (Berlin), 194, 97, 1994.

96. Kleinman, M.T., Leaf, D.A., Kelly, E., Caiozzo, V., Osann, K., and O'Niell, T. Urban angina in the mountains: effects of carbon monoxide and mild hypoxemia on subjects with chronic stable angina, *Arch. Environ. Health*, 53, 388, 1998.

97. Kapoor, R.K., Kumar, A., Chandra, M., Misra, P.K., Sharma, B., and Awasthi, S., Cardiovascular responses to treadmill exercise testing in anemia, *Indian Pediatr.*, 34, 607, 1997.

98. Koskolou, M.D., Roach, R.C., Calbet, J.A., Radegran, G., and Saltin, B., Cardiovascular responses to dynamic exercise with acute anemia in humans, *Am. J. Physiol.*, 273, H1787, 1997.

99. Braden, D.S., Covitz, W., and Milner, P.F., Cardiovascular function during rest and exercise in patients with sickle-cell anemia and coexisting alpha thalassemia-2, *Am. J. Hematol.*, 52, 96, 1996.

100. Roach, R.C., Koskolou, M.D., Calbet, J.A., and Saltin, B., Arterial O_2 content and tension in regulation of cardiac output and leg blood flow during exercise in humans, *Am. J. Physiol.*, 276, H438, 1999.

101. Marthan, R., Castaing, Y., Manier, G., and Guenard, H., Gas exchange alterations in patients with chronic obstructive lung disease, *Chest*, 87, 470, 1985.

102. Maynard, R.L. and Waller, R., Carbon monoxide, in *Air Pollution and Health*, Holgate, T.S., Samet, J.M., Koren, H.S., and Maynard, R.L., Eds., Academic Press, London, 1999, 33.

103. U.S. Environmental Protection Agency, Integrative summary and conclusions, *in Air Quality Criteria for Carbon Monoxide*, U.S. EPA, Office of Research and Development, Washington, D.C., 2000, chap. 7.

104. Mennear, J.H., Carbon monoxide and cardiovascular disease: an analysis of the weight of evidence, *Reg. Toxicol. Pharmacol.*, 17, 77, 1993.

105. Hinderliter, A.L., Adams, K.F., Price, C.J., Herbst, M.C., Koch, G., and Sheps, D.S., Effects of low-level carbon monoxide exposure on resting and exercise-induced ventricular arrhythmias in patients with coronary artery disease and no baseline ectopy, *Arch. Environ. Health*, 44, 89, 1989.

106. Chaitman, B.R., Dahms, T.E., Byers, S.L., Carroll, L.W., Younis, L.T., and Wiens, R.D., Carbon monoxide exposure of subjects with documented cardiac arrhythmias, *Res. Rep. Health Effects Inst.*, 52, 1, 1992.

107. Dahms, T.E., Younis, L.T., Wiens, R.D., Zarnegar, S., Byers, S.L., and Chaitman, B.R., Effects of carbon monoxide exposure in patients with documented cardiac arrhythmias, *J. Am. Coll. Cardiol.*, 21, 442, 1993.

108. Sheps, D.S., Herbst, M.C., Hinderliter, A.L., Adams, K.F., Ekelund, L.G., O'Neil, J.J., Goldstein, G.M., Bromberg, P.A., Dalton, J.L., and Ballenger, M.N., Production of arrhythmias by elevated carboxyhemoglobin in patients with coronary artery disease, *Ann. Intern. Med.*, 113, 343, 1990.

109. Sheps, D.S., Herbst, M.C., Hinderliter, A.L., Adams, K.F., Ekelund, L.G., O'Neil, J.J., Goldstein, G.M., Bromberg, P.A., Ballenger, M.N., Davis, S.M., and Koch, G., Effects of 4 percent and 6 percent carboxyhemoglobin on arrhythmia production in patients with coronary artery disease, *Res. Rep. Health Effects Inst.*, 41, 1, 1991.

110. Sheps, D.S., Adams, K.F., Bromberg, P.A., Goldstein, G.M., O'Neil, J.J., Horstman, D., and Koch, G., Lack of effect of low levels of carboxyhemoglobin on cardiovascular function in patients with ischemic heart disease, *Arch. Environ. Health*, 42, 108, 1987.

111. Allred, E.N., Bleecker, E.R., Chaitman, B.R., Dahms, T.E., Gottlieb, S.O., Hackney, J.D., Hayes, D., Pagano, M., Selvester, R.H., and Walden, S.M., Acute effects of carbon monoxide exposure on individuals with coronary artery disease, *Res. Rep. Health Effects Inst.*, 1, 1989.

112. Allred, E.N., Bleecker, E.R., Chaitman, B.R., Dahms, T.E., Gottlieb, S.O., Hackney, J.D., Pagano, M., Selvester, R.H., Walden, S.M., and Warren, J., Short-term effects of carbon monoxide exposure on the exercise performance of subjects with coronary artery disease, *N. Engl. J. Med.*, 321, 1426, 1989 (published erratum appears in *N. Engl. J. Med.*, 322 (14), 1019, 1990).

113. Kleinman, M.T., Davidson, D.M., Vandagriff, R.B., Caiozzo, V.J., and Whittenberger, J.L., Effects of short-term exposure to carbon monoxide in subjects with coronary artery disease, *Arch. Environ. Health*, 44, 361, 1989.

114. Allred, E.N., Bleecker, E.R., Chaitman, B.R., Dahms, T.E., Gottlieb, S.O., Hackney, J.D., Pagano, M., Selvester, R.H., Walden, S.M., and Warren, J., Effects of carbon monoxide on myocardial ischemia, *Environ. Health Perspect.*, 91, 89, 1991.

115. Adams, K.F., Koch, G., Chatterjee, B., Goldstein, G.M., O'Neil, J.J., Bromberg, P.A., and Sheps, D.S., Acute elevation of blood carboxyhemoglobin to 6% impairs exercise performance and aggravates symptoms in patients with ischemic heart disease, *J. Am. Coll. Cardiol.*, 12, 900, 1988.

116. Farber, J.P., Schwartz, P.J., Vanoli, E., Stramba Badiale, M., and De Ferrari, G.M., Carbon monoxide and lethal arrhythmias, *Res. Rep. Health Effects Inst.*, 1, 1990.

117. Verrier, R.L., Mills, A.K., and Skornik, W.A., Acute effects of carbon monoxide on cardiac electrical stability, *Res. Rep. Health Effects Inst.*, 1, 1990.

6 The Interacting Effects of Altitude and Carbon Monoxide

James J. McGrath

CONTENTS

6.1 INTRODUCTION

The atmosphere is an envelope of air made up of a mixture of gases (nitrogen, oxygen, carbon dioxide, and water vapor) that surrounds the Earth and permits human beings and other life-forms to exist. Life can exist only in the lower regions of this envelope where variations in its thickness, and therefore its weight, are expressed as changes in atmospheric pressure. Atmospheric pressure varies with altitude; at sea level the weight of the atmosphere, expressed as barometric pressure (P_B), is 760 mm Hg. As people ascend to altitudes above sea level, they rise higher into the atmosphere and consequently experience lower P_B.

0-8493-2065-8/00/$0 00+$.50
© 2000 by CRC Press LLC

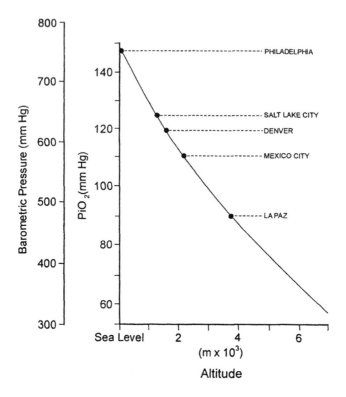

FIGURE 6.1   Barometric pressure and inhaled PO_2 (PiO_2) in population centers at various altitudes.

The total atmospheric or P_B is made up of the sum of the partial pressures of its constituent gases oxygen, (PO_2), nitrogen (PN_2), carbon dioxide (PCO_2), and water vapor (PH_2O). With the exception of PH_2O, which varies with temperature, the relative proportions of the partial pressures of the constituent gases remain about the same at different attitudes. At higher elevations, total P_B, as well as the partial pressures of the constituent gases, is reduced. Of these gases, it is oxygen (O_2) that is of greatest physiological importance and as people ascend to higher elevations the PO_2 of inhaled air is reduced (Figure 6.1).

In recent years, there has been a massive expansion of tourism and sports at altitude including snowmobiling, hiking, paragliding, skiing, biking, and mountaineering. As people travel into the high-altitude areas of the world, exposure to carbon monoxide (CO) may be exacerbated because CO emissions at altitude are often higher. This is due both to the incomplete nature of combustion and to the cold conditions prevailing at altitude. Wood-burning fireplaces, woodstoves, and space heaters (both kerosene and gas) are used extensively in mountainous communities. Moreover, CO emissions from light-duty vehicles (passenger cars, light-duty trucks, vans, and sports utility vehicles) are highest during "cold" starts, which are more frequent at altitude.

The potential health problems associated with exposure to CO at high altitude have not received a great deal of study. Yet it has been estimated that there are over

35 million people living or visiting in mountainous regions at altitudes above 1524 m (5000 ft) in the United States.[1] Worldwide, it has been estimated that nearly 40 million people live permanently at elevations above 2424 m (8000 ft) and as many as another 40 million visit high-altitude regions.[2] This chapter will explore the potential effects of exposure to CO on humans at altitude. The focus will be on the newcomer to altitude who, for a number of physiological reasons, may be more vulnerable than the long-term resident to CO exposure. The physiological changes that enable the long-term resident (human or animal) to adapt to high-altitude hypoxia are beyond the scope of this chapter.

6.2   OXYGEN TRANSPORT

Oxygen is transported from the lungs to the tissues in reversible combination with hemoglobin, a conjugated protein molecule in which the protein, globin, is joined with heme, an iron-porphyrin moiety. The hemoglobin molecule is remarkable; it reversibly binds O_2 in the lungs, where the concentration of O_2 is high and releases it in the tissues where the O_2 is low. The amount of O_2 transported from the lungs to the tissues as oxyhemoglobin depends on the PO_2 in the inhaled air, the amount of functional hemoglobin, and the resulting percentage of hemoglobin saturated with O_2. The relationship between PO_2 and hemoglobin saturation with oxygen is depicted by the sigmoidal-shaped oxyhemoglobin dissociation curve (Figure 6.2). Two regions of the normal curve are of particular interest. The first is the flatter region between O_2 tensions of 70–100 mm Hg. In this region there is little change in the oxyhemoglobin saturation or arterial blood oxygen content with changes in PO_2; a decrease in PO_2 from 100 to 70 only decreases the hemoglobin saturation to about 93%. This feature enables people to experience a moderate reduction in PO_2 without much of a reduction in the amount of O_2 transported by the blood to the tissue. The second region of interest is the steeper region between O_2 tensions of 10 and 40 mm Hg. This is the PO_2 range found in metabolically active tissues. Because the amount of O_2 that hemoglobin can bind depends on the local PO_2, oxyhemoglobin dissociates in this region and releases O_2 to the metabolizing tissue. At a PO_2 of 40 mm Hg, hemoglobin saturation is only 75%. If the PO_2 is decreased further to 10, hemoglobin saturation decreases to about 13%. In this way, the flat, upper part of the curve protects the body by enabling blood to load O_2 in the lungs over a broad range of PO_2 values, whereas the steep middle and lower parts protect the metabolizing tissues by enabling large amounts of O_2 to be unloaded to the tissues with relatively small decreases in PO_2. In healthy subjects at sea level it is only the upper portion of the O_2 dissociation curve that is used in O_2 unloading; the lower part may be viewed as a reserve that is drawn on only during exercise or in pathological conditions. The oxyhemoglobin dissociation curve is also not fixed; it is changed by several factors, the most important of which for our consideration here are pH and CO (Figure 6.2).

With a normal oxyhemoglobin dissociation curve, 5 ml of O_2 per 100 ml of blood is delivered to the metabolizing tissue at a mixed venous PO_2 of 40 mm Hg (Point V, Figure 6.2). With 50% anemia, where the O_2-carrying capacity is reduced to 10 ml/100 ml (rather than the normal 20 ml/100 ml), 5 ml of O_2 per 100 ml of

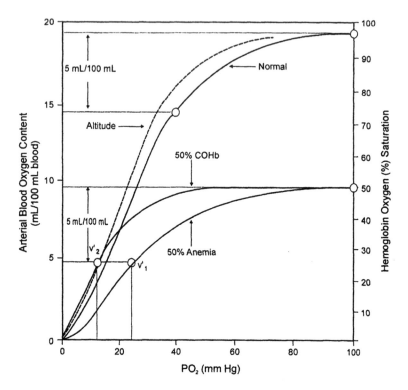

FIGURE 6.2   The oxyhemoglobin dissociation curve of normal human blood, blood at altitude, blood containing 50% COHb, and of blood with 50% hemoglobin caused by anemia.

blood is delivered at a PO$_2$ less than 40 (Point Vl1). With 50% carboxyhemoglobin (COHb), 5 ml of O$_2$ per 100 ml blood is delivered at a PO$_2$ of less than 20 (Point Vl2) because of the leftward shift of the curve and the decreased oxygen-carrying capacity caused by COHb formation. With altitude, 5 ml of O$_2$ per 100 ml blood is delivered at a somewhat lower PO$_2$ than 40 (the PO$_2$ depends on the altitude) because of the leftward shift of the curve caused by altitude-induced respiratory alkalosis.

6.3   EFFECTS OF ASCENT TO ALTITUDE

When an individual ascends to altitude, several meteorological changes are experienced: atmospheric pressure falls and the O$_2$ tension in the ambient air is reduced; it is usually cooler in the summer and extremely cold during the winter; it usually is drier; the solar radiation is more intense and the ultraviolet light is increased. It is, however, the decrease in O$_2$ tension or hypoxia and the ensuing hypoxemia that elicit the most-pronounced physiological response to altitude ascent.

There are numerous population centers at high altitude in the world in which humans are subject to reduced atmospheric pressure and reduced PO$_2$ (Figure 6.1). Atmospheric pressure decreases nearly logarithmically with increasing elevation; at an altitude of 1500 m (5000 ft), the atmospheric pressure is approximately 630 mm Hg, a decrease of about 17% of the standard sea level value. The diurnal variation in atmospheric pressure at a given altitude is normally on the order of 1 to 3%. At Keystone, CO (2840 m; 9300 ft), the average change of atmospheric pressure between good weather and a storm is 20 to 40 mm Hg.[3] This corresponds to an altitude change of 550 to 1000 ft and a maximum change of about 2000 ft.

The reduced P_B and consequently reduced PO_2 experienced at higher altitudes reduce the percentage of hemoglobin saturated with O_2 and reduces O_2 transport to the tissues. However, physiological mechanisms are activated that serve to reduce the tissue hypoxia, so a clear distinction must be drawn between the newly arrived visitor (sojourner) and the resident at altitude. The sojourner will be more hypoxic than the fully adapted resident for several reasons.[4] Initially, the sojourner will exhibit a hypoventilation relative to the adapted resident, particularly during sleep, because ventilatory adaptation requires several days at altitude. Hypoventilation results in a lowering of arterial blood PO_2, O_2 saturation, and O_2 content and, for reasons to be discussed, an increase in COHb from endogenous sources. The ensuing hypoxemia stimulates the sympathetic nervous system and increases heart rate, myocardial contractility, and arterial blood pressure. The resulting increased cardiac work requires an increase in coronary blood flow.

While the ventilatory response of the sojourner at altitude is lower than the fully adapted resident, it is higher than that of the sea level resident. It is this higher-than-normal sea level ventilation or hyperventilation that "blows off" carbon dioxide and reduces the concentration of carbonic acid in the blood; this, in turn, reduces blood acidity and produces respiratory alkalosis. The loss of carbon dioxide and the decreasing hydrogen ion concentration cause a conformational change in the hemoglobin molecule that increases the affinity of hemoglobin for O_2 and diminishes its release to the tissues. This corresponds to a leftward shift of the oxyhemoglobin dissociation curve (Figure 6.2). An extreme example of this phenomenon whereby the respiratory alkalosis accompanying hyperventilation increases the affinity of hemoglobin for O_2 and facilitates loading of O_2 in the pulmonary capillaries is provided by West.[5] The calculated arterial blood pH of two climbers who scaled Mt. Everest (8848 m; 28,028 ft) was between 7.7 and 7.8; the barometric pressure and inspired PO_2 at the summit were, respectively, 253 and 43 mm Hg.

After several days at altitude, a number of compensating mechanisms serve to reduce the initial impact of the hypoxic stress. Ventilation increases progressively; this increases arterial blood O_2 tension, saturation, and content. Plasma volume decreases; this increases the hematocrit ratio (hemoconcentration) and results in an increase in the O_2-carrying capacity of the blood. The reasons for the reduction in plasma volume with ascent to altitude are complex. The stimulus is hypoxia, which appears to result in a decrease in total body water, intracellular water, and plasma volume but without a significant change in extracellular fluid. Confounding factors appear to be excessive fluid loss though the respiratory tract and diuresis, blunted thirst, and inadequate fluid intake at altitude.[3] The polycythemia occurring at this time further increases the arterial blood O_2 content as compared with sea level values.

Although sympathetic activity is increased, cardiac beta-receptor responsiveness decreases, which mitigates the initial tachycardia. These factors combined with a decreased stroke volume cause cardiac output to return to normal (or even subnormal) levels.[4] There is also compensation for the initial respiratory alkalosis and an increase in the concentration of 2,3-diphosphoglycerate within the red cells. These changes return blood pH and hemoglobin affinity for O_2 toward preascent levels and facilitate the release of O_2 to the tissues, an effect that more than offsets the slight decrease in arterial O_2 saturation. These responses are particularly important for the heart because they remove the demands for increased coronary blood flow at moderate altitude.[6]

For the long-term resident adapted to altitude, systemic blood pressure returns to normal or falls below values normal for sea level. Cardiac output remains at (or below) levels normal for sea level. Tissue capillary density may increase, thereby decreasing diffusion distances and enhancing O_2 delivery. Consequently, demands on the coronary circulation are not increased. An absolute polycythemia develops (i.e., total red cell mass attains levels greater than at sea level). As a consequence, the normal turnover of this greater mass of red cells results in an increase in the endogenous production of CO[7] and elevated COHb levels.[8,9]

6.4   EFFECTS OF CARBON MONOXIDE EXPOSURE

The relationship between CO and its effect on oxygen transport, termed the Haldane effect, has been analyzed by Guttierrez.[10] CO, by binding with hemoglobin, not only reduces the total O_2-carrying capacity of hemoglobin, but it also shifts the oxyhemoglobin curve toward the left. The higher the percent COHb, the further the curve is shifted to the left and the less efficient is hemoglobin as an oxygen transporter.[11] This effectively places the steep or unloading part of the curve in the region where PO_2 is below that usually found in tissues (Point VI2, Figure 6.2). The consequence of the leftward shift is that although hemoglobin binds O_2 readily in the lungs, it is released less readily to the tissues; this increases tissue hypoxia. By shifting the curve to the left, the CO effect differs from the effects of anemia produced by reducing blood hemoglobin by 50%

It is for these reasons that the newly arrived visitor to altitude is considered to be at greatest potential risk from CO exposure. By binding with hemoglobin, CO further reduces arterial O_2 content and exacerbates the effect of altitude-induced alkalosis by further increasing the affinity of hemoglobin for O_2; these actions may impair O_2 delivery to an even greater degree.

If the visitor remained long enough at altitude to complete physiological adaption, the initial risk from CO exposure would decline progressively with time. The period of increased risk may be prolonged in the elderly because adaptation to high altitude proceeds more slowly with increasing age[12] and because the rate at which COHb dissociates following CO exposure decreases with age.[13]

6.5   CARBON MONOXIDE SOURCES AT ALTITUDE

CO poisoning is a common problem encountered in a variety of outdoor settings such as tents and cabins heated with woodstoves and other devices and may pose a special danger at altitude. Foutch and Henrichs[14] reported on the fatal CO exposure of two young, healthy mountain climbers who succumbed to fumes generated by a small cookstove in the enclosed space of their tent at 4300 m (14,200 ft) on Mount McKinley in Alaska. Turner et al.[15] measured CO levels in various shelters (tents, igloos, and snow caves) produced by heaters used for cooking and melting snow on a climb of Denali (Mount McKinley). Measurements were made during the climb at between 2000 and 5200 m (6600 and 17,160 ft). Mean CO concentrations in the shelters usually exceeded the 35 ppm 1-h limit set by the EPA, and a mean value of 165 ppm, with a maximum of 190 ppm, was reached in one snow cave.

CO sources, including light-duty vehicles, trucks, and heating devices such as space heaters and fireplaces, at altitude are increasing as humans travel into mountain resort areas. Other sources of CO exposure in mountain recreational communities include hibachi and charcoal grills, other heating and cooking appliances, snowmobiles and other small engines, and camping equipment.

Emissions of CO are much less in newer automobiles and light-duty trucks because they are manufactured to conform to all altitude standards. This has been accomplished by a combination of electronic controls, oxygen and pressure sensors, and the catalytic converter. Although the CO emissions of newer vehicles has decreased dramatically, factors such as tampering with emission control devices, numerous "cold" starts, the age of the vehicle fleet, and the increased numbers of automobiles and sports vehicles being driven in mountain communities all contribute to CO in the ambient air at altitude.

Numerous factors exacerbate ambient CO levels in mountain recreational communities.[16] Older vehicles tuned for mountain driving emit 1.8 times more CO at 2424 m (8000 ft) than at 1600 m (5280 ft) in Denver, whereas such vehicles tuned for sea level driving conditions emit almost four times more CO at altitude. CO emissions from vehicles are increased at altitude by driving at reduced speeds along steep grades under poor driving conditions and by the cold conditions prevailing at altitude. CO emissions from automobiles and light-duty trucks increase dramatically during cold weather. This is because more fuel is needed to start at cold temperatures and because some emission control devices (such as oxygen sensors and catalytic converters) operate less efficiently when they are cold. Because automobiles and light-duty trucks emit more CO and other air pollutants at high altitude, especially older vehicles if they are not tuned for such driving, influxes of tourists into high-altitude resort areas may drastically increase pollution levels in general and CO levels in particular. Finally, geographic factors, such as population growth concentrated along valley floors combined with the reduced mixing of air at altitude as well as the reduced volume of air available for pollutant dispersal, lead to accumulation of pollutants, including CO, in mountain valleys. These factors are all exacerbated in cities at altitude in the developing countries.

As humans trek into mountainous regions throughout the developing world, they are subject to increasing exposures to CO. This is especially true in cities in developing countries such as Mexico with large fleets of older vehicles lacking emission controls (Table 6.1). Fernandez-Bremauntz et al.[18] investigated the CO exposure experienced by street vendors working on busy roads in Mexico City. CO measurements taken at street level in selected avenues in the city center were compared with the concurrent concentrations measured at the nearest fixed-site monitoring station. Short-term street-level CO concentrations ranged from 2.0 to 70 ppm with a mean

TABLE 6.1
Net Mean Carbon Monoxide Concentration Ranges Inside Vehicles

Travel Mode	Washington, D.C. Net Mean[a] CO Conc. (ppm)	Mexico City, Mexico (elev. 2241 m)
Automobile	7–12	37–47
Diesel bus	2–6	14–27
Rail transit	0–3	9–13

[a] Net mean CO concentration = mean in-vehicle CO concentration minus mean ambient CO concentration.

Source: Adapted from U.S. Environmental Protection Agency.[17]

TABLE 6.2
The 8-h Air Quality Standards for Carbon Monoxide in Different Regions

Region	ppm	Agency
United States	9	U.S. EPA
Lake Tahoe Basin	6	CA Air Resources Board
Mexico	13	PAHO
Europe	10	World Health Organization

concentration of 26 ppm. There was a significant positive correlation between street level and fixed-site CO concentrations. CO concentrations measured at the pavement were consistently higher than the concurrent fixed-site monitor levels; the average ratio of street/fixed-site CO concentrations was 2.2. There were more than 1000 street vendors working in the surveyed avenues. More than 80% of the vendors reported that they work at least 6 days a week, with an average working shift of 10 h per day. The results of this study indicate that street vendors, as well as tourists visiting in Mexico City, are exposed to CO concentrations well above national and international air quality standards.

Fernandez-Bremauntz and Ashmore[19] compared measurements of CO made concurrently inside vehicles and at fixed-site monitoring stations in Mexico City. During the study period, ambient CO concentrations were very high, in excess of the U.S. (9 ppm) and the Mexican (13 ppm) 8-h standards for CO (Table 6.2). The in-vehicle concentrations of CO for all modes of transportation were always higher than the concurrent ambient concentrations measured at the fixed-site monitors. Average in-vehicle/ambient CO ratios for each mode of transportation were: automobile, 5.2; minivan, 5.2; minibus, 4.3; bus, 3.1; trolleybus, 3.0; and metro, 2.2. Ambient CO concentrations at selected stations varied from 3 to 28 ppm during peak commuting hours and, depending on the locations, CO levels were increased above ambient in automobiles by 37 to 47 ppm and in minivans by 29 to 52 ppm.

Snowmobiling can be a significant source of CO exposure at altitude. The typical snowmobile is not equipped with pollution control equipment and is powered by a two-stroke engine, which emits high levels of CO. CO emissions from snowmobiles are remarkable; CO emissions from a typical snowmobile ranged from 9.9 g/mile at 10 mph to 19.9 g/mile at 40 mph. Comparable figures for CO emissions from a modern automobile are 0.01 to 0.04 g/mile. During the winter of 1993 to 1994, over 87,000 tourists traveled by snowmobile in Yellowstone National Park[20] where altitudes range from 1600 to 3442 m (5282 to 11,358 ft). Near the West Yellowstone entrance, a point where over 1000 snowmobiles a day enter the park, 1-h air samples exceeded 35 ppm CO in the winter of 1994 to 1995.

Snowmobilers typically travel for several hours in large groups along narrow trails that are frequently located at high altitude. One such trail is the Continental Divide Snowmobile Trail,[21] a 250-mile trail that crosses the Continental Divide at 2909 m (9600 ft). Snook and Davis[20] studied the CO exposure of a snowmobiler while traveling in the wake of a lead snowmobile on a 2 to 3 mile straight trail in Grand Teton National Park, WY. Average exposure measurements taken at distances 25 and 125 ft behind the lead snowmobile were as high as 23 ppm with individual measurements as high as 45 ppm.

6.6 CARDIOVASCULAR EFFECTS OF CARBON MONOXIDE AT ALTITUDE

Earlier studies compared the effects of CO with those of altitude, but there are relatively few studies of the effects of CO at altitude. Forbes et al.[22] reported that CO uptake was increased during light activity at an altitude of 4877 m (16,000 ft). This was shown to be caused by altitude hyperventilation stimulated by a decreased arterial O_2 tension. An increased pulse rate in response to the combined stresses of altitude and CO exposure was reported by Pitts and Pace.[23] The subjects were 10 healthy men who were exposed to simulated altitudes of 2121, 10,000, and 4545 m (7000, 10,000, and 15,000 ft) with COHb levels of 0, 6, and 13%, respectively. The mean pulse rate during the first 5 min after exercise had the greatest correlation with the change in COHb level or inspired PO_2. The authors reported that the response to a 1% increase in COHb level was equivalent to that obtained by elevating a healthy group of men 335 ft in altitude. This relationship was stated only for a range of altitudes from 7000 to 10,000 ft and for increases in COHb up to 13%.

Weiser et al.[24] studied the effects of low-level CO exposure on aerobic work at an altitude of 1810 m (5973 ft). The young subjects inhaled a bolus of 100% CO to achieve COHb levels of 5%. CO impaired work performance at altitude to the same extent as at sea level. Because these subjects were Denver residents and, presumably, fully adapted to altitude, they would have had an arterial O_2 concentration the same as at sea level (about 20 ml O_2/100 ml). Hence, 5% COHb would lower arterial O_2 concentration about the same amount at both altitudes and impair work performance at altitude to the same extent as at sea level. There were small but significant submaximal exercise changes in cardiorespiratory function during CO exposure. The working heart rate increased and the postexercise left ventricular

TABLE 6.3
The Effects of Exposure to Carbon
Monoxide and Altitude on Time to Angina

Condition	Reduction in Time to Angina, %
Sea level	0
Altitude[a]	11
Sea level + CO[b]	9
Altitude + CO	18

[a] 2100 m simulated.
[b] 3.9% COHb.

Source: Data from Kleinman et al.[26]

ejection time shortened, but not to the same extent as when clean air was breathed. Exposure to CO resulted in a lower anaerobic threshold, the energy level at which the exercising subject switches to anaerobic metabolism and lactate accumulates in the bloodstream. CO exposure also caused a greater minute ventilation at work rates higher than the anaerobic threshold.

Environmental exposure to inhaled CO has been reported to increase the risk of coronary artery disease. Sudden cardiac death, a frequent manifestation of coronary artery disease, is usually a result of ventricular dysrhythmia. Leaf and Kleinman[25] investigated the effect of exposure to CO (COHb = 3.9%) at sea level and at a simulated high altitude of 2100 m (6930 ft) on the incidence of cardiac ectopy in subjects with coronary artery disease. In the study, 17 men with documented coronary artery disease and stable angina pectoris performed cardiopulmonary exercise stress tests after random exposure to either CO or clean air at sea level or at a simulated altitude of 2100 m. The percentage of O_2 saturation in each subject's arterial blood was reduced from a baseline level of 98% to approximately 94% after CO or simulated altitude and to approximately 90% after CO at simulated altitude. The average incidence of exercise-induced ventricular ectopy was approximately doubled after all exposures (CO, altitude, CO and altitude), and a significant trend of increased ectopy with decreased O_2 saturation in arterial blood was observed. These investigators concluded that exposure to increased levels of hypoxemia, resulting from hypoxic and/or CO exposures, increased the susceptibility to ventricular ectopy during exercise in individuals with stable angina pectoris; however, this risk was nominal for those without ectopy.

Kleinman et al.[26] reported on the effects of combined exposure to altitude and CO in patients with stable angina who resided at or near sea level (Table 6.3). The subjects performed cardiopulmonary exercise stress tests at sea level or at a simulated altitude of 2100 m (6930 ft) while being exposed to either clean air or CO (3.9% COHb). Compared with sea level, the time to onset of angina was reduced by simulated altitude (11%), by breathing CO at sea level (9%) and by breathing CO at simulated altitude (18%). Other cardiopulmonary parameters were also adversely affected by concomitant CO–altitude exposure. The authors conclude that high altitude exacerbates the effects of exposure to CO in unacclimatized individuals with coronary artery disease.

Song et al.[27] compared the effects of exposure to CO (28.6 mg/m³) in a population of workers at 2300 m (7590 ft) with effects in a control group of local residents. They reported that the same CO concentrations had a greater effect on the workers at altitude than at sea level and the workers were more affected by CO at altitude than the local inhabitants. The symptoms observed included headache, vertigo, fatigue and weakness, memory impairment, insomnia, palpitation, and neurobehavioral function.

6.7 VISION EFFECTS OF CARBON MONOXIDE AT ALTITUDE

There are a number of older studies on the effects of CO at altitude on the visual performance of aviators. Changes in visual sensitivity were reported to occur at COHb concentrations of 5% or at a simulated altitude of approximately 2438 m (8000 ft).[28] Later, McFarland[29] expanded on these original observations and stated that a pilot flying at 1818 m (6000 ft) while breathing 0.005% CO is at an altitude physiologically equivalent to approximately 12,000 ft. McFarland states that the sensitivity of the visual acuity test is such that the effects of CO absorbed from a cigarette are demonstrable. In subjects inhaling smoke from three cigarettes, the saturation of blood with CO was equal to that at an altitude of approximately 7500 ft. The loss of arterial oxygen saturation resulting from the decreased PO_2 in the inspired air was approximately 4%. The absorption of a similar amount of CO at 2273 m (7500 ft) caused a combined loss of visual sensitivity equal to that which occurs at 3030 to 3333 m (10,000 to 11,000 ft). The initial report[28] was confirmed by Halperin et al.[30], who also reported that recovery from the detrimental effects on visual sensitivity lagged behind CO elimination from the blood.

Combined exposure to altitude and CO has been reported to decrease flicker fusion frequency (FFF), i.e., the crucial frequency in cycles per second at which a flickering light appears to be steady.[31] Whereas mild hypoxia, that occurring at 2727 to 3637 m (9000 to 12,000 ft) alone impaired FFF, COHb levels of 5 to 10% decreased the altitude threshold for onset of impairment to 1515 to 1818 m (5000 to 6000 ft).

These studies have been thoroughly reviewed and criticized by Benignus[32] on methodological and statistical grounds and have been characterized as nonreplicable. Nevertheless, the visual effects of breathing CO at high altitude may present a particular hazard in high-performance aircraft.[33] As described earlier, the increased ventilation caused by ascent to altitude could result in an increase in blood pH and a leftward shift of the oxyhemoglobin dissociation curve. Although such a small shift would probably have little or no physiological consequences under normal conditions, it may take on physiological importance for aviators required to fly and perform tedious tasks involving a multitude of cognitive processes under a variety of operational conditions. The leftward shift of the oxyhemoglobin dissociation curve may be further aggravated by preexisting alkalosis caused by hyperventilation resulting from anxiety. The potential for this effect has been observed in studies reporting that respiratory minute volume may be increased by 110% during final landing approaches requiring night vision devices.[34] Thus, the hypoxia-inducing effects of CO inhalation could accentuate the cellular hypoxia caused by stress- and altitude-induced hyperventilation.

6.8   COMPARTMENT SHIFTS OF CARBON MONOXIDE AT ALTITUDE

The possibility that CO may pose a special threat at altitude is suggested by the studies of Luomanmaki and Coburn.[35] They reported that during hypoxia, CO shifts out of the blood and into the tissues of anesthetized dogs. In experiments using [14]CO, they observed that radioactivity did not change when arterial PO_2 tension was increased from 50 to 500 mm Hg; however, when arterial PO_2 was decreased to less than 40 mm Hg, [14]CO activity decreased to levels as low as 50% of control. When arterial PO_2 was returned to normal, the [14]CO reentered the blood. The possibility that the [14]CO had been sequestered in the spleen was excluded by determining that there was no significant difference between splenic and central venous [14]CO radioactivity, either before or after the [14]CO shift.

These workers also studied the shift of CO out of the blood during hypoxia by measuring the rate of increase of blood COHb when CO was administered into the rebreathing system at a constant rate. They reasoned that if the partition of CO between vascular and extravascular stores remained constant, the increase in blood COHb should be proportional to the amount of CO administered. They found that COHb increased at a constant rate up to a saturation of 50%. With additional CO, there was a decrease in the rate at which COHb increased, suggesting that proportionally greater quantities of CO were going into extravascular stores. The rate of COHb buildup became nonlinear at a COHb level of 50%, which corresponds to an arterial PO_2 of 80 mm Hg.[36]

In a study designed to estimate intracellular PO_2 during exercise, Clark and Coburn[37] measured the effects of hypoxia on compartmental shifts of CO between blood and muscle. These studies were conducted in human subjects under conditions where the total-body CO stores remained constant. COHb decreased about 6% during a bout of acute maximum exercise at a PO_2 equivalent to 3300 to 3600 m (11,000 to 12,000 ft).

The shift of CO out of the blood during bouts of hypoxia has been confirmed in studies conducted on both men and women undergoing maximal aerobic capacity tests at altitudes of 55, 1524, 2134, and 3058 m (182, 5030, 7640, and 10,100 ft) and CO concentrations of 0, 50, 100, and 150 ppm.[38] CO shifted into extravascular spaces during maximum work but returned to the vascular space within 5 min after exercise stopped (Figure 6.3). This liberation of CO was related to the concentration of COHb achieved as noted by the regression equation:

$$y = 0.0017 + 0.3047x \qquad (6.1)$$

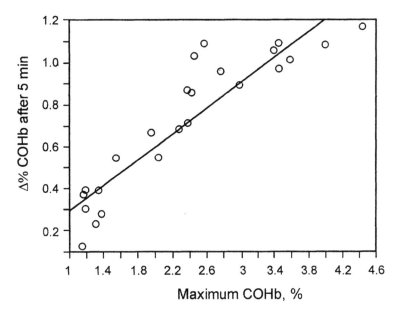

FIGURE 6.3   Relationship between the increase in COHb ($\Delta\%$ COHb) observed at the end of a 5-min recovery period and COHb concentration present at exhaustion after attainment of maximum aerobic capacity (maximum COHb, %).[1]

where x is the COHb concentration at exhaustion with a correlation coefficient, r, of 0.9.

The observations were further extended by Dahms and Baudendistel[39] who investigated whether mixed venous blood with an O_2 tension greater than 40 mm Hg could facilitate loss of CO from the vascular space. Using chronically instrumented goats and a closed rebreathing system, these workers observed that small changes in mixed venous O_2 tension resulted in small changes in tissue CO levels and noted that there is not a threshold value of O_2 below which CO begins to move into the tissue. They concluded that, at any measured COHb concentration, both sojourners and residents at altitude would potentially have a greater risk relative to similar exposure at sea level. Moreover, the effects during exercise would be greater than those observed during rest.

Agostoni et al.[40] presented a theoretical model supporting these observations; they developed a series of equations predicting that with decreased venous PO_2, CO moves out of the vascular compartment and into skeletal and heart muscle, increasing the formation of carboxymyoglobin (COMb) in the tissues.

Thus, the available evidence suggests that during hypoxemia, CO moves into the extravascular compartment causing the COMb/COHb ratio to increase. During heavy smoking (COHb levels of 10%), as much as 30% of the cardiac myoglobin may be saturated with CO.[36] Presumably, this situation would be worsened during altitude exposure because of the attendant arterial hypoxemia.

6.9 ENDOGENOUS PRODUCTION OF CARBON MONOXIDE

A small amount of CO, approximately 0.4 ml/h, is produced endogenously by catabolism of hemoglobin and other hemoproteins.[41] Transport of this endogenously produced CO by the blood results in an average physiological background level of 0.4 to 0.7% COHb at sea level. Certain drugs, chemical exposures, and disease states (e.g., hemolytic anemias) can increase CO production within the body and may cause COHb levels as high as 4 to 6%.[41,42] Typical COHb levels reported for urban nonsmokers are 1.8% for Milwaukee and 2.5% for Chicago. These levels represent a rise above background levels and, presumably, the additional COHb derives from exposure to exogenous CO in the ambient environment. Body stores of CO are influenced by endogenous as well as exogenous CO, and the level of alveolar air CO is directly related to the concentration of COHb in the blood.[43]

6.10 CARBOXYHEMOGLOBIN LEVELS AT ALTITUDE

With ascent to altitude, PO_2 is reduced and, assuming PCO remains constant, the concentration of COHb in the blood is increased. These results are in accord with the predictions of Collier and Goldsmith,[44] who transformed and rearranged the Coburn–Foster–Kane (CFK) equation,[45] and derived an equation expressing COHb in terms of endogenous and exogenous sources of CO.

$$COHb = [FICO (P_B - 47) / 10^6 K] + \underline{V}CO \, Z \, / \, K \qquad (6.2)$$

where

$$
\begin{aligned}
COHb &= COHb \ (\%) \\
FICO &= \text{fraction inspired CO (ppm)} \\
P_B &= \text{barometric pressure (torr)} \\
\underline{V}CO &= \text{rate of CO production (ml/min STPD)} \\
K &= P_C O_2/(M \times SO_2) \\
Z &= 1/D_L CO + (P_B - 47/\underline{V}_A) \\
P_C O_2 &= \text{mean partial pressure of pulmonary capillary } O_2 \text{ (torr)} \\
M &= \text{Haldane constant} \\
SO_2 &= O_2 Hb \ (\%) \\
D_L CO &= \text{carbon monoxide diffusing capacity (ml/min/torr)} \\
(\underline{V}_A) &= \text{alveolar ventilation (ml/min STPD)}
\end{aligned}
$$

The model predicts that FICO will produce a higher COHb at high altitude where P_B and, consequently, $P_C O_2$ and SO_2 are lower than at sea level. According to this relationship, a given partial pressure of CO will result in a higher percent COHb at high altitudes (where PO_2 is reduced). Thus, Collier and Goldsmith[44] calculate that humans breathing 8 ppm CO will have equilibrium COHb levels (Table 6.4) of 1.4% at sea level and 1.6, 1.8, and 1.8%, respectively, at 1530, 3050, and 3660 m (5049,

TABLE 6.4

Calculated Equilibrium Values of Percent Carboxyhemoglobin and Percent Oxyhemoglobin in Humans Exposed to Ambient Carbon Monoxide at Various Altitudes

Ambient CO	Actual Altitude							
	Sea Level		1530 m		3050 m		3660 m	
(ppm)	COHb	O_2Hb	COHb	O_2Hb	COHb	O_2Hb	COHb	O_2Hb
0	0.2	97.3	0.26	93.6	0.35	82.4	0.37	73.3
4	0.8	96.8	0.9	93	1.1	·82.1	1.1	73.1
8	1.4	96.2	1.6	92.5	1.8	81.7	1.8	72.9
12	2.1	95.6	2.3	91.9	2.5	81.3	2.5	72.7
16	2.7	95.1	2.9	91.4	3.2	80.9	3.2	72.5

Notes: The table is for unacclimatized, sedentary individuals at one level of activity (VO_2 = 500 ml/min).

Source: U.S. Environmental Protection Agency.[1]

10,065, and 11,090 ft). Moreover, these workers predict an increase in COHb at altitude even in the absence of inhaled CO (due to endogenous production of CO). COHb levels will be higher at altitude even when FICO is zero because of the relationship between $P_C - O_2$ and SO_2; $P_C - O_2$ decreases in a linear fashion with P_B while SO_2 decreases in a sigmoidal fashion that is described by the oxyhemoglobin dissociation curve. The net effect of changes in these two terms is to increase COHb. However, because diffusing capacity (DLCO) and alveolar ventilation (V_A) both tend to increase with altitude,[46] changes in these two parameters will tend to decrease COHb when FICO is zero.

COHb levels have been shown to increase significantly in humans and laboratory animals at altitude even in the absence of exposure to exogenous CO. In healthy laboratory rats breathing ambient air, COHb levels increased from 0.68% at 1000 m (control) to 1.16 and 1.68%, respectively, at 3030 and 4545 m.[8] In healthy human subjects also breathing ambient air, COHb levels increased from 0.790 and 0.795% (Table 6.5), in males and females, respectively, at College Station, TX (elevation 100 m) to 0.947 and 0.945% after about 20 h at Hoosier Pass, CO (elevation 3497 m).[9] These results indicate that sojourners in high-altitude areas throughout the world (i.e., Denver, Lake Tahoe, Mexico City, La Paz) experience a small increase in body stores of CO that is independent of local CO levels, and suggest that this is a result of the physiological processes by which the body excretes endogenously produced CO. Based on an equivalency of 6 ppm CO to 1% COHb, this effect would be comparable to an exposure to one additional part per million of CO in the atmosphere at Hoosier Pass, CO. These results are also consistent with studies showing elevated rates of CO production in residents of Leadville, CO (elevation 3100 m; 10,230 ft).[7]

CO uptake is also enhanced at altitude in humans breathing CO.[9] In subjects breathing 9 ppm CO for 1 h, COHb increased from 0.78 to 1.464% at 100 m and

TABLE 6.5
COHb (%) Levels in Humans Breathing
Ambient Air at Different Altitudes

Altitude (ft)	Sex	COHb[a] (%)
330	Males (10)	0.790 ± 0.052
330	Females (10)	0.795 ± 0.038
330	Males + females (20)	0.792 ± 0.044
11,540	Males (10)	0.947 ± 0.076[b]
11,540	Males (10)	0.945 ± 0.075[b]
11,540	Males and females (10)	0.946 ± 0.074[b]

[a] Values are mean ± SD.
[b] Significantly different from 330 ft ($p < 0.05$).

TABLE 6.6
COHb (%) Levels in Humans Breathing
Ambient Air or CO at Different Altitudes

Altitude (ft)	Treatment	COHb[a] (%)
330	Air	0.780 ± 0.052
330	9 ppm CO[b]	1.464 ± 0.123[c]
11,540	Air	0.959 ± 0.089[c]
11,540	9 ppm CO[b]	1.982 ± 0.209[c,d,e]

[a] Values are mean ± SD.
[b] 2% COHb.
[c] Significantly different from 330 ft air ($p < 0.05$).
[d] Significantly different from 330 ft CO ($p < 0.05$).
[e] Significantly different from 11,540 ft air ($p < 0.05$).

from 0.959 to 1.982% at 3497 m (Table 6.6). It would be expected that in subjects breathing CO, COHb concentrations would be higher at high altitude than at sea level. These observations are in accord with a model presented by Benignus[47] to predict COHb levels following inhalation of 200 ppm CO at two different barometric pressures and two inhaled O_2 concentrations. In general, the model provides support for COHb being elevated when P_B is reduced at a constant inhaled O_2 concentration and when O_2 is reduced at constant P_B. COHb increased from 7.59 to 8.60%, respectively, when P_B was reduced from 760 to 456 mm Hg and increased from 7.59 to 10.43% when O_2 was reduced from 20.93 to 12%.

Thus, humans sojourning at altitude have an increased body burden of CO and a significantly higher level of endogenously derived COHb. Because of the higher endogenously derived COHb levels, humans breathing CO at altitude will more quickly attain the COHb level of 2% associated with 1-h National Ambient Air

Quality Standard (NAAQS) for CO of 9 ppm. It is expected that for individuals exposed to low levels of CO such as found in urban air pollution in cities at altitude, COHb levels, if measured precisely, will be marginally higher than predicted from studies conducted at sea level. However, these differences will be small compared with the range of COHb levels known to produce adverse health effects at sea level.

6.11 MEASURING CARBON MONOXIDE AT ALTITUDE

Because gaseous air pollutants such as CO are measured in ambient air, changes in atmospheric pressure directly affect their volume. According to Boyles law, if the temperature of a gas is held constant, the volume occupied by the gas varies inversely with the pressure (as pressure decreases, volume increases). The major effect of altitude on CO measurements is that the P_B prevailing at different elevations causes differences in the gas volume measurements. These differences in volume caused by changes in P_B must be considered when measuring CO concentrations.

The two units most commonly used in reporting CO concentrations are parts per million (ppm), a volumetric measurement, and micrograms per cubic meter ($\mu g/m^3$), a gravimetric measurement. When $\mu g/m^3$ units are used, measurements are corrected, for regulatory purposes, to a reference temperature of 25°C and a reference pressure of 760 torr. When ppm units are used, volume corrections are not necessary because the ppm unit is not dependent on pressure effects.

6.11.1 VOLUMETRIC MEASUREMENTS

The measurement of CO in ppm (volume per volume) units is widely used in the air pollution scientific community. It is dimensionless and is numerically equivalent to such ratios as microliters per liter, molecules per million molecules, and micromoles per mole.

Because the ppm unit is a ratio of two volumes, the measurement is independent of pressure and temperature. Both gases (CO and the air sample) closely follow the ideal gas law (Boyle's law), where volume changes proportionately in response to changes in pressure; thus, their ratio remains unchanged. This means that a CO analyzer, properly calibrated for the altitude of use, can measure concentrations in ppm under conditions of varying pressure without error due to the resulting changes in volume.

The primary disadvantage of the use of ppm units is that while the ppm unit does not vary with pressure, the mass of a gaseous pollutant per unit volume does. For example, air containing 1 ppm of CO at a high altitude such as Denver, CO, will still contain 1 ppm of CO if that air is brought down to sea level. However, the mass per unit volume of air will increase with the decreasing altitude.

This relationship can be demonstrated as follows (recall that parts per million can be defined as microliters (μl) of gaseous pollutant per liter (l) of air):

$$ppm = CO\ (\mu l)\ /\ air\ (l) \qquad (6.3)$$

If gas behavior is assumed to be ideal, the ppm concentration unit can be converted into moles of CO per liter of air, a unit useful in considering health effects caused by exposure to air pollution. The concentration of CO in moles per liter of air can be derived from the ppm unit according to Equation 6.4[48]:

$$\text{moles of CO / liter of air} = \text{ppm} \times 1.604 \times 10^{-8} \times P / T \qquad (6.4)$$

where

P = atmospheric pressure at sampling site, torr
T = temperature at sampling site, K

(Note that the reference temperature, 298 K or 25°C is used by the EPA as more appropriate to ambient air measurement conditions than the conventional 273 K or 0°C)

It is evident from Equation 6.4 that constant parts per million contain differing numbers of moles of CO per unit volume of air as pressure (and temperature) changes. For this reason exposure to identical parts per million CO concentration at differing pressures does not expose the receptor to identical numbers of moles of CO per unit volume of air. Although CO concentrations reported in ppm units are not directly affected by altitude differences, the effective mass per unit volume differences should be considered when interpreting data taken from different altitudes. It is important to note that if concentration results, in ppm, are to be converted to moles per liter air or to $\mu g/m^3$, the pressure at which the sample was taken must be known.

6.11.2 GRAVIMETRIC MEASUREMENTS

The mass per cubic meter measurement is used in setting air quality standards primarily for nongaseous pollutants. When CO concentrations are reported in mg/m^3, the volume of air sampled must be corrected to a reference pressure and temperature (temperature effects are not considered in this discussion) if the comparisons of data from different attitudes are to be made. If not, the calculation of the mass of CO per unit volume of air sampled will be a function of pressure, and the concentration of CO obtained will vary with the altitude of the sampling site.

For example, consider a situation in which 40 mg/m^3 of CO is measured without correction at an altitude of 1576 m (5200 ft) (atmospheric pressure of about 625 torr). If a cubic meter of that air is pressurized to 760 torr (sea level reference conditions), it would be compressed to about 80% of its original volume, and the correct CO concentration per cubic meter would be correspondingly higher, according to the following calculation:

$$V_2 = P_1 V_1 / P_2 \qquad (6.5)$$

where

V_2 = volume at reference conditions
P_2 = 760 torr
V_1 = 1 m^3
P_1 = 625 torr

and, therefore,

$$V_2 = 625 \text{ torr} \times 1 \text{ m}^3 / 760 \text{ torr} = 0.82 \text{ m}^3$$

thus, the CO concentration corrected to sea level would be

$$40 \text{ mg} / 0.82 \text{ m}^3 = 48.8 \text{ mg} / \text{m}^3$$

Thus, for purposes of comparison, all CO measurements reported in mg/m³ should be corrected to the reference pressure of 760 torr.

6.11.3 CONVERSION OF UNITS

The conversion factor to convert ppm units of CO to mg/m³ at 25°C and 760 mm Hg is **1.15**. Thus, a CO concentration of 35 ppm is equivalent, approximately, to 40 mg/m³. Conversely, the factor to convert mg/m³ units of CO to ppm is 0.87. Thus, 10 mg/m³ CO is equivalent, approximately, to 9 ppm.

6.12 CONCLUSIONS

The potential effects of exposure to CO at altitude have not received a great deal of study. Available evidence indicates that the effects are dose dependent, not only in terms of the dose of CO, but also the "dose" or level of altitude. A distinction must be made between the sojourner at altitude and the long-term resident with the sojourner being more sensitive to CO. A number of physiological factors combine to produce a greater sensitivity to CO in the sojourner at altitude. These factors include:

1. A leftward shift of the oxyhemoglobin dissociation curve at altitude caused by respiratory alkalosis;
2. A leftward shift of the oxyhemoglobin dissociation curve by the Haldane effect and a reduction in hemoglobin saturation caused by formation of COHb;
3. Higher endogenous COHb levels at altitude caused by hypoxia;
4. A higher CO body burden at altitude at any measured COHb level produced by the movement of CO out of the blood and into the tissue as PO_2 is reduced;
5. A greater uptake rate of ambient CO at altitude.

These physiological changes combined with the emission of higher levels of CO produced by less efficient combustion processes at altitude cause the newcomer to altitude to be at increased risk from exposure to CO.

ACKNOWLEDGMENTS

The author gratefully acknowledges the helpful advice and comments provided by Drs. Gary Hatch, Robert Chapman, and James Raub.

REFERENCES

1. U.S. Environmental Protection Agency, Air Quality Criteria for Carbon Monoxide, EPA/600/8-90/045F, Office of Health and Environmental Assessment, Environmental Criteria and Assessment Office, Research Triangle Park, NC, 1991. (Available from NTIS, Springfield, VA, PB93-167492.)
2. Moore, L.G., Altitude-aggravated illness: examples from pregnancy and prenatal life, *Ann. Emerg. Med.*, 16, 965, 1987.
3. Hultgren, H., *High Altitude Medicine*, Hultgren Publications, Stamford, CA, 1997.
4. Grover, R.F., Weil, J.V., and Reeves, J.T., Cardiovascular adaptation to exercise at high altitude, *Exercise Sport Sci. Rev.*, 14, 269, 1986.
5. West, J.B., The 1988 Stevenson Memorial lecture, physiological responses to severe hypoxia in man, *Can. J. Physiol. Pharmacol.*, 67, 173, 1989.
6. Grover, R.F., Lufschanowski, R., and Alexander, J.K., Alterations in the coronary circulation of man following ascent to 3,100 m altitude, *J. Appl. Physiol.*, 41, 832, 1976.
7. Johnson, R.L., Jr., Rate of Red Cell and Hemoglobin Destruction after Descent from High Altitude, U.S. Air Force School of Medicine, Brooks Air Force Base, TX, 1968.
8. McGrath, J.J., Effects of altitude on endogenous carboxyhemoglobin levels, *J. Toxicol. Environ. Health*, 35, 127, 1992.
9. McGrath, J., Schreck, R., and Lee, P., Carboxyhemoglobin levels in humans: effects of altitude, *Inhalation Toxicol.*, 5, 241, 1993.
10. Guttierrez, G., Carbon monoxide toxicity, in *Air Pollution — Physiological Effects*, McGrath, J.J. and Barnes, C.D., Eds., Academic Press, New York, 1982, 354.
11. Roughton, F.J.W., Transport of oxygen and carbon dioxide, in *Handbook of Physiology*, Vol. 1, Fenn, W.O. and Rahn, H., Eds., American Physiological Society, Washington, D.C., 1964, chap. 31.
12. Robinson, S., Dill, D.B., Ross, J.C., Robinson, R.D., Wagner, J.A., and Tzankoff, S.P., Training and physiological aging in man, *Fed. Proc.*, 32, 1628, 1973.
13. Pace, N., Strajman, E., and Walker, E., Influence of age on carbon monoxide desaturation in man, *Fed. Proc.*, 7, 89, 1948.
14. Foutch, R.G. and Henrichs, W., Carbon monoxide poisoning at high altitudes, *Am. J. Emerg. Med.*, 6, 596, 1988.
15. Turner, W.A., Cohen, M.A., Moore, S., Spengler, J.D., and Hackett, P.H., Carbon monoxide exposure in mountaineers on Denali, *Alaska Med.*, 30, 85, 1988.
16. Kirkpatrick, L.W. and Reeser, W.K., Jr., The air pollution carrying capacities of selected Colorado mountain valley ski communities, *J. Air Pollut. Control Assoc.*, 26, 992, 1976.
17. U.S. Environmental Protection Agency, Air Quality Criteria for Carbon Monoxide [External Review Draft], EPA/600/P-99/001, National Center for Environmental Assessment, Research Triangle Park, NC, 1999.
18. Fernandez-Bremauntz, A.A., Ashmore, M.R., and Merritt, J.Q., A survey of street sellers' exposure to carbon monoxide in Mexico City, *J. Exposure Anal. Environ. Epidemiol.*, 3, 23, 1993.
19. Fernandez-Bremauntz, A.A. and Ashmore, M.R., Exposure of commuters to carbon monoxide in Mexico City II, comparison of in-vehicle and fixed-site concentrations, *J. Exposure Anal. Environ. Epidemiol.*, 5, 497, 1995.
20. Snook, L.M. and Davis, W.T., An investigation of driver exposure to carbon monoxide while traveling in the wake of a snowmobile, paper no. 97-RP143.02, presented at 90th Annual Meeting & Exhibition of the Air & Waste Management Association, Toronto, Ontario, Canada, Air & Waste Management Association, Pittsburgh, PA, 1997.

21. Wilkinson, T., Snowed under, *Natl. Parks*, 69, 32, 1995.
22. Forbes, W.H., Sargent, F., and Roughton, F.J.W., The rate of carbon monoxide uptake by normal men, *Am. J. Physiol.*, 143, 594, 1945.
23. Pitts, G.C. and Pace, N., The effect of blood carboxyhemoglobin concentration on hypoxia tolerance, *Am. J. Physiol.*, 148, 139, 1947.
24. Weiser, P.C., Morrill, C.G., Dickey, D.W., Kurt, T.L., and Cropp, G.J.A., Effects of low-level carbon monoxide exposure on the adaptation of healthy young men to aerobic work at an altitude of 1,610 meters, in *Environmental Stress: Individual Human Adaptations*, Folinsbee, L.J., Wagner, J.A., Borgia, J.F., Drinkwater, B.L., Gliner, J.A., and Bedi, J.F., Eds., Academic Press, New York, 1978, 101.
25. Leaf, D.A. and Kleinman, M.T., Urban ectopy in the mountains: carbon monoxide exposure at high altitude, *Arch. Environ. Health*, 51, 283, 1996.
26. Kleinman, M.T., Leaf, D.A., Kelly, E., Caiozzo, V., Osann, K., and O'Niell, T., Urban angina in the mountains: effects of carbon monoxide and mild hypoxemia on subjects with chronic stable angina, *Arch. Environ. Health*, 53, 388, 1998.
27. Song, C.P. et al., Health effects on workers exposed to low concentration carbon monoxide at high altitude, *Zhonghua Yufang Yixue Zazhi*, 27, 81, 1993.
28. McFarland, R.A., Roughton, F.J.W., Halperin, M.H., and Niven, J., The effects of carbon monoxide and altitude on visual thresholds, *J. Aviat. Med.*, 15, 381, 1944.
29. McFarland, R.A., The effects of exposure to small quantities of carbon monoxide on vision, in *Biological Effects of Carbon Monoxide*, Coburn, R. F., Ed., *Ann. N.Y. Acad. Sci.*, 174, 301, 1970.
30. Halperin, M.H., McFarland, R.A., Niven, J.I., and Roughton, F.J.W., The time course of the effects of carbon monoxide on visual thresholds, *J. Physiol.* (London), 146, 583, 1959.
31. Lilienthal, J.L., Jr. and Fugitt, C.H., The effect of low concentrations of carboxyhemoglobin on the "altitude tolerance" of man, *Am. J. Physiol.*, 145, 359, 1946.
32. Benignus, V.A., Behavioral effects of carbon monoxide exposure: results and mechanisms, in *Carbon Monoxide*, Penney, D.G., Ed., CRC Press, Boca Raton, FL, 1996, 211.
33. Denniston, J.C., Pettyjohn, F.S., Boyter, J.K., Kelliher, J.C., Hiott, B.F., and Piper, C.F., The Interaction of Carbon Monoxide and Altitude on Aviator Performance: Pathophysiology of Exposure to Carbon Monoxide, Rep. no. 78-7, U.S. Army Aeromedical Research Laboratory, Fort Rucker, AL, 1978. (Available from NTIS, Springfield, VA, AD-A055212.)
34. Pettyjohn, F.S., McNeil, R.J., Akers, L.A., and Faber, J.M., Use of Inspiratory Minute Volumes in Evaluation of Rotary and Fixed Wing Pilot Workload, Rep. no. 77-9, U.S. Army Aeromedical Research Laboratory, Fort Rucker, AL, 1977.
35. Luomanmaki, K. and Coburn, R.F., Effects of metabolism and distribution of carbon monoxide on blood and body stores, *Am. J. Physiol.*, 217, 354, 1969.
36. Coburn, R.F., The carbon monoxide body stores, in *Biological Effects of Carbon Monoxide*, Coburn, R.F., Ed., *Ann. N.Y. Acad. Sci.*, 174, 11, 1970.
37. Clark, B.J. and Coburn, R.F., Mean myoglobin oxygen tension during exercise at maximal oxygen uptake, *J. Appl. Physiol.*, 39, 135, 1975.
38. Horvath, S.M., Agnew, J.W., Wagner, J.A., and Bedi, J.F., Maximal Aerobic Capacity at Several Ambient Concentrations of Carbon Monoxide at Several Altitudes, Rep. no. 21, Health Effects Institute, Cambridge, MA, 1988.
39. Dahms, T.E. and Baudendistel, L.J., The effect of hypoxia on the movement of carbon monoxide out of the vascular compartment, paper no. 91-138.5, presented at 84th Annual Meeting and Exhibition of the Air & Waste Management Association, June, Vancouver, British Columbia, Canada, Air & Waste Management Association, Pittsburgh, PA, 1991.

40. Agostoni, A., Stabilini, R., Viggiano, G., Luzzana, M., and Samaja, M., Influence of capillary and tissue PO_2 on carbon monoxide binding to myoglobin: a theoretical evaluation, *Microvasc. Res.*, 20, 81, 1980.
41. Coburn, R.F., Blakemore, W.S., and Forster, R.E., Endogenous carbon monoxide production in man, *J. Clin. Invest.*, 42, 1172, 1963.
42. Stewart, R.D., The effect of carbon monoxide on humans, *J. Occup. Med.*, 18, 304, 1976.
43. Stewart, R.D., Stewart, R.S., Stamm, W., and Seelen, R.P., Rapid estimation of carboxyhemoglobin level in fire fighters, *J. Am. Med. Assoc.*, 235, 390, 1976.
44. Collier, C. and Goldsmith, J., Interactions of carbon monoxide at altitude, *Atmos. Environ.*, 17, 723, 1983.
45. Coburn, R.F., Forster, R.E., and Kane, P.B., Considerations of the physiological variables that determine the blood carboxyhemoglobin concentration in man, *J. Clin. Invest.*, 44, 1899, 1965.
46. Hurtado, A., Animals in high altitude: resident man, in *Handbook of Physiology: Adaptation to the Environment*, American Physiological Society, Washington, D.C., 1964, 843.
47. Benignus, V.A., A model to predict carboxyhemoglobin and pulmonary parameters after exposure to O_2, CO_2, and CO, *Aviat. Space Environ. Med.*, 66, 369, 1995.
48. U.S. Environmental Protection Agency, Altitude as a Factor in Air Pollution, EPA/600/9-78-015, Office of Research and Development, Environmental Criteria and Assessment Office, Research Triangle Park, NC, 1978. (Available from NTIS, Springfield, VA, PB-285645.)

7 Interactions Among Carbon Monoxide, Hydrogen Cyanide, Low Oxygen Hypoxia, Carbon Dioxide, and Inhaled Irritant Gases

David A. Purser

CONTENTS

0-8493-2065-8/00/$0.00+$.50
© 2000 by CRC Press LLC

7.1   INTRODUCTION

7.1.1   GENERAL

An enormous research effort has been applied over many years to understanding the mechanisms and clinical consequences of carbon monoxide (CO) poisoning, yet there are still many unresolved issues in this field. The effects of hypercapnia and low oxygen hypoxia have also been extensively studied, particularly in the context of respiratory disease and also with respect to effects occurring during diving or at altitude. Compared with the effects of these gases, those of exposure to hydrogen cyanide and of inhaled irritants have received relatively little attention, and the effects of exposure to all these gases in mixtures containing a variety of concentration combinations are poorly characterized. The lack of research in this area is surprising based on the knowledge that many thousands of people are injured by exposures to such complex gas mixtures every year and many hundreds die as a result. These people are fire victims. In the United Kingdom the annual number of injuries caused by exposure to smoke and toxic gases from fires is approximately 7000 and the number of deaths is approximately 450.[1]

Interest in this problem was stimulated in the 1970s as a result of concerns with respect to the toxicity of combustion products. In the United Kingdom, studies of fire death and injury statistics revealed a fourfold increase in fire deaths and injuries caused by exposure to toxic smoke since the late 1950s,[2] while research in the United States revealed examples of highly toxic combustion products evolved from certain polymeric materials.[3,4] Previously, almost all fire deaths and injuries not resulting from burns were attributed to CO poisoning, but concerns then arose that other toxic gases might also be important. This led to a variety of research approaches to the toxic and physiological effects of exposure, both to pedigree gases and to combustion product atmospheres containing complex fire effluent mixtures.[5] The primary aim of these studies was to understand the mechanisms of toxicity and to identify materials yielding particularly toxic effluent. Other work involved clinical and pathological studies of fire victims to understand causes of injury and death.[6] More recently, attempts have been made to develop calculation models — known as N-gas or fractional effective dose (FED) models — to predict the effects of exposure to known combinations of toxic gases in combustion product atmospheres.[7,8]

The extent to which this research enables the effects of exposure to mixtures of asphyxiant and irritant gases occurring in fire effluent to be understood and modeled depends very much on the type of study and the study aims. The majority of combustion toxicity studies have involved simple lethal toxic dose exposure studies of rats or mice, usually for 30-min exposure periods.[7,9] The aim of many of these studies has been to rank materials in terms of toxic potency. Other work, such that conducted at the National Institute of Standards and Technology (NIST)[7] and Utah,[3] was aimed at measuring lethal toxic potency of mixed combustion atmospheres from individual materials in rats, but measurements were made of the concentrations of the major toxic gases during the experiments, so that some attempt could be made at identifying the main contributory agents. This was followed by work, mainly at NIST[7] and SWRI,[9] in which rats were exposed to individual asphyxiant and irritant gases, and certain gas mixtures, to measure the lethal toxic exposure doses of each gas to rats, and to determine some possible interactions in gas mixtures. Based upon this work, N-gas and FED models were developed to predict the effects of exposure to toxic gas mixtures in terms of time and dose to lethality for rodents.[7,9-11] The predictions of these models could then be compared with the results of small-scale combustion toxicity tests involving rats to determine the extent to which the lethal exposure doses in the tests could be explained in terms of the models.[8,13]

In the United Kingdom a somewhat different approach was taken.[5] The position taken in there was that all fires are likely to prove fatal to an exposed subject remaining for a sufficient period in the fire environment, due to the effects of heat or toxic gases. What is more important in terms of survival is how quickly an exposure causes incapacitation, preventing the occupant from escaping. Once an occupant is incapacitated, death is very likely to occur. A research program was therefore undertaken using primates (cynomologus monkeys) to examine the physiological effects of exposure to each of the main toxic gases known to occur in fires (CO, HCN, CO_2, and low O_2). Exposure periods of up to 30 min were used over a range of concentrations during which effects on respiratory, cardiovascular, neurophysiological, and behavioral variables were measured and, in particular, times and doses required to cause incapacitation.[8,14] Further work involved sublethal exposures to combustion product atmospheres, to measure exposure times and doses producing incapacitation and to determine the extent that the effects of exposure to combustion product atmospheres could be explained in terms of the effects of the measured concentrations of the major toxic gases present.[5,15,16] Similar work was later undertaken at Southwest Research Institute using baboons.[17] Unfortunately, although both of these programs produced valuable data on the incapacitating exposure doses of individual pedigree gases, and upon the agents responsible for causing incapacitation in fire atmospheres, they were both discontinued before detailed studies could be made of the interactions between components of gas mixtures.

Despite these limitations, it has been possible to develop predictive models for time to incapacitation in humans for gas mixtures typical of fire atmospheres.[8] The models also consider to some extent the effects of metabolic work. These models are based upon the following sources of data:

1. Reasonably detailed data on time and dose to incapacitation for individual gases based upon data obtained from experimental exposure in humans and other primates.
2. A limited amount of experimental data on time and dose to incapacitation for some gas mixtures, such as CO and CO_2, CO_2 and low O_2, CO and low O_2, in humans and other primates.
3. Some experimental data on time and dose to incapacitation in primates for combustion product atmospheres containing mixtures of gases such as CO, HCN, and CO_2.
4. Reasonably detailed experimental data on time and dose to death in rats for the most important gas mixtures, which establishes to some extent the degree of additivity or synergism between individual gases in terms of lethal toxic exposure doses.

Based upon these data, models have been developed that are currently being standardized for application to fire safety engineering and fire hazard analysis. Although these models are physiologically based, they are still derived largely from empirically observed effects and interactions. A true physiological model could be envisaged which would enable predictions of tissue oxygen saturation or rates of oxygen metabolism based upon the known physiology, pharmacokinetics, and metabolism of complex toxic gas mixtures. Such models, although theoretically possible, are yet to be developed.

7.1.2 Effects of Species Differences

Whereas primates might be considered a reasonably good direct model for the physiological effects of exposure to asphyxiant and irritant gases in humans, care needs to be exercised in using results obtained from rodents. Rats are generally more resistant to hypoxia and can survive prolonged exposure to lower oxygen concentrations than humans (rat 30-min LC_{50} 5.4% O_2).[7] Rats are also more tolerant of hypercapnia (30-min LC_{50} 47% CO_2).[7] Also, when restrained rats or mice are exposed to respiratory irritants, they show a marked sensory irritant reflex decrease in breathing rate and minute volume, which may decrease the rate of uptake of asphyxiant gases present in some situations.[8,18] In humans and other primates, a brief period of respiratory rate depressions is followed by an increase in respiration resulting from pulmonary irritancy.[5,8] However, the evidence suggests that the increased respiration does not obviously increase the rate of CO uptake in primates, possibly due to ventilation/perfusion distribution changes during the inhalation of irritant gases.[8] In rodents, the small body size and higher metabolic rate result in a much greater volume of air inhaled per gram body weight (or per milliliter of blood) than in large animals and humans, so that the rate of uptake of gases such as CO is much more rapid in rodents and equilibrium is reached much more quickly.

Despite these differences, it is considered in general that rat LC_{50} data for mixed combustion product atmospheres derived from materials decomposed in small-scale combustion toxicity tests provide a reasonable indication of likely lethal toxic potency to humans, and that physiological interactions between different asphyxiant gases in mixtures should be reasonably indicative of likely effects in humans, provided allowances are made for the known species differences.

7.1.3 FIRE ATMOSPHERES

An important consideration with respect to modeling toxic and physiological effects in occupants of buildings during fires is the range of concentrations of different asphyxiant gases likely to be encountered. This depends upon the position of the occupant relative to the fire, the chemical composition of the fuel, the combustion conditions in the fire, and the rate of fire development. Some basic points are considered here to put limits on the concentrations of individual gases and typical concentration ratios in gas mixtures likely to be encountered in typical fire scenarios.

In terms of the fire chemistry, the mixture of asphyxiant gases produced, the basic scenarios confronting building occupants, and the hazard development, fires can be classified into three basic types[8]:

1. Smoldering/nonflaming fires;
2. Early, well-ventilated fires;
3. Ventilation-controlled flaming fires:
   a. Small, vitiated flaming fires and
   b. Postflashover, ventilation-controlled fires.

1. Smoldering/Nonflaming Fires

These occur when an object is exposed to a heat source and is heated sufficiently to decompose. Examples might include a lighted cigarette dropped onto a sofa or when a chair or garment is placed too close to a heater. Fires often start this way and, providing flaming does not occur, the development of the hazard is slow. The main hazard is to sleeping occupants of small enclosures such as domestic rooms or houses. Such fires produce high yields of partially oxidized organic compounds (approximately 50% of the mass decomposed) which are highly irritating to the respiratory tract. Inorganic acids as gases provide a further source of irritants if the appropriate elements (halogens, sulfur, nitrogen, or phosphorus) are present in the decomposing materials.

During smoldering/nonflaming fires the O_2 concentration in the fire room usually remains near ambient and CO_2 is less than 0.2%. Most nitrogen-containing materials do not produce significant amounts of HCN under such conditions. The main asphyxiant agent is therefore CO, which may be present at concentrations in the approximate range 100 to 2000 ppm for several hours. The other major hazard is exposure to pulmonary irritants. There may be some limited degree of interaction between the respiratory effects of exposure to irritants and the rate of uptake and toxicity of CO.

2. Early, Well-Ventilated Fires

These fires occur when an object burns with a flame, when the fire is supplied with plentiful fresh air, and when the fire is small compared with the size of the enclosure in which it burns. Under these conditions, the ratio of fuel to air is low so that combustion is most efficient. For most non-fire-retarded materials, the main products are CO_2, water, and heat, and the yields of smoke and toxic products tend to be low initially. As the fire develops, ventilation tends to become more restricted compared with the size of the fire and a more hazardous mixture of products may be formed. CO and CO_2 can be significant toxic products, and many inorganic products may be released as acid gases. Some materials (particularly if treated with some fire retardants) are unable to burn efficiently, producing high yields of CO and organic products even under well-ventilated conditions. As the fire grows and uses up the available oxygen inside a building, it tends to become ventilation controlled.

For a fire to remain well ventilated, the O_2 concentration in the air entering the flames cannot be less than 15%. The rising fire plume entrains large volumes of air and is usually considerably diluted by the time it reaches the breathing zone of a room occupant. The gas mixture to which building occupants are exposed under these conditions would contain up to approximately 5000 ppm CO, 5% CO_2, and 15% O_2 (approximately 5% O_2 depletion). If the fuel contains nitrogen, then up to 500 ppm NO_x and 100 ppm HCN may be present. Other acid gases may also be present, usually at concentrations below approximately 1000 ppm, depending upon the composition of the burning materials. Under such conditions, the most important asphyxiant is likely to be CO, but the hyperventilatory stimulus from CO_2 can have a significant effect on rates of CO uptake.

3. Ventilation-Controlled Flaming Fires

During the period when most occupants are at risk, the majority of fires in small enclosed buildings (such as houses, apartments, hotels, boarding houses, cellular offices, and small shops) are likely to be restricted-ventilation (vitiated) preflashover fires. However, if external windows or doors are open, or (as often happens) are opened at a later stage of the event, then flashover may occur, where all surfaces are ignited in high-temperature (often as high as 1000°C) conflagrations. In either case the air supply is restricted compared with the fuel available for combustion. Most fires in buildings become ventilation controlled after a few minutes. Ventilation-controlled fires, both pre- and postflashover are the main threat to building occupants. The restricted ventilation results in high yields of CO, CO_2, hydrogen cyanide, organic products, smoke, and inorganic acid gases. Ventilation-controlled fires therefore tend to be a worst case for toxicity, since they produce large amounts of effluent containing high yields of toxic products.

By the time fires become ventilation restricted, the effluent from the fire tends to be less diluted than at the well-ventilated stage, rapidly filling building enclosures, although there tends to be more dilution further away from the seat of the fire. The effluent atmosphere to which a building occupant is exposed can contain up to 5%

CO (typically 5000 ppm 5% CO) so that victims almost always have high blood carboxyhemoglobin (COHb) concentrations. However, O_2 concentrations in the undiluted plume can be as low as 1% (typically 1 to 12%) and CO_2 concentrations as high as 15% (typically 5 to 15%). In mixed fuel materials typical of furnished buildings, the nitrogen content may be sufficient to provide HCN concentrations of up to 3000 ppm, since vitiated combustion conditions favor the formation of HCN rather than NO_x. High concentrations of organic and inorganic irritants are also likely to be present.

People may therefore be exposed to different combinations of asphyxiant and irritant gases at different relative concentrations depending upon the fire conditions, their proximity to the fire, and the nature of the burning materials.

7.1.4 CASE STUDIES

Indirect evidence for interactions between fire gases can be obtained from pathology studies of fire victims. A number of data sets exist for COHb concentrations in fire victims, and in nonfire victims of CO poisoning.[6,19] Most fire victims dying from exposure to toxic smoke (rather than from burns) exhibit substantial COHb concentrations, and in general the distribution of COHb concentrations in subjects considered to have died as a result of CO poisoning is similar to that of toxic smoke victims.[19] It therefore seems likely that for many fire victims, CO inhalation is the major cause of death. However, there are differences in the distributions in that a larger proportion of fire fatalities occur at the lower end of the COHb concentration distribution (<50% COHb), while many more nonfire fatalities achieve concentrations above 80% COHb. These differences have been interpreted in a number of ways, but one possibility is that the differences reflect the contribution to lethality made by other toxicants as would be predicted from a knowledge of the composition of fire atmospheres and the results of the animal toxicity studies. A difficulty with this kind of study is that other asphyxiant fire gases such as CO_2 and low reduced O_2 do not leave stable residues in the body as does CO in the form of COHb. The other major candidate, HCN, does leave cyanide in the blood and body tissues, but the residues are very unstable, so that cyanide concentrations measured in tissues some time after exposure provide a poor indication of the concentrations present at the time of death. In a study where blood samples were taken from fire victims at the fire scene during the fire incidents, very high blood cyanide concentrations were obtained.[20]

Another difficulty with pathology studies is that they do not provide information on the causes of incapacitation in fires. Based upon the animal studies, it is quite possible for incapacitation to be caused principally by exposure to HCN in some fires and for the subject to then continue to load CO while unconscious until death occurs (see Section 7.2.3.2). Also, based upon experiments in human volunteers and laboratory animals, and upon case reports, it is likely that escape attempts can be considerably inhibited or slowed by exposure to optically dense and irritant smoke,[8] with the result that occupants become trapped and then overcome by asphyxiant gases.

7.2 PHYSIOLOGICAL EFFECTS OF EXPOSURE TO INDIVIDUAL TOXIC GASES AND INTERACTIONS WITH CARBON MONOXIDE

7.2.1 GENERAL

Although data on the concentration/time/dose relationships of the incapacitating and lethal asphyxiant effects of individual gases in humans are limited, they are adequate for the development of usable predictive incapacitation models. Information on interactions is also limited, but it is possible to predict likely degrees of interaction based upon physiological data for individual gases and such experimental data for gas combinations as do exist. In the following sections, the basic effects of each gas alone, its interaction with CO, and its interactions with other gases are considered in terms of physiological mechanisms. Effects are considered in terms of relationships between exposure concentration and time to incapacitation for humans and the effects on lethality in rodents. Also considered are exposure doses (concentration to which a subject is exposed multiplied by exposure time) causing incapacitation and the relationship between exposure concentrations and incapacitating exposure doses.

To predict the effects of gas mixtures containing CO on time and exposure dose to incapacitation in humans, or exposure dose for lethality in rats, expressions are needed to predict the effects of CO alone. A number of models with varying degrees of sophistication have been derived from human experimental data, for predicting time to attain a particular COHb concentration. The effects of CO and its interactions considered here represent emergency situations such as fires, where the inhaled CO concentration is well in excess of equilibrium with the COHb concentration. In such cases the departure from a linear uptake rate is not great for a constant CO concentration, so that a reasonable prediction of COHb concentration can be obtained from the Stewart equation:[21]

$$\%COHb = (33.7 \times 10^{-5})\ (ppm\ CO)^{1.036}\ (RMV)\ (t)$$

where

ppm CO = CO concentration (ppm)
RMV   = minute volume (l/min)
t = exposure time (min.)

To predict the time to incapacitation, the %COHb achieved after a time t is expressed as a fraction of the %COHb causing incapacitation. This constitutes the FED. As a default condition, it is assumed that subjects are engaged in light activity (minute volume 25 l/min) and that incapacitation (loss of consciousness) occurs when a COHb concentration of 30% is achieved. Loss of consciousness was observed to occur in actively moving primates carrying out an operant conditioning paradigm at 30% COHb.[8,21] The FED for incapacitation (FED_{Ico}) at time t is then given by

$$FED_{Ico} = (8.2925 \times 10^{-4} \times ppm\ CO^{1.036}) \times t/30 \tag{7.1}$$

The FED can then be correlated with the FEDs for other gases depending upon the nature of the toxic interactions between CO and the other gases present.

7.2.2 EFFECTS OF EXPOSURE TO CARBON DIOXIDE AND INTERACTIONS WITH CARBON MONOXIDE, HCN, LOW OXYGEN HYPOXIA, AND IRRITANTS

7.2.2.1 Hyperventilation

The most obvious effect of inhaled CO_2 is a hyperventilatory drive and respiratory acidosis. The hyperventilatory drive is important because it increases the rate of uptake of other asphyxiant gases, particularly CO and HCN. For a given exposure concentration of either or both of these gases, the presence of CO_2 should decrease the time to incapacitation, or the time required for the uptake of a lethal dose in proportion to the increase in ventilation. This should occur for subjects at rest, and should be even more marked for exercising subjects.

The ventilatory response to CO_2 varies among individuals and there are some variations in reported data.[8] To estimate the extent of the effect of inhaled CO_2 on the inhalation of other gases, an average curve for the ventilatory response to inhaled CO_2 has been constructed from data given in four sources.[23–26] A regression curve has been fitted to these data[8] and, based upon an assumed resting minute volume of 6.8 l/min, a curve has been constructed of the extent to which the resting minute volume is multiplied at different inhaled %CO_2 concentrations (Figure 7.1). A simplified expression for this multiplication factor (VCO_2) has been derived as follows:

$$VCO_2 = [CO_2]/4 \ \text{l/min} \tag{7.2}$$

where $[CO_2] = \%CO_2$.

Although this provides a reasonable estimate of the ventilatory response to inhaled CO_2, the increase in the rate of uptake of CO resulting from this hyperventilation will be somewhat less than the increase in minute volume, due to the inefficiencies resulting from the effects of dead space, CO diffusion capacity, and other factors. In practice, the efficiency of uptake decreases as ventilation increases. The Coburn–Forster–Kane equation[27,28] for CO uptake takes these factors into account, and provides somewhat lower prediction for the increase in uptake rate of CO and other gases with increased minute volume. This relationship has been used to derive a modified uptake expression and also a slightly higher value (7.1 l/min) has been used for the resting minute volume. Based upon these considerations, the modified expression for the multiplicatory effect of increases in inhaled %CO_2 above ambient is as follows:

$$VCO_2 = [CO_2]/5 \ \text{l/min} \tag{7.3}$$

This expression predicts an approximate doubling of the rate of uptake of CO at 4% inhaled CO_2 and an increase by approximately a factor of 3 at 6% inhaled CO_2.

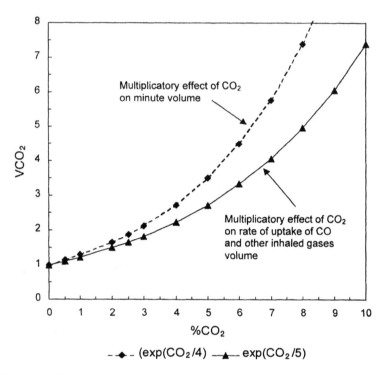

FIGURE 7.1   Effect of inhaled CO_2 concentration on minute volume (data from References 23 to 26) and rate of uptake of CO and other inhaled gases (previous data modified by uptake efficiency factor taken from the Coburn–Forster–Kane equation[27,28]). (From Purser, D.A., Hazards to life in fire, *Hartford Environmental Research*, 1999. With permission.)

7.2.2.2 Toxic Effects of Hypercapnia

At normal O_2 concentrations, very high inhaled concentrations of CO_2 cause respiratory stimulation, distress, acidosis, headache, decrement in cognitive and psychomotor ability, increased circulating catecholamines, and loss of consciousness. Up to concentrations of 5%, CO_2 is well tolerated, although from approximately 3 to 6% there is gradually increasing respiratory distress. This becomes severe at approximately 5 to 6% with clinical comments from subjects such as "breathing fails to satisfy intense longing for air" or "much discomfort, severe symptoms impending" with headache and vomiting also occurring.[28] Due to the gradual equilibrium process these signs tend to worsen during exposure but they are unlikely to be severe over periods of 30 or even 60 min at inhaled concentrations of 5 to 6%. Exposure to 4% can be tolerated for days although above 6.5% detectable slowing of mental function occurs after approximately 20 min.[29] In some experiments subjects have tolerated 7% CO_2 in normal oxygen for 30 to 40 min,[30] but stupor and loss of consciousness have been reported to occur within a few minutes at concentrations between 7 and 10%.[26,31] Loss of consciousness is likely within 2 min at 10% inhaled CO_2 in humans.[26] Intoxication by CO_2 dose not follow Haber's rule, in that a short exposure to a high concentration causes rapid intoxication while tolerance is much prolonged at lower concentrations.[8,26] An exposure dose for incapacitation is therefore not constant ($C \times t \neq K$) (Ct for 10% = 20%·min, Ct for 5% = 175%·min). An expression for predicting approximate time to incapacitation for humans (t_{Ico2}) has been estimated as follows[8]:

$$t_{Ico2} = \exp (6.16 - 0.52 \times \% \, CO_2) \tag{7.4}$$

Levin conducted experiments on a wide range of individual gases and gas mixtures in rats and has reported them to be very resistant to toxicity by CO_2, with a 30-min LC_{50} concentration of 47% CO_2. No deaths occurred in rats exposed to 26% CO_2 for 30 min.[12]

On the basis of these experiments in humans and rodents, it is considered that in mixed atmospheres of toxic gases the main effect of CO_2 will be to increase the rate of uptake of other toxic gases, but above approximately 6.5% CO_2 the direct toxic effects are likely to become increasingly important. These considerations are important because in fire effluent the CO_2 concentration can be up to 15%, although the effluent is likely to be diluted somewhat before reaching the breathing zone of a building occupant. CO_2 is also used in certain inert gas fire-extinguishing systems which may be released in occupied spaces. When discharged, these produce CO_2 concentrations of around 5%, with O_2 concentrations of around 12%. During fires these gases are mixed with the CO_2 and other gases produced by the fire.

7.2.2.3 Effects of Carbon Dioxide on Low Oxygen Hypoxia

In both fires and releases of inert gas extinguishing systems CO_2 is always present with reduced O_2 concentrations. Inhaled CO_2 has long been known to have beneficial effects during low O_2 hypoxia so that time to incapacitation is increased and the degree of incapacitation at different maintained oxygen concentrations is reduced.[32,33]

The deleterious effects of inhaling low O_2 mixtures result partly from effects on blood pH and PCO_2. A reduction in arterial O_2 concentration stimulates respiration and increases cerebral blood flow, but this results in a reduction in blood CO_2 concentration (PCO_2). Since CO_2 itself has a powerful effect on both respiration and cerebral blood flow, the reduced PCO_2 and resultant alkalosis mask the peripheral chemoreceptor respiratory stimulation by hypoxia and also reduce the respiratory drive by respiratory centers in the brain. The lowered respiration and cerebral blood flow result in an increased hypoxic effect.

The hyperventilatory effect of CO_2 increases the rate of oxygen uptake and there is also a benefit from the rightward shift in the O_2 dissociation curve caused by CO_2. This improves the delivery of oxygen to the tissue, counteracting the respiratory alkalosis that otherwise occurs.[32,33] The beneficial effects of exposure to 5% CO_2 during exposures to 10 to 12% O_2 for periods of up to an hour have been demonstrated in experiments on human volunteers.[34] Six male human volunteers were exposed for 20 min to an atmosphere produced by mixing an inert gas mixture containing CO_2 with air to obtain an O_2 concentration of 12 to 13% and a CO_2 concentration of 3 to 4%. Minute respiratory volume was doubled, but there were no adverse clinical signs, and tests of cognitive function indicated that the subjects remained alert during exposure. At even lower O_2 concentrations, 5% CO_2 has been shown to prolong time of useful consciousness at O_2 concentrations of 10, 8, 6, 4, and even 2%.[32]

In rats, when effects on lethality were studied, CO_2 was found to increase the toxic effects of hypoxia. Thus, the 30-min LC_{50} for rats exposure to low O_2 alone was found to be 5.4% O_2, while the addition of 5% CO_2 gave an LC_{50} concentration of 6.4% O_2.[12]

7.2.2.4   Effect of Carbon Dioxide on Carbon Monoxide Toxicity

When considering time to incapacitation (loss of consciousness) in humans, it is considered that the main effect of CO_2 will be to increase the rate of uptake of CO and thus reduce time to loss of consciousness on a pro rata basis. Compared with this, it is likely that other interactions will have relatively minor effects. It is possible that the presence of increased CO_2 may somewhat counteract the leftward shift of the O_2 dissociation curve caused by CO so that deleterious effects on oxygen delivery to the tissues may be somewhat reduced.[35] On the other hand, the combination of a respiratory acidosis induced by CO_2 with a metabolic acidosis induced by CO may have some deleterious effect.

The latter certainly seems to be the case when exposures are continued to lethal levels in rats. The effect on the uptake and toxicity of CO has been confirmed in rats by Levin et al.,[7,12] where exposure to CO_2 not only increased the rate of uptake of CO but also caused a greater lethality for a given COHb concentration. In these experiments the synergistic effects of CO_2 on CO lethality increased up to 5% CO_2. Above 5% CO_2 the toxicity of CO reverted back to the toxicity of CO itself. During exposure to 5% CO_2 with CO the rate of uptake of CO was increased during the first 15 min of a 30-min exposure period, at which point the rats became comatose. Deaths occurred both during and some time after exposure when CO_2 + CO mixtures were used, but during exposure only when CO alone was used. During these experiments the degree of acidosis induced by 5% CO_2 and 2500 ppm was considerably greater than for 2500 ppm CO alone (Figure 7.2).[36] Recovery from this acidosis required more than 90 min in surviving rats. It is likely that the increased mortality resulted from a combination of prolonged tissue hypoxia with an increased metabolic and respiratory acidosis, although it is not clear why the synergism decreased at higher CO_2 concentrations.

To model the lethal effects in rats of CO_2 in combination with CO and other gases, Levin et al.[7,12] developed terms for input into their N-gas equation. These terms m and b represent the slope and intercept of the line for the relationship between CO_2 concentration and lethal CO concentrations for a 30-min exposure period. These are –18 and 122,000, respectively, for CO_2 concentrations of 5% or less, or 23 and –38,600 for CO_2 concentrations above 5%. It is assumed that CO_2 had no effect on the toxicity of other inhaled gases.

Based upon the same data, Purser[8] developed a somewhat different approach. At 5% CO_2 it was assumed that the enhanced lethality was partly due to the increased rate of uptake of CO and partly due to acidosis. A multiplicatory factor was developed for the hyperventilatory effect of CO_2 on the uptake of CO (and any other toxic gases present) and an additive term for the acidosis effect as follows:

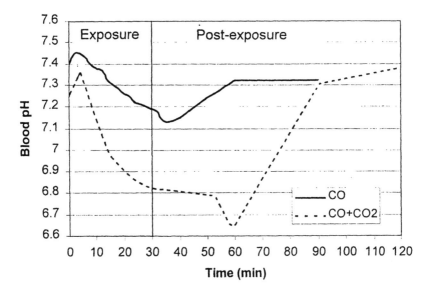

FIGURE 7.2   Changes in blood pH in rats during and after exposure to 2500 ppm CO alone and 2500 ppm CO + 5% CO_2. (From Levin, B. et al., Rep. NBSIR 88-3753, U.S. National Bureau of Standards, Gaithersburg, MD, 1988, 368.)

$$\text{Multiplicatory } CO_2 \text{ factor for rat lethality} = 1 + \{\exp(0.14 \times [CO_2]) - 1\}/2 \quad (7.5)$$

$$\text{Additive acidosis factor for rat lethality} = [CO_2] \times 0.05 \quad (7.6)$$

At 5% CO_2 the effect is to increase the inhaled CO dose by 50% with an additive acidosis factor of 0.25. Thus, for rats with an LC_{50} concentration for 30 min of 5700 ppm, 50% lethality is predicted following a 30-min exposure to half this concentration.

7.2.2.5   Effects of Carbon Dioxide on Cyanide Toxicity

In the absence of experimental data, it is considered that the effects of CO_2 on HCN toxicity will be similar to those of CO_2 on CO toxicity, particularly in terms of rate of uptake and time to incapacitation.

7.2.2.6   Effects of Carbon Dioxide on Toxicity of Inhaled Irritants

When rats and mice are exposed to irritants, the reflex depression of breathing tends to decrease the rate of uptake of other toxic gases.[8,18] If this respiratory depression is counteracted to some degree by exposure to CO_2, then an enhanced toxicity is likely to occur. This has been found experimentally when rats inhaled irritant acid gases in the presence of CO_2, with increased deaths possibly caused by postexposure acidosis and increased lung damage.[9,12,36] Exercise also causes a CO_2-driven hyperventilation, and there is evidence that this may also cause death when rodents are exposed to irritants at normally sublethal concentrations. This is of particular concern with respect to occupants attempting to escape from fires, and fire victims often suffer from postexposure lung inflammation and edema, which is sometimes fatal.

7.2.3 TOXICITY OF CYANIDE AND INTERACTIONS BETWEEN CARBON MONOXIDE AND CYANIDE

7.2.3.1 Toxicity of Hydrogen Cyanide

HCN is commonly encountered with CO in fire atmospheres so that an understanding of its direct effects and interactions with CO are important. Like CO, cyanide causes a chemical asphyxia, the main target organs being the brain and heart. This results in loss of consciousness with respiratory and cerebral depression, arrhythmias, and ECG changes, somewhat similar to those induced by CO.[5,14–16] The pattern of toxicity during the early stages of intoxication is very different from that induced by CO. The physiological changes occuring during a 30-min exposure of a primate are shown in Figure 7.3. After a few minutes of exposure, depending upon the exposure concentration, there is a marked period of hyperventilation, during which the minute volume is increased by a factor of up to 4. Within 1 to 5 min of this hyperventilatory episode loss of consciousness occurs, accompanied by a significant decrease in heart rate with arrhythmia, T-wave abnormalities characteristic of cardiac hypoxia, and EEG signs of severe cerebral depression. The minute volume then decreases to below-normal levels and there may be a partial recovery of consciousness followed by a slow decline if exposure continues. Once exposure ceases, recovery occurs within a few minutes.

Another important aspect of the effects of CO and HCN is the difference between the time–concentration–effect relationships for these two gases. This is illustrated in Figure 7.4. It demonstrates the relationship between time to incapacitation and concentration for the two gases. In these experiments, when active animals were exposed to CO, incapacitation (loss of consciousness) occurred at approximately 30% COHb. Since CO was present in the inhaled atmosphere at well above equilibrium levels, uptake was approximately linear, so that incapacitation occurred at a constant exposure dose. In this case approximately 27,000 ppm·min. It did not make any difference whether a high concentration of CO was inhaled for a short time or an equivalent low concentration for a long time. Incapacitation (loss of consciousness) occurred when an exposure dose of 27,000 ppm·min was achieved, as shown in the smooth curve illustrated. The relationship therefore follows Haber's rule, which states that time and concentration are equivalent. For HCN the curve has a very different shape. At low concentrations of HCN (up to around 100 ppm) time to incapacitation is long (>20 min), but at higher concentrations (approaching 200 ppm) the time to incapacitation is very short, around 2 min. This means that while exposure to low cyanide concentrations is relatively safe, at around 200 ppm and above incapacitation occurs very rapidly. HCN is therefore a very rapid "knock-down" gas in fires. Of course, this effect does not show in cadavers after the event.

From the primate exposure data an expression has been developed for the prediction of time to incapacitation (loss of consciousness) at different inhaled HCN

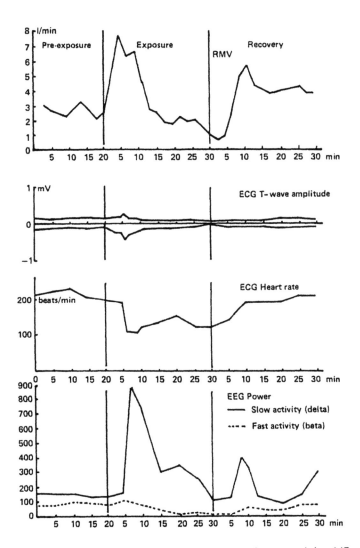

FIGURE 7.3 Physiological effects of exposure to an atmosphere containing 147 ppm HCN in the cynomolgus monkey. (From Purser, D.A. et al., *Arch. Environ. Health*, 39, 394, 1984.)

concentrations. It is considered that this might be somewhat conservative for direct comparison with a resting adult human but could be reasonably predictive of time to incapacitation in children or exercising adults.

For a constant HCN concentration time to incapacitation (t_{Icn}) is given by

$$t_{Icn} = 220 \, / \, \exp \, ([CN] \, / \, 43) \tag{7.7}$$

so that

$$FED_{Icn} = \{ \exp \, ([CN])43) \} \, t/220 \tag{7.8}$$

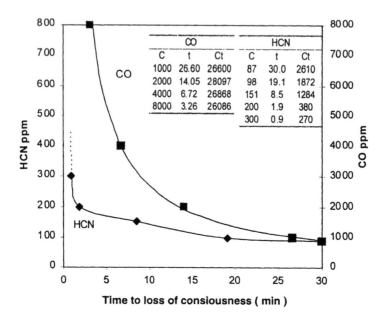

FIGURE 7.4 Relationship between time to incapacitation and concentration for HCN and CO exposures in primates. Time and concentration are equivalent for CO; for HCN a small increase in concentration causes a large decrease in time to incapacitation. (From Purser, D.A., Hazards to life in fire, *Hartford Environmental Research*, 1999. With permission.)

Despite the differences between exposure doses producing incapacitation at different HCN concentrations, for simple rat lethality estimations, a 30-min exposure time is assumed and the LC_{50} concentration is 165 ppm. In rats, less-marked differences in exposure doses for incapacitation and lethality were observed at different exposure concentrations.[7,12]

7.2.3.2 Interactions between Carbon Monoxide and Cyanide

Although both CO and HCN cause tissue hypoxia, it is not to be expected that much of an additive or synergistic interaction would occur between these gases. This is because the primary mechanism of action of CO is usually considered to be an impairment of the transport and delivery of O_2 to the tissue, while the primary mechanism of action of cyanide is impairment of the use of O_2 at the cellular level. It is therefore to be expected that either one or the other gas would determine the rate-limiting step in O_2 supply and utilization.[14] When animals are exposed to HCN, there is usually excess O_2 in the venous blood. However, the results of experimental exposures in rats to CO and HCN mixtures show some degree of additivity,[7] and experiments with primates have shown that time to incapacitation by HCN is reduced by the presence of near toxic concentrations of CO.[15] There are several reasons this might be the case. One reason is the hyperventilation resulting from HCN exposure (Figure 7.3), which further increases the rate of uptake of cyanide and of any CO present. Another possible reason derives from a proposed mechanism of CO toxicity. It has been suggested that CO may exert direct tissue toxicity in addition to its effects on the blood. Particularly when the uptake of CO is rapid, it has been suggested that sufficient dissolved CO may reach the tissues to exert direct toxic effects. *In vitro*, CO has been shown to exert an inhibitory effect on O_2 metabolism. Although the affinity of cytochrome oxidase for CO is considerably less than that of cyanide, it is possible that some degree of additive inhibition of O_2 metabolism may occur at the tissue level.

Whatever the reason for the observed effects, it is therefore safest to assume that these gases are directly additive in terms of exposure doses for incapacitation and death, so that an end point will be reached when the fractions of the toxic doses for each individual gas add up to unity.

The effects in an accidental exposure will therefore depend to some extent on the relative concentrations of the two gases present. In flaming domestic fires, typical HCN concentrations range from 0 to 3000 ppm, while CO concentrations may range from 1000 to as high as 50,000 ppm. When the nitrogen content of the fuel is low, so that the HCN concentration remains below approximately 100 ppm and the fire becomes sufficiently vitiated to produce CO in excess of 5000 ppm, then the main toxic effects are likely to be due to CO, with some minor additive contribution from HCN. When the nitrogen content of the fire is higher, such as with vitiated fires involving upholstered furniture, then HCN often exceeds 200 ppm when CO is less than 5000 ppm. In this case, incapacitation is predicted from HCN before a large dose of CO can be inhaled. Once a subject has become unconscious as a result of the effects of cyanide, breathing continues for some time, during which there is a continued uptake of CO until death from asphyxia occurs. It is quite possible in this situation for a subject to achieve high or even fatal COHb concentrations before death even though the initial collapse might have been caused by HCN exposures. This is illustrated in Figure 7.5, which shows the FEDs for HCN and CO in two fire experiments. The fires both consisted of single armchair fires in an open domestic lounge. The first (Figure 7.5a) involved one chair only; the second involved a fully furnished room. Both fires evolved high concentrations of HCN and CO. Figure 7.5a shows the situation in the fire room, while Figure 7.5b shows the situation in the upstairs bedroom for a different fire. For Figure 7.5a incapacitation is predicted from HCN after 5 min. Occupants collapsing at this point would continue to load CO and would be predicted to reach 50% COHb if they survived a further 3 min. For Figure 7.5b incapacitation due to HCN is predicted at 4.5 min, with 50% COHb being predicted 1.5 min later. Both exposures would almost certainly prove fatal to an occupant remaining after 5 min in the first case and after 4.5 min in the second.

Caution is also necessary with the interpretation of blood cyanide concentrations in fire victims postmortem. When a primate was exposed to a high concentration of HCN (approximately 250 ppm) for 3 min, the arterial blood cyanide concentration increased to a very high level (18.1 µg/ml), but within a further 5 min breathing air the concentration had decreased to 3.5 µg/ml.[37] It is therefore likely that when fire victims are exposed to high HCN concentrations for short periods in fires, loss of consciousness occurs as a result of transient high plasma cyanide concentrations. HCN uptake is then reduced when the subject loses consciousness (or dies) and the cyanide in the plasma disperses throughout the body fluids. Also, cyanide decomposes

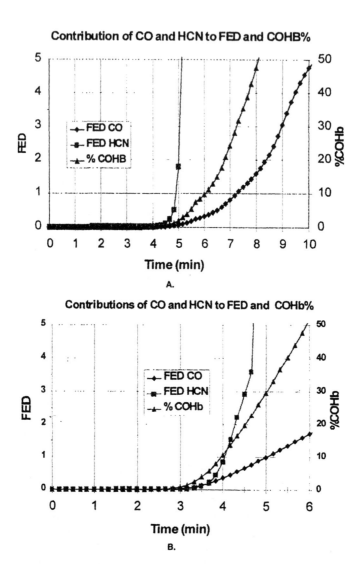

FIGURE 7.5   Relative contributions of CO and HCN to FED for incapacitation and predicted % COHb from furniture fires in a domestic lounge. Upper panel (A): toxic hazards in a lounge from a fire involving an armchair with fire-retarded acrylic covers and combustion-modified polyurethane foam filling. Lower panel (B): toxic hazards in a bedroom from a fire involving an armchair with fire-retarded cotton covers and combustion-modified polyurethane foam filling within a fully furnished room. (From Purser, D.A., Hazards to life in fire, *Hartford Environmental Research*, 1999. With permission.)

rapidly in cadavers,[38] by approximately 50% in 1 to 2 days, and may subsequently decrease further in stored blood. For these reasons, blood cyanide concentrations measured in fire victims are often relatively low, but when samples are obtained immediately after exposure, considerably higher, toxicologically significant or life-threatening levels are detected.[20]

7.2.4 Hypoxia and Interactions between Carbon Monoxide and Low Oxygen Hypoxia

7.2.4.1 Low Oxygen Hypoxia

The effects of exposure to lowered inspired O_2 concentrations depend upon a number of factors including the inhaled CO_2 concentration, the physical activity of the subject, and the health status of the subject. Lowered O_2 in the inspired air, or lowered O_2 in the lungs or blood resulting from exercise, can be tolerated to some extent, due to compensatory mechanisms. These include increased cerebral blood flow and more efficient unloading of oxygen into the tissues from oxyhemoglobin at low oxygen tensions.[39] However, a point is reached where these compensatory mechanisms fail and a marked cerebral depression occurs. In primates, cerebral depression occurred as a result of inhaling 10% O_2 at normal pressure,[14] and similar effects have been reported in human volunteers breathing 12% O_2 for periods of 15 min or more[39] (consisting of lethargy and impaired consciousness). The effects of hypoxia induced by reduced atmospheric pressure at altitude are well known and are basically similar to those of exposure to equivalent reduced O_2 concentrations at sea level pressure. When a subject becomes equilibriated to different equivalent sea level O_2 concentrations, the effects are as follows:

1. 3000 to 4500 m equivalent to 11.8 to 14.4% O_2 at sea level: Compensated phase with relatively mild effects, although the ability to carry out aerobic work is reduced.
2. 4500 to 6000 m equivalent to 9.6 to 11.8% O_2 at sea level: Manifest hypoxia, with degradation of higher mental processes and neuromuscular control, loss of volition and dulling of the senses, and a marked increase in cardiovascular and respiratory activity.
3. >6000 m equivalent to <9.6% O_2 at sea level: Rapid deterioration in judgment and comprehension leading to unconsciousness followed by cessation of respiration and finally of circulation at death.

Two important considerations in relation to the effects of sudden hypoxia are, first, that the severity of the effects does not increase linearly with the degree of O_2 deprivation. Second, they do not occur immediately, but require a certain time to develop depending upon the inhaled concentration and the time for equilibrium to be established between the air, the lungs, the blood circulation, and the tissues.

With regard to the severity of the effects, the body is designed to cope with a certain degree of hypoxia as a result of situations that may occur naturally and has mechanisms to compensate for a decreased blood O_2, such as increases in respiration and in cerebral blood flow. However, once a point is reached where these mechanisms are no longer able to compensate, decline in function can be dramatic. The critical point for hypoxia at rest appears to be around 10 to 12% O_2. Above 12%, effects are relatively minor but, below 10%, loss of consciousness is likely to occur. When primates were exposed to 10% O_2 at rest for 30 min,[14] they remained conscious but became very lethargic, with clear signs of cerebral depression (increased slow wave brain activity, impaired reflexes, ventricular extrasystoles, reduced peripheral nerve conduction velocity, and effects on auditory evoked potential). Another problem in low O_2 environments is reduced exercise capability. At 11 to 12.6% O_2 maximal O_2 consumption is reportedly decreased by 24 to 35%, with a greatly reduced endurance time (78% reduction).[40]

When a subject is suddenly introduced to a low-pressure or low-oxygen atmosphere, loss of consciousness occurs when the partial pressure of oxygen in the cerebral venous blood falls below 20 mm Hg. Due to the effects of the compensatory mechanisms, to residual O_2 in the lungs, and to O_2 stores available from the blood, a certain period of time elapses before the O_2 tension of the venous blood declines to this critical level. The time taken for this depletion depends upon the level to which the O_2 concentration falls, but also on the activity level of the subject (which affects O_2 demand) and the minute volume (which determines the time for the air in the lungs to reach equilibrium with the inspired O_2 concentration). Measurements of these effects are made frequently on human subjects being exposed to sudden decompression to different simulated altitudes for the measurement of time of useful consciousness. Based upon the results of such studies, an expression has been developed for time to loss of consciousness (t_{lo}) for a subject exposed to a hypoxic environment, as follows:

$$(t_{lo}) \text{ min} = \exp \left[(8.13 - 0.54 \,(20.9 - \%O_2)) \right] \qquad (7.9)$$

so that FED_{lo} (the fraction of an incapacitating dose of low O_2 hypoxia) is given by:

$$FED_{lo} = t/\exp \left[8.13 - 0.54 \,(20.9 - \%O_2) \right] \qquad (7.10)$$

7.2.4.2 Interactions Between Carbon Monoxide and Low Oxygen Hypoxia

It is to be expected that CO and low oxygen hypoxia would be additive, since both reduce the percentage oxygen saturation of arterial blood, and CO also impairs the delivery of oxygen to the tissues by causing a leftward shift of the O_2 dissociation curve.[35] It is possible that during the early stages of CO exposure in hypoxic subjects the CO occupies the upper, oxygen-free, part of the O_2 dissociation curve and, therefore, has little effect. At altitude, subjects at rest have been reported to remain symptom free at low levels of CO saturation.[41] However, more severe CO exposures have been reported to be additive with the effects of altitude.[42,43] Also, the hyperventilation resulting from exposure to low O_2 concentrations will lead to some increase in the rate of CO uptake.

7.2.5 INTERACTIONS BETWEEN CARBON MONOXIDE AND INHALED IRRITANTS

The effects of inhaled irritants on respiration depend upon the aqueous solubility of gases, and the extent to which they are associated with respirable particles, as in smoke atmospheres. Fire effluent contains irritant acid gases such as HCl, HBr, HF,

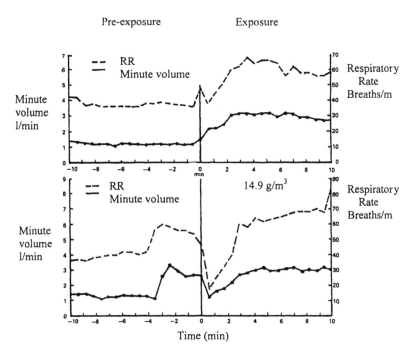

FIGURE 7.6   Respiratory effects of exposure to an atmosphere containing organic irritants from polypropylene decomposed under nonflaming oxidative conditions in the cynomolgus monkey.

SO_2, and NO_2, with organic irritants such as acrolein, formaldehyde, and phenol. There are also considerable differences between the responses of different animal species. In particular, rats and mice show a marked sensory (upper respiratory tract) irritant response, consisting of a marked reduction in both breathing rate and tidal volume.[8,18] Mice in particular can reduce their respiratory rate and minute volume by up to 90% for prolonged periods when exposed to irritant atmospheres if they are at rest.[18] During such episodes, they reduce their peripheral circulation, appearing to elicit a kind of diving reflex. The effect of this is that the rate of uptake of other toxic gases such as CO can be considerably reduced, leading to apparent reductions in CO toxicity on an exposure concentration basis. The effect appears to depend upon the combination of irritant and CO to which the animals are exposed.

In humans and primates the effects are very different. The initial effects of inhaled irritant gases or smokes is sensory irritation and a (trigeminal nerve) reflex decrease in breathing rate, as in the rodents.[8,44] After a short period with breath holding or respiratory pauses at the end of the inspiratory phase, the incidence of pauses decreases and a hyperventilatory response occurs, resulting from the stimulation of vagal lung irritant receptors. The effect is illustrated in Figure 7.6, which shows the respiratory effects occurring when primates were exposed to a smoke containing organic irritants, produced by decomposing polypropylene under nonflaming oxidative conditions.[8] At the lower exposure concentration, the hyperventilatory effect occurs alone. At high concentrations, it is preceded by 2 to 3 min of respiratory depression. When primates or humans are exposed to irritant gases with a low aqueous solubility, such as NO_2, there may be little sensory irritant response, so that lung irritation occurs from the beginning of the exposure.

The effect of this hyperventilatory response might be expected to increase the rate of uptake of other toxic gases such as CO, but during exposure of primates to smoke atmospheres containing both CO and organic irritants there was no evidence of an increased rate of uptake of CO despite a measured increase in minute volume. During such exposures, the respiratory pattern was disturbed and the breathing labored. It is considered that bronchoconstriction and ventilation–perfusion changes may have reduced the rate of CO uptake, despite the increased minute volume. Indeed, it is considered that in humans, the deleterious effects of inhaled irritants on lung function might well result in some degree of hypoxia which might be additive to some extent with the effects of inhaled CO.

7.2.6 INTERACTIONS BETWEEN INHALED NITROGEN DIOXIDE AND HYDROGEN CYANIDE

Inhaled NO_2, apart from acting as a pulmonary irritant, converts hemoglobin to methhemoglobin. This can have hypoxic consequences if a significant percentage of hemoglobin is converted. At the concentrations of NO_2 usually present in fire gas mixture with <250 ppm, this effect should be relatively minor. Experiments with rats have demonstrated additive effects between NO_2 and either CO or low O_2 hypoxia. However, nitrite is a known antidote to cyanide poisoning, so it is to be expected that NO_2 might have some protective effect against HCN inhalation. This is indeed the case as demonstrated by exposures to NO_2 + HCN mixtures in rats.[45] The degree of protection appears to be approximately that 1 ppm of NO_2 protects against 1 ppm of HCN so that the toxicity of a gas mixture in terms of asphyxiant effects is given approximately by subtracting the NO_2 concentration from the HCN concentration. (In practice three molecules of NO_2 are able to convert two molecules of oxyhemoglobin to methemoglobin.) The toxicity of the HCN excess is then approximately equivalent to the effect of the same concentration of HCN administered alone.

In fire effluent, well-ventilated fires tend to produce NO and NO_2, while vitiated fires tend to produce HCN, so that high concentrations of NO_2 and HCN seldom occur in the same atmosphere.[8]

7.3 MODELS FOR THE EFFECTS OF ASPHYXIANT GAS MIXTURES CONTAINING CARBON MONOXIDE, HYDROGEN CYANIDE, LOW OXYGEN, AND IRRITANTS

7.3.1 GENERAL

Having considered the effect of individual gases and their interactions with CO and with each other, the global interactions likely to occur in complex gas mixtures can be considered. This is important because in many fire atmospheres all the gases are likely to be present in various combinations. In theory, it should be possible to develop a physiological model capable of taking into account all the physiological and biochemical effects of inhaled gas mixtures. This would then predict the tissue oxygen tension and its metabolic availability. In practice, models developed so far are based upon empirical observations of interactions between individual pairs of gases and a few experiments on three component mixtures, together with adjustments for some known basic physiological effects. There are two types of models available. One, developed by Purser, is intended to predict time to incapacitation (and death) in humans for exposure to gas mixtures found in fire atmospheres. The other type is primarily intended to predict death in rats (LC_{50} concentrations) following a 30-min exposure to mixed combustion product atmospheres generated in laboratory-scale toxicity tests. Two examples of such models are presented here, the N-gas model developed by Levin et al.[7,12,45] and the lethal FED model developed by Purser.[8,11] It is possible to verify the predictions of these rat lethality models to some extent by reference to experiments in which rats have been exposed to combustion product atmospheres containing gas mixtures. In general, a good agreement has been obtained when the main toxic components of the atmospheres have been identified and measured.

Because the concentrations of individual gases in fire atmospheres, and sometimes in small-scale tests, vary with time during exposure, the models are all FED models. In these models it is assumed that the fractions of an incapacitating or lethal dose acquired during each unit of time during an exposure can be integrated to provide an indication of the summed exposure dose acquired. This is expressed as a fraction of the exposure dose required to cause incapacitation or death. When the summed exposure dose reaches unity, then the end point (incapacitation or death) is predicted to occur.

7.3.2 Purser Human Incapacitation Model

The Purser human incapacitation model is based upon the following precepts:

1. In lightly active subjects (minute volume 25 l/min) incapacitation due to CO occurs when a COHb concentration of 30% has been achieved (this can be modified under different circumstances such as resting or heavy work).
2. For HCN, the relationship between exposure concentration and time to incapacitation in humans is similar to that observed in other primates, and the FED of HCN is directly additive with that for CO. The FED for HCN may be corrected for the deleterious effect of any additional nitriles present and for the protective effect of NO_2. (Acetonitrile, acrylonrile, propionitrile, and benzonitrile are usually present in fire atmosphere with HCN and have an approxmately equivalent toxicity on a molar basis.)
3. A term is included for the asphyxiant effects of inhaled irritants on respiratory function. This term represents a predicted FED for lethality due to pulmonary irritation and is included here to model some additional effects on asphyxiation.

4. The rate of uptake of CO and HCN is directly proportional to the hyperventilatory drive produced by inhaled CO_2 (corrected for uptake inefficiency factors).
5. The FED for low O_2 hypoxia is directly additive with that from CO and HCN. (The deleterious effect of CO_2-induced hyperventilation on time to equilibrium with a reduced O_2 atmosphere is ignored, as are the beneficial effects of CO_2 on moderate hypoxia.)
6. For situations where CO_2 is present at concentrations in excess of 5%, an expression for the direct incapacitating effects of hypercapnia is included as an "or" term.

On the basis of the above, the fractional dose equation for asphyxia is

$$FED_{IN} = (FED_{Ico} + FED_{Icn} + FLD_{irr}) \times VCO_2 + FED_{Io} \text{ or } FED_{Ico2} \quad (7.11)$$

where

FED_{IN} = fraction of an incapacitating dose of all asphyxiant gases
FED_{Ico} = fraction of an incapacitating dose of CO
FED_{Icn} = fraction of an incapacitating dose of HCN (and nitriles, corrected for NO_2)
FLD_{irr} = fraction of an irritant dose contributing to hypoxia
VCO_2 = multiplication factor for CO_2-induced hyperventilation
FED_{Io} = fraction of an incapacitating dose of low oxygen hypoxia
FED_{Ico2} = fraction of an incapacitating dose of CO_2

Each individual term in the FED equation is itself the result of the following equations, which give the FED for incapacitation for each gas and the multiplication factor for CO_2, where t is the exposure time at a particular concentration in minutes. The FEDs acquired over each period of time during the fire are summed until the total FED_{IN} reaches unity, at which point incapacitation (loss of consciousness) is predicted. To allow for differences in sensitivity and to protect susceptible human subpopulations, a target factor of 0.1 to 0.5 FED is suggested for design purposes to allow for the safe escape of nearly all exposed individuals. Death is predicted at approximately two to three times the incapacitating dose.

$$FED_{Ico} = (8.2925 \times 10^{-4} \times ppm \ CO^{1.036}) \times t/30 \quad (7.1)$$

$$FED_{Icn} = (\exp([CN]/43)) \ t/220 \quad (7.8)$$

where [CN] represents the concentration of cyanide corrected for the presence of other nitriles besides HCN and for the protective effect of NO_2. [CN] can be calculated as [CN] = [HCN] + [total organic nitriles] − [NO_2].

$$FLD_{irr} = FLD_{HCl} + FLD_{HBr} + FLD_{HF} + FLD_{SO_2} + FLD_{NO_2} + FLD_{CH_2CHO}$$

$$+ FLD_{HCHO} + \Sigma \ FLD_x \quad (7.12)$$

TABLE 7.1
Lethal Exposure Doses of Irritants Contributing to Asphyxia and Lung Damage

Gas	Exposure Doses Predicted to Be Lethal to Half the Population (ppm·min)
HCl	114,000
HBr	114,000
HF	87,000
SO_2	12,000
NO_2	1,900
CH_2CHO (acrolein)	4,500
HCHO (formaldehyde)	22,500

TABLE 7.2
Simplified Lookup Table for Solutions to Individual Toxic Gas FED Equations for Incapacitating Exposure Doses over a 1-min Exposure Time

ppm HCN	FED_{lcn}	FED_{lco} = CO ppm/25,000 %CO_2	VCO_2	%O_2	FED_{lo2}
0–50	0	0–2	1.0	21–13	0.00
50–100	0.05	2–3	1.5	13–12	0.02
100–125	0.10	3–4	2.0	12–11	0.05
125–150	0.15	4–5	2.5	11–10	0.08
150–200	0.50	5–6	3.0	10–9	0.15
200 +	1.00	6–7	3.5	9–8	0.20
		7–8	4.5	8–7	0.40
		8–10	4.8	7–6	0.70

where $\Sigma\, FLD_x = FLD_s$ for any other irritants present and where the FED for each irritant gas is equal the exposure dose acquired divided by the lethal exposure dose. Table 7.1 shows the lethal exposure doses for irritant gases commonly found in fire atmospheres.

$$VCO_2 = \exp\,([CO_2]/5) \qquad (7.3)$$

$$FED_{Io} = t/\exp\,[8.13 - 0.54\,(20.9 - \%O_2)] \qquad (7.10)$$

Table 7.2 shows a simplified lookup table of FEDs for incapacitation for each gas, illustrating the relationship between concentration and FED for asphyxiant gases. For CO the FED is simply related to the exposure dose; for HCN and decreased O_2 the FED acquired each minute increases considerably with exposure concentration, as does the multiplication factor VCO_2.

Table 7.3 shows a worked example of the application of the Purser human incapacitation model to the profile of asphyxiant and irritant gases occurring during a furniture fire in a domestic room. In this example, although some of the data are

TABLE 7.3
Life Threat Analysis for the First 6 min of a Furniture Fire

	1	2	3	4	5	6
Gas Conc. Each Minute						
HCl (ppm)	10	50	150	200	250	200
Acrolein (ppm)	0.4	0.8	2.0	6.0	12.0	14.0
Formaldehyde (ppm)	0.6	1.2	3.0	9.0	18.0	21.0
CO (ppm)	0	0	500	2000	3500	6000
HCN (ppm)	0	0	50	150	250	300
CO_2 (%)	0	0	1.5	3.5	6.0	8.0
O_2 (%)	20.9	20.9	19.0	17.5	15.0	12.0
Fractional Lethal Dose (irritants)						
FLD_{HCl}	0.00	0.00	0.00	0.00	0.00	0.00
$FLD_{acrolein}$	0.00	0.00	0.00	0.00	0.00	0.00
FLD_{form}	0.00	0.00	0.00	0.00	0.00	0.00
ΣFLD_{Irr}	0.00	0.00	0.00	0.00	0.00	0.00
Fractional Asphyxiant Dose						
FED_{Ico}	0.00	0.00	0.02	0.07	0.13	0.23
FED_{Icn}	0.00	0.00	0.01	0.15	1.52	4.87
FLD_{irr}	0.00	0.00	0.00	0.00	0.01	0.04
VCO_2	1.00	1.00	1.35	2.01	3.32	4.95
FED_{Io2}	0.00	0.00	0.00	0.00	0.01	0.04
FED_{IN} (asphyxiants)	0.00	0.00	0.04	0.45	5.50	25.29
FED_{IN}	0.00	0.00	0.04	0.50	**6.00**	31.29

Note: The end point, incapacitation (for asphyxiant gases), is reached when the FED value reaches 1. Limiting values are emboldened. Lethal values are approximately two to three times incapacitating levels for dose-related parameters.

taken from a real experiment, HCl, acrolein, formaldehyde, and HCN were not measured, so likely concentrations of these gases have been added for the purpose of illustrating the calculation method. The gas concentrations and acquired FEDs are shown for 1-min intervals. The upper section shows the gas concentrations increasing as the fire grows over 6 min. Asphyxiant gases, HCN and CO and CO_2 reach toxicologically significant concentrations during the fourth minute. Irritant gases, although present at concentrations likely to cause painful sensory irritation, are not at sufficiently high concentrations to cause deep lung damage over a few minutes. This is shown in the second section where the accumulated lethal doses of each irritant and their summed effects are too small to register over the time period examined. The third section shows the results of the analysis for asphyxiants. During the fourth minute the FEDs for CO and HCN, combined with the hyperventilatory stimulus from CO_2 are sufficient to provide an FED of 0.5 when added to the small FED from the previous minute. During the fifth minute, when the HCN concentration reaches 250 ppm, this provides an FED of 1.52 and the total summed FED reaches 6.

It is therefore predicted that a room occupant breathing this atmosphere would lose consciousness during the fifth minute and is likely to have received a lethal asphxyiant dose during the fifth and sixth minutes.

Figure 7.7 shows the results of an FED analysis applied to a more typical domestic fire. The fire involved flaming ignition of a single armchair made from a combustion-modified polyurethane foam containing melamine and phosphate fire retardants covered with fire-retarded acrylic fabric. Both of these contain a significant amount of organic nitrogen. The fire was conducted in the ground-floor (U.S. first-floor) lounge of a typical British domestic house. The lounge door was left open so that the fire effluent could escape into the hall and stairway and fill the upstairs landing and an open bedroom. The outer doors and windows were closed. This is a typical fire scenario in which deaths commonly occur. In this situation the fire grew initially as shown by the gas concentration curves presented in the upper part of the figure. After 6 min, the combustion conditions became vitiated when the O_2 concentration in the fire room dropped to around 15%, eventually decreasing to 11% O_2. The fire, which was still quite small, self-extinguished after 8 min having consumed around 3 kg of material from the chair. These conditions produce high yields of CO from carbon in the fuel and high yields of HCN from nitrogen in the fuel, resulting in a highly toxic asphyxiating atmosphere. The peak HCN concentration was over 1100 ppm and the peak CO concentration over 1%.

The predicted effect on a room occupant is presented in the lower part of Figure 7.7, which shows the contributions to asphyxia made by each gas individually and in combination. Incapacitation (loss of consciousness) is predicted when any line exceeds an FED of 1. The picture is dominated by the extremely high concentration of HCN, which is predicted to cause incapacitation at the end of the fourth minute, as the concentration exceeds 400 ppm. The FED for CO alone exceeds unity at around 9.5 min. The FED for low O_2 hypoxia is predicted to remain quite small, reaching only 0.1 after 10 min. This is because the O_2 concentration does not fall below 11% and then increases to 16% as the gases in the house become mixed after the fire extinguishes. Similarly, irritants are predicted to have only a minor asphyxiant and lung inflammatory effect over this timescale. The hyperventilatory effect of CO_2 becomes significant from around 5 min. The effect that this might have on the CO toxicity if HCN is removed from the analysis is shown by the open diamonds (asphyxia, no HCN). This line rises quite steeply from 5 min, reaching an FED of unity after 7 min. It is therefore predicted that in a similar fire involving non-nitrogen-containing fuels, the presence of CO_2 would significantly increase the rate of uptake of CO, reducing the time to incapacitation by CO from 9.5 to 7 min. Since all the FED curves are rising steeply, particularly the HCN curve, it is considered that uncertainties in the precision of the model will have little effect on time to incapacitation.

7.3.3 Models for Predicting Lethality (LC$_{50}$ Concentrations) in Rats Following a 30-min Exposure

Models for predicting lethality in rats were primarily developed to reduce the need for animal experiments in evaluating the lethal toxic potency of combustion product

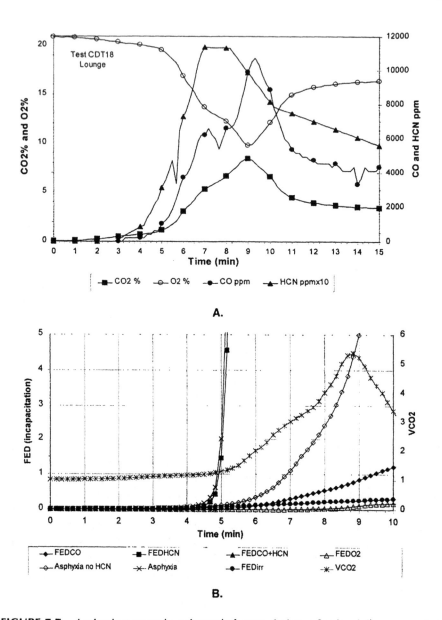

FIGURE 7.7 Asphyxiant gases in a domestic lounge during a fire involving an armchair with fire-retarded acrylic covers and combustion-modified polyurethane foam filling and FEDs for effects of asphyxiant gases for a room occupant. Upper panel (A): asphyxiant gases in lounge during an armchair fire in a domestic lounge. Lower panel (B): FED for incapacitation for individual asphyxiant gases, combinations of gases, and VCO_2, for gases shown in the upper part of the figure. (From Purser, D.A., Hazards to life in fire, *Hartford Environmental Research*, 1999. With permission.)

atmospheres evolved from materials in small-scale toxicity tests. However, it is also possible to apply such models to full-scale fire atmospheres and to make some assessment of exposure doses likely to cause incapacitation (at approximately half to a third of the lethal exposure dose). When such methods are applied to fire hazard analysis, it is assumed that an exposure dose lethal to rats would be approximately the same as a lethal exposure dose to humans. With asphyxiant gases, this does imply some margin of safety, since rodents tend to load asphyxiant gases more rapidly than humans, due to their small body size and high metabolic rate (low body mass/surface area ratio).

The main concepts in these models are similar, but they differ somewhat regarding which factors are considered and the ways the various interactions are handled.[10] The models estimate the FED for lethality in terms of the concentration of each toxic gas present, expressed as a fraction of the LC_{50} concentration. The basic concept is that fractions of lethal doses of almost all gases are directly additive. Thus, if half the lethal 30-min concentration of CO is present with half the 30-min lethal concentration of HCN, then exposure of rats to this mixture for 30 min will on average result in the deaths of half the animals. Deaths are considered both during exposure and over a postexposure observation period of up to 14 days. It is considered that in terms of overall lethality, the effects of asphyxiant gases (CO, HCN, and low O_2 hypoxia) are additive with the effects of irritant gases (acid gases such as HCl and NO_2 and organic irritants such as formaldehyde). It is recognized that CO_2 modifies the toxicity of other gases and this is treated differently in different models. It is also recognized that NO_2 has some protective effect against HCN toxicity due to methemoglobin formation.

For use in the models, the currently accepted values for the 30-min LC_{50} concentration of gases commonly occurring in fire atmospheres are shown in Table 7.4.

7.3.3.1 Levin et al.[45] N-Gas Model for Predicting FED for Lethality in Rats (30-min exposure + 14 days postexposure observation period)

The N-gas model developed by Levin at al.[45] is based on a large number of LC_{50} experiments on individual asphyxiant and irritant gases and on experiments on various concentration mixtures of two or even three gases.[7,12,45]

The key precepts of the model are as follows:

1. Fractions of lethal doses of almost all gases are directly additive.
2. Based upon combination experiments with CO and CO_2, it is considered that CO_2 enhances the lethal toxicity of CO, with a maximum effect at a concentration of 5% CO_2. Above 5% CO_2 the enhancement of CO toxicity decreases. This effect is handled in the N-gas equation by the use of constants m and b, which represent the slope and intercepts of the combination gas toxicity curve.[7]
3. In a recent version of the model, the corrective terms are added for the protective effect of NO_2 on HCN toxicity.[45] Versions of the model are also considered for predicting deaths during exposure and combined deaths during and after exposure.

TABLE 7.4
Currently Accepted 30-min LC_{50}
Concentrations for Common Fire Gases

Gas	Conc. (ppm)
CO	5700
HCN	165
HCl	3800
HBr	3800
HF	2900
SO_2	400–1400
NO_2	170
Acrolein	150
Formaldehyde	750

The current version of the N-gas model[45] for total deaths (during and after exposure) is as follows:

$$FED = (m[CO]/[CO_2] - b) + (21 - [O_2] / 21 - LC_{50} \, O_2) + \{([HCN] / LC_{50} \, HCN)$$

$$\times (0.4 \, [NO_2]/LC_{50} \, NO_2) + (0.4 \, [NO_2] / LC_{50} \, NO_2)\} + ([HCl] / LC_{50} \, HCl)$$

$$+ ([HBr] / LC_{50} \, HBr) \tag{7.13}$$

where

For $[CO_2] \le 5\%$, $m = -18$ and $b = 122{,}000$
For $[CO_2] > 5\%$, $m = 23$ and $b = -38{,}600$

The N-gas equation works reasonably well and is based upon some very comprehensive experimental data obtained from rats for key gas interactions. However, although CO_2 has been shown to exert synergistic effects on the toxicity of several other gases, a CO_2 correction is applied only to CO, on the basis that CO is likely to be the major toxic gas present. Also, the correction of HCN toxicity for the protective effect of NO_2 as expressed in the equation works only for the case of 200 ppm NO_2. As the term is currently expressed, a low level of NO_2 or no NO_2 would negate HCN toxicity, which is incorrect, while a level higher than 200 ppm would enhance it, which is also incorrect. Another problem is that in fire gas atmospheres organic irritants can be major causes of lung inflammation and death, and these are omitted from the analysis.

7.3.3.2 Purser Model for Predicting FED for Lethality in Rats
(30-min exposure + 14 days postexposure observation period)

The Purser rat lethality model[8] is based upon the same general concepts as the N-gas model and makes use of the rat LC_{50} data for individual gases and gas interactions obtained mainly by Levin et al.[7,12,45] and Hartzell et al.[9]

The key precepts of the model are as follows:

1. Fractions of lethal doses of all gases except CO_2 are directly additive.
2. The main effect of CO_2 is considered to be a multiplicatory effect on the rate of uptake of other gases depending upon the extent of CO_2-driven hyperventilation. In addition, it is considered that once animals are incapacitated, CO_2-induced respiratory acidosis enhances the metabolic acidosis already present, providing an additive toxicity factor.
3. Low O_2 hypoxia is usually a minor factor in small-scale rodent toxicity experiments and can be ignored unless O_2 concentrations are allowed to decrease below 12%. At low levels an additive term can be used.
4. A correction for the protective effect of NO_2 on HCN toxicity can be made if necessary and for the additive effect of other nitriles present.
5. It is considered important to make allowance for the effect of all inorganic acid gases present and for organic irritants.

The estimated FED for lethality in rats for a 30-min exposure to any defined set of toxic gas concentrations is calculated as follows:

$$FED = \{([CO] / LC_{50}\ CO) + ([CN] / LC_{50}\ HCN)$$

$$+ ([each\ acid\ gas] / LC_{50}\ each\ gas) + ([each\ org.\ irritant] / LC_{50}\ each\ org.\ irritant)\}$$

$$\times VCO_2 + A + (21 - [O_2] / 21 - LC_{50}\ O_2) \tag{7.14}$$

where

VCO_2 = multiplication factor for CO_2-driven hyperventilation
 = $1 + (\exp(0.14 \times [CO_2]) - 1)/2$
A = an acidosis factor = $[CO_2] \times 0.05$

[CN] represents the concentration of cyanide corrected for the presence of other nitriles besides HCN and for the protective effect of NO_2. [CN] can be calculated as

$$[CN] = [HCN] + [total\ organic\ nitriles] - [NO_2] \tag{7.15}$$

Where data on organic irritant concentrations are absent, it is recommended that a contribution to the overall FED should be derived from an estimate of the total yield of organic products. The FED component for organic irritants (FED_{org}) is then estimated as

$$FED_{org} = mass\ loss\ concentration\ of\ organic\ material \times$$

$$\%\ carbon\ present\ as\ organic\ carbon \div 25 \tag{7.16}$$

FED_{org} is then substituted for the organic irritant term in Equation 7.14.

TABLE 7.5
30-min LC_{50} Analysis Using the Purser Method of Materials Decomposed under Early, Well-Ventilated Flaming Conditions in the NBS Cup Furnace

Material	Douglas Fir, 485°C		Flexible Polyurethane Foam and Polyester, 52°C	
Mass loss exposure concentration g/m³	39.8		39.0	
	Conc.	FED	Conc.	FED
CO	3400 ppm	0.60	2270 ppm	0.40
HCN	0 ppm	0.00	63 ppm	0.38
FED presumed due to irritants		0.05		0.05
Total FED uncorrected		0.65		0.83
CO_2	3.71%		3.36%	
× VCO_2		1.34		1.30
+ A		0.19		0.17
Total FED corrected ·		1.06		1.25
LC_{50} calculated	37.5 g/m³		31.2 g/m³	
LC_{50} observed	39.8 g/m³		39.0 g/m³	

If these data are unavailable, an FED_{org} of 0.5 should be used for nonflaming decomposition and 0.15 for vitiated or inefficient combustion, with 0.05 used for well-ventilated combustion at the LC_{50} mass loss concentration for the material or product under test.

If the concentrations of the irritants present and their lethal exposure doses are known, then the equation can be solved fully. Where unknown irritants are present, the equation enables the maximum LCt_{50} to be predicted based upon the asphyxiant gases and any known irritants.

Table 7.5 shows an example of lethal FED calculations based upon gas effluent concentrations for experiments conducted using the National Bureau of Standards (NBS) cup furnace method, for which the LC_{50} was measured using rats.[12] The table shows the toxic gas concentrations at the LC_{50} concentrations for each material. From these, the actual LC_{50} exposure concentrations have been calculated according to Equation 7.14 and these can be compared with the estimated LC_{50} exposure concentrations calculated according to the Purser FED equation (Equation 7.14). This shows data for a well-ventilated flaming case. In this instance it is possible to account for all the observed toxic potency in terms of asphyxiant gases. The calculation formula slightly overestimates the toxic potency in these examples, particularly if a term is added for a notional small contribution from organic irritants.

7.4   CONCLUSIONS

Based upon experimental work in human volunteers, other primates, and rodents, and upon basic physiology, it is possible to derive reasonably good expressions for calculating the effects of inhaling individual asphyxiant and irritant gases in terms of time and exposure dose to incapaciation and death for humans. Sufficient data also exist to determine the main interactions between pairs of asphxyiant gases, and even for more complex mixtures. The main interactions are that CO, HCN, and low O_2 hypoxia are more or less directly additive in terms of fractional effective doses required for incapacitation or death. Inhaled irritants also show some degree of additivity in exacerbating asphyxiant effects. CO_2 is most important as a cause of hyperventilation, increasing the rate of uptake of CO, HCN, and irritants, and thereby hastening the point where an incapacitating exposure dose has been inhaled. There is also evidence that CO_2 exerts an additive effect with asphyxia, at least under near-lethal conditions, possibly resulting from the combination of metabolic and respiratory acidosis and other factors. CO_2 has a beneficial effect on low O_2 hypoxia, both in terms of time of useful consciousness at very low O_2 concentrations, and in terms of maintaining mental function during prolonged exposure periods at 10 to 12% O_2. However, there is evidence that CO_2 enhances the lethal effects of extreme hypoxia.

Based upon these considerations, it has been possible to develop calculation models for time and exposure dose to incapacitation in humans, and for estimating lethal exposure doses in rats.

REFERENCES

1. Home Office/The Scottish Office, Fire Statistics, United Kingdom, Government Statistical Service, 1993 (summarized annually).
2. Bowes, P.C., Smoke and toxicity hazards of plastics in fires, *Ann. Occup. Hyg.*, 17, 143, 1974.
3. Petajan, J.H., Voorhees, K.J., Packham, S.C., Baldwin, R.C., Einhorn, I.N., Grunnet, M.L., Dinger, B.G., and Birky, M.M., Extreme toxicity from combustion products of a fire-retardant polyurethane foam, *Science*, 187, 742–744, 1975.
4. Levin, B.C., Fowell, A.J., Birky, M.M., Paabo, M., Stolte, A., and Malek, D., Further Development of a Test Method for the Assessment of the Acute Inhalation Toxicity of Combustion Products, Report NBSIR 82-2532, National Bureau of Standards, Washington, D.C., 1982.
5. Purser, D.A. and Woolley, W.D., Biological studies of combustion atmospheres, *J. Fire Sci.*, 1, 118–144, 1983.
6. Anderson, R.A., Watson, A.A., and Harland, W.A., Fire deaths in the Glasgow area: 1. General conclusions and pathology, *Med. Sci. Law*, 21, 175, 1981.
7. Levin, B., Gurman, J., Paabo, M., Baier, L., and Holt, T., Toxicological effects of different time exposures to fire gases: carbon monoxide or hydrogen cyanide or to carbon monoxide combined with hydrogen cyanide or carbon dioxide, in *Proceedings of the 9th Joint Panel Meeting of the UNJR Panel on Fire Research and Safety*, Rep. NBSIR 88-3753, U.S. National Bureau of Standards, Gaithersburg, MD, 1988, 368.
8. Purser, D.A., Toxicity assessment of combustion products, in *SFPE Handbook of Fire Protection Engineering*, DiNenno, P.J., Ed., National Fire Protection Association, Quincy, MA, 1995, chap. 8, Section 2, 85–146.
9. Hartzell, G.E., Grand, A.F., and Switzer, W.G., Modelling of toxicological effects of fire gases. VI. Further studies of the toxicity of smoke containing hydrogen chloride, in *Advances in Combustion Toxicology*, Vol. 2, Hartzell, G.E., Ed., Technomic, Lancaster, PA, 1989, 285–308.

10. International Standards Organisation. International Standard, Determination of the lethal toxic potency of fire effluents, ISO, 13344, 1996.

11. Purser, D.A., The evolution of toxic effluents in fires and the assessment of toxic hazard, *Toxicol. Lett.*, 64/65, 247, 1992.

12. Levin, B.C., Paabo, M., Gurman, J.L., and Harris, S.E., Effects of exposure to single or multiple combinations of the predominant toxic gases and low oxygen atmospheres produced in fires, *Fundam. Appl. Toxicol.*, 9, 236, 1987.

13. Purser, D.A., The harmonization of toxic potency data for materials obtained from small and large scale tests and their use in calculations for the prediction of toxic hazard in fire, presented at First International Fire and Materials Conference, September 24–25, 1992, Washington, D.C., Proceedings: Interscience Communications, London, 1992, 179–200.

14. Purser, D.A., A bioassay model for testing the incapacitating effects of exposure to combustion product atmospheres using cynomolgus monkeys, *J. Fire Sci.*, 2, 20, 1984.

15. Purser, D.A. and Grimshaw, P., The incapacitative effects of exposure to the thermal decomposition products of polyurethane foams, *Fire Mater.*, 8, 10, 1984.

16. Purser, D.A. Grimshaw P., and Berrill, K.R., Intoxication by cyanide in fires: a study in monkeys using polyacrylonitrile, *Arch. Environ. Health*, 39, 394, 1984.

17. Kaplan, H.L., Grand, A.F., Switzer, W.G., Mitchell, D.S., Rogers, W.R., and Hartzell, G.E., Effects of combustion gases on escape performance of the baboon and the rat, *J. Fire Sci.*, 3, 228, 1985.

18. Hirschler, M.M. and Purser, D.A., Irritancy of the smoke (non-flaming mode) from materials used for coating wire and cable products, both in the presence and absence of halogens in their chemical composition, *Fire and Mater.*, 17, 7, 1993.

19. Nelson, G.L., Carbon monoxide and fire toxicity: a review and analysis of recent work, *Fire Technol.*, 34, 39–57, 1998.

20. Baud, F.J., Barriot, P., Toffis, V., Riou, B., Vicaut, E., Lecarpentier, Y. Bourdon, R., Astier, A., and Bismuth, C., Elevated blood cyanide concentrations in victims of smoke inhalation, *N. Engl. J. Med.*, 325, 1761–1766, 1991.

21. Stewart, R.D., Peterson, J.E., Fisher, T.N., Hosko, M.J., Baretta, E.D., Dodd, H.C., and Hermann, A.A., Experimental human exposure to high concentrations of carbon monoxide, *Arch. Environ. Health*, 26, 1, 1973.

22. Purser, D.A. and Berrill, K.R., Effects of carbon monoxide on behaviour in monkeys in relation to human fire hazard, *Arch. Environ. Health*, 39, 308, 1983.

23. Lambertson, C.J., Carbon dioxide and respiration in acid-base homeostasis, *Anaesthesiology*, 21, 642–651, 1960.

24. Comroe, C.J., Forster, R.E., Dubois, A.B., Briscoe, W.A., and Carlsen, E., in *The Lung*, Year Book Medical Publishers, Chicago, 1962.

25. Altman, P. and Ditter, D.S., Eds., *Environmental Biology*, Federation of American Societies for Experimental Biology, Bethesda, MD, 1966.

26. King, B.G., High concentration–short time exposures and toxicity, *J. Ind. Hyg. Toxicol.*, 31, 365, 1949.

27. Coburn, R.F., Forster, R.E., and Kane, P.B., Consideration of the physiological variables that determine the blood carboxyhaemoglobin concentration in man, *J. Clin. Invest.*, 44, 1899, 1965.

28. Peterson, J.E. and Stewart, R.D., Predicting the carboxyhaemoglobin levels resulting from carbon monoxide exposures, *J. Appl. Physiol.*, 4, 633, 1975.

29. Clark, J.M., Sinclair, R.D., and Lenox, J.B., Chemical and nonchemical components of ventilation during hypercapnic exercise in man, *J. Appl. Physiol.*, 48, 1065, 1980.

30. Brackett, N.C., Jr., Cohen, J.J., and Schwartz, W.B., Carbon dioxide titration curve of normal man. Effect of increasing degrees of acute hypercapnia on acid–base equilibrium, *N. Engl. J. Med.*, 272, 6, 1965.

31. Schulte, J.E., Sealed environments in health and disease, *Arch. Environ. Health*, 8, 437–452, 1964.

32. Gibbs, F.A., Gibbs, E.L., Lennox, W.G., and Nims, L.F., The value of carbon dioxide in counteracting the effects of low oxygen, *J. Aviat. Med.*, 14, 250, 1943.

33. Karl, A.A., McMillan, G.R., Ward, S.L., Kissen, A.T., and Souder, M.E., Effects of increased ambient CO_2 on brain tissue oxygenation and performance in the hypoxic rhesus, *Aviat. Space Environ. Med.*, 49, 984–989, 1978.

34. Lambersten, C.J., Inergen: Summary of Relations; Physiologic Factors and Fire Protection Engineering Design, Environmental Biomedical Research Data Center Rep. 3-30-93, Institute for Environmental Medicine, University of Pennsylvania, Philadelphia, PA, 1993.

35. Root, W.S., Carbon monoxide, in *Handbook of Physiology*, Section 3: Respiration, American Physiological Society, Washington, D.C., 1965, 1087.

36. Levin, B.C., Paabo, M., Gurman, J.L., and Harris, S.E., Toxicological effects of the interactions of fire gases, in Proceedings of the Smoke/Obscurants Symposium X, Vol. 2, Tech. Rep. AMCPM-SMK-T-001-86, Harry Diamond Laboratories, Adelphi, MD, April 22–24, 1986.

37. Purser, D.A., Determination of blood cyanide and its role in producing incapacitation in fire victims, presented at Royal Society of Chemistry (Analytical Division) Meeting, Analytical Aspects of Biological Safety Evaluation, Huntingdon, U.K., June 6, 1984.

38. Ballantyne, B., The forensic diagnosis of cyanide poisoning, in *Forensic Toxicology*, Wright, Bristol, U.K., 1974, 99–113.

39. Luft, U.C., Aviation physiology — the effects of altitude, in *Handbook of Physiology*, Section 3: Respiration, American Physiological Society, Washington, D.C., 1965, 1099.

40. Gleser, M.A. and Vogel, J.A., Effects of acute alterations of + VO_2 max. on endurance capacity of men, *J. Appl. Physiol.*, 34, 443, 1973.

41. Von Leggenhager, K., New data on the mechanisms of carbon monoxide poisoning. *Acta Med. Scand.*, 196 (Suppl. 563), 1, 1974.

42. Heim, J.W., The toxicity of cabon monoxide at high altitude, *J. Aviat. Med.*, 10, 211–215, 1939.

43. McFarland, R.A., Roughton, F.J.W., Halperin, M.H., and Niven, J.I., The effects of carbon monoxide on altitude and visual thresholds, *J. Aviat. Med.*, 15, 381, 1944.

44. Purser, D.A. and Buckley, P., Lung irritance and inflammation during and after exposure to thermal decomposition products from polymeric materials, *Med. Sci. Law*, 23, 142, 1983.

45. Levin, B.C., New research avenues in toxicology: 7-gas N-gas model, toxicant suppressans, and genetic toxicology, *Toxicology*, 115, 89, 1996.

8 Carbon Monoxide Poisoning and Its Management in the United States

Neil B. Hampson

CONTENTS

8.1   SIGNIFICANCE OF CARBON MONOXIDE POISONING IN THE UNITED STATES

Carbon monoxide (CO) poisoning is a significant health problem in the United States, killing an estimated 3700 people annually.[1] It is the single most common cause of poisoning death in the country. During the decade of the 1980s, there were approximately 1100 deaths each year due to accidental CO exposure and an additional 2600 suicidal deaths.[1] The death rate from unintentional CO poisoning declined in the 1980s, but this was offset by an equal rise in the suicidal death rate.

Nonfatal CO poisoning is even more common. For many years, it has been widely quoted that CO intoxication causes approximately 10,000 affected individuals to seek medical attention or miss at least 1 day of normal activity annually. However, this estimate is decades old, first published the medical literature in the early 1970s,[2] and derived by the U.S. Public Health Service from limited data in the late 1960s.[3] A recent study of hospitals in the Pacific Northwest found an emergency department visit rate for CO poisoning of 18.1/100,000 population per year in a three state region and estimated over 40,000 visits for CO poisoning annually in the entire country.[4]

The actual number of nonfatal cases in the United States annually is likely to be significantly larger for several reasons. Even this higher estimate of disease incidence includes only emergency department visits for recognized CO poisoning. The signs and symptoms of CO poisoning are nonspecific[5] and underdiagnosis of CO poisoning is well described.[6,7] Additionally, not all patients are treated in emergency departments. Those treated in medical offices or clinics would not be represented. Finally, patients may attribute the nonspecific symptoms of CO poisoning (e.g., headache, nausea) to alternative causes such as viral illness, staying home from work or school but not seeking medical evaluation.

8.2 TEMPORAL AND GEOGRAPHIC EPIDEMIOLOGY OF CARBON MONOXIDE POISONING

CO poisoning has a seasonal distribution, being much more common in winter than in summer.[1,8–17] In the case of accidental exposures, this is often attributed to the fact that home furnace accidents are more common during the indoor heating season. Severe winter storms can play a role, as well, resulting in CO poisoning when individuals without power use alternative energy sources for heating and/or cooking or when they are trapped in motor vehicles by snow.[18–22] Suicidal CO deaths have been reported to be more common in the United States during the months of March through May.[1] A recent report found an increased number of patients treated in Seattle for intentional CO poisoning in the months of March, April, and October, correlating with the amount of rain on the days prior to a suicide attempt.[23]

CO poisoning also has a geographic distribution in the United States, again related at least in part to climate and home heating. The cold and high-altitude states have the highest accidental death rate due to CO poisoning, while warmer states have the lowest rates.[1] The CO-related death rate in Alaska (2.72/100,000) is approximately 50-fold that reported in Hawaii (0.05/100,000). California has the second lowest accidental death rate (0.25/100,000), felt to be due to stringent automobile emission standards, in addition to the relatively warm climate. Few data have been published on the geographic distribution of nonfatal CO poisoning. One survey of U.S. hyperbaric oxygen (HBO) treatment facilities found that the largest number of treatments for CO poisoning was performed in Minnesota during the year studied.[16]

8.3 DEMOGRAPHICS OF INDIVIDUALS WITH CARBON MONOXIDE POISONING

Although individuals of all ages are victims of CO poisoning, it is most common among those of middle age. U.S. Consumer Product Safety Commission (CPSC) data indicate that accidental non-fire-, non-automobile-related CO poisonings are most common in the 25 to 44 year age group.[15] In a survey of U.S. hyperbaric facilities, the age group most commonly referred for treatment of severe CO poisoning was also 25 to 44 years.[16] Case series of consecutive patients reported from individual treatment facilities confirm this, with average patient ages of 34 years in Illinois,[24] 34 years in North Carolina,[25] 35 years in Utah,[14] and 34 years in Washington

State.[25] The source of CO may have a relationship to victim age. Patients treated at a hyperbaric facility in Seattle for CO poisoning due to indoor burning of charcoal briquets were significantly younger than those who sustained poisoning during recreational boating activities (mean ages 28 vs. 37 years, respectively; $p < 0.02$).[10,13]

As middle-aged individuals are most frequently poisoned, it is not surprising that they account for the greatest number of deaths from CO poisoning.[1,9,12,15] However, the age-specific death rate is highest among the elderly, suggesting an increased susceptibility to the toxic effects of CO. A Centers for Disease Control study of U.S. deaths due to accidental CO poisoning in the 1980s found the highest death rate to be among those over 75 years of age.[1] The death rate was reported to be significantly higher among persons 65 years of age and older in studies of unintentional CO poisoning in both Colorado and Michigan.[9,11]

CO poisoning has a preponderance for the male sex. Males accounted for 56% of unintentional nonfatal CO poisonings in Colorado from 1986 to 1991.[11] Among various series of patients treated for accidental and/or intentional CO intoxication at U.S. hyperbaric facilities, males have been reported to account for 63 to 73% of patients.[14,16,24,26] The source of CO poisoning may play a role with regard to gender of individuals poisoned, as was previously noted with patient age. Among 79 consecutive patients treated at a Seattle hyperbaric facility for CO poisoning resulting from indoor use of charcoal briquets, for example, a small majority (52%) were female.[10]

Mortality has been more common among males than females in all series reported. In a national study of deaths from accidental CO poisoning of all causes in the 1980s, the death rate was almost three times greater for males than for females (0.78 vs. 0.26/100,000).[1] Of accidental U.S. non-fire-, non-automobile-related CO deaths from 1990 to 1994, 70% were male.[15] Similar findings have been described in a variety of studies on individual states. Males have accounted for 72% of unintentional CO deaths in California,[17] 81% in Michigan,[9] 74% in New Mexico,[12] and 72% in Colorado.[11] In the latter study, the death rate was 2.6 times greater for males than for females (1.3 vs. 0.5/100,000).

In addition to age and sex, race is also a risk factor for CO poisoning. Members of minority races are more likely both to be treated for and to die from unintentional CO intoxication. While a number of reports provide data on the racial and/or ethnic composition of the population studied, only a few compare this with the general population to allow calculation of poisoning rates or an estimation of risk. A study of all patients accidentally poisoned with CO and treated with HBO in Washington State from 1987 to 1997 found that the relative risk of accidental poisoning compared with whites was 3.96 for Hispanics and 2.91 for blacks.[27] In New Mexico, the annual death rate from unintentional CO poisoning per 100,000 persons has been reported as 2.41 for blacks, 0.83 for Native Americans, 0.47 for Hispanics, and 0.46 for whites.[12] A national study of accidental CO deaths found race-specific death rates to be more than 20% higher for blacks than for whites.[1]

Possible explanations for higher death rates among minorities could include excess exposure to CO within these groups, enhanced susceptibility to CO intoxication, and/or poorer access to medical care. The relative contributions of these factors is unknown. Limited data exist to support the concept of excess exposure to

CO in certain racial or ethnic groups. For example, in a study of CO poisoning due to the indoor burning of charcoal briquets, it was found that the incidence among Asians and Hispanic whites was far in excess of their representation among the general population.[10] Common reasons for indoor charcoal use were found to be cooking and home heating, apparently a continuation of ethnic customs following immigration to the United States. Enhanced susceptibility to the effects of CO poisoning among specific racial groups has never been reported and, therefore, seems unlikely to be the explanation for higher death rates. It is also difficult to implicate lack of access to health care as the reason when individual minority groups receive a disproportionate number of HBO treatments for poisoning.[27]

8.4 SOURCES OF CARBON MONOXIDE INVOLVED IN POISONINGS

CO poisoning has been reported to occur from exposure to virtually every form of combustion that exists in society. While specific sources of CO have been reported to predominate at certain times of the year and/or among select subpopulations (e.g., specific racial groups), some studies do provide information on CO sources for large unselected populations of poisoned individuals. Among 807 nonfatal unintentional poisonings in Colorado, the CO source was a furnace in almost half the cases (Table 8.1).[11] The list of sources of CO was much more heterogeneous in a series of 631 accidental, severe CO poisonings referred for HBO treatment (Table 8.1).[25] Results from a national survey of U.S. hyperbaric treatment centers, obtaining information from 51% of facilities, found that an indoor gas appliance was the most common source of CO (33%) among patients treated from 1994 to 1995.[16] Among 295 CO-poisoned patients with CO poisoning treated at a Utah hyperbaric facility, the most common sources of CO were internal combustion engines (50%) and furnaces (37%).[14] The U.S. CPSC tracks hospital emergency department visits for non-fire-, non-automobile-related cases of CO poisoning resulting from consumer products.[15] From their sample of 1110 cases reported from 1992 to 1996, consumer products implicated in CO poisonings included heating systems (71%), gas ranges/ovens (6%), grills (4%), portable generators and pumps (4%), fuel-powered tools (3%), gas water heaters (2%), and gas clothes dryers (1%). Differences in CO sources among the various series are likely related to differences in the population studied.

The most common source of CO resulting in unintentional CO-related death in the United States is motor vehicle exhaust, comprising 57% of such fatalities in a national study.[1] The majority are associated with stationary automobiles. CPSC show that the consumer products most likely to be sources of CO for non-fire-, non-automobile-related deaths are heating systems (73%), charcoal grills (10%), gas water heaters (5%), camp stoves/lanterns (5%), and gas ranges/ovens (5%).[15]

Series of cases from individual states show some variability in these figures, again likely related to geographic and population influences. Among 444 accidental CO deaths in California, common sources of CO were heating and cooking appliances (40%), motor vehicles (31%), and charcoal grills (13%).[17] In a series of 74 unintentional CO deaths from New Mexico, the most frequent sources of CO were home heating equipment (50%) and motor vehicles (46%).[12] Finally, the most

TABLE 8.1
Sources of Carbon Monoxide in Cases of Unintentional Poisoning

˙CO Source	Ref. 11 (n = 807)	Ref. 25 (n = 631)
Furnace	345 (42.7%)	65 (10.3%)
Motor vehicle	178 (22.1%)	108 (17.1%)
Fire	53 (6.6%)	77 (12.2%)
Indoor charcoal	Not reported	79 (12.5%)
Gas-powered electrical generator	Not reported	59 (9.4%)
Other gas-powered motor	73 (9.0%)	Not reported
Other indoor appliance	63 (7.8%)	Not reported
Boat	Not reported	42 (4.7%)
Other/unknown	95 (11.8%)	201 (31.9%)

common sources of CO in 174 unintentional CO fatalities in Colorado were fire (36%), motor vehicles (34%), and furnaces (10%).[11]

8.5 CIRCUMSTANCES OF CARBON MONOXIDE POISONING

CO poisoning may be accidental or intentional. Among series of patients referred to HBO facilities for treatment of severe poisoning, the proportion of accidental cases has been 76% in North Carolina,[25] 82% in Utah,[14] and 72% in Washington State.[25] Among fatalities from CO poisoning, only 31% are unintentional.[1] Males account for 71 to 76% of suicidal CO deaths and motor vehicles are the source of CO in 97% of cases.[1,25]

Simultaneous exposure of multiple individuals is relatively common in incidents of CO poisoning. Among cases of CO poisoning due to indoor burning of charcoal briquets treated at a Seattle HBO facility, 69% of incidents involved poisoning sufficiently severe to require HBO treatment of more than one individual.[10] This likely relates to the indoor use of charcoal in a family setting for heating or cooking. In a series of accidental California CO fatalities, most vehicular exposures resulted in a single death (94%), while multiple deaths were more common when heating and cooking appliances were the source of CO (33%).[17] Among cases of CO poisoning collected by the CPSC (non-fire, non-automobile consumer product related), 22% of fatal incidents involved more than one death and 45% of nonfatal incidents resulted in more than one person poisoned.[15]

The location of the individual at the time of poisoning is obviously related to the source of CO. In the case of nonvehicular poisonings, the majority occur in a residential setting, in either a home or garage.[12,15,17,28] Other relatively common locations include campers, tents, boats.[13,15] In one series of 80 accidental poisonings occurring in residential settings, 39 (49%) occurred while the individual was asleep.[28] This has obvious implications for prevention with CO detectors/alarms.

Prior consumption of ethanol is common among those with CO poisoning, presumably affecting judgment, altering consciousness, and predisposing to exposure which would not otherwise have occurred. Alcohol has been reported to be involved in 31 to 42% of unintentional CO deaths in series of poisoning from all causes.[12,17,28] The rate of alcohol use appears to vary depending upon the source of CO.[17,28] In one study, alcohol was detected in 80% of adult fire victims, all of whom had elevated carboxyhemoglobin (COHb) levels.[29] Blood alcohol levels were higher among victims discovered in bed as compared with those found near an exit. Among deaths from unintentional CO poisoning from motor vehicle exhaust, 47 to 68% have been reported to have blood alcohol concentrations of at least 0.10 g/dl.[9,12,30] In a series of 59 patients dying from accidental CO poisoning from charcoal briquets, 37% involved alcohol consumption.[31] In a report of 16 deaths from CO poisoning due to faulty home heating systems, only 5 individuals (31%) had a positive blood or liver ethanol test, and none was over 0.10 g/dL.[32]

While alcohol undoubtedly increases risk for CO exposure, there are limited data that suggest it might actually be protective in CO poisoning. One clinical study of fatal CO poisonings noted that blood COHb levels were higher among those who also had the highest blood alcohol levels, raising the possibility that longer exposure to CO is tolerated before death in individuals with higher blood ethanol levels.[33] One laboratory study demonstrated that pretreatment of rats with ethanol increased both tolerance and survival at various levels of CO exposure.[34]

8.6 MANAGEMENT OF CARBON MONOXIDE POISONING IN THE UNITED STATES

As mentioned previously, the death rate from accidental CO poisoning has been declining in the United States. Proposed explanations for this observation have included (1) disease prevention, related to factors such as automobile emission control regulations, (2) more stringent occupational exposure standards, and (3) public education.[1] Additional possibilities include improvement in diagnosis and medical management of the disease. While underdiagnosis of CO poisoning has been described,[6,7] it is possible that physician recognition of CO poisoning is improving. Approximately 400 articles on CO poisoning were published in the English-language medical literature from 1985 to 1998.[35] It is not unreasonable to expect that this degree of availability of information about the disease would enhance knowledge and awareness among medical practitioners.

Another possible explanation for the declining death rate from accidental CO poisoning is improved treatment of the disease. All agree that appropriate management of acute CO poisoning includes removal of the individual from the source of exposure and administration of supplemental oxygen to enhance clearance of CO from the body. Some degree of disagreement exists in the U.S. medical community as to the exact roles for normobaric vs. HBO therapy in management of the CO-poisoned patient.[36,37] However, this is an area that is the subject of intense research, the majority of which supports HBO treatment in at least some subgroups of CO-poisoned patients.

As of January 1999, six prospective clinical trials have been reported comparing normobaric oxygen and HBO in the treatment of patients with acute CO poisoning. Three of these have been published in peer-reviewed form[38–40] and three only in abstract form to date.[41-43] Among the five trials in which the treatment groups have been unblinded, three showed statistically superior clinical outcomes among patients treated with HBO,[39,40,43] while two showed equivalent outcomes with normobaric oxygen and HBO.[38,42]

Two of the trials have been performed in the United States. In one, which is still ongoing,[41] patients with acute CO poisoning are being randomized to three HBO treatments in 24 h (50 min oxygen at 3.0 atmospheres absolute, or atm abs, followed by 60 min oxygen at 2.0 atm abs) or three sham HBO treatments with 100% oxygen breathing of the same duration at 1.0 atm abs. Interim analysis of results after enrollment of 49 patients demonstrated a 32% (8/25) rate of neurologic sequelae in one treatment arm and 17% (4/24) in the other. As the sequelae rates were not statistically different ($p = 0.0538$), the study was not unblinded and continues to accrue patients.

The second U.S. trial randomized 60 patients with mild CO poisoning, excluding those with history of unconsciousness or cardiac compromise, to treatment with HBO vs. normobaric oxygen at 1.0 atm abs until asymptomatic.[40] The HBO protocol utilized included 30 min of oxygen breathing at 2.8 atm abs, followed by 90 min at 2.0 atm abs. Patients were followed with serial neuropsychological testing in an attempt to detect development of delayed neurologic sequelae (DNS). DNS developed in 7 of 30 patients (23%) treated with normobaric oxygen and in no patients following treatment with HBO ($p < 0.05$). Among those developing DNS, impairment persisted for an average of 6 weeks and often interfered with activities of daily life.

Data collected by the Maryland Institute of Emergency Medical Services System demonstrate that the number of HBO treatments performed annually in the United States for CO intoxication has increased steadily over the past two decades.[44] In a 1992 study of North American HBO facilities, it was found that 51 multiplace and 90 monoplace facilities in the United States utilized their hyperbaric chamber for treatment of acute CO poisoning.[45] In that year, multiplace hyperbaric chambers were used to treat 1117 CO-intoxicated patients, with individual facilities treating 0 to 161 patients (mean 22 patients per facility). Monoplace chambers treated a total of 1240 patients in 1992, with individual facilities treating 0 to 112 patients (mean 14 patients per facility). Combining data from monoplace and multiplace facilities found that 2357 total patients were treated at 141 centers, averaging 17 patients per facility. In the case of multiplace facilities, this represented a 34% increase in patients treated annually as compared with figures from 2 years earlier.[46]

With regard to patient selection for treatment with HBO, recommendations are provided by the Hyperbaric Oxygen Therapy Committee of the Undersea and Hyperbaric Medical Society.[47] In their 1999 report, the committee recommends that CO-intoxicated patients with transient or prolonged unconsciousness, neurological signs, cardiovascular dysfunction, or severe acidosis be referred for HBO therapy irrespective of their COHb levels. The same report notes that the role of neuropsychological testing in patient selection for HBO is unclear. Finally, the committee suggests that treatments be performed at a pressure of 2.4 to 3.0 atm abs. In patients with persistent neurologic dysfunction after the initial treatment, retreatment may be performed once or twice daily until there is no further improvement in cognitive functioning, to a maximum of five HBO treatments.

TABLE 8.2
Proportion of North American Hyperbaric Unit Medical Directors Utilizing
Various Criteria to Determine Need for HBO Therapy in Patients with
CO Poisoning Presenting Less Than 3 h after Exposure[45]

Criteria	Percent
Patient arrives at emergency department unconscious with COHb = 9.5%	98
Initially unconscious upon CO exposure, arriving at the emergency department awake and asymptomatic, with normal neurologic examination and COHb = 9.5%	77
History of CO exposure, no loss of consciousness, COHb = 9.5% and	
No signs or symptoms	7
Headache and dizziness only	48
Electrocardiogram suggesting acute myocardial ischemia	91
Focal neurologic abnormality on physical examination	94
Abnormal psychometric testing	91
No loss of consciousness, presenting with headache, nausea, COHb = 40%, and normal neurological examination, electrocardiogram, and neuropsychiatric testing	92

While these recommendations may appear clear, there exists tremendous variation in clinical HBO practice in North America with regard both to patient selection
criteria and treatment protocols utilized for this disease. When each area is examined
in detail, it is seen that consensus exists in only selected aspects of the management
of the patient with CO poisoning.

Medical directors of hyperbaric chamber facilities in the United States and
Canada were recently surveyed to determine patient selection criteria utilized for
application of HBO in CO poisoning.[45] Approaches were found to be most similar
when dealing with the more severely poisoned patient (Table 8.2). A significant
majority of medical directors administer HBO to CO-poisoned patients with coma,
focal neurologic deficits, ischemic changes on electrocardiogram, abnormal neuropsychiatric testing, or transient loss of consciousness, despite a relatively low COHb
level. The approach to patients with milder degrees of poisoning is less clear. Only
a minority of hyperbaric facility medical directors surveyed in that study would
utilize HBO for the patient with a slightly elevated COHb level and either no
symptoms or headache and dizziness only (Table 8.2).

The appropriate role of the COHb level in determining which cases of CO
poisoning warrant HBO treatment remains undefined. Nearly all North American
medical directors (92%) would administer HBO to a poisoned patient with COHb
of 40%, headache, and nausea (Table 8.2). Only two thirds, however, identify
COHb level as an independent criterion for the HBO treatment of an asymptomatic
patient.[45] Therefore, while the majority use HBO to treat the patient with a specified
minimum COHb level irrespective of clinical signs, some also require symptomatic

manifestation of the poisoning before recommending HBO treatment. Based upon the results of that study, it would appear that manifestations of headache or nausea are considered sufficient symptoms by these physicians to administer HBO therapy.

When the ÇOHb level is applied as an independent indication for HBO therapy, the range of COHb values utilized is quite wide.[45] A COHb level of 25% is identified most often, but this value is used by only half of those applying COHb level as a sole criterion for HBO treatment. This variability may result from the fact that it is not possible to draw firm conclusions from the published clinical literature with regard to the role of COHb in determining need for HBO therapy.

The importance placed upon the delay from CO exposure to medical evaluation as a factor in determining appropriateness for HBO treatment is also quite variable among hyperbaric facility medical directors.[45] Previous studies have demonstrated that effectiveness of HBO therapy decreases with the duration of delay to treatment.[48,49] Delayed treatment is associated with an increased incidence of residual neurologic deficits after treatment, as well as increased mortality. Neither these nor other studies, however, have precisely defined time limits beyond which HBO therapy for CO poisoning will be ineffective and should therefore be withheld. This lack of information is apparent from the time limits utilized by HBO physicians. One half of North American medical directors use a time limit to deny HBO treatment to a patient with only transient loss of consciousness.[45] When time limits are applied in such instances, intervals ranging from 6 to 48 h are most commonly used, but delays of 1 to 2 months are allowed by some physicians. In the CO-poisoned patient presenting with focal neurologic findings, time limits are applied by only one quarter of directors to determine eligibility for HBO treatment.

Related to the issue of temporal delay is the patient presenting with delayed development of neurologic or neuropsychiatric sequelae after CO poisoning. A majority of medical directors in the United States and Canada utilize HBO to treat such patients.[45] Published data regarding the efficacy of such treatment are contradictory.[50-52]

Interestingly, management of acute CO poisoning in pregnancy remains a topic of controversy in North America. Only 74% of American and Canadian hyperbaric facilities have treated or would treat pregnant patients with CO intoxication.[46] One quarter do not use HBO for pregnant CO-poisoned patients despite a lack of data demonstrating increased risk from such treatment and recommendations from authors in both the United States and Europe that such patients be treated.[53,54]

As noted above, there is greatest agreement among hyperbaric physicians regarding use of HBO for the most severely poisoned patients. This seems to be reflected in hyperbaric medicine practice. A recent study found that 6.9% of those evaluated for CO poisoning in emergency departments in Washington, Idaho, and Alaska were referred for HBO therapy.[4] The report also estimated the number of emergency department visits nationally for CO poisoning (42,890). Using a 1992 figure for the number of CO-poisoned individuals treated with HBO in the United States (2355), a nationwide hyperbaric treatment rate of 5.7% was calculated. Thus, while HBO is recommended for treatment of CO poisoning, it is generally reserved for a select population of patients.

Once the decision has been made to utilize HBO for CO poisoning, physician opinions regarding hyperbaric treatment protocols are quite varied. In 1990, 1023 cases of acute CO poisoning were treated in 42 multiplace hyperbaric chamber facilities in North America, with 38 U.S. facilities treating 832 patients and four Canadian facilities treating 191 patients.[46] A total of 18 different hyperbaric protocols were used at those facilities for primary treatment of acute CO poisoning. These include 3 protocols with a maximum pressure of 3.0 atm abs, 13 protocols with a maximum pressure of 2.8 atm abs, and 2 protocols with a maximum pressure of 2.4 to 2.5 atm abs. The oxygen dose delivered by these protocols (calculated by multiplying the minutes of 100% oxygen breathing by atm abs pressure) differs by a factor of over threefold. In the year studied, 28% of patients were treated at facilities utilizing 3.0 atm abs, 55% at facilities utilizing 2.8 atm abs, and 17% at facilities utilizing 2.4 to 2.5 atm abs. While this might suggest that a consensus exists for treatment at 2.8 atm abs, it should be recognized that the slight majority of patients treated at that pressure were divided among 13 protocols.

The most frequently identified multiplace treatment protocol in North America utilizes a maximum pressure of 3.0 atm abs and is commonly known as the "U.S. Air Force" protocol, developed and applied by the U.S. Air Force for treatment of CO poisoning.[55] While this protocol was identified as the primary treatment protocol by more multiplace facilities than any other, it is utilized by only 33% of facilities. Furthermore, just 15% of patients treated in multiplace hyperbaric chambers in North America in 1990 were managed by this protocol, attesting to the lack of consensus in this area.

There are no published prospective studies comparing outcome of patients with acute CO poisoning treated with different HBO protocols. While the relative benefit of different protocols has not been directly compared, side effects of some treatment protocols have been evaluated. A large study reviewed 300 patients treated at each of three hyperbaric pressures to define the incidence of central nervous system (CNS) oxygen toxicity associated with treatment at various partial pressures of oxygen.[25] It found that CNS toxicity, as manifest by grand mal seizure activity, was significantly more common among patients treated at 2.80 or 3.00 atm abs, as compared with 2.5 atm abs.

8.7   CONCLUSIONS

CO intoxication is a common health problem in the United States. The mortality rate from accidental CO poisoning appears to be decreasing, probably due to improvements in both disease prevention and treatment. While a consensus does exist among North American HBO medical directors with regard to many issues in CO poisoning, discrepancy still persists about several aspects of patient selection and HBO treatment protocols. Despite these discrepancies, there are suggestions that the benefits seen with HBO therapy in clinical trials are indeed impacting the outcome of the disease nationally. When one compares the increasing number of HBO treatments performed in the United States for CO poisoning with the declining number of accidental CO-related deaths, a statistically significant correlation is seen. It is hoped that future refinements in patient selection criteria and hyperbaric treatment protocols for CO intoxication will result in improved management of this common form of poisoning.

REFERENCES

1. Cobb, N. and Etzel, R.A., Unintentional carbon monoxide-related deaths in the United States, 1979 through 1988, *J. Am. Med. Assoc.*, 266, 659–663, 1991.
2. Schaplowsky, A.F., Oglesbay, F.B., Morrison, J.H., Gallagher, R.E., and Berman, W., Jr., Carbon monoxide contamination of the living environment: a national survey of home air and children's blood, *J. Environ. Health*, 36, 569–573, 1974.
3. Hampson, N., Incidence of carbon monoxide poisoning in the United States [letter], *Undersea Hyperb. Med.*, 26, 47–48, 1999.
4. Hampson, N.B., Emergency department visits for carbon monoxide poisoning, *J. Emerg. Med.*, 16, 695–698, 1998.
5. Hampson, N.B., Ball, L.B., and Macdonald, S.C., Are symptoms helpful in the evaluation of patients with acute carbon monoxide poisoning [abstract]? *Undersea Hyperb. Med.*, 26 (Suppl.), 50, 1999.
6. Grace, T.W. and Platt, F.W., Subacute carbon monoxide poisoning: another great imitator, *J. Am. Med. Assoc.*, 246, 1698–1700, 1981.
7. Baker, M.D., Henretig, F.M., and Ludwig, S., Carboxyhemoglobin levels in children with nonspecific flu-like symptoms, *J. Pediat.*, 113, 501–504, 1988.
8. Hampson, N.B., Dunford, R.G., and Norkool, D.M., Temporal epidemiology of emergency hyperbaric treatments [abstract]. *Undersea Biomed. Res.*, 18 (Suppl.), 85, 1991.
9. Anonymous, Unintentional deaths from carbon monoxide poisoning — Michigan, 1987–1989, *MMWR*, 41, 881–882, 1992.
10. Hampson, N.B., Kramer, C.C., Dunford, R.G., and Norkool, D.M., Carbon monoxide poisoning from indoor use of charcoal briquets, *J. Am. Med. Assoc.*, 271, 52–53, 1994.
11. Cook, M., Simon, P.A., and Hoffman, R.E., Unintentional carbon monoxide poisoning in Colorado, 1986 though 1991, *Am. J. Public Health*, 85, 988–990, 1995.
12. Moolenaar, R.L., Etzel, R.A., and Parrish, R.G., Unintentional deaths from carbon monoxide poisoning in New Mexico, 1980 to 1988: a comparison of medical examiner and national mortality data, *West. J. Med.*, 163, 431–434, 1995.
13. Silvers, S.M. and Hampson, N.B., Carbon monoxide poisoning among recreational boaters, *J. Am. Med. Assoc.*, 274, 1614–1616, 1995.
14. Howe, S., Hopkins, R.O., and Weaver, L.K., A retrospective demographic analysis of carbon monoxide poisoned patients [abstract]. *Undersea Hyperb. Med.*, 23 (Suppl.), 84, 1996.
15. Ault, K., Estimates of non-fire-related carbon monoxide poisoning deaths and injuries, Memorandum to Elizabeth Leland, U.S. Consumer Product Safety Commission, December 10, 1997.
16. Dodson, W.W., Santamaria, J.P., Etzel, R.A., Desautels, D.A., and Bushnell, J.D., Epidemiologic study of carbon monoxide poisoning cases receiving hyperbaric oxygen treatment [abstract], *Undersea Hyperb. Med.*, 24 (Suppl.), 38, 1997.
17. Girman, J.R., Chang, Y.-L., Hayward, S.B., and Liu, Y.-S., Causes of unintentional deaths from carbon monoxide poisoning in California, *West. J. Med.*, 168, 158–165, 1998.
18. Glass, R.I., O'Hare, P., and Conrad, J.L., Health consequences of the snow disaster in Massachusetts, February 6, 1978, *Am. J. Public Health*, 69, 1047–1049, 1979.
19. Faich, G. and Rose, R., Blizzard morbidity and mortality: Rhode Island 1978, *Am. J. Public Health*, 69, 1050–1052, 1979.
20. Geehr, E.C., Salluzzo, R., Bosco, S., Braaten, J., Wahl, T., and Wallenkampf, V., Emergency health impact of a severe storm, *Am. J. Emerg. Med.*, 7, 598–604, 1989.

21. Anonymous, Carbon monoxide poisonings associated with snow-obstructed vehicle exhaust systems — Philadelphia and New York City, January 1996, *MMWR*, 45, 1–3, 1996.

22. Houck, P.M. and Hampson, N.B., Epidemic carbon monoxide poisoning following a winter storm, *J. Emerg. Med.*, 15, 469–473, 1997.

23. Geltzer, A.J., Geltzer, A.M., Dunford, R.G., and Hampson, N.B., Effects of weather on incidence of attempted suicide by carbon monoxide (CO) poisoning in the Seattle area [abstract], *Undersea Hyperb. Med.*, 25 (Suppl.), 48, 1998.

24. Sloan, E.P., Murphy, D.G., Hart, R., Cooper, M.A., Turnbull, T., Barreca, R.S., and Ellerson, B., Complications and protocol considerations in carbon monoxide-poisoned patients who require hyperbaric oxygen therapy: report from a ten-year experience, *Ann. Emerg. Med.*, 8, 629–634, 1989.

25. Hampson, N.B., Simonson, S.G., Kramer, C.C., and Piantadosi, C.A., Central nervous system oxygen toxicity during hyperbaric treatment of patients with carbon monoxide poisoning, *Undersea Hyperb. Med.*, 23, 215–219, 1996.

26. Norkool, D.M. and Kirkpatrick, J.N., Treatment of acute carbon monoxide poisoning with hyperbaric oxygen: a review of 115 cases, *Ann. Emerg. Med.*, 14, 1168–1171, 1985.

27. Ralston, J.D. and Hampson, N.B., Carbon monoxide poisoning: are minority races at increased risk [abstract]? *Undersea Hyperb. Med.*, 25 (Suppl.), 48, 1998.

28. Yoon, S.S., Macdonald, S.C., and Parrish, R.G., Deaths from unintentional carbon monoxide poisoning and potential for prevention with carbon monoxide detectors, *J. Am. Med. Assoc.*, 279, 685–687, 1998.

29. Barillo, D.J., Rush, B.F., Jr., Goode, R., Reng-Lang, L., Freda, A., and Anderson, E.J., Is ethanol the unknown toxin in smoke inhalation injury? *Am. Surg.*, 52, 641–645, 1986.

30. Baron, R.C., Backer, R.C., and Sopher, A.M., Unintentional deaths from carbon monoxide in motor vehicle exhaust: West Virginia, *Am. J. Public Health*, 79, 328–330, 1989.

31. Liu, K-S., Girman, J.R., Hayward, S.B., Shusterman, D., and Chang, Y.-L., Unintentional CO deaths in California from charcoal grills and hibachis, *J. Exposure Anal. Environ. Epidemiol.*, 3 (Suppl.), 143–152, 1993.

32. Caplan, Y.H., Thompson, B.C., Levine, B., and Masemore, W., Accidental poisonings involving carbon monoxide, heating systems, and confined spaces, *J. Forensic Sci.*, 31, 117–121, 1986.

33. King, L.A., Effect of ethanol in fatal carbon monoxide poisonings, *Hum. Toxicol.*, 2, 155–157, 1983.

34. Sharma, P. and Penney, D.G., Effects of ethanol in acute carbon monoxide poisoning, *Toxicology*, 62, 213–226, 1990.

35. *MEDLINE Database*, United States National Library of Medicine, Washington, D.C., 1999.

36. Tibbles, P.M. and Perotta, P.L., Treatment of carbon monoxide poisoning: a critical review of human outcome studies comparing normobaric oxygen with hyperbaric oxygen, *Ann. Emerg. Med.*, 24, 269–276, 1994.

37. Van Meter, K.W., Weiss, L., Harch, P.G., Andrews, L.C., Jr., Simanonok, J.P., Staab, P.K., and Gottlieb, S.F., Should the pressure be off or on in the use of oxygen in the treatment of carbon monoxide-poisoned patients? *Ann. Emerg. Med.*, 24, 283–288, 1994.

38. Raphael, J.C., Elkharrat, D., Jars-Guincestre, M.C., Chastang, C., Chasles, V., Vercken, J.B., and Gajdos, P., Trial of normobaric and hyperbaric oxygen for acute carbon monoxide intoxication, *Lancet*, 2, 414–419, 1989.

39. Ducasse, J.L., Celsis, P., and Marc-Vergnes, J.P., Non-comatose patients with acute carbon monoxide poisoning: hyperbaric or normobaric oxygenation? *Undersea Hyperb. Med.*, 22, 9–15, 1995.
40. Thom, S.R., Taber, R.L., Mendiguren, I., Clark, J.M., Hardy, K.R., and Fisher, A.B., Delayed neuropsychologic sequelae after carbon monoxide poisoning: prevention by treatment with hyperbaric oxygen, *Ann. Emerg. Med.*, 25, 474–480, 1995.
41. Weaver, L.K., Hopkins, R.O., Larson-Lohr, V., and Haberstock, D., Double-blind, controlled, prospective, randomized clinical trial (RCT) in patients with acute carbon monoxide (CO) poisoning: outcome of patients treated with normobaric oxygen or hyperbaric oxygen — an interim report [abstract]. *Undersea Hyperb. Med.*, 22 (Suppl.), 14, 1995.
42. Scheinkestel, C.D., Jones, K., Cooper, D.J., Millar, I., Tuxen, D.V., and Myles, P.S., Interim analysis — controlled clinical trial of hyperbaric oxygen in acute carbon monoxide poisoning [abstract], *Undersea Hyperb. Med.*, 23 (Suppl.), 7, 1996.
43. Mathieu, D., Wattel, F., Mathieu-Nolf, M., Durak, C., Tempe, J.P., Bouachour, G., and Sainty, J.M., Randomized prospective study comparing the effect of HBO versus 12 hours NBO in noncomatose CO poisoned patients [abstract], *Undersea Hyperb. Med.*, 23 (Suppl.), 7–8, 1996.
44. Maryland Institute of Emergency Medical Services System, Baltimore.
45. Hampson, N.B., Dunford, R.G., Kramer, C.C., and Norkool, D.M., Selection criteria utilized for hyperbaric oxygen treatment of carbon monoxide poisoning, *J. Emerg. Med.*, 13, 227–231, 1995.
46. Hampson, N.B., Dunford, R.G., and Norkool, D.M., Treatment of carbon monoxide poisonings in multiplace hyperbaric chambers, *J. Hyperb. Med.*, 7, 165–171, 1992.
47. Hyperbaric Oxygen Therapy Committee, Hyperbaric Oxygen Therapy: 1999 Committee Report, Hampson, N.B., Ed., Undersea and Hyperbaric Medical Society, Kensington, MD, 1999.
48. Goulon, M., Barois, A., Rapin, M., Nouailhat, F., Grosbuis, S., and Labrousse, J., Intoxication oxycarbonee et anoxie aigue par inhalation de gaz de charbon et d'hydrocarbures, *Ann. Med. Intern.* (Paris), 120, 335–349, 1969. (English translation: *J. Hyperb. Med.*, 1, 23–41, 1986.)
49. Skeen, M.B., Massey, E.W., Moon, R.E., Shelton, D.L., Fawcett, T.A., and Piantadosi, C.A., Immediate and longterm neurological sequelae of carbon monoxide intoxication [abstract], *Undersea Biomed. Res.*, 18 (Suppl.), 36, 1991.
50. Myers, R.A.M., Snyder, S.K., and Emhoff, T.A., Subacute sequelae of carbon monoxide poisoning, *Ann. Emerg. Med.*, 14, 1163–1167, 1985.
51. Hopkins, R.O. and Weaver, L.K., Does late repetitive hyperbaric oxygen improve delayed neurologic sequelae associated with carbon monoxide poisoning [abstract]? *Undersea Biomed. Res.*, 18 (Suppl.), 34, 1991.
52. Coric, V., Oren, D.A., Wolkenberg, F.A., and Kravitz, R.E., Carbon monoxide poisoning and treatment with hyperbaric oxygen in the subacute phase, *J. Neurol. Neurosurg. Psychiatr.*, 65, 245–247, 1998.
53. Van Hoesen, K.B., Camporesi, E.M., Moon, R.E., Hage, M.L., and Piantadosi, C.A., Should hyperbaric oxygen be used to treat the pregnant patient for acute carbon monoxide poisoning? *J. Am. Med. Assoc.*, 261, 1039–1043, 1989.
54. Elkharrat, D., Raphael, J.C., Korach, J.M., Jars-Guincestre, M.C., Chastang, C., Harboun, C., and Gajdos, P., Acute carbon monoxide intoxication and hyperbaric oxygen in pregnancy, *Intensive Care Med.*, 17, 289–292, 1991.
55. U.S. Air Force, Hyperbaric chamber operations, Air Force Pamphlet, 161-27 79-81, July 5, 1983.

9 Death by Suicide Involving Carbon Monoxide Around the World

Pierre Baume and Michaela Skopek

CONTENTS

9.1 INTRODUCTION

Although a relatively rare event, suicide ranks among the top 10 causes of death in most Western nations.[1] However, in some countries such as Australia, it ranks as the leading cause of death by injuries, ahead of motor vehicle accidents and well ahead of homicides which have been gradually falling to a 60-year low.[2] Indeed, Australia now has one of the highest suicide rates for young males, where a fourfold increase has taken place in the last three decades or so.[3] A great proportion of these suicide

0-8493-2065-8/00/$0.00+$.50
© 2000 by CRC Press LLC

deaths are associated with only relatively few methods. These include, but are not limited to, hanging, firearms, drug overdose, and deliberate exposure to motor vehicle exhaust emissions or carbon monoxide (CO) poisoning, which together account for about 85% of all suicides.[4]

Suicide methods are usually influenced by a complex constellation of social, cultural, psychological, environmental, and physical factors, which precede the individuals' decisions to end their lives. A number of studies have shown that the popularity of particular methods of suicide changes within and between countries over time.[5-12] In the Australian context, the increase in suicide rates especially in males has been associated with a nearly fivefold increase in suicide by hanging and a quadrupling in suicide by CO poisoning. Fortunately, the increases in those two methods have been somewhat offset by a substantial decrease in suicides by overdose over the last 25 years and firearms for the last 10 years.[2,4]

This chapter examines the impact of suicide by CO poisoning. Although the focus is centered on the Australian experience, this discussion takes an international perspective. American and European studies examining profiles of suicide completers and attempters as well as suicide prevention initiatives in these continents are discussed.

9.1.1 Defining Terms

In this chapter CO poisoning refers to all deaths relating to CO from car exhaust (ICD 9 Coding E 952, suicide and self-inflicted poisoning by other gases and vapors). The category of suicide by CO consists mainly of poisoning from motor vehicle exhaust gas. CO poisoning accounts for nearly all suicide deaths by gas in Australia and overseas and generally occurs in association with a motor vehicle.[13,14] In 1997 only four suicides from poisoning by gases in domestic use were recorded in Australia. This category, ICD 9 coding E 951 (suicide and self-inflicted poisoning by gases in domestic use) will not be used in this discussion.[2] Although there is a code specific to motor vehicle exhaust gas suicide (E925.0), data to this fourth digit has only been available from the Australian Bureau of Statistics (ABS) since 1979. Therefore, E952 is preferred for consistency.

9.2 BACKGROUND TO THE PROBLEM

CO poisoning has become one of the principal means of suicide in many affluent countries. Individuals at risk of suicide are usually fully conversant that this approach is both lethal and painless.[15] In most advanced economies, where automobiles are an everyday utility, deliberate CO poisoning ranges from 5 to 20% of all suicides.[16] There are some Western countries, however, where the incidence of suicide by this method is rare, such as Japan and Hong Kong, despite a large proportion of car ownership.[17,18]

In European countries suicide by exposure to motor vehicle exhaust has been reported to be especially high.[10,19,20] In Sweden, for example, CO poisoning is the fourth leading method of suicide, accounting for 15% of all male and 3% of all female suicides.[21] These trends have been found in a number of other European countries such as Ireland, Finland, and Britain, as well as in the United States and

Australia.[8,11,14,22-26] In Finland, Ohberg et al.[24] reported a significant increase in CO suicides from 1965 to 1991 for young men, although this increase was specifically severe in those aged 20 to 24 years, but very rare in women. In Britain, the number of CO suicides increased from around 350/year in 1974 to over 1200 in 1991, being the second most common method of suicide for men and the third most common method for women. For both genders, suicide rates for this method increased between 1983 and 1991 and then declined.[14] Since their peak in 1991, rates for both men and women have fallen by just one third. Suicide rates reached a peak at the beginning of the 1990s and then fell dramatically in 1993. Indeed, the decrease in suicide rates from CO poisoning accounted for most of the overall fall in suicide rates for men in England and Wales since 1991.[14]

In Australia, a recently published federal report "Access to Means of Suicide by Young Australians"[27] highlights concerns regarding increasing rates of CO poisoning over the last two decades. The data show a greater than twofold rise in fatal CO poisoning in young males and that this method was the second leading method of suicide by young people of both genders (males 36.1% and females 36.8%) behind hanging. Early speculation suggested that the rise in CO poisoning might have been due to substitution and may have resulted from the recent decline in firearm suicides. As the decline in male firearm suicides has been only one third of the magnitude of the rise in CO and hanging suicides, this suggested explanation appears to have limited validity.[4]

Suicide methods vary considerably in terms of "lethality." Some methods, especially CO poisoning, rank very highly in terms of the probability of death. In one study, the authors reported a firearms lethality rate of 85%, 80% for hanging, and 77% for CO poisoning.[28] Lethality is partly dependent on the length of time that elapses between the suicide event and death.[29] CO poisoning may take several minutes to hours depending on the effectiveness of the setup and whether or not the vehicle is equipped with a catalytic converter, as well as how much alcohol is consumed at the time.[11,26] Site of death may also be an important consideration. If the person chooses to drive away to a remote and/or concealed site, it may be more difficult to reduce the potential for a lethal outcome. In Australia the great majority of people who die by suicide do so at their place of residence.[3]

The rise in deliberate exposure to motor vehicle exhaust gas as a method of suicide may be related in part to the proportion of cars available in circulation.[24,26] In a recent review of ownership of cars in 28 countries, nations with more cars per capita had significantly higher suicide rates from CO poisoning.[8,23] Environmental factors such as climate may also play a role. Tanney[30] reported that a significant proportion of suicides due to CO poisoning in Canada were attributable to the northerly climate where enclosed garages for cars provided a suitable "gas chamber."[30] In Australia the rise in suicide by this method has also been partly attributed to method substitution associated with the introduction of nontoxic domestic gas, which resulted in a marked decline in suicides by that method.[7]

Geographic location has also been reported to be a possible factor in the selection of a suicide method. A number of recent studies have reported sharp increases in rural suicides with firearms, hanging, and CO poisoning, especially in young people.[31-33] Figure 9.1 highlights the rise of CO suicides by residence

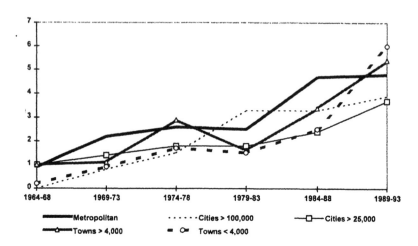

FIGURE 9.1   Car exhaust suicide rates per 100,000 by residence, in Australia 1964 to 1993, Males 15 to 24 years.

between 1964 and 1993 for males aged 15 to 24 years. There are complex reasons for these trends. In city areas, firearm ownership is not as high as in rural areas, and the publicity surrounding the rural suicide epidemic may have persuaded metropolitan families to secure their firearms. The availability of and need for firearms in small rural areas, a lack of safety in storage, and the use of firearms in dwindling populations may combine to elevate rural firearm suicide rates in young people, especially when combined with high impulsivity and alcohol consumption. Firearm availability is potentially significant for younger males completing suicide, who may be more prone to impulsivity and aggression and may be unwilling or unable to access mental health services. Given the frequency, impulsivity, and low intent of many suicide attempts, the high lethality of guns may convert many attempts at suicide into deaths by the presence of a firearm.[5]

Multiple etiological factors contribute to a suicide outcome in an individual, but it has been suggested that suicide modeling, especially within young people's social networks may be responsible at least in part for the increasing suicide rates in males of this age group.[34,35] A large number of reports have supported the hypothesis that there may be a significant relationship between the reporting of a suicide and subsequent suicides. These include those that have examined the link between media reports about suicide and subsequent suicides.[11,36–40] A number of authors have reported clustering of suicides,[34,41,42] which suggests that imitation occurs following media coverage or personal contact with the suicide event.[43–46] Press and television reports have been demonstrated to influence significantly young people to complete suicide.[47] More recently, the Internet has been identified as a new medium of communication with a possible influence on suicide rates. A study by Baume et al.[48] reported that the Internet could be influencing individuals in their selection of a particular method. A number of sites appeared not only to list the details of every step that should be taken, as well as the methods available, but also provided interactive forums to encourage active suicidal acts.[48] This evidence supports the view that if suicide is discussed and described explicitly in a public forum, vulnerable individuals may consider it a viable option. Hence, people who are struggling with seemingly insurmountable personal, interpersonal, or family problems may be the most vulnerable when exposed to the news of the suicide of a public person or a close friend.[5,16,49]

These arguments appear to be supported in part in relation to CO poisoning in two recent studies. The first, carried out in Finland by Ohberg et al.,[24] described a sharp rise in suicide by car exhaust after the showing of a movie *Force Play* in which the method was explicitly described. More recently during an Australian study of suicide attempts by CO poisoning in New South Wales (NSW), a prominent male citizen died by CO suicide. The case was covered extensively by the media and was cited by 20% of individuals (all male) as having influenced their choice of method. General media portrayal of suicide by exposure to motor vehicle exhaust was cited by a further 25% as having influenced their method choice. The number of survivors of attempted suicide presenting for treatment to the study center was 40% higher in the 6 months following the prominent citizen's suicide compared with the rate 12 to 24 months later.[11]

As well as the mortality associated with deliberate CO poisoning, there is also a significant morbidity associated with a suicide attempt by this method. The NSW Centre for Mental Health performed the most recent investigation of suicide attempts and suicide mortality. Estimates suggested that 22,000 to 30,000 people in NSW made attempts at self-harm of some kind in 1992 and there were 741 deaths by suicide in that year. There was a clear distinction between the three methods most likely to be fatal, and the two methods most likely to be used. Hanging, firearms, and motor vehicle exhaust gas were the methods used in 14% of these attempts, and resulted in 68% of deaths. There were 203 attempts at suicide with motor vehicle exhaust in NSW in 1992; 127 of these died without hospitalization, 76 were hospitalized, 4 died in hospital, and 72 survived. Thus, the fatality rate was 66% and the hospitalization rate was 37%. This compared with the fatality rate for hanging being 82% and the hospitalization rate of 19%, while the fatality rate of suicide attempts with firearms was 75% and the hospitalization rate was 31%.[50]

A limited number of studies have also examined the psychopathology and psychosocial profile of individuals using exposure to motor vehicle exhaust gas as a suicide method. A recent Swedish study investigated 194 cases of completed suicide by this method using a psychological autopsy methodology. A higher incidence of completed suicide was seen in rural regions, with the majority being middle-aged males (88%). Most died in a car outdoors with a vacuum cleaner tube connected from the exhaust to the passenger compartment. Psychiatric morbidity was present in 61% of cases and 37% had made a previous attempt. Suicide notes were found in 40% of cases. Blood alcohol was detected in 51%. The study concluded that reducing access to this method of suicide was of primary importance in future suicide prevention strategies.[21]

Exploration of the cultural context and meaning of exposure to motor vehicle exhaust gas has been reported, although in a limited way, for both indigenous people and immigrants. In the only study of this kind conducted in Australia, the authors found that the method of suicide for migrants from English-speaking countries was very similar to that of the Australian born. However, males from English-speaking countries had a slightly lower usage of firearms and a corresponding higher usage of poisonous substances. They found that for the period 1970 to 1992, CO suicide for females accounted for 11.6% for those born in Australia. This was compared with 12.3% for those born in other English-speaking countries, 11% for those born in Western Europe, 4.4% for those born in Southern Europe, and less than 5% for those born in Eastern Europe, the Middle East, or Asia. For males, the Australian average was 21.3% for the above period. This compared with 27% for those males who migrated from English-speaking countries. While 6% from those who originated from Southern Europe, 19.9% from Western Europe, 8.8% from Eastern Europe, 14.9% from Northern Europe, and about 8% for those whose origin was either the Middle East or Southern or Northern or Southeast Asia, used this method.[51]

Car exhaust as a means of suicide is widely available and accessible in developed countries. Individuals choosing this method rate it as relatively easy and painless.[11] It is not considered to have the violent visual repercussions associated with firearms, jumping, and/or even hanging. Individuals appear to give consideration to who will find them and how they will appear to this person and at times want to lessen the distress experienced by their survivors as they lessen their own distress through suicide. Although the question of cultural acceptability of this method has not previously been addressed, it can be hypothesized, however, that males tend to identify more readily with their motor vehicles and thus find it a more acceptable method.

Given the frequency of CO poisoning as a suicide method, it is surprising how few studies have addressed this issue. Changes in the frequency of CO poisoning as a suicide method cannot generally be explained by changes in availability, although knowledge about the availability of the method could influence rates. Fluctuations of suicides by CO poisoning within the same country over time, as well as between countries at any single point in time, more likely reflect changes or differences in the cultural acceptability of this method.

In summary, a disproportionate increase in suicides in males in recent years suggests a gradual but significant change in method, which could be linked, to a changing social context. The rise in deaths by car exhaust suicides in males has not been confined to any one location, suggesting that, in general, males have a continuing high lethality. The wide media coverage in Australia of death in custody by hanging in the mid-1980s and the suicide by CO poisoning of certain high-profile individuals, combined with the language of tabloid newspapers and other media, may have installed in the mind of some vulnerable individuals an association between despair and suicide.[47] CO poisoning and hanging are now understood to refer almost always to suicide. Therefore, an action (CO) may have become synonymous with an intention (suicide).[5] Furthermore, the availability of the means of suicide may influence rates of suicide and must therefore be considered in any suicide prevention initiative.

A number of themes emerge out of this background review, namely, that suicide rates and method selection are influenced by sociocultural factors. These vary from region to region, and generally males are more likely to choose this method. The next sections will now discuss the changes over time for this method across the life span, as well as the psychosocial profile of individuals who attempt and/or complete suicide using this method.

9.3 SUICIDE AND PARASUICIDE BY CAR EXHAUST GAS IN AUSTRALIA

This section examines the trends in CO suicide for Australia as a whole as well as for states and territories for the period 1960 to 1997. The information used in this section is derived primarily from raw data provided by the ABS[2] and the Centre for Injury Surveillance at Flinders University.[52]

9.3.1 METHODS

Mortality rates were derived from the number of deaths recorded each year for males and females with Australia or states and territories as the usual place of residence, across all age groups in the ABS Mortality Tabulations for 1960 to 1997. Rates were calculated by dividing the number of deaths in each category in each year by the number of males and females in the same age group estimated by the ABS to be living in Australia or in individual states or territories in that year.[2] The mortality rates reported here are for suicides by gender per 100,000 population of the same gender in the following age groups per annum: 15 to 24 years, 25 to 34 years, 35 to 44 years, 45 to 54 years, 55 to 64 years, 65 to 74 years, and 75 years and over.

The categories of interest for methods of suicide were self-inflicted injury by hanging, strangulation, and suffocation (E953), by firearms and explosives (E955), by poisoning by other gases and vapors (E952), and by poisoning by solid or liquid substance (E950). Coding is according to the World Health Organization, *International Classification of Diseases* (ICD), 7th revision from 1960 to 1973, 8th revision from 1974 to 1978, and 9th revision from 1979 to 1997.[2]

9.3.2 RESULTS

As previously discussed, there are only relatively few methods used in Australia as a means of suicide. These are hanging, CO, firearms, and overdose. Together these represent over 85% of all suicides in the country. Figures 9.2 and 9.3 outline the changes in rates of suicide for these four methods from 1960 to 1997 for males and females, respectively. As can be seen, there have been some welcome decreases in suicide rates by overdose for both genders and by firearms for males, which continue to show a declining trend. However, a significant increase in both hanging and CO poisoning is noted, where CO is now the second most common method of suicide in Australia for both genders combined. Both hanging and CO have risen in parallel and more sharply since the early 1980s and show no sign of plateauing or decreasing at this time. For males, CO suicide rates are now at the highest rate for the study period and are now twice as common as firearm suicides, whose rates are now lower than overdose deaths. Although the rise was gradual but constant until 1984, a sharp rise ensued thereafter in parallel with hanging until the end of 1997. The sharper rise for CO poisoning for males starts almost a decade before the decline in firearm suicides. These trends appear to lend very little support for the idea of method substitution.

For females, the trends for the four main methods are similar but the rates are far smaller. A very sharp decline in poisoning by solids and liquids is observed since

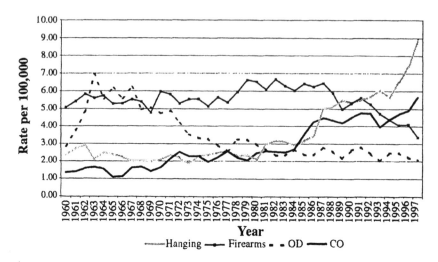

FIGURE 9.2   Methods of suicide in Australia 1960 to 1997, males all ages.

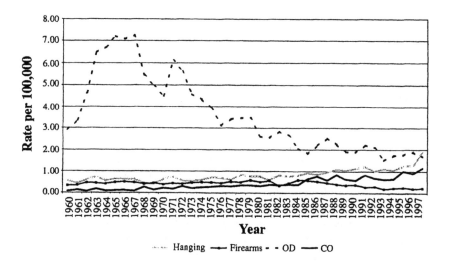

FIGURE 9.3   Methods of suicide in Australia 1960 to 1997, females all ages.

the mid-1960s. Figure 9.3 outlines the changes in trend in the other three methods, which seem to parallel those of males; namely, that both hanging and CO suicides begin to rise more sharply since the mid-1980s. However, firearm suicides appear to start declining at approximately the same time but from a very low baseline. In 1997, for the first time in Australia, the most common method of suicide for females is hanging. Assuming continuing trends, CO will soon become the second most common method for females, ahead of overdoses. These trends are of concern for this group, as females have traditionally chosen less active and less violent methods of suicide.

The number of people in Australia who choose CO as a form of suicide has gradually increased from 74 in 1960 to 629 in 1997. This represents an 8.5-fold

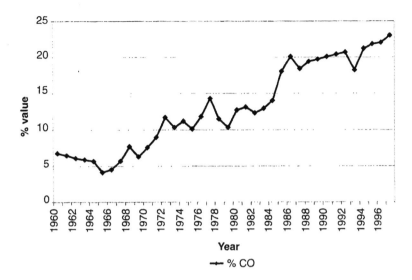

FIGURE 9.4   Percent automobile exhaust suicides as a proportion of all suicides, in Australia, all ages, 1960 to 1997.

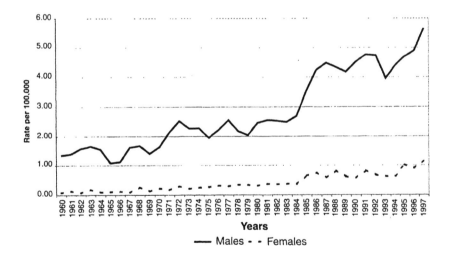

FIGURE 9.5   Automobile exhaust suicides, in Australia, all ages, 1960 to 1997.

increase in raw terms. In 1960, CO suicides represented 6.8% of all suicides. This proportion has now increased to 23% of all suicides, or 19.2% for males and 4% for females in 1997 (Figure 9.4). Accordingly, the rates for this method have risen from 0.08/100,000 in 1960 for females and 1.35/100,000 for males in the same year to 1.16/100,000 for females and 5.65/100,000 for males in 1997 (Figure 9.5).

In terms of age group distribution, CO suicide, unlike hanging which is more common in young people and firearms in older populations, is generally more prevalent for both males and females in the adult populations. Figure 9.6 shows the distribution of CO suicides for males and females in 1997. As can be observed, the

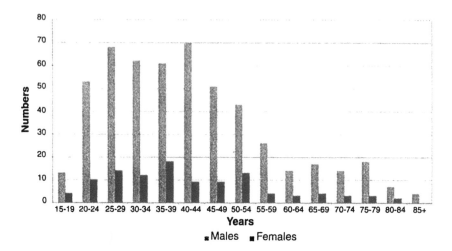

FIGURE 9.6   Automobile exhaust suicide cases per 5-year groups, in Australia, 1997.

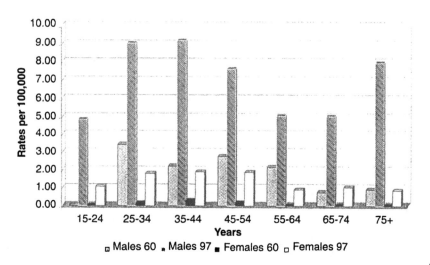

FIGURE 9.7   Automobile exhaust suicides, both genders, in Australia, in two time periods, 1960 and 1997.

peak for males is in those aged 40 to 44 years, followed closely by those aged 25 to 29 years. For females the largest numbers are found in those aged 35 to 39 years.

Figure 9.7 shows the changes in age groups that have taken place in the period 1960 to 1997. For females in the 1960 period, CO was very uncommon across all age groups. The age group where rates are highest was 35 to 44 years. Nearly four decades later, CO suicides have become common across all age groups for females, with the highest rates in those aged 25 to 54 years. Increases have already been reported for males between those two age periods. Males aged between 25 and 54 years have the highest rates, similar to the female trend. However, there are some significant changes during those two time periods. Males aged 15 to 19 years had

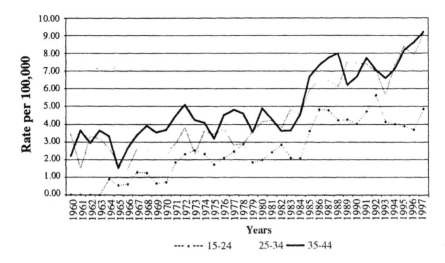

FIGURE 9.8   Automobile exhaust suicides, males 15 to 24, 25 to 34, 35 to 44 years old, in Australia 1960 to 1997.

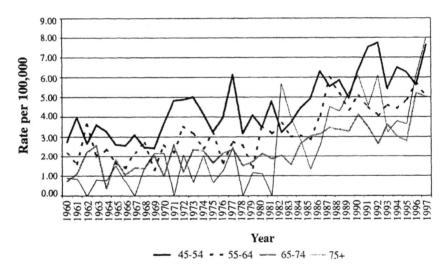

FIGURE 9.9   Automobile exhaust suicides, males 45 to 54, 55 to 64, 65 to 74, 75+ years old, in Australia 1960 to 1997.

insignificant rates in 1960 and those rates are now similar to those aged 55 to 74 years. There has been a trend reversal in older age groups, where those 75 years and over now have the highest rates (7.98/100,000).

When rates for males are compared over time as outlined in Figures 9.8 and 9.9, gradual increases have been witnessed for all age groups. Those who are least likely to use this method are aged 15 to 24 years. The sudden and more significant increase noted earlier for ages combined appears to be present for all age groups. This indicates that all age groups tend to change their method of suicide. Whatever the

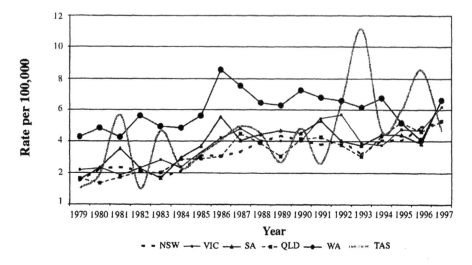

FIGURE 9.10   Automobile exhaust suicides, all states (NT and ACT excluded), Australia 1979 to 1997.

factors that influence selection of this method of suicide, they do not appear to have affected one group more than another.

9.3.3   VARIATIONS BETWEEN STATES AND TERRITORIES

A rise in the use of CO for suicide has been noted in every Australian state and territory (Figure 9.10). The national distribution of these suicide deaths, however, is not uniform and there are some marked differences between regions. Because of the size of the population concerned, these data will not be broken down by age groups for the purpose of comparisons. Furthermore, only data from 1979 to 1997 will be used. Because of the small population of Northern Territory and the Australian Capital Territory (ACT), these two states were excluded. Although Western Australia (WA) did not have the highest rate of CO suicide in 1997, it has consistently done so for most of the period under consideration.

Table 9.1 shows the proportion of CO suicide in comparison with other methods and the proportion of all methods in each state compared with the national trend for 1997. In that year, South Australia recorded the highest percentage followed closely by WA. The lowest percentage of CO suicides of any state was in Queensland. Car exhaust suicide was the second most common method of suicide in all states and territories for 1997.

9.3.4   PARASUICIDE

In addition to the burden of death associated with this method, a small number of attempted suicides resulting in admission to hospital have been noted. As can be expected because of the lethality of this method, only a few suicide attempts are recorded as CO poisoning. Attempts using this method are uncommon when compared with overdoses or cutting and piercing incidents.[53] However, the morbidity

TABLE 9.1
Percentage of Suicides by Methods in all States and for Australia, 1997

	NSW	VIC	QLD	SA	WA	AUS
CO	21.5	25.5	19.8	28.9	27.5	23
Hanging	37.1	31.5	39.5	35.5	37.6	36.2
Guns	11.4	11.5	15.9	8.6	8.6	12.1
Overdose	13	13.3	11.8	11.7	13.7	12.7

Note: NSW = New South Wales, VIC = Victoria, QLD = Queensland, SA = South Australia, WA = Western Australia, AUS = Australia.

Source: Raw data from Reference 2.

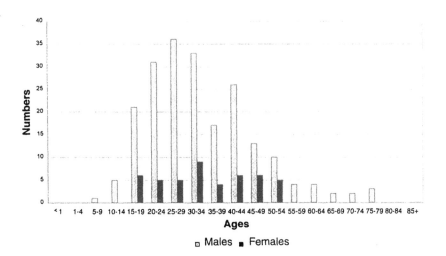

FIGURE 9.11   Suicide attempts by CO, both genders, in Australia 1996 to 1997.

associated with this method is severe, with approximately 70% of cases initially requiring hospitalization. The average length of stay for such individuals is 6 days. Many exhibit prolonged neuropsychiatric sequelae.[11] National data indicate that in 1993/94 there were 263 males and 44 females who required hospitalization for CO poisoning. These represented 1.5% of all admissions for suicide attempts. For the period 1994/95, there were 195 males and 42 females, or 1% of all suicide attempts. In the 1995/96 period there were 376 males and 68 females, or 1.75% of all attempts. In the last period where data is available, i.e., 1996/97, there were 254 males and 46 females, or 1.5% of all reported cases.

   The age distribution for suicide attempts by CO poisoning is shown in Figure 9.11. The distribution of age groups for this method is similar to all suicide attempt profiles where the tendency is for a younger population to engage in the behavior. In 1996/97, the most frequent presentations were found in those aged 25 to 29 years with some cases reported in those aged as young as 10 to 14 years. The number of attempts by this method is very low for the older age group and is not reported for females over the age of 54 years. The most common age group for females 30 to 34 years.

9.4 PROFILES OF INDIVIDUALS WHO COMPLETE AND ATTEMPT SUICIDE WITH CAR EXHAUST

This section is divided into four parts. Section 9.4.1 presents the profiles of CO suicides in an Australian population from a recent study by Baume et al.[3] The sample includes a total 532 CO suicides from a population of 2867 completed suicides. Section 9.4.2 describes the findings of a recent prospective study of attempted suicide by CO in Australia by Skopek and Perkins.[11] Section 9.4.3 compares the profiles of a group of suicide attempters and completers matched for the time period during which the CO poisoning occurred, and Section 9.4.4 describes data collected from the individuals described in Section 9.4.2 from a two year follow-up.

9.4.1 PROFILES OF CARBON MONOXIDE SUICIDES IN AN AUSTRALIAN POPULATION

The variables that are discussed, and for which information was available, for a group of 530 CO suicides include the following: age, gender, cultural background, geographic location, past psychiatric history, the place and time of death, previous suicide attempts, and the presence of alcohol at the time of death. These data are derived from a study originally undertaken by Baume et al.[3]

The majority of those who completed suicide were males (86%), aged 30 to 54 years, with the highest percentage (36%) in those aged 35 to 44 years. In terms of cultural background, the great majority were Caucasian, with less than one 1% coming from indigenous and Asian populations combined. Nearly half of the sample, or 46%, occurred in metropolitan areas, while 37% were reported in provincial towns (population of about 100,000) and about 17% in rural areas. Approximately 22% had a history of psychiatric disorders; however, only 50% of the raw data contained information about psychiatric history. In terms of place of death, 51% died at home, whereas the rest died in their car away from their place of residence. For all methods combined, the Queensland study reported that almost 75% of completed suicides occurred at home. This tends to indicate that individuals who chose exposure to car exhaust were less likely to die at home than those using other methods. A total of 15% reported a previous suicide attempt before the death (55% of the sample did not have information available). About 42% of those who selected this method had a blood alcohol level at the time of death and were more likely to have died in the summer months (28.5%) as compared with fall (23%), winter (24.8%), or spring (23.7%).

9.4.2 PROFILE OF SUICIDE ATTEMPTERS BY CARBON MONOXIDE IN AN AUSTRALIAN POPULATION

Between June 1996 and February 1997, a prospective, longitudinal pilot study of 30 consecutive cases of deliberate exposure to motor vehicle exhaust gas accepted for hyperbaric oxygen (HBO) treatment at a major public hospital in Sydney was conducted by Skopek and Perkins.[11] These cases represented risk of serious morbidity. Criteria for HBO therapy at the center included a period of loss of consciousness, and/or a COHb level of greater than 25%, and/or the presence of neurologic signs, all signifying exposure to near-lethal levels of CO. The psychosocial profile as well as factors influencing method choice and circumstances of the act were examined. Psychiatric assessment was performed prior to or following the first HBO treatment after medical stabilization and assessment by an anesthetist.

The demographic characteristics of this population (n = 30) revealed that the majority, 25 patients, were male (83.3%) and that 5 patients were female. The age range was 22 to 73 years, and 26 patients (76.7%) were between the ages of 20 and 50. The mean age for the group was 39.6 years — for males 38.2 years and for females 46.2 years. Of the patients, 20 (66.7%) were married or in a *de facto* relationship; 16 (53.3%) were employed, the vast majority in unskilled jobs; 28 patients (93.3%) resided in urban areas, and none of the patients belonged to an indigenous group.

Of the patients, 50% had a psychiatric history; 40% had made suicide attempts in the past; 70% reported regular alcohol intake; 33% reported regular illicit drug use; and 30% (all male) had been charged with police offenses in the past.

An outline of events immediately preceding the suicidal act was obtained by self-report. A minority of patients reported more than one precipitant. Of the patients, 60% stated that an argument with their partner had precipitated their suicidal act. Other named precipitants included psychiatric symptoms (23.3%) such as low mood, anxiety, and sleep disturbance; job related stress (10%); charges or a pending court case (10%); the anniversary of a loss event (6.7%); and arguments with other relatives (6.7%). One patient could not name a precipitant.

The self-reported impact of stressors in the 6 months prior to the suicidal act revealed that the majority of patients (70%) had experienced marital or relationship discord with a significant other.

Up to six reasons for choice of this method were given by each patient. Most (63%) stated the method's ease, painlessness, and availability influenced their decision. In all, 60% could name a specific person as being a model for their behavior; 40% of these stated that they personally knew someone who had used this method to commit suicide in the past and the remainder (20%) named the media representation of the suicide of a prominent male citizen during the period of the study as having influenced their decision. Interestingly, all who were influenced by this representation were male.

Some (23%) stated that the idea for the method had come from media portrayal in general. Together with those influenced by the above-mentioned case, 43% in total were influenced by media representation of suicide. Further, 27% stated that the known lethality of the method had influenced their decision. Only one patient (3.3%) did not give a reason for choice of this method.

Two thirds of the patients admitted to spending minimal if any time planning their suicidal act: 23% denied taking any time to plan the act immediately following a precipitating event, and 43% had spent less than 24 hours in planning. In contrast, 17% had been planning the act for weeks to months, and 17% did not recall how long they had spent planning.

Most (90%) of patients connected a hose from the exhaust pipe to the passenger compartment; the remaining patients sat in a closed garage with the engine running. Almost half (47%) admitted to using alcohol at the time of exposure to car exhaust and 20% took an overdose of tablets, benzodiazepines in most cases, at the time of exposure.

Carboxyhemoglobin (COHb) levels were taken by the referring service in 90% of cases: 20% had COHb levels below 20%, and 70% had levels above 20%. The highest COHb level in the group was 50%. As delays in obtaining a blood sample for COHb measurement and the administration of oxygen prior to blood sampling has an effect on COHb level, neuropsychiatric signs and symptoms, and eventual outcome, the COHb level was not used as a prognostic indicator or the sole determinant for administration of HBO.

Mental state examination found 63% of patients euthymic in mood and reactive in affect, 33% admitted to depressed mood, and 80% denied further suicidal ideation. Of the patients, 37% regretted their suicide act, while 53% remained ambivalent and 7% regretted failure of the method. Only 26% of patients appeared moderately or severely depressed as measured by the Beck Depression Inventory. The objective component of the Beck Suicide Intent Scale was used to assess the intensity of the wish of the individual to terminate his or her life prior to exposure to car exhaust. In the majority of cases the scores were consistent with low intent, suggesting impulsivity, relative spontaneity, or ambivalence. Adjustment disorder with depressed mood was the most prevalent psychiatric diagnosis (77% cases) based on history obtained from the patient and significant others and repeated mental state examinations. Major depression could only be diagnosed in 17% of cases.

The discrepancy between the choice of a highly lethal method, the lack of evidence for high suicide intent on objective measurement, and denial of intent after the act by the majority of patients suggests involvement of important patient variables. These may include deficient problem-solving skills and help-seeking behavior, personality factors such as proneness to anger and impulsivity, and substance abuse. In the presence of an adverse life event, usually a relationship breakup, subsequent consumption of substances may have contributed to negative cognitions, anger, feelings of helplessness, isolation, and the decision to "end it all" with minimal time spent planning. The outcome of access to a painless, accessible, yet lethal method was determined in the majority not by patient variables but rather by extraneous factors such as unexpected discovery, inadequate fuel, or use of a car with a warm engine and efficient catalytic converter. Individuals were unaware of these factors when selecting their method. Low suicide intent may not be sufficient to determine survival when a potentially lethal method is used. This population described the method as highly acceptable and accessible, emphasizing the urgent need for reduction of access to this means of suicide. While psychiatric intervention with survivors may have some impact on further self-harming behavior and needs to be standardized across services, the reluctance of these individuals to seek help prior to their initial act stresses the primary importance of targeting access to this method of suicide.

TABLE 9.2
Comparisons of Suicide Attempters and Completers
(*t*-test signif.: *p* < 0.001)

	Completed Suicides	Attempted Suicides	Significance
Age	44.1	39.6	No
Gender	25 male/5 female	25 male/5 female	No
CO Level	66.6	27.9	Yes
Geographical	50.8	50.8	No
Place	49.1	50.9	No
Alcohol	47.4	52.6	No
Note	49.2	50.8	No
Attempts	42.3	57.7	No
Threats	41.2	58.8	Yes
Employment	50.8	49.2	No
Psychiatric history	49.1	50.9	No
Psychiatric diagnosis	41.7	58.3	Yes
Stress	39.6	60.4	Yes
Separation	44.6	55.6	No

9.4.3 COMPARISON OF SUICIDE COMPLETERS AND SUICIDE ATTEMPTERS BY CARBON MONOXIDE POISONING

A total of 30 individuals who had attempted and 30 individuals who had completed suicide during the period between June 1996 and February 1997 in Australia were compared. The 30 individuals who attempted suicide have been described above. A random number of individuals who died by suicide during that period were selected for comparison. The variables used and which were available for comparison for both groups included age, gender, CO level at the time of death or attempt, geographic location, place of death, alcohol consumption, presence of a suicide note, previous suicide attempts, threats of suicide, employment status, psychiatric history, psychiatric diagnosis, stressful precipitating events, and relationship status.

The results are outlined in Table 9.2. In terms of age and gender, the two groups were similar. Individuals completing suicide were, as expected, more likely to have higher COHb (66%) — the value achieved statistical significance. Equally, there was no significant difference between the groups in terms of geographic location, i.e., metropolitan vs. rural, or place of death, i.e., at residence vs. away from the home. Blood alcohol concentration was similar in both groups. The presence of a suicide note was also investigated and there was no difference between the groups.

The rate of previous attempts was similar between the groups, but those who attempted were more likely to have made threats than those who completed. Employment status was similar between the groups. While the history of psychiatric disorders was not significant between the groups, the presence of a psychiatric diagnosis was significant toward those who attempted. Those who attempted reported lower incidence of depressed mood after the act; however, care needs to be taken in the interpretation of these findings. Those who attempted were subject to a rigorous psychiatric assessment, while those who completed were not, as information for psychological autopsy study was reported by nonpsychiatric observers. A level of significance was obtained in terms of a stressful precipitating event being experienced. This was in the direction of the suicide attempts group and may be due to the unique access to self-report in this group. There was no difference between the groups in terms of marital status or history of separation.

While there are some differences between the groups, the sample size and reporting procedure in the case of completers differed from that of attempters. Caution needs to be exercised in the interpretation of these results. In summary, the similarities between the two groups are significant. The age and gender distribution is similar to what was found in the larger sample[3] and in the national data set discussed above. Because of the similarity between the groups, notwithstanding the limitations previously discussed, one could hypothesize that the suicide intent of individuals in both attempter and completer groups may be similar and that the survivors were probably saved because of poor planning on their part rather than because of a low level of suicide intent.

9.4.4 FOLLOW-UP OF SURVIVORS OF ATTEMPTED SUICIDE BY EXPOSURE TO CAR EXHAUST

The 30 individuals who attempted suicide and who were discussed in the pilot study described above were followed up 2 years later. This was done by way of a telephone interview.[54] Changes in place of abode, marital status, employment, psychosocial stressors, alcohol and drug use were examined. The rate of subsequent episodes of deliberate self-harm was investigated as was current mental state including mood and suicidal ideation. Presence of regret for the attempt was examined, as was regular contact with a physician as well as current involvement in medical and/or psychiatric follow-up or treatment.

Of the initial 30 individuals seen at time 1 (T1), 16 (53.3%) were able to be contacted at 2-year follow-up (Time 2 = T2). They included 12 of the original 25 males and 4 of the original 5 females. One patient (male age 28) had completed suicide (by hanging) since T1. Of the 16 individuals contacted, 36% (n = 6) had changed residence since T1. The 13 individuals not able to be contacted at T2 were no longer residing at the address given at T1. In total, 63.3% of subjects had moved since T1.

Six of the 16 subjects (37.5%) had changed their marital status. Nine (56%) subjects remained in the same employment, but of the six whose employment status had changed, five (31%) had been placed on a disability support pension since their suicide attempt. Four of these subject's disabilities were related to psychiatric morbidity associated with CO toxicity and the fifth subjects' disability was cardiovascular illness exacerbated by CO exposure. Ten of the 16 (63%) subjects remained in contact with a general practitioner (GP), two had acquired GPs, but four fell out of contact with their GP. Seven (44%) reported psychosocial stressors at T2. These included relationship discord (n = 1), social isolation (n = 3), psychiatric symptoms (n = 3), symptoms of medical illness (n = 3), gambling (n = 1), and work-related stress (n = 1).

Two individuals attempted suicide after $T1$ (12.5%). One completed suicide by hanging and the other survived multiple attempts including electrocution, stabbing, and a motor vehicle accident. All of these attempts were related to intravenous drug use. Since commencing a detoxification program, methadone maintenance, and attending regular counseling, the subject reported diminished suicidal ideation, improvement in mood, and no further self-harming behavior. All of the sample denied suicidal ideation at $T2$.

Four of the 16 (25%) subjects reported regular alcohol consumption, but all denied problem drinking. Illicit drug use at $T2$ included cannabis ($n = 2$). Of the five subjects using cannabis at $T1$, only one continued use at $T2$. There was no other reported illicit drug use. Three (19%) subjects at $T2$ reported residual cognitive impairment, predominantly short-term memory difficulties. All had been discharged to an inpatient rehabilitation program at $T1$ with significant cognitive impairment. None of the 16 subjects had experienced delayed neuropsychological sequelae after a period of recovery.

At $T2$, 15 (94%) subjects regretted their previous suicidal behavior. The last individual accepted the outcome that was due to setup failure, "the only reason I'm talking to you is because I ran out of fuel (gasoline), no one could have helped me," but did not regret his decision.

All subjects at $T1$ were referred for psychiatric and medical follow-up in their local area. At $T2$ three (19%) stated they had never complied with the arrangements, six (38%) attended follow-up for some period but were no longer attending any service, and seven (44%) remain in current follow-up. Of these, five attended psychiatric follow-up, one attends a rehabilitation clinic, and one attends regular GP counseling.

All 30 subjects at $T1$ did not notify others of their immediate suicide intent. When questioned about this at $T2$, only 2 of the 16 offered reasons for their lack of help-seeking behavior. The responses included "no one could help," "I didn't want to suffer loss of face," "I wanted to preserve my anonymity and didn't know how," "I thought of taking time off but it seemed a cop out or a sign of weakness."

In summary, the loss to follow-up is significant in this population and has been previously described.[55] The majority of individuals seen at $T1$ had changed residence by $T2$. This may suggest a lack of secure attachments or the inability to form or maintain attachments in these individuals. This could in part be explained by personality features, lack of social skills and problem-solving ability, but also by social stratification and psychiatric/psychological symptoms. These features may also result in the lack of effective help-seeking behavior characteristic of these individuals.

9.5 DISCUSSION AND CONCLUDING REMARKS

Suicide by exposure to motor vehicle exhaust gas represents a significant and preventable public health problem. This chapter has shown that the completed suicide rate by this method exceeds the rate of attempted suicide (at least in Australia) by a factor of 12. Therefore, CO poisoning is a highly effective way of taking one's life. This is different from most other commonly used suicidal poisoning methods, where attempts greatly outnumber the number of deaths and have generally been decreasing as a proportion of all suicides. During the last four decades, this method of suicidal poisoning has claimed thousands of lives in most countries where suicides are recorded. In Australia, unlike in most other countries, this method has been increasing both in terms of rates and as a proportion of all suicides.

In Australia this method is predominantly a male phenomenon and appears more frequent in urban areas. This differs significantly from previous European studies that found it to be primarily a rural problem.[14,21,24] The majority of those who choose this method are aged between 30 and 50 years for both genders. In 1960, this method accounted for about 6% of all suicides and by 1997 this had increased to 23%, the highest proportion on record. The rate of increase has been noted for all age groups and for both genders between 1960 and 1997. About 50% of those individuals die at their place of residence and high proportions (55%) have detectable blood alcohol levels. They are more likely to die during the summer months. The media appear to have a significant impact on the choice of this method.

The high morbidity in this population is also noted and can in part be measured by the need for disability support. Thus, the financial burden of CO poisoning to the community is significant, as well as the lowering of social stratification and the concomitant "stigma" experienced by the individual.

Australia has introduced a number of measures to reduce pollution levels resulting from burned fuels. Motor vehicle emissions are controlled by catalytic converters (ADR 37-00; Australian Standard 2877). The purpose of such a mechanism is to convert CO and associated pollutants into the by-products of carbon dioxide and water. Australia, unlike the United States, has been late in introducing stringent regulations for the reduction of CO concentration emanating from the combustion of fuels in motor vehicles. These regulations were introduced over several years and, as such, permissible CO levels have been reduced from 24.3 g/km in July 1976 to 9.3 g/km in 1986 for new passenger vehicles, to 2.1 g/km from 1998 on for all passenger vehicles. The latter CO level was introduced in the United States in 1981.[26] The effectiveness of catalytic converters to reduce suicide rates is not only dependent on emission reduction, but also how quickly the car fleet can be replaced. While catalytic converters fitted to cars manufactured since 1986 reduce the CO in exhaust gas, they do not eradicate it completely and modern cars remain a common means of suicide. Catalytic converters in the bulk of the Australian car fleet do not meet the most recent design standards, and thus morbidity and mortality from deliberate car exhaust exposure remain high.

The Australian automobile fleet is on average 11 years old compared with 7 years in the United States.[26] This means that even with the introduction of more stringent regulations, which should prove effective if the evidence from other countries is an indication, it will be a slow process. Hence, there is a need in the meantime for short- and medium-term solutions.

Although empirically limited, the follow-up pilot study presented above should caution clinicians in terms of morbidity outcome following a suicide attempt. There should be hypervigilance in the collection of at least three contact addresses/telephone numbers, not only for the subject, but also for an immediate relative and a significant other. There should be the establishment of a strong therapeutic alliance at $T1$, which is crucial if one is to affect the prognosis of these individuals. Finally, further scientific research into the management of deliberate CO poisoning should be encouraged, as it is associated with significant morbidity both to the individual and to the community.

Most studies confirm the positive correlation of availability of a method with its use for suicide.[56,57] Whether people switch from a restricted method to another method has only been a matter of interest. Certainly, the most famous example in the literature of the effect of restriction is that of the detoxification of domestic gas in the United Kingdom in the 1960s.[58] In Northern Ireland, suicide rates by car exhaust gas increased in 1960 to 1988 coincident with car ownership, with no corresponding decreases in the use of other methods.[8]

There is a strong correlation between the availability and the use of a method of suicide. Evidence suggests that restrictions on the availability of a method reduce its use for suicide but do not necessarily affect the overall suicide rate. For example, the decrease in firearm suicide, but a continued rise in overall suicide rates partly attributable to increased CO suicide, suggests that one method tends to be replaced by other methods. The United Kingdom may, however, be an exception. A fall in overall suicides may be explained to some extent by new legislation in the United Kingdom, which, from January 1993, required all new gasoline-powered vehicles to be fitted with catalytic converters, which reduce CO emissions. Of automobiles licensed at the end of 1994, 93% were gasoline vehicles. Of these, 14% were first registered in 1993 or later.[59] By the end of 1996, 90% of licensed automobiles were gasoline vehicles, of which 28% were first registered in 1993 or later. Taking into consideration, that even before the introduction of the legislation some cars with catalytic converters already existed, the steadily growing number of cars with catalytic converters may go some way to explain the decreasing suicide rate for this method since the early 1990s. The increased popularity of diesel cars, which emit much lower levels of CO, may also have contributed to this decrease.[14] This reduction has also been reported in the United States, where studies on suicide by car exhaust suggest a decline from 8.9% of all suicides in 1970 to 5.6% in 1991 following the introduction of stricter guidelines for catalytic converters.[23,60] Motor vehicle ownership is not a conventional risk factor. In Britain a recent study showed that suicide rates were lower in those who owned a car than for those who did not.[61] A number of studies have now demonstrated the efficacy of catalytic converters provided the CO emission is lower than 2.1 g/km.[61]

However, because suicide is often an impulsive action of a person preoccupied with it, reducing opportunities for suicide by restrictions on availability should be considered to be a means of prevention. In this case, reducing the availability of cars is inappropriate, but restricting the availability of a lethal gas via more stringent catalytic converters could prove very useful.

Reduction of access to means of CO suicide has recently been addressed in Australia. Recommendations recently forwarded to the Commonwealth Government to address the problem of the existing fleet include the introduction of passenger compartment gas sensors incorporating an alarm system and engine cut-out mechanism. Other proposals debated but not widely supported as a means of suicide prevention included alteration of the shape of existing exhaust pipes to preclude insertion of hoses, addition of malodorous substances to fuel, and garage alarms.

In summary, while CO suicides appear to be a problem in many Western nations, recent studies are encouraging and indicate a major decline in this method following the introduction of strict guidelines for catalytic converters.[14,24,62-64] This, however, is not the case in every country.

REFERENCES

1. Diekstra, R.F., Epidemiology of suicide, *Encephale*, 22, 15–18, 1996.
2. Australian Bureau of Statistics, Cause of Death Australia. Catalogue No. 3303.0. Australian Government Publishing Service, Canberra, 1998.
3. Baume, P.J.M., Cantor, C.H., and McTaggart, P., *Suicides in Queensland: A Comprehensive Study: 1990–1995*. Australian Academic Press, Brisbane, 1998.
4. Cantor, C.H. and Baume, P.J.M., Access to methods of suicide: what impact? *Aust. N. Z. J. Psychiatr.*, 32, 8–14, 1998.
5. Baume, P.J.M. and Clinton, M., Social and cultural patterns of suicide in young people in rural Australia, *Aust. J. Rural Health*, 5, 115–120, 1997.
6. Bille-Brahe, U. and Jessen, G., Suicide in Denmark, 1922–1991: the choice of method, *Acta Psychiatr. Scand.*, 90, 91–96, 1994.
7. Burvill, P.W., Changing patterns of suicide in Australia, 1910–1977, *Acta Psychiatr. Scand.*, 62, 258–268, 1980.
8. Curran, P.S. and Lester, D., Trends and methods used for suicide in Northern Ireland, *Ulster Med. J.*, 60, 58–62, 1991.
9. Farmer, R. and Rhode, J., Effect of availability and acceptability of lethal weapons on suicide mortality: an analysis of some international data, *Acta Psychiatr. Scand.*, 62, 436–466, 1984.
10. Pounder, D.J., Changing patterns of male suicide in Scotland, *Forensic Sci. Int.*, 51, 79–87, 1991.
11. Skopek, M. and Perkins, R., Deliberate exposure to motor vehicle exhaust gas: the psychosocial profile of attempted suicide, *Aust. N. Z. J. Psychiatr.*, 32, 830–838, 1998.
12. Snyder, M.L., Methods of suicide used by Irish and Japanese samples: a cross-cultural study from 1964 to 1979, *Psychol. Rep.*, 74, 127–130, 1994.
13. Routley, V., Motor Vehicle Exhaust Gassing Suicides in Australia; Epidemiology and Prevention, Monash University Accident Research Centre Rep. No. 139. Monash University Accident Research Centre, Sept., 1998.
14. Hawton, K., Why has suicide increased in young males? *Crisis*, 19, 119–124, 1998.
15. Rhyne, C., Templer, D., Brown, L., and Peters, N., Dimensions of suicide: perceptions of lethality, time, and agony, *Suicide Life Threatening Behav.*, 25, 373–380, 1995.
16. Lester, D., Alcoholism and drug abuse, in *1992 Assessment and Prediction of Suicide*, Maris, R.W., Berman, A.L., Maltsberger, J.T., and Yufit, R.I., Eds., Guildford Press, New York, 1992, 321–336.
17. Takahashi, Y., Culture and suicide from a Japanese psychiatric perspective, *Suicide Life Threatening Behav.*, 27, 137–145, 1996.
18. Yip, P., Teenage attempted suicide in Hong Kong, *Crisis*, 19, 67–72, 1998.
19. McClure, G.M.G., Suicide in children and adolescents in England and Wales 1960–1990, *Br. J. Psychiatr.*, 165, 510–514, 1994.
20. Pounder, D.J., Why are the British hanging themselves, *Am. J. Forensic Med. Pathol.*, 14, 135–140, 1993.

21. Ostrom, M., Thorson, J., and Eriksson, A., Carbon monoxide suicide from car exhausts, *Soc. Sci. Med.*, 42, 447–451, 1996.

22. Ohberg, A., Lonnqvist, J., Sarna, S., Vuori, E., and Penttila, A., Trends and availability of suicide methods in Finland: proposals for restrictive measures, *Br. J. Psychiatr.*, 166, 35–43, 1995.

23. Lester, D., Car ownership and suicide by car exhaust in nations of the world, *Percept. Mot. Skills*, 79, 898, 1994.

24. Ohberg, A., Lonnqvist, J., Sarna, S., and Vuori, E., Violent methods associated with high suicide mortality among the young, *J. Am. Acad. Child Adolescent Psychiatr.*, 35, 144–153, 1996.

25. Burvill, P.W., The changing pattern of suicide by gassing in Australia, 1910–1987: the role of natural gas and motor vehicles, *Acta Psychiatr. Scand.*, 81, 178–184, 1989.

26. Routley, V. and Ozanne-Smith, J., The impact of catalytic converters on motor vehicle exhaust gas suicides, *Med. J. Aust.*, 168, 65–68, 1998.

27. Cantor, C.H., Turrell, G., and Baume, P.J.M., Access to means of suicide by young Australians, *Australian Government Publishing Service (AGPS)*, Canberra, 1996.

28. Kleck, G., *Point Blank: Guns and Violence in America*, Aldine de Gruyter, New York, 1992.

29. McIntosh, J.L., Methods of suicide, in *1992 Assessment and Prediction of Suicide*, Maris, R.W., Berman, A.L., Maltsberger, J.T., and Yufit, R.I., Eds., Guildford Press, New York, 1992, 381–397.

30. Tanney, B., Suicide prevention in Canada: a national perspective highlighting progress and problems, *Suicide & Life Threatening Behavior*, 25, 105–122, 1995.

31. Dudley, M., Waters, B., Kelk, N., and Howard, J., Youth suicide in New South Wales: urban rural trends, *Med. J. Aust.*, 156, 83–88, 1992.

32. Dudley, M., Cantor, C., and de Moore, G., Jumping the gun: firearms and the mental health of Australians, *Aust. N. Z. J. Psychiatr.*, 30, 370–381, 1996.

33. Cantor, C.H. and Slater, P.J., The impact of firearm control legislation on suicide in Queensland: preliminary findings, *Med. J. Aust.*, 162, 583–585, 1995.

34. Gould, M. and Davidson, L., Suicide contagion among adolescents, in *Advances in Mental Health*, Vol. III: *Depression and Suicide*, Stillman, A. and Feldman, R., Eds., JAI Press, Greenwich, 1988.

35. Gould, M.S., Suicide clusters and media exposure, in *Suicide over the Life Cycle*, Blumenthal, S.J. and Kupfer, D.J., Eds., American Psychiatric Press, Washington, D.C., 1990.

36. Phillips, D., The influence of suggestion on suicide: substantial and theoretical implications of the Werther effect, *Am. Sociol. Rev.*, 39, 340–354, 1974.

37. Phillips, D., Motor vehicle fatalities increase after publicized suicide stories, *Science*, 196, 1464–1465, 1977.

38. Phillips, D., Airplane accident fatalities increase just after stories about murder and suicide, *Science*, 201, 748–750, 1978.

39. Wasserman, I., Imitation and suicide: a re-examination of the Werther effect, *Am. Sociol. Rev.*, 49, 427–436, 1984.

40. Gould, M. and Shaffer, D., The impact of suicide in television movies, *N. Engl. J. Med.*, 315, 690–694, 1986.

41. Berman, A., Fictional depiction of suicide in television films and imitation effects, *Am. J. Psychiatr.*, 145, 982–986, 1988.

42. Martin, G., Adolescent suicide 3: imitation and clustering phenomenon, *Youth Stud. Aust.*, 11, 28–32, 1992.

43. Church, I. and Phillips, J., Suggestion and suicide by plastic bag asphyxia, *Br. J. Psychiatr.*, 144, 100–101, 1984.
44. Goldney, R., A spate of suicide by jumping, *Aust. J. Soc. Issues*, 21, 119–125, 1986.
45. Niemi, T., The time–space distances of suicides committed in the lock-up in Finland in 1963–67, *Psychiatr. Fenn.*, 267–270, 1975.
46. Shaffer, D., Suicide in childhood and early adolescence, *J. Child Psychol. Psychiatr.*, 15, 275–291, 1974.
47. Hassan, R., Effects of newspaper stories on the incidence of suicide in Australia: a research note, *Aust. N. Z. J. Psychiatr.*, 29, 480–483, 1995.
48. Baume, P.J.M., Cantor, C., and Rolfe, A., Cybersuicide: the role of interactive suicide notes on the Internet, *Crisis*, 18, 71–79, 1997.
49. Shaffer, D., Garland, A., Gould, M., Fisher, P., and Trautman, P., Preventing teenage suicide: a critical review, *J. Am. Acad. Child Adolescent Psychiatr.*, 27, 675–687, 1988.
50. Sayer, G., Stewart, G., and Chipps, J., Suicide Attempts in NSW: Associated Morbidity and Mortality, Public Health Bull., NSW Department of Health, June, 1996.
51. McDonald, B. and Steel, Z., *Immigrants and Mental Health: An Epidemiological Analysis*, Sydney Transcultural Mental Health Centre, Sydney, 1997.
52. Bordeaux, S., Centre for Injury Prevention, Flinders University, unpublished data, 1998.
53. Cassell, E., Routley, V., and Ozanne-Smith, J., Prevention of suicide–reducing car exhaust gas poisoning, Hazard, 25, Victorian Injury Surveillance System, Monash University Accident Research Centre, 1995.
54. Skopek, M. and Baume, P., The prevention of carbon monoxide poisoning suicide, *Australian Government Publishing Service (AGPS)*, 1999 (in press).
55. Raphael, J.-C., Elkharrat, D., Jars-Guincestre, M.-C. et al., Trial of normobaric and hyperbaric oxygen for acute carbon monoxide intoxication, *Lancet*, August 19, 1989.
56. Lester, D., Changing rates of suicide by car exhaust in men and women in the United States after car exhaust was detoxified, *Crisis*, 10, 164–168, 1989.
57. Lester, D. and Frank, M.L., The use of motor vehicle exhaust for suicide and the availability of cars, *Acta Psychiatr. Scand.*, 79, 238–240, 1989.
58. Kreitman, N., The coal gas story, *Br. J. Prev. Soc. Med.*, 30, 86–93, 1976.
59. Kelly, S. and Bunting, J., Trends in suicide in England and Wales 1982–96, *Popul. Trends*, Summer, 29–41, 1998.
60. Clarke, R.V. and Lester, D., Toxicity of car exhausts and opportunity for suicide: comparison between Britain and the United States, *J. Epidemiol. Comm. Health*, 41, 114–20, 1987.
61. Lewis, G., Hawton, K., and Jones, P., Strategies for preventing suicide, *Br. J. Psychiatr.*, 171, 351–354, 1997.
62. Risser, D. and Schneider, B., Carbon monoxide-related deaths from 1984 to 1993 in Vienna, Austria, *J. Forensic Sci.*, 40, 368–371, 1995.
63. La Harpe, R., Suicide in the Geneva canton (1971–1990): An analysis of the forensic medicine autopsy sample, *Arch. Kriminol.*, 195, 65–74, 1995.
64. Lester, D., The toxicity of car exhaust and its use as a method for suicide, *Psychol. Rep.*, 77, 1090, 1995.

10 Carbon Monoxide as an Unrecognized Cause of Neurasthenia: A History

Albert Donnay

CONTENTS

10.1 INTRODUCTION

The chronic, polysymptomatic, multiorgan, and highly variable condition of neurasthenia, meaning literally "nervous weakness," was first defined in the medical literature by Dr. Eugene Bouchot of the Faculté de Medecine de Paris in a seminal

0-8493-2065-8/00/$0.00+$.50

lecture on *nervosisme* published in 1858.[1] It gradually spread to become one of most commonly diagnosed and written about disorders in urban medicine throughout the industrialized world from the 1880s through 1920s, with over 1900 medical papers and books on the subject referenced in the *Index Catalog of the Library of the* [U.S. Army] *Surgeon General's Office* and the *Quarterly Cumulative Index Medicus* during this period.

Although originally promoted and remembered in the United States as a uniquely American disorder, 55% of the medical literature from this period was in languages other than English, especially French, German, and Russian. The number of papers published globally per year on neurasthenia peaked in the 1900s before gradually declining to zero in 1957. Articles on the disorder began to reemerge in the 1960s, and since the 1970s the English-language literature has steadily averaged about seven papers per year, while the nonEnglish-language literature has been declining again since the breakup of the Soviet Union in the late 1980s and is now less common.

Researchers have proposed many endogenous and exogenous causes of neurasthenia and especially its higher prevalence among urban dwellers, women, and "brain workers." These included genetic, technological, societal, psychiatric, and even iatrogenic factors, but no single cause was ever agreed upon or established.[2] It remains today a disease of unknown but presumed psychiatric etiology and is coded as a neurotic disorder (F48.0) in the World Health Organization's latest *International Classification of Diseases* (ICD), 10th edition.[3]

This chapter examines the previously unexplored hypothesis that carbon monoxide (CO) poisoning from the indoor use of illuminating gas in the 19th century and the very similar utility gas made in the first half of the 20th century may have been a major unrecognized cause of neurasthenia. The history of neurasthenia and illuminating gas are reviewed and the putatively causal role of CO is assessed according to the 10 Unified Criteria for causation of disease (an expanded form of the Henle–Koch postulates) developed by Alfred Evans of the World Health Organization.[4] These allow one to measure the strength of any hypothesized causal association based on the number of criteria it fulfills. The hypothesis that CO poisoning from manufactured gas was a cause of neurasthenia fulfills all 10, with available evidence supporting consistent associations with (1) prevalence, (2) exposure, (3) incidence, (4) temporality, and (5) a spectrum of (6) measurable host responses, including (7) evidence of experimental reproduction from either controlled or natural experiments showing a decline in the disorder with (8) elimination or modification of the putative cause and (9) prevention or modification of the host response. The last criterion simply requires that the causative association (10) makes biological and epidemiological sense.

The hypothesis also fulfills all of the nine more-open-ended qualitative scales recommended by Sir Austin Bradford Hill[5] for assessing causal associations with environmental factors: it is a relatively strong, consistent, and specific association; it is temporally correct; it demonstrates some biological gradient of response; it is plausible, coherent, and supported by experimental evidence, and also can be supported by analogy. The gradually shifting upper- to lower-class prevalence of illuminating gas poisoning, for example, is similar to that of lead poisoning in ancient Rome, albeit on a different timescale, which also was first recognized as a common affliction of the urban upper classes but is now most common among the urban poor.

Given the toxicological, temporal, geographic, and occupational associations that suggest CO poisoning played a causative role in 19th-century neurasthenia, CO also may play a role in some of the late-20th-century disorders to which neurasthenia has been compared and by which its many symptoms are now more commonly and often multiply diagnosed. These overlapping disorders include chronic fatigue syndrome (CFS, also ME), defined primarily by fatigue and weakness;[6,7] fibromyalgia syndrome (FMS), defined by muscle pain and weakness;[8] multiple chemical sensitivity (MCS), defined by sensitivity or intolerance to previously tolerated levels of sensory stimuli (predominantly chemicals and odors, but also lights, sounds, foods, drugs and alcohol, touch, hot or cold, etc.);[9] and Gulf War Syndrome (GWS), an ill-defined and all-encompassing diagnosis for Gulf War veterans with multiple medically unexplained symptoms[10] that was first refined by factor analysis in 1997 into three subsyndromes quite similar to CFS, FMS, and MCS.[11]

The unproven etiologies of all these disorders are still hotly debated, with the main dividing line today — just as with neurasthenia a century ago — being between those who favor psychogenic theories and those who favor organic causes.[12] All these disorders also are commonly reported as undistinguished by any consistent signs or objective diagnostic tests, but neither do any of the protocols developed for their screening and diagnosis suggest evaluation for CO poisoning, even though the symptoms of chronic low-level exposure to CO are so similar to those of CFS, FMS, MCS, and GWS. All these disorders may begin suddenly or gradually (CO poisoning and CFS literature both commonly refer to a "flu-like onset"), and involve long bouts of persisting or recurring weakness and fatigue with difficulty sleeping, memory loss, concentration difficulties, muscle pains and cramps, headaches or migraines, dizziness, lack of coordination, nausea, vomiting, diarrhea, chest pain, tachycardia, dyspnea, and changes in sensory tolerance or sensitivity, especially to noises, lights, and odors.

10.2   NEURASTHENIA

10.2.1   RECOGNITION AND DEFINITION OF NEURASTHENIA

The first English-language description of neurasthenia-like symptoms in the context of illuminating gas poisoning was written by Edgar Allan Poe in his 1832 story, "A Decided Loss" (later renamed by Poe as "Loss of Breath").[13] It appeared 25 years before neurasthenia was first defined in French medical literature, and 37 years before its American definition. Poe's story is a sharp satire in which one of the main characters, Mr. Windenough, dies "gloriously while inhaling gas," while the narrator, saying, "every writer should confine himself to matters of experience," details his own "sensations upon the gallows," which read very much like a description of CO poisoning:

I heard my heart beating with violence — the veins in my hands and wrists swelled to nearly bursting ... and I felt that my eyes were starting from their sockets. Yet when I say that in spite of all this my sensations were not absolutely intolerable, I will not be believed.

There were noises in my ears — first like the tolling of huge bells — then like the beating of a thousand drums — then, lastly, like the low, sullen murmurs of the sea. ... Although, too, the powers of my mind were confused and distorted, yet I was — strange to say! — well aware of such confusion and distortions.

Memory, which, of all other faculties, should have first taken its departure, seemed on the contrary to have been endowed with quadrupled power. [Note this is clearly meant to be satirical: the phrasing shows Poe recognizes that memory was usually the first faculty affected in gas poisoning.]

A rapid change was now taking place in my sensations. The last shadows of concentration flitted away from my meditations. A storm — a tempest of ideas, vast, novel, and soul-stirring, bore my spirit like a feather afar off. Confusion crowded upon confusion like a wave upon a wave.

During the brief passage to the cemetery [i.e., outdoors, in fresh air] my sensations, which for some time had been lethargic and dull, assumed, all at once, a degree of intense and unnatural vivacity for which I can in no manner account. I could distinctly hear the rustling of the plumes — the whispers of the attendants — the solemn breathings of the horses of death. ... I could distinguish the peculiar odor of the coffin — the sharp acid smell of the steel screws. I could see the texture of the shroud as it lay close against my face.

Poe next described these same symptoms — tachycardia, tinnitus, confusion, memory loss, and multisensory sensitivity — and many more in "The Fall of The House of Usher," first published in 1839.[14] Most critics believe this story is only a figment of Poe's fertile imagination, and many imaginative allegorical interpretations have been proposed.[15] Only David Sloane[16] considered the possibility that Poe may have been describing a specific medical condition of his time — nervous fever, which supposedly was caused by overexposure to the bad air or miasmas arising from the rot and decay of animal and vegetable matter (i.e., swamp gas or sewer gas) — but this analysis does not suggest Poe suffered from the condition, only that he may have borrowed from a contemporary home medical book, The Family Physician, in describing it.

While nine of the 30 specific symptoms Poe described (Table 10.1) are suggestive of "nervous fever," all 30 also may be read as a literal description of the effects of CO poisoning, as jointly experienced by the soul, mind, and body — specifically, the narrator and Roderick (who together may be interpreted as representing the "Us" in Usher) and Roderick's sister Madeline (the "her").

A 1999 survey of people with multiple chemical sensitivity whose illness began after months or years of chronic CO poisoning ($n = 10$) found they reported experiencing an average of 27 of these 30 symptoms in the prior month, compared with just 2 experienced by healthy controls ($n = 10$) (Donnay, unpublished). The extraordinary apparent sensitivity and specificity of such a survey for screening of both CO poisoning and MCS deserves further study in the general population.

While Poe's many descriptions of CO poisoning symptoms have been read by millions worldwide over the last 150 years, only the French poet Charles Baudelaire appears to have recognized them as descriptions of Poe's own illuminating gas

TABLE 10.1
Symptoms Described by Edgar Allan Poe in "The Fall of the House of Usher" 1839

1. "ghastly pallor of the skin ... "
2. "miraculous lustre of the eye"
3. "gossamer texture" [of hair]
4. "nervous agitation"
5. "alternately vivacious and sullen"
6. "voice varied from tremulous indecision"
7. "... to that species of energetic concision — abrupt, weighty, unhurried, and hollow-sounding enunciation — that leaden, self-balanced, and perfectly modulated guttural utterance, which may be observed in the lost drunkard"
8. "it was, he said, a constitutional and a family evil, and one for which he dispaired to find a remedy — a mere nervous affection, he immediately added, which would soon pass off"
9. "it displayed itself in a host of unnatural sensations"
10. "he suffered much from a morbid acuteness of the senses"
11. "insipid food was alone endurable"
12. "could wear only garments of certain texture"
13. "odors of all flowers were oppressive"
14. "eyes were tortured by even a faint light"
15. "there were but peculiar sounds, and these from stringed instruments, which did not inspire him with horror"
16. "phantasmagoric conceptions ... wild fantasies"
17. "fear"
18. "without having noticed my presence"
19. "he arrested and overawed attention ... an intensity of intolerable awe"
20. "radiation of gloom"
21. "painted an idea ... pure abstractions"
22. "intense mental collectedness and concentration ... observable only in particular moments"
23. "roamed from chamber to chamber with hurried, unequal, and objectless step"
24. "sleep came not near my couch"
25. "gazing upon vacancy for long hours, in an attitude of the profoundest attention, as if listening to some imaginary sound"
26. "hysteria in his whole demeanor"
27. "struggled to reason off the nervousness which had dominion over me"
28. "irrepressible tremor gradually pervaded my frame"
29. "there sat upon my heart an incubus of utterly causeless alarm"
30. "overpowered by an intense sentiment of horror, unaccountable yet unendurable"

Source: Extracted from Poe, E.A., *The Unabridged Edgar Allan Poe*, Running Press, Philadelphia, 1983.

poisoning. In the preface to his 1856 translation of Poe's *Histories Extraordinaires*, Baudelaire wrote:

> All the documents I have read led me to the conviction that for Poe the United States was nothing more than a vast prison which he traversed with the feverish agitation of a being made to breathe a sweeter air — nothing more than a great gas lighted nightmare — and that his inner, spiritual life, as a poet or even as a drunkard, was nothing but a perpetual effort to escape the influence of this unfriendly atmosphere.[17]

Baudelaire also shares with Poe a unique facial "twist" not previously recognized in the annals of neurology in which one eye is lower than the other, while the mouth slants downward on the other side. The same neurological abnormality is seen in some people with MCS who associate the onset of their illness with a chronic exposure to CO, but not in those who attribute their MCS to other causes such as pesticide exposures.[18]

The first physician to write about the disorder appears to have been Dr. Eugene Bouchot, a distinguished physician at the Faculté de Medecine de Paris, who in an 1858 paper[1] and then in an 1860 book[19] (without reference in either to Poe or Baudelaire) distinguished two types of what he called acute and chronic *nervosisme*. He distinguished these from both hypochondriasis and hysteria, the two most common psychiatric diagnoses of the era, and noted consistent improvement with extended visits to the country. The name *nervosisme* died out of French completely by the 20th century, however, and apparently had little influence on physicians in America, where the same syndrome was fleetingly described in the 1860s as a "hyperesthetic state" by Silas Weir Mitchell and colleagues[20] and "nervous asthenia" by Fordyce Barker,[21] again distinguished from hysteria.

The U.S. military first confronted a similar disorder of nervous weakness and chronic fatigue in the Civil War. Known initially as Da Costa's syndrome,[22] by World War I most U.S. and other medical literature had redefined its symptoms as "war neurasthenia"[23] and "battleship neurasthenia."[24] The same neurasthenic disorder was then rediscovered and renamed repeatedly after each major U.S. battle experience, starting with Soldier's Heart after World War I[25] and most recently as Gulf War Syndrome.[26]

In the United States, neurasthenia was first defined by this name — without reference to either Poe or Bouchot — in two independently published papers that just happened to both appear in April 1869. One was by Dr. George Miller Beard,[27] an influential neurologist in New York City, and the other, by Edward Van Deusen,[28] an insane-asylum director in Michigan. Both suggest they have discovered a new disorder, but their descriptions of its symptoms and history match closely with each other and with what was then known about similar disorders such as hysteria and neuralgia. Beard went on to become a widely published and quoted promoter of the neurasthenia diagnosis until his death in 1883, after which it gained even greater popularity. While Beard's name remains closely associated with the diagnosis, the names of Bouchot and Van Deusen remained obscure and rarely cited.

Bouchot, Van Deusen, and Beard all described neurasthenia as a chronic disorder of nervous weakness with such extensive and variable symptomatology as to defy any single consistent or comprehensive classification. Beard, for example, cited 71 common symptoms in his popular 1881 book, on *American Nervousness*,[29] but said even this long list (Table 10.2) was only representative and not complete. Numerous subtypes were defined in the late 19th and early 20th century to address this variability, starting with "sexual neurasthenia" published posthumously by Beard in 1886,[30] followed by "neurasthenia in children"[31] similar to attention hyperactivity deficit disorder today, "neurasthenia in women,"[32] "syphilitic neurasthenia," "angioparalytic or pulsating neurasthenia,"[33] "senile neurasthenia,"[34] "abdominal and gastric neurasthenia"[35] similar to irritable bowel syndrome, "tropical neurasthenia,"[36] "endocrine neurasthenia"[37] similar to hypothyroidism, "ocular neurasthenia,"[38] "traumatic neurasthenia"[39] similar

TABLE 10.2
Symptoms of Neurasthenia Described by George Beard, 1881

1. Insomnia
2. Flushing
3. Drowsiness
4. Bad dreams
5. Cerebral irritation
6. Dilated pupils
7. Pain, pressure, and heaviness in head
8. Changes in the expression of the eye [one or both eyes look different]
9. Eye fatigue, discomfort or tearing
10. Noises in the ears [ringing, etc.]
11. Atonic voice [abnormal tone]
12. Mental irritability [short temper]
13. Tenderness of the teeth and gums
14. Nervous dyspepsia [upset stomach]
15. Desire for stimulants and narcotics
16. Abnormal dryness of the skin, joints and mucous membranes
17. Sweaty hands and feet with redness
18. Fear of lightning [or surprising loud noises like thunder]
19. Fear of responsibility
20. Fear of open or closed places
21. Fear of society
22. Fear of being alone
23. Fear of contamination
24. Fear of fears
25. Fear of everything
26. Deficient mental control [spacey]
27. Lack of decision in minor matters
28. Feeling of hopelessness
29. Deficient thirst and/or deficient capacity for assimilating fluids
30. Abnormal secretions [from any of eye, nose, mouth, etc.]
31. Excess salivation
32. Tenderness of the spine
33. Sensitivity to cold or hot water
34. Sensitivity to changes in weather
35. Coccyodynia [pain in "tail bone"]
36. Back pain
37. Heaviness of the loins and limbs
38. Shooting pains
39. Foot pain
40. Cold hands and feet
41. Localized numbness in extremities
42. Hypersensitive to touch
43. Tremulous heartbeat or pulse
44. Rapid, strong heartbeat or pulse
45. Intolerance of certain foods, medicines and external irritants
46. Local [isolated] muscle spasms
47. Difficulty swallowing

TABLE 10.2 (continued)

48. Convulsive limb or body movements, especially when going to sleep
49. Cramps
50. A feeling of profound exhaustion unaccompanied by positive pain that comes and goes
51. Ticklishness [increased or excess]
52. Vague pains and fleeting neuralgias [throbbing or stabbing nerve pain]
53. Itching: all over or localized
54. Chills or hot flashes
55. Attacks of temporary paralysis
56. Pain in perineum [between legs]
57. Irritability of the prostatic urethra [the urinary canal]
58. Excessive gaping and yawning
59. Rapid decay and irregularities of teeth
60. Vertigo or dizziness
61. Headache that feels like an explosion in the brain at the back of the neck
62. Dribbling and incontinence of urine
63. Frequent urination
64. Spasmodic twitching or jerking of different parts of the body
65. Trembling of muscles or portions of muscles in different parts of the body
66. Exhaustion after defecation or urination
67. Dry hair or hair loss
68. Slow reaction of skin to pressure [takes long time to clear redness]
69. [Men]: Involuntary ejaculation
70. [Men]: Impotence
71. [Women]: Sexual dysfunction
72. [Women]: Menstrual dysfunction

Source: Extracted from Beard, G.M., *American Nervousness, Its Causes and Consequences,* William Wood & Co., New York, 1881.

to fibromyalgia, and "war" or "battleship" neurasthenia seen only in soldiers and sailors[23,24] similar to Gulf War syndrome. None of these subclassifications, however, was associated explicitly with illuminating gas exposure or, in the military, with exposure to CO from firing and exploding shells.

Not surprisingly in this Victorian era, sexual neurasthenia attracted the most medical attention. It was the subject of 122 articles from 1880 to 1930 according to the index catalogs of the U.S. Army Surgeon General's library,[40] from which the U.S. National Library of Medicine was later created. Neurasthenia in women was the subject of another 55 articles during this period, and together these two categories exceeded all the other subtypes combined, which no doubt laid an important foundation for America's rapid acceptance of sexually oriented Freudian psychotherapy in the early 1900s. Neurasthenia in children, who like women were most likely exposed to manufactured gas in their homes and schools, ranked a close third with 48 articles.

Stedman's Medical Dictionary lists several other neurasthenic subtypes of unspecified origin: cardiac or cardiovascular, cerebral, and spinal.[41] All had disappeared from the medical literature by 1957, when for the first time in 100 years not a single medical journal article on neurasthenia was indexed in any language.

But the diagnosis of neurasthenia has been included in every edition of the *International Classification of Diseases*, and its description there is remarkably consistent with CO poisoning. The 10th edition, published by the World Health Organization, lists neurasthenia as follows in Section F48 on "Other Neurotic Disorders":[3]

F48.0 Neurasthenia:

Considerable cultural variations occur in the presentation of this disorder; two main types occur, with substantial overlap. In one type, the main feature is a complaint of increased fatigue after mental effort, often associated with some decrease in occupational performance or coping efficiency in daily tasks. The mental fatigability is typically described as an unpleasant intrusion of distracting associations or recollections, difficulty in concentrating and generally inefficient thinking. In the other type, the emphasis is on feelings of bodily or physical weakness and exhaustion after only minimal effort, accompanied by a feeling of muscular aches and pains and inability to relax. In both types, a variety of other unpleasant physical feelings, such as dizziness, tension headaches, and a sense of general instability, is common. Worry about decreasing mental and bodily well-being, irritability, anhedonia, and varying minor degrees of both depression and anxiety are all common. Sleep is often disturbed in its initial and middle phases but hypersomnia may also be prominent.

Diagnostic Guidelines:

Definite diagnosis requires the following:

a) either persistent and distressing complaints of increased fatigue after mental effort, or persistent and distressing complaints of bodily weakness and exhaustion after minimal effort;
b) at least two of the following:
   — feelings of muscular aches and pains
   — dizziness
   — tension headaches
   — sleep disturbance
   — inability to relax
   — irritability
   — dyspepsia;
c) any autonomic or depressive symptoms present are not sufficiently persistent and severe to fulfill the criteria for any of the more specific disorders in this classification.
   Includes: fatigue syndrome
   Excludes: asthenia NOS (R53)
               burn-out (Z73.0)
               malaise and fatigue (R53)
               post-viral fatigue syndrome (G93.3)
               psychasthenia (F48.8)

10.2.2   Etiologic Theories of Neurasthenia

Bouchot, Beard, and Van Deusen all defined neurasthenia as an organic brain disorder and blamed it on the many energy-draining influences of their "modern" fast-paced urban society. Beard specifically ascribed its cause to the unnatural and overstimu-lating influence of the "steam engine, the telegraph, the printing press and the higher education of women."[29]

Neurasthenia quickly became a very fashionable diagnosis among upper-class women and "brain workers" of both sexes. Its supposed organic etiology as a neuro-logical disorder was embraced by patient and physician alike as much preferable to the more stigmatized psychiatric alternatives of hysteria and madness. Numerous lifestyle, societal, racial, and other inherited factors were invoked to explain its many variations, accommodating everyone from hardworking "brain workers" and delicate upper-class women to lazy and degenerate factory workers. But neurasthenia gradually fell from favor around 1900 as the demographics of the disorder shifted from the upper to lower classes and as, in the absence of any evidence of organic lesions or other objective signs, it was increasingly reinterpreted by physicians as psychiatric.[42]

The concept of a "psychosomatic neurasthenia" was first formally proposed in 1890 by Charles Dana, another prominent New York City neurologist.[43] He later introduced the term *phrenasthenias* to distinguish the neuroses that he said accounted for half of all neurasthenia.[44] Although Dana recognized all the same mental and physical symptoms as Beard, he claimed the physical manifestations were only psychosomatic, caused by depression and anxiety. Freud's seminal 1895 paper distinguishing anxiety neurosis from neurasthenia contributed to this shift, although Freud himself maintained that all the symptoms of both — which he acknowledged often occurred together — were caused by purely physical factors: the repression of sexual fulfillment in cases of anxiety neurosis and excessive masturbation in neurasthenia.[45]

In 1906, Janet[46] introduced the concept of a mutually exclusive distinction between neurasthenia and psychasthenia, with the later due to phobias, obsessions, and compulsions. This distinction is still in place today, as codified in the 10th edition of the *International Classification of Diseases*.[3] The distinction was and still is not rigorously followed in clinical practice, however, as clinicians recognized a great deal of overlap between the two.

One of the most perceptive interpretations, not surprisingly, was that of William Osler. In the first edition of his famous textbook on *The Principles and Practice of Medicine*,[47] he recognized "hyperosmia" and "hyperesthesia" as both distinct con-ditions and symptoms of an organic neurasthenia, about which he succinctly noted, "The entire organism reacts with unnecessary readiness to slight stimuli." Unfortu-nately, later 20th century editions redefined neurasthenia in psychiatric terms.

As the medical debate over the etiology of neurasthenia waxed and waned, many physicians gave up on the diagnosis entirely, viewing it as no more than a dumping ground for many common disorders.[48]

The derision and decline of neurasthenia continued throughout the first half of the 20th century until the subject disappeared entirely from the medical literature in 1957, at least as tracked by *Index Medicus*. Physicians and medical historians now consider 19th-century neurasthenia to have been a diagnostic garbage can for a variety of societally influenced complaints for which no objective evidence or cause could be found.[7,42,49-51]

Missing from both the 19th- and 20th-century medical debates over the etiology of neurasthenia is any discussion of the possible role of CO or illuminating gas exposure. While industrial medicine specialists wrote extensively about both CO and manufactured gases after 1900, they never connected either with neurasthenia. The famous 19th-century German chemist Justus von Liebig may have recognized the connection. Decades after his death, von Oettingen credits him with coining the term *neurasthenia chemicorum* to describe the syndrome caused by exposure to the manufactured gases used in chemical laboratories,[52] but no reference to this term has been found in Liebig's writings or any other 19th-century medical literature.

As discussed below, illuminating gas was obtained primarily from burning coal and contained from 4 to 50% CO, depending on the method of manufacture. With gas flow rates of 2 to 5 ft³ of gas/h,[53] gas lines and fixtures leaking even small amounts of CO at these high concentrations could quickly build up to lethal levels in poorly ventilated indoor environments. Given the crude quality of 19th-century gas piping and gas fixtures, lethal leaks were common.

While the deadly effects of acute exposure to high levels of illuminating gas were well recognized and reported in the 19th century, there was little recognition or discussion of the effects of chronic low-level exposures.[54] The nonlethal effects of gaslights that received the most medical and public attention were the stifling heat and great amount of moisture given off by illuminating gas when it burned, and the "sewer gas" odor of hydrogen sulfide in coal gas.

Even though over 1400 medical books and articles were written on CO poisoning before 1945 — with dozens of them warning about the dangers of illuminating gas and dozens more arguing about the existence of a unique syndrome associated with chronic low-level CO exposure — none recognized the possibility that exposure to CO in manufactured gas may have been a cause of neurasthenia.[52] Only in 1952 did Theron Randolph, considered the father of modern clinical ecology or environmental medicine, first publish an abstract noting the strong association between domestic exposure to utility gas and the symptoms of neurasthenia.[55] But given the psychiatric connotations of neurasthenia, he had already long recommended diagnosing its symptoms as "allergic toxemia" in all cases in which a causative toxic agent could be identified.[56] This term was first proposed by Albert Rowe[57] in 1930 to diagnose food intolerances accompanied by migraines but it died out in the 1950s.

10.2.3 OVERLAP WITH MULTIPLE CHEMICAL SENSITIVITY, CHRONIC FATIGUE SYNDROME, FIBROMYALGIA SYNDROME, AND GULF WAR SYNDROME

Patients presenting with neurasthenic symptoms are now more commonly and multiply diagnosed with a variety of recently defined and putatively unique disorders, including MCS, CFS, FMS, and GWS. The striking similarity of all these disorders, but primarily CFS, with 19th-century neurasthenia has been noted by many psychiatrically minded critics, who contend that each is just another culturally determined variant of the same supposedly psychogenic original.[7,58–68] Indeed, none of the current definitions of any of these disorders is sufficiently unique to distinguish them.[69] Almost all those who meet the current Centers for Disease

Control (CDC) criteria for CFS, for example, also meet the current ICD-10 criteria for neurasthenia.[70]

Fewer than 5% of the papers on CFS, FMS, or MCS published in the 1990s addresses any of their many overlaps with each other or with neurasthenia.[12] Only one discusses the similarity of CO poisoning with MCS, but even in this case report of a woman with both, the author fails to see any causal connection and dismisses the MCS as psychogenics.[60] The similarity with CFS also has been reported in only one case of CO poisoning, but dismissed as a misdiagnosis.[70a] No papers on CFS, MCS, or FMS have ever mentioned CO poisoning as a possible cause. Most are more interested in pursuing recently proposed theories. In the case of MCS, for example, these include limbic kindling,[71] neurogenic inflammation,[72] porphyrin disorders,[73] immune dysfunction,[74] and toxicant-induced loss of tolerance.[75]

Although CO poisoning is not usually considered a cause of CFS, FMS, or MCS, and carboxyhemoglobin (COHb) levels are routinely assessed in their diagnosis, there are anecdotal reports of MCS[76] and CFS[77] patients responding well to repeated sessions of normobaric and hyperbaric oxygen treatment, respectively.

MCS was first defined by this name in 1987 by Mark Cullen from an occupational medicine perspective.[78] The hallmark symptom is increased sensitivity to or reduced tolerance for previously tolerated levels of various sensory stimuli, most notably inhaled and ingested substances such as chemical odors, irritants, drugs, and alcohol, but also any combination of sound, light, touch, pain, hot and/or cold weather, electromagnetic fields, and physical, mental, or social stress.

While the etiology of MCS is still unknown, the American Lung Association, American Medical Association, Consumer Product Safety Commission, and Environmental Protection Agency issued a joint statement in 1994 declaring that the "current consensus is that in claimed or suspected cases of MCS, complaints are not to be dismissed as psychogenic, and a thorough workup is essential."[79]

A more recent consensus, published in 1999 by 34 clinicians and researchers, established the first formal criteria for the diagnosis of MCS:[80]

1. The symptoms are reproducible with [repeated chemical] exposure.
2. The condition is chronic.
3. Low levels of exposure [lower than previously or commonly tolerated] result in manifestations of the syndrome.
4. The symptoms improve or resolve when the incitants are removed.
5. Responses occur to multiple chemically unrelated substances.
6. Symptoms involve multiple organ systems.

MCS is the most common diagnosis among U.S. veterans in the Gulf War Registry maintained by the U.S. Department of Veterans Affairs[81] and the majority of both veteran and civilian cases also meet the current diagnostic criteria for CFS and/or FMS.[82]

CFS was first defined "for research purposes" by a consensus of clinicians and researchers in 1988[83] and redefined in 1994.[84] The diagnosis requires fatigue lasting 6 months or more that causes a substantial reduction in previous levels of activity (50% or more), is unrelieved by rest, and includes at least four of the following eight symptoms:

1. Short-term memory loss or concentration difficulties severe enough to cause a substantial reduction in previous levels of activity;
2. Sore throat;
3. Tender lymph nodes (glands) in neck or armpit;
4. Muscle pain;
5. Pain in two or more joints without swelling or redness;
6. Headaches of a new type, pattern, or severity;
7. Not feeling refreshed after sleep;
8. Still feeling fatigued more than 24 h after physical exertion.

The symptoms of CFS should not be explained by other diseases but specific exemptions are allowed for the concurrent diagnosis of FMS, MCS, neurasthenia, somatoform and anxiety disorders, and nonpsychotic depression. Unfortunately, these potential overlaps are rarely considered in the differential diagnosis of CFS.

FMS also was defined by a consensus of clinicians and researchers as a chronic disorder, but one that need only have lasted 3 months.[85] The diagnosis requires evidence of widespread pain and tenderness on both left and right sides, above and below the waist, and anywhere along the spine, as demonstrated by hypersensitivity to at least 11 of 18 specific trigger points.

While neurasthenia was defined in its day by many more specific symptoms than CFS, FMS, or MCS, all are extremely variable and polysymptomatic, sharing literally dozens of symptoms. Of note is that all their cardinal symptoms — chronic fatigue, widespread pain and tenderness, and sensory intolerance — are included in the detailed descriptions of neurasthenia noted above, including those of Poe, Bouchot, von Deusen, and Beard.

10.3 CARBON MONOXIDE IN MANUFACTURED GAS

10.3.1 History and Chemistry of Manufactured Gas

The 19th century saw the commercial development of many types of manufactured gas for illuminating and so-called utility purposes[53,86]:

1. Tar Gas: Made from coal tar, the first choice of the first American illuminating gas company, used for street lighting in Baltimore from 1817 to 1822.
2. Coal Gas: Made from destructive distillation of coal in coke ovens, yielding gas that was 40% methane and 4 to 6% CO; first used for street lighting in the United States in Baltimore in 1822, quickly became the most commonly used illuminating gas and remained so until overtaken by water gas plants in 1907.
3. Natural Gas or "Swamp" Gas: Commonly over 90% methane with varying minor percentages of other light hydrocarbons; first used for gas lighting in Fredonia, NY in 1824.
4. Water Gas: Made by spraying water vapor over incandescent coke ($H_2O + C$) yielding hydrogen and CO with no hydrogen sulfide or other noxious impurities. Invented in France in 1810, patented in the United States

in 1854, first produced in Philadelphia in 1858, and gradually adopted by almost all U.S. gas companies, which usually mixed water gas with other gases to dilute its high CO content and increase its illuminating power.

5. Carburetted Water Gas: Made from mixing steam with water gas and oil gases (ethane, ethylene, acetylene, and benzene), yielding up to 36% hydrogen, 30% CO, and 26% methane.

6. Carburetted Air Gas: Made by forcing air through a carburetor so that it becomes saturated with (10 to 17%) gasoline vapors; commonly used to illuminate country houses and hotels located outside the range of municipal gasworks.

7. Oil Gas: Made in iron retorts from heavy petroleum until improved with "light oils" (naphtha) in Saratoga, NY in 1870. Oil gas could be burned by itself, but was usually mixed with water gas or coal gas to increase illuminating power.

8. Acetylene Gas: Produced for lighting purposes from 1862 by heating charcoal with an alloy of zinc and calcium and later by adding water to calcium chloride.

9. Mixed Gas: Commonly a mixture of coal and water gas yielding a final CO concentration of 25 to 35%, but the term was applied to any combination of the above types, and many combinations were tried in the effort to increase gas efficiency.

Production methods were crude, involving relatively low temperatures and pressures resulting in highly variable gases whose composition varied greatly day to day, season to season, and city to city.[87] Leakage rates also were high, with a 19th-century trade publication citing leakage rates averaging 20% nationwide in 1862.[53]

Toxic chemicals of primary concern in manufactured gas included CO (4.5 to 7.5% in coal gas, up to 25% in carbureted water gas, 50% in pure water gas, and commonly 25 to 35% in mixed gas), methane at 4 to 43%, and hydrogen sulfide at 1 to 3% — but absent from pure water gas.[88]

Various "illuminants" including paraffin, gasoline, and petroleum oils were commonly added to increase the proportion of brightly burning methane-, ethylene-, and benzene-series hydrocarbons. These usually comprised 2 to 16% of the final gas and consisted primarily of ethylene and benzene, with much smaller percentages of acetylene, butylene, toluene, xylene, crontonylene, propylene, amylene, and mesitylene.[87]

Unwanted "impurities" were a significant problem in both 19th- and 20th-century production of coal gas and its derivatives. Significant percentages of hydrogen sulfide, carbon disulfide, ammonia, and tar were only partially removed by various filtering, scrubbing, and washing techniques, leaving a strong odor of "sewer gas." Illuminating and later utility gas, were distributed at such low pressures (an average $1/2$ in. of water) and variable temperatures that the saturated light hydrocarbons in the gas often condensed in pools inside low points of distribution lines and fixtures. This reduced both the flow of gas and its illuminating power, causing gas lights to flicker dimly and, in worst cases, be extinguished with often fatal results.

With the gradual switch to electric lighting from the 1880s through the rural electrification programs of the 1930s, manufactured gases were designed and used less for lighting and more as "utility gas" for space and water heating, refrigeration, cooking, and industrial power applications.

Production methods shifted to focus on maximizing heat output, the practice of adding illuminants was abandoned, and the final product was renamed "utility gas" to highlight its multiple uses. It was the primary form of domestic gas sold in the United States in the first half of the 20th century, and was only gradually displaced by natural gas with the introduction of transcontinental pipelines in the 1950s.[53] The gradual shift away from using manufactured gases for lighting greatly reduced the extent of gas piping, fixtures, and leaks in most homes and offices, but led to concentration of the gases in kitchens and laundry rooms, were women were disproportionately exposed.

10.3.2 SYMPTOMS OF CHRONIC CARBON MONOXIDE POISONING

"Chronic carbon monoxide poisoning" refers not to the chronic neurological sequelae of acute CO poisoning from a single high level exposure but to an even more variable syndrome associated with repeated or continuous exposure to small doses of CO. The first medical descriptions appeared long before CO was identified as the culprit. Bernardino Ramazzini, in his seminal 1713 book on *The Diseases of Workers*, devoted a chapter to the "serious complaints such as headaches, pains in the eyes and even severe dyspnea" that affect confectioners, due to their continuous work standing over sugar pans heated by coal braziers."[89a]

The symptoms of occupational coal gas exposure were first attributed to CO in 1879 by Benjamin Richardson, a fellow of the Royal College of Physicians in England. He described how lace-frame workers placed coal stoves with glowing coals beneath their looms in winter to warm their fingers:

> There is no visible smoke from the fuel, and yet there is a dangerous exhalation. From the chafer is evolved carbonic acid gas [CO_2], and worse still, carbonic oxide [CO]. The first of these gases is suffocative, and is injurious, if so little as one percent of it be present in the air breathed. The second is directly more hurtful, and in much smaller proportion. It causes headache, nausea, giddiness, and irregular muscular power, with palpitation of the heart.[97]

The etiology of chronic CO poisoning has been the subject of almost continuous medical controversy since the disorder was first recognized as an occupational disease in the U.S. in 1908 by Dr. George Kober, a professor of hygiene at George-town University. Kober chaired a federal Committee on Social Betterment, whose final report identified the symptoms of chronic exposure to CO among workers in the gas industry as "headache, dizziness, slow pulse, anemia, general debility, and diseases of respiratory and digestive organs."[89] But industry physicians claimed the symptoms were either psychiatric or feigned by malingering workers seeking compensation. They pointed to the lack of correlation between the symptoms of chronic CO poisoning and normal carboxyhemoglobin levels, and the absence of any other objective abnormalities or biomarkers. Even Alice Hamilton questioned the very existence of chronic CO poisoning as a distinct disorder in her classic textbook on

Industry Toxicology, saying its many symptoms "cannot be distinguished from the symptoms of ill health which come from factory work itself."[89b]

Neither supporters nor critics of the diagnosis of chronic CO poisoning, however, ever equated it directly with neurasthenia. Dr. George Apfelbach came close, saying its symptoms "simulate hystero-neurasthenia."[90] He wrote:

> Persons who do not acquire a tolerance to the gas will become anaemic, irritable, neurasthenic, and complain of headaches, anorexia, loss of weight, backaches, vertigo and gastro-intestinal disorders. To this list Watkins,[91] who investigated the conditions in the steel industry, adds: nausea, vomiting, lack of coordination, muscular weakness, rapid fatigue, palpitation and cardiac distress, irregular pulse, neuritis with ensuing paralysis, and melancholia or hallucinations. ... From the point of view of the employer: mental alertness is gone, while judgment, strength and coordination are impaired. ...
>
> Wilmer[92] gives in great detail the symptoms and impressions recorded by a very intelligent patient whose whole family was subjected to slow poisoning by the gas, escaping in an old house into which they had moved which was said to be "haunted" due, plainly, to hallucinations of sight and hearing which all occupants experienced, and which turned out eventually to be due to CO poisoning. There were, also, peculiar changes in temperament so that the persons became just the opposite of what they were before. The picture looked like hysteria. Wilmer describes the optic neuritis scotoma and other disturbances in the field of vision of two of the children, one of whom had not entirely recovered after a period of four years. ...
>
> Georgine Luden[93] of the Mayo Clinic gives, in great detail, the symptoms in herself and several others which were due to a faulty furnace in the residence and caused disturbances extending over a year or so. These symptoms she divides into three periods. ... Those of the *second period* were many and included varied subjective symptoms, but also the following objective symptoms: sudden drowsiness; shivering attacks; muscular spasm (short); extreme thirst (satisfied with lots of tea); great craving for sugar; transient bronzing; pinched face for 24–36 hours (one looks old); low blood pressure with normal pulse pressure; transitory "uncomfortable" heart action; irregular flattening pulse; respiratory oppression after the intake of food, sometimes with slight shivering attacks; and low blood-sugar values. *Third Period:* Hypersensitization to minute amounts of CO, as a whiff of exhaust gas from a passing auto, smoke from a chimney; transient, acute nausea; numbness of feet; scalp felt too tight, etc. Veins in scalp seem to bulge in certain areas; "coal-gas headache" for several hours which is characteristic, localized, sore to touch, intense, no effects from the usual remedies (aspirin, phenacetin, pyramidin, etc.); dull headache at times.

The first in-depth study of both acute and chronic CO poisoning was conducted in 1921 for the U.S. Department of Labor by Dr. Alice Hamilton, the mother of modern occupational medicine.[86] On the subject of chronic CO poisoning, she wrote:

> The whole question of chronic-monoxide poisoning is very obscure as to its nature, its prevalence, and its diagnosis. It can not be regarded as caused by an accumulation of carbon monoxide in the body, which reaches a certain degree and then produces symptoms, because that could occur only if a person were breathing carbon-monoxide contaminated air all the time, and this is never true.

There is a marked variation in the way different individuals are affected by long exposure to small quantities of carbon monoxide. One person will suffer from ill-health, constant headache, neuralgic pains, perhaps albumen in the urine, after a few months' work in a room where others have passed several years with no trouble at all.

There is great difficulty in diagnosing this form of occupational poisoning, because so rarely are the symptoms at all characteristic. Usually there is only complaint of headache, palpitation of the heart, breathlessness, general nervousness, and disturbed sleep. Indigestion (especially for solid food), loss of appetite, and acid eructations two or three hours after meals are very common. There may be sleeplessness at night and drowsiness in the daytime. Often this drowsiness is so great as to lead to the taking of stimulants. The French attribute the alcoholism of male cooks to the effect of gas from the stoves, and say that they suffer from loss of energy and great irritability.

A great number of industrial workers are exposed to small quantities of carbon monoxide in the course of their work, and may suffer from poisoning, although there is almost no positive evidence that this is actually the cause of the ill-health, so often complained of in these trades. Those who are most in danger are thought to be the following:

a. Laundry workers, especially ironers.
b. Pressers in tailor shops.
c. Furnace tenders.
d. Gas workers, especially men making connections and reading meters.
e. Painters working in rooms in which salamanders are kept burning to dry the walls.
f. Printers, especially linotypists, and men working at melting kettles heated with gas.
g. Molders of metal where gas is used for heat.
h. Men working in canneries on soldering machines.
i. Miners, especially in mines where blasting is done continuously.
j. Garage workers, especially repairers.

Note that Dr. Hamilton does not mention illuminating gas exposure, which by the 1920s had been almost completely replaced by electric lighting. But the symptoms she describes from these various other sources of CO exposure are consistent with earlier descriptions of neurasthenia.

10.4 CARBON MONOXIDE AS A PUTATIVE CAUSE OF NEURASTHENIA — EVANS'S 10 UNIFIED CRITERIA FOR DISEASE CAUSATION

The three original Henle–Koch postulates, first suggested by Jakob Henle[94] in 1840 and refined by his pupil Robert Koch[95] in 1890, were developed to establish causation of infectious disease. These were modified and expanded by Alfred Evans, based on his review of a dozen other algorithms designed to assess the cause of other types of diseases, into a set of 10 "Unified Criteria" for assessing the causation of any acute or chronic disease (Table 10.3).[96] The hypothesis that CO poisoning from exposure to illuminating and utility gas was a cause of neurasthenia meets all of the 10 Unified Criteria, as reviewed below.

TABLE 10.3
Criteria for Causation: A Unified Concept

1. Prevalence of the disease should be significantly higher in those exposed to the putative cause than in case controls not so exposed ... The putative cause may exist in the external environment or in a defect in host response.

2. Exposure to the putative cause should be present more commonly in those with the disease than in controls without the disease when all risk factors are held constant.

3. Incidence of the disease should be significantly higher in those exposed to the putative cause than in those not so exposed as shown in prospective studies.

4. Temporally, the disease should follow exposure to the putative agent with a distribution of incubation periods on a bell-shaped curve.

5. A spectrum of host responses should follow exposure to the putative agent along a logical biologic gradient from mild to severe.

6. A measurable host response following exposure to the putative cause should regularly appear in those lacking this before exposure (i.e., antibody, cancer cells) or should increase in magnitude if present before; this pattern should not occur in persons so exposed.

7. Experimental reproduction of the disease should occur in higher incidence in animals or man appropriately exposed to the putative cause than in those not so exposed; this exposure may be deliberate in volunteers, experimentally induced in the laboratory, or demonstrated in a controlled regulation of natural exposure.

8. Elimination or modification of the putative cause or of the vector carrying it should decrease the incidence of the disease (control of polluted water or smoke or removal of the specific agent).

9. Prevention or modification of the host's response on exposure to the putative causes should decrease or eliminate the disease (immunization, drug to lower cholesterol, specific lymphocyte transfer factor in cancer).

10. The whole thing should make biologic and epidemiologic sense.

Source: Evans, A.S., *Yale J. Biol. Med.*, 49, 175–195, 1976. With permission.

10.4.1 PREVALENCE CRITERIA

The relative prevalence of neurasthenia was significantly higher in those exposed to the putative cause — manufactured gas — than others not so exposed, whether analyzed by occupation, social class, or geography. Neurasthenia was first described as a disease of "brain-workers" and upper-class women.[29] It did not afflict farmers, construction workers, and others who labored outdoors in fresh air, all of whom were generally considered healthier than average.[97] As use of gas lighting spread, neurasthenia also became a disorder of factory workers, cooks, bakers, laundresses, tailors, printers, painters, and others who were multiply exposed to gas used for lighting, cooking, heat, and/or power. A 1911 study of 7000 garment workers in St. Louis, for example, reported 25% suffered from neurasthenia,[98] but their disease was blamed on other ills such as neglect of personal hygiene, low wages, and irregularity of employment.[98a] Not until the 1920s was their illness first described in the medical literature as chronic CO poisoning, but by then no reference was made to its earlier characterization as neurasthenia.[86]

According to a prevalence map published by Beard[29] in 1881, neurasthenia also was concentrated geographically in the gaslit urban areas of the Northeast, the industrial Midwest, and mid-Atlantic states, which hosted the highest number of manufactured gas plants, while few cases were reported in the West and Southwest, which had few gas plants.

Although now rarely diagnosed in the United States, neurasthenia is still common in China and Russia (two countries that still manufacture gas from coal for domestic and industrial purposes), but there, too, it is now widely considered psychiatric. A nationwide epidemiological study in China put the prevalence at 1.3% of the general population,[99] and a study of over 1700 Chinese-Americans in the Los Angeles area found 6.4% met the WHOs ICD-10 criteria for neurasthenia.[100] The same WHO criteria were used in a prevalence study of young adults in Switzerland, which found only 0.9% whose symptoms had lasted the 3 months required for diagnosis but an incredible 12% whose symptoms had lasted at least 1 month.[101]

Neurasthenia clearly caused by toxic exposures — so-called organic or toxic neurasthenia — has been well documented, but only in a few European studies of people occupationally exposed to toluene and other organic chemicals used in jet fuel,[102,103] car paint,[104] paint production,[105] printing,[106] and the textile industry.[107] The diagnosis of toxic neurasthenia also has been proposed to explain the chronic psychiatric and neurological symptoms afflicting Vietnam War veterans who were exposed to Agent Orange and other pesticides.[108]

The prevalence of MCS, FMS, and CFS in the general population today is estimated at up to 6.3% for MCS,[109] 2 to 4% for FMS,[110] and up to 2.6% for CFS,[111] while the total prevalence of those suffering from CO poisoning at any time is unknown. The MCS data come from a statewide survey by the California Department of Health Services, which the found the prevalence of MCS higher only among women (adjusted Odds Ratio = 1.63, 95% Confidence Interval = 1.23 to 2.17) and Hispanics (OR = 1.82, 95%CI = 1.21 to 2.73). Both groups would be expected to have greater-than-average exposure to CO based on the greater time they spend around domestic gas appliances and consumer products containing dichloromethane (DCM). DCM is a common solvent and spray propellant that, whether inhaled or ingested, is metabolized directly to CO. This endogenous CO can increase COHb to fatal levels,[112] and even at sublethal levels the COHb it forms has a half-life twice that of the COHb formed from directly inhaled CO.[113]

10.4.2 Exposure Criteria

Whether examined by geography, occupation, or social class, exposure to CO from illuminating or utility gas was much more common in those with neurasthenia than in those without. As noted above, neurasthenia was widely recognized in its day as a disease of people who spent a lot of time indoors (in gaslit environments), and was relatively rare among those who worked out-of-doors or lived in rural areas, among whom exposure to gas lighting was rare. And while the same classes and most of the same occupations that suffered from neurasthenia in the eastern United States existed in the West, the disease was uncommon in those states without manufactured gas plants.

Neurasthenia was a common and popular diagnosis in its early years among upper-class men and women who could afford gas lighting in their homes, offices, and shops,[114-116] but by the turn of the century the disorder was more common among lower-class factory workers who could less easily afford to upgrade to electric lighting and who also tended to live in older, gaslit, urban housing.[98,117,118] Electric lighting, like gas lighting before it, was quickly adopted by the upper classes in the 1880s and

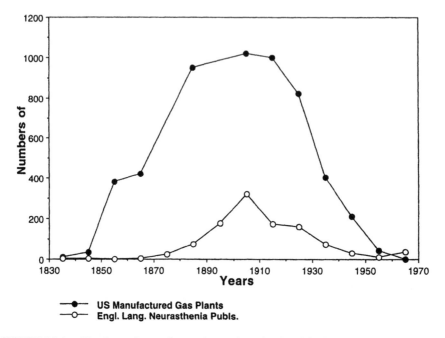

FIGURE 10.1   Numbers of manufactured gas plants in the United States and the English-language literature on neurasthenia, 1830s to 1960s, by decade. 1830s and 1840s are all E. A. Poe. The medical literature starts after "water gas" begins being produced in 1858.

1890s, resulting in noticeably decreased patient load and income for those physicians whose practices had specialized in treating neurasthenia among the wealthy.

10.4.3   INCIDENCE CRITERIA

Unfortunately, no controlled prospective studies of neurasthenia were done in the 19th century. But the natural incidence of neurasthenia over the long term, like its preva-lence, can be reasonably inferred from the number of English-language medical papers on the subject that were published annually. This clearly peaked and fell in step with the peak and fall of manufactured gas plants in the U.S. (Figure 10.1), although the neurasthenia literature did not begin to appear until approximately 40 years after the introduction of coal gas. Only after the introduction of water gas in 1873 — which, at up to 50% CO, was much more toxic than coal gas — does the medical literature begin to take off. The same geographic, occupational, and social class associations seen between the use of manufactured gas and the prevalence of neurasthenia over so many decades also are true of its incidence.

Prospective studies of acutely CO-poisoned patients done in the 20th century show a much higher incidence of chronic neurological sequelae typical of neuras-thenia compared with normal controls, with a lower incidence among those treated within 6 h of CO exposure by hyperbaric oxygen.[119]

No good data are available on the current annual incidence of CO poisoning, although an estimated 10,000, or more, people are seen annually for this reason in emergency rooms. CFS and FMS likewise lack good incidence data, but that on MCS comes from a random survey of 4000 adults conducted by the California State Department of Health in 1995 and replicated again in 1996 using the Behavioral Risk Factor Surveillance Survey methodology of the U.S. CDC. Looking at the last 5 years, the study calculated the average annual incidence of new MCS cases to be 0.36%.[109]

10.4.4 TEMPORALITY CRITERIA

No 19th-century studies associating exposure to manufactured gas and the onset of neurasthenia symptoms are known, but again the rise, peak, and fall of neurasthenia in the medical literature closely matches the number of manufactured gas plants operating in the United States (Figure 10.1). Some evidence for temporality in individual cases is provided by anecdotal reports of cases that improved while away from their gaslit homes and/or workplaces for extended periods but who promptly worsened upon their return.[92] In the case of Edgar Allan Poe, whose first description of gas poisoning symptoms appeared in 1832,[13] he was most likely first chronically exposed for a few months in 1829, while living at his foster father's expense in the Indian Queen Hotel in downtown Baltimore.[18]

10.4.5 SPECTRUM OF RESPONSE CRITERIA

A spectrum of host responses following a logical biological gradient from mild to severe has been reported in response to both acute and chronic CO exposures. Both the immediate symptoms of sublethal CO poisoning[52,86,120,121] and their delayed neurological sequelae[122-125] are consistent with 19th- and 20th-century descriptions of neurasthenia. These symptoms do not correlate well with COHb, however, the most commonly measured biomarker of CO exposure. That COHb does not show a consistent dose–response relationship with and is not predictive of chronic CO symptoms, especially at low levels[126] and in cases involving the chronic neurological sequelae typical of neurasthenia,[127] may be explained in part by the delayed effects of brain lipid peroxidation seen after tissue reoxygenation[128] and the phenomenon of habituation seen in some cases of continuous long-term exposure, such as among cigarette smokers and those living at high altitudes.[129,130]

COHb also may be increased by chronic exposure to any sufficiently stressful stimulus, as all stressors tested so far — physical, biological, chemical, and social — induce heme-oxygenase-1, a universal stress protein found in all red-blooded mammals, to break down heme into CO, biliverdin, and iron.[131] A different spectrum of response is evident in these habituated individuals who can tolerate relatively high levels of CO exposure compared with those with little CO exposure who are not habituated and therefore more sensitive to the effects of CO at low levels.

10.4.6 MEASURABLE HOST RESPONSE CRITERIA

Except in cases of habituation or adaptation to repeatedly high levels of CO, increases in COHb and CO in breath are measurable after each significant new acute exposure to CO.[132] Other CO-induced changes consistent with the symptoms of neurasthenia can be seen in decreased plasma volume,[133] decreased 2,3-diphosphoglycerate at high levels of CO, impaired oxygen dissociation,[134] and increased levels of glucose[135,136] and erythrocytosis.[137] Other measurable changes seen with increasing CO exposure include impairment of cerebrovascular circulation,[138] cardiovascular function,[139] and cognitive function.[140]

10.4.7 EXPERIMENTAL REPRODUCTION CRITERIA

Neurasthenia occurred at a higher rate among people exposed to manufactured gas than it did among those not so exposed. As discussed above, this is true whether exposure is assessed geographically,[29,88] temporally (Figure 10.1), by social class, or by occupation. The history of the natural experiment is sufficient to fulfill the criteria, which also may be met by evidence from deliberate exposure of animals or human volunteers under real or laboratory conditions. While no attempts were made to reproduce neurasthenia by chronic exposure to manufactured gas lighting under controlled laboratory conditions, numerous natural experiments involving chronic exposure to CO have shown that both the physical and mental symptoms of neurasthenia are associated with exposure to CO from various sources, including illuminating gas.[140a]

10.4.8 ELIMINATION OR MODIFICATION OF CAUSE CRITERIA

The decline in the use of illuminating gas after 1900, as reflected in the declining number of U.S. gas plants in operation, was matched by a decline in the diagnosis of neurasthenia, as reflected by the declining number of papers published on the subject (Figure 10.1). Theron Randolph, considered the father of modern clinical ecology or environmental medicine, was the first to identify chronic exposure to utility gas as a cause of systemic neurasthenic symptoms in 1952,[55] and he later noted that in over 400 cases the only cure had been to remove all the gas piping from the dwelling.[141]

10.4.9 PREVENTION OR MODIFICATION OF RESPONSE CRITERIA

Haldane[143] first showed in 1895 that concurrently exposing mice to high levels CO and hyperbaric oxygen at 2 atm (absolute) prevented their unconsciousness and death. A hundred years later, hyperbaric treatment at 2.8 atm began within 6 hours of moderate CO poisoning (involving symptomatic but not unconscious cases) was shown in a prospective randomized trial to eliminate the risk of chronic neurological sequelae in humans completely, the incidence of which was 23% in those treated only with normobaric oxygen ($p < 0.05$).[144] Similar effects have been shown in rats. Although both acute and chronic symptoms of CO poisoning respond well to hyperbaric oxygen treatment, they may be prevented entirely if exogenous CO exposures are avoided and endogenous sources are kept within normal limits (through stress reduction techniques, for example, and the avoidance of exposure to products such as propellant spray cans that contain DCM, which is metabolized to CO).

10.4.10 MAKES-SENSE CRITERIA

That CO poisoning from exposure to manufactured gas could have been a cause of neurasthenia makes both epidemiological sense, in terms of the limited data available about the higher prevalence of neurasthenia in those states, occupations, and social classes with the greatest exposure to manufactured gas, and biological sense, in terms of what is now known about the biological functions of CO. CO is not just an air pollutant and potential asphyxiant that binds competitively with oxygen on hemoglobin and other heme proteins, but a natural breakdown product of these same heme compounds produced in response to stress of any kind. It is also a neurotransmitter involved in the control of numerous functions also disturbed in neurasthenia, including respiration, heart rate, circulation, learning, memory, vision, olfaction, and sensory habituation to lights, sounds, and odors.[145]

10.5 CONCLUSIONS

The hypothesis that the syndrome of neurasthenia may have been caused in the late 19th and early 20th century by carbon monoxide poisoning from exposure to illuminating and utility gas satisfies Evan's ten Unified Criteria for assessing the causation of disease. Given the overlap of neurasthenia symptoms with those of multiple chemical sensitivity, chronic fatigue syndrome. Given the overlap of neurasthenia symptoms with those of multiple chemical sensitivity, chronic fatigue syndrome, fibromyalgia and Gulf War Syndrome, it is possible that these other disorders also may be caused in some cases by carbon monoxide poisoning.

Unfortunately, carbon monoxide poisoning is not routinely screened for in hospital emergency rooms or private practices, and it is well recognized that many cases remain undiagnosed. Given the ease, speed, and accuracy with which endogenous CO levels can be screened with hand-held breath analyzers, breath testing for CO is recommended not just whenever CO poisoning is suspected but as part of any evaluation of neurasthenia, CFS, FMS, MCS, or GWS. Knowing the parts per million (ppm) of CO in breath, which must be corrected to account for the level of CO in ambient air, one can then make a rough estimate of the carboxyhemoglobin (COHb) in blood: the percent COHb in health resting adults, smokers or nonsmokers, is approximately 1/6th the level of CO in their breath.[146] Additional blood testing is recommended to measure the percent carboxyhemoglobin and the difference in the partial pressure of oxygen in arterial and venous blood.

As with so many public health crises, however, the high rate of CO poisoning in the U.S. is best countered by prevention and early detection. Attached garages and all combustion appliances, including gas ranges and ovens, should be vented. Just as almost every residence in the U.S. now has a smoke detector, every residence also should have at least one high-level CO alarm or low-level CO monitor. The latter are recommended for pregnant women, children, the elderly, and anyone with chronic disease at greater risk from CO exposure.

REFERENCES

1. Bouchot, E., De l'etat nerveux dans sa forme aigue et chronique, *Bull. Acad. Med.*, 23, 980, 1857–1858.
2. Chatel, J.C. and Peele, R., A centennial review of neurasthenia, *Am. J. Psychiatr.*, 126, 1404–1413, 1970.

3. World Health Organization, *The ICD-10 Classification of Mental and Behavioral Disorders: Clinical Descriptions and Diagnostic Guidelines*, World Health Organization, Geneva, 1992.

4. Evans, A.S., Causation and disease: the Henle–Koch postulates revisited, *Yale J. Biol. Med.*, 49, 192, 1976.

5. Hill, A.B., The environment and disease: association or causation? [President's Address from January, 14, 1965], *Proc. R. Soc. Med. Sect. Occup. Med.*, 295–302, 1965.

6. Wessely, S., Old wine in new bottles: neurasthenia and "ME," *Psychol. Med.*, 20, 35–53, 1990.

7. Abbey, S.E. and Garfinkel, P.E., Neurasthenia and chronic fatigue syndrome: the role of culture in the making of a diagnosis, *Am. J. Psychiatr.*, 148, 1638–1646, 1991.

8. Smythe, H., Fibrositis syndrome: a historical perspective, *J. Rheumatol.*, 19 (Suppl.), 2–6, 1989.

9. Corrigan, F.M., MacDonald, S., Brown, A., Armstrong, K., and Armstrong, E.M., Neurasthenic fatigue, chemical sensitivity and GABAa receptor toxins, *Med. Hypotheses*, 43, 195–200, 1994.

10. Wolfe, J. Proctor, S.P., Davis, J.D. et al., Health symptoms reported by Persian Gulf War veterans two years after return, *Am. J. Ind. Med.*, 33, 104–113, 1998.

11. Haley, R.W., Kurt, T.L., and Hom, J., Is there a Gulf War syndrome? Searching for syndromes by factor analysis of symptoms, *J. Am. Med. Assoc.*, 277, 215–222, 1997.

12. Donnay, A., On the recognition of multiple chemical sensitivity in medical literature and government policy, *Int. J. Toxicol.*, 18, 383–392, 1999.

13. Poe, E.A., A decided loss [Loss of breath], in *The Unabridged Edgar Allan Poe*, Running Press, Philadelphia, 1983.

14. Poe, E.A., The Fall of the House of Usher, in *The Unabridged Edgar Allan Poe*, Running Press, Philadelphia, 1983.

15. Carlson, E.W., Ed., *The Recognition of Edgar Allan Poe*, Ann Arbor Paperbacks, University of Michigan Press, Ann Arbor, 1970.

16. Sloane, D.E.E., Usher's nervous fever: the meaning of medicine in Poe's *The Fall of the House of Usher*, in *Poe and His Times, the Artist and His Milieu*, Fisher, B.F., Ed., The Edgar Allan Poe Society, Baltimore, 1990, 147–153.

17. Carlson, E.W., Ed., *A Companion to Poe Studies*, Greenwood Press, Westwood, CT, 1996, 31.

18. Donnay, A., Poisoned Poe: evidence that Edgar Allan Poe may have suffered from neurasthenia (a.k.a. multiple chemical sensitivity and chronic fatigue syndrome) as a result of exposure to illuminating gas, presented at International Edgar Allan Poe Conference, Richmond, VA, October 10, 1999.

19. Bouchot, E., *Du nervosisme aigu et chronique et des maladies nerveuses*, Librairie J.-B. Bailliere et Fils, Paris, 1860.

20. Mitchell, S.W., Morehouse,G., and Keen, W.W., *Gunshot Wounds and Other Injuries of the Nerves*, Lippincott, Philadelphia, 1864.

21. Flint, A., *A Treatise on the Principles and Practice of Medicine; Designed for the Use of Practitioners and Students of Medicine*, Henry C. Lea, Philadelphia, 1866, 640–641.

22. Da Costa, J.M., On irritable heart; a classical study of a form of functional cardiac disorder and its consequences, *Am. J. Med. Sci.*, 71, 52, 1871.

23. Burton-Fanning, F.W., Neurasthenia in soldiers of the home forces, *Lancet*, 907–911, 1916.

24. Evans, S.G., Battleship neurasthenia, *Mil. Surg.* (Richmond VA), 24, 32–36, 1909.
25. MacCurdy, J.T., *War Neurosis*, Cambridge University Press, Cambridge, 1918, 113.
26. Hyams, K.C., Wignall, F.S., and Roswell, R., War syndromes and their evaluation: from the U.S. Civil War to the Persian Gulf War, *Ann. Intern. Med.*, 125, 398–405, 1996.
27. Beard, G.M., Neurasthenia, or nervous exhaustion, *Boston Med. Surg. J.*, 3, 217–221, 1869.
28. von Deusen, E.H., Observations on a form of nervous prostration, neurasthenia, culminating in insanity, *J. Insanity*, Suppl. to the Asylum's Annual Report for 1867–1868, 25, 445–461, 1869.
29. Beard, G.M., *American Nervousness, Its Causes and Consequences*, a supplement to Nervous Exhaustion (Neurasthenia), William Wood & Co, New York, 1881.
30. Beard, G.M., Sexual Neurasthenia (Nervous Exhaustion); Its Hygiene, Causes, Symptoms and Treatment, with a chapter on diet for the nervous, Rockwell, A.D., Ed., E.B. Trent & Co, New York, 1886.
31. Hermann, H.W., On certain forms of nervousness in children, *St. Louis Polyclin.*, 141–147, 1889.
32. Deale, H.B. and Adams, S.S., Neurasthenia in young women, *Am. J. Obstet.*, 24, 190–195, 1894.
33. Dana, C.L., On a new type of angioparalytic or pulsating neurasthenia, *J. Am. Med. Assoc.*, 25, 110–112, 1895.
34. M'Dounell, W.C., Case of senile neurasthenia with paresis, *Aesculap. Soc. Abstr. Tr.* 1895–1898 (London), 6, 68, 1899.
35. MacCallum, H.A., Gastric neurasthenia, *Lancet*, 695–700, 1906.
36. King, W.W., Tropical neurasthenia, *J. Am. Med. Assoc.*, 47, 1518, 1906.
37. Levi, L. and de Rothschild, N., Neurasthenie throidienne, *Rev. Neurol.* (Paris), Vol. 47, 82, 1907.
38. Woods, H., Ocular neurasthenia, *J. Am. Med. Assoc.*, 48, 211–215, 1907.
39. Monguzzi, U., La reazione miastenica nella neurasthenia traumatica, *Gazz. Med. Lomb.* (Milan), Vol. 48, 58–60, 1907.
40. U.S. Army, *Index Catalog of the Library of the Surgeon General's Office*, Government Printing Office, Washington, D.C., Series I (1880–1895), II (1896–1916), and III (1918–1932).
41. Stedman, T.L., *Stedman's Medical Dictionary*, 26th ed., Williams & Wilkins, Baltimore, 1995.
42. Bassoe, P., The origin, rise and decline of the neurasthenia concept, *Wisc. Med. J.*, 27, 11–14, 1928.
43. Dana, C.L., Neurasthenia, *Post-Graduate* (N.Y.), 6, 26–38, 1890/91.
44. Dana, C.L., The partial passing of neurasthenia, *Boston Med. Surg. J.*, 339–344, 1904.
45. Freud, S., On the grounds for detaching a particular syndrome from neurasthenia under the description "anxiety neurosis," in *Standard Edition*, J. Strachey, Ed., Hogarth Press, London, 1895, 87–115.
46. Janet, R., *Les Obsessions et la Psychasthenie*, Felix Alcan, Ancienne Librarie Germer Bailliere et Cie, Paris, 1903.
47. Osler, W., *The Principles and Practice of Medicine*, D. Appleton and Company, New York, 1892, 971–980.
48. Buzzard, F., The dumping ground of neurasthenia, *Lancet*, 1–4, 1930.
49. Shorter, E., *From Paralysis to Fatigue: A History of Psychosomatic Illness in the Modern Era*, Free Press, New York, 1992.

50. Gosling, F., *Before Freud: Neurasthenia and the American Medical Community: 1970–1910*, University of Illinois Press, Springfield, 1987.

51. Chatel, J. and Peele, R., A centennial review of neurasthenia, *Am. J. Psychol.*, 126, 48–57, 1970.

52. von Oettingen, W.F., Carbon monoxide: its hazards and the mechanism of its action, Public Health Bull. 290, U.S. Government Printing Office, Washington, D.C., 1944.

53. Anonymous, Diary of an industry, *Am. Gas J.*, 22–52, Oct., 1959.

54. Morton, H., *Illuminating Water Gas Proved a Deadly Poison: The Predictions of Professor Henry Morton*, pamphlet, Henry Morton, New York, 1882.

55. Randolph, T.G., Sensitivity to petroleum including its derivatives and antecedents, *J. Lab. Clin. Med.*, 40, 931–932, 1952.

56. Randolph, T.G., Fatigue and weakness of allergic origin (allergic toxemia) to be differentiated from nervous fatigue or neurasthenia, *Ann. Allergy*, 3, 418–430, 1945.

57. Rowe, A.H., Allergic toxemia and migraine due to food allergy, *Calif. West. Med.*, 33, 785–793, 1930.

58. Gothe, C.J., Molin, C., and Nilsson, C.G., The environmental somatization syndrome, *Psychosomatics*, 36, 1–11, 1995.

59. Greenberg, D.B., Neurasthenia in the 1980s: chronic mononucleosis, chronic fatigue syndrome, and anxiety and depressive disorders, *Psychosomatics*, 31, 129–37, 1990.

60. Hartman, D.E., Missed diagnoses and misdiagnoses of environmental toxicant exposure. The psychiatry of toxic exposure and multiple chemical sensitivity, *Psychiatr. Clin. North Am.*, 21, 659–670, 1998.

61. Hickie, I., Hadzi-Pavlovic, D., and Ricci, C., Reviving the diagnosis of neurasthenia, *Psychol. Med.*, 27, 989–994, 1997.

62. Leitch, A.G., Neurasthenia, myalgic encephalitis or cryptogenic chronic fatigue syndrome? *Q. J. Med.*, 88, 447–450, 1995.

63. Massey, R.U., Neurasthenia, psychasthenia, CFS, and related matters, *Conn. Med.*, 60, 627–628, 1996.

64. Shorter, E., Multiple chemical sensitivity: pseudodisease in historical perspective, *Scand. J. Work Environ. Health*, 23 (Suppl. 3), 35–42, 1997.

65. Simpson, M., Bennett, A., and Holland, P., Chronic fatigue syndrome/myalgic encephalomyelitis as a twentieth-century disease: analytic challenges, *J. Anal. Psychol.*, 42, 191–199, 1997.

66. Stewart, D.E., The changing faces of somatization, *Psychosomatics*, 31, 153–158, 1990.

67. Ware, N.C. and Kleinman, A., Culture and somatic experience: the social course of illness in neurasthenia and chronic fatigue syndrome, *Psychosom. Med.*, 54, 546–560, 1992.

68. Wessely, S., History of postviral fatigue syndrome, *Br. Med. Bull.*, 47, 919–941, 1991.

69. Hyams, K.C., Developing case definitions for symptom-based conditions: the problem of specificity, *Epidemiol. Rev.*, 20, 148–156, 1998.

70. Farmer, A., Jones, I., Hillier, J., Llewelyn, M., Borysiewicz, L., and Smith, A., Neuraesthenia revisited: ICD-10 and DSM-III-R psychiatric syndromes in chronic fatigue patients and comparison subjects, *Br. J. Psychiatr.*, 167, 503–506, 1995.

70a. Knobloch, L. and Jackson, R., Recognition of chronic carbon monoxide poisoning, *Wisc. Med. J.*, 98, 26–29, 1999.

71. Bell, I.R., Miller, C.S., and Schwartz, G.E., An olfactory-limbic model of multiple chemical sensitivity syndrome: possible relationships to kindling and affective spectrum disorders, *Biol. Psychiatr.*, 32, 218–242, 1992.

72. Bascom, R., Meggs, W.J., Frampton, M., Hudnell, K., Kilburn, K., Kobal, G., Medinsky, M., and Rea, W., Neurogenic inflammation: with additional discussion of central and perceptual integration of nonneurogenic inflammation, *Environ. Health Perspect.*, 105, 531–537, 1997.

73. Ziem, G. and McTamney, J., Profile of patients with chemical injury and sensitivity, *Environ. Health Perspect.*, 105, 417–436, 1997.

74. Levin, A.S. and Byers, V.S., Environmental illness: a disorder of immune regulation, *Occup. Med.*, 2, 669–681, 1987.

75. Miller, C.S., Toxicant-induced loss of tolerance — an emerging theory of disease? *Environ. Health Perspect.*, 105, 445–453, 1997.

76. Rea, W., Oxygen therapy as an adjunct to treating the chemically sensitive [Abstract], presented at 1998 Annual Meeting of the Am. Acad. of Environ. Med., Baltimore, November 7, 1998.

77. Kent, H., Customers lining up for high-cost hyperbaric therapy, *Can. Med. Assoc. J.*, 160, 1043, 1999.

78. Cullen, M.R., The worker with multiple chemical sensitivities: an overview, *Occup. Med.*, 2, 655–661, 1987.

79. American Lung Association, American Medical Association, U.S. Consumer Product Safety Commission, and U.S. Environmental Protection Agency, Indoor Air Pollution, An Introduction for Health Professionals, U.S. Government Printing Office, Washington, D.C., 1994, 20.

80. Bartha, L., Baumzweiger, W., Buscher, D.S. et al., Multiple chemical sensitivity: a 1999 consensus, *Arch. Environ. Health*, 54, 147–149, 1999.

81. Fukuda, K., Nisenbaum, R., Stewart, G. et al., Chronic multisymptom illness affecting Air Force veterans of the Gulf War, *J. Am. Med. Assoc.*, 280, 981–988, 1998.

82. Donnay, A. and Ziem, G., Prevalence and overlap of chronic fatigue syndrome and fibromyalgia syndrome among 100 patients with multiple chemical sensitivity, *J. Chronic Fatigue Syndr.*, 5, 71–80, 1999.

83. Holmes, G. et al., Chronic fatigue syndrome: a working case definition, *Ann. Intern. Med.*, 108, 387–389, 1988.

84. Fukuda, K., Straus, S.E., Hickie, I., Sharpe, M.C., Dobbins, J.G., and Komaroff, A., The chronic fatigue syndrome: a comprehensive approach to its definition and study. International Chronic Fatigue Syndrome Study Group, *Ann Intern. Med.*, 121, 953–959, 1994.

85. Wolfe, F., Smythe, H.A., Yunus, M.B. et al., The American College of Rheumatology 1990 criteria for the classification of fibromyalgia, *Arthritis Rheum.*, 33, 160–172, 1990.

86. Hamilton, A., Carbon-Monoxide Poisoning, Bull. of the U.S. Bureau of Labor Statistics, Industrial Accidents and Hygiene Series 291, December, 1921, U.S. Government Printing Office, Washington, D.C., 1922.

87. Cowdery, E.G., Principles of manufacture and distribution of gas, with particular reference to lighting, in *Lectures on Illuminating Engineering Delivered at the Johns Hopkins University*, October and November 1910, under the Joint Auspices of the University and the Illuminating Engineering Society, Vol. 1, The Johns Hopkins Press, Baltimore, 1911, 277–385.

88. Brown, E.C., *Brown's Directory of American Gas Companies: Gas Statistics*, Press of Progressive Age Publishing Co., Philadelphia, 1887, 74–75.

89. Kober, G.M., Industrial and Personal Hygiene: A Report of the Committee on Social Betterment, The President's Homes Commission, Washington, D.C., 1908, 45.

89a. Ramazzini, B., *De Morbis Artificum*, 1713. Translated by Wright, W.C. in *Diseases of Workers*, Hafner Pub. Co., New York, 1964, 427.

89b. Hamilton, A. and Hardy, H.L., *Industrial Toxicology*, 2nd ed., Paul B. Hoeber (Medical Book Dept. of Harper & Brothers), New York, 1949, 243.

90. Apfelbach, G.L., Carbon monoxide poisoning, in *Industrial Health*, Kober, G.M. and Hayhurst, E.R., Eds., P. Blakiston's Son & Co., Philadelphia, 1924. [Reprinted with revisions by Hayhurst from *Diseases of Occupation and Vocational Hygiene*, Kober, G.M. and Hanson, W.C., Eds., P. Blackiston's Son & Co., Philadelphia, 1916.]

91. Watkins, J.A., Carbon Monoxide in the Steel Industry, Tech. Paper 156, U.S. Bureau of Mines, Washington, D.C., 1917.

92. Wilmer, W.H., Effects of carbon monoxide upon the eye, *Am. J. Ophthal.*, 4, 73–90, 1921.

93. Luden, G., Chronic carbon monoxide poisoning, *Mod. Med.*, 3, 27, 1921.

94. Henle, J., *On Miasmata and Contagie*, translated with an introduction by George Rosen, Johns Hopkins Press, Baltimore, 1938.

95. Koch, R., Ueber bakteriologische Forschung, in *Verh. X. Int. Med. Congr.*, Berlin, 1890, 1892, 35.

96. Evans, A.S., Causation and disease: the Henle–Koch postulates revisited, *Yale J. Biol. Med.*, 49, 175–195, 1976.

97. Richardson, B.W., *Manuals of Health on Health: and Occupation*, Pott, Young & Co., New York, 79, 1879.

98. Schwab, S., Neurasthenia among the garment workers, *Bull. Am. Econ. Assoc.*, 4th Series, no. 2, 265–270, 1911.

98a. Apfelbach, G.L., Tailors, garment and laundry workers, in *Industrial Health*, Kober, G.M. and Hayhurst, E.R., Eds., P. Blakiston's Son & Co., Philadelphia, 1924, 283.

99. Epidemiological Study Group of Mental Disorders in 12 Areas in China, Twelve areas epidemiological survey of neuroses, *Chin. J. Neuropsychiatr.*, 19, 87–91, 1986.

100. Zheng, Y.P., Lin, K.M., Takeuchi, D., Kurasaki, K.S., Wang, Y., and Cheung, F., An epidemiological study of neurasthenia in Chinese-Americans in Los Angeles, *Compr. Psychiatr.*, 38, 249–259, 1997.

101. Merikangas, K. and Angst, J., Neurasthenia in a longitudinal cohort study of young adults, *Psychol. Med.*, 24, 1013–1024, 1994.

102. Knave, B., Mindus, P., and Struwe, G., Neurasthenic symptoms in workers occupationally exposed to jet fuel, *Acta Psychiatr. Scand.*, 60, 39–49, 1979.

103. Struwe, G., Knave, B., and Mindus, P., Neuropsychiatric symptoms in workers occupationally exposed to jet fuel — a combined epidemiological and casuistic study, *Acta Psychiatr. Scand.*, 303 (Suppl.), 55–67, 1983.

104. Husman, K., Symptoms of car painters with long-term exposure to a mixture of organic solvents, *Scand. J. Work Environ. Health*, 6, 19–32, 1980.

105. Risberg, J. and Hagstadius, S., Effects on the regional cerebral blood flow of long-term exposure to organic solvents, *Acta Psychiatr. Scand.*, 303 (Suppl.), 92–99, 1983.

106. Orbaek, P. and Nise, G., Neurasthenic complaints and psychometric function of toluene-exposed rotogravure printers, *Am. J. Ind. Med.*, 16, 67–77, 1989.

107. Predescu, V., Nica, S.T., Meiu, G., Prica, A., Cucu, I., Damian, N., Popovici, I., Roman, I., Grigoroiu, M., and Curelaru, S., Observations on neurasthenia and neurasther ''' syndromes in a group of women working in textile industry, *Neurol. Psych   ^45–252, 1976.

108. Hall,   ''terans suffer from toxic neurasthenia? *Aust. N. Z*

109. Kreutzer, R., Neutra, R.R., and Lashuay, N., Prevalence of people reporting sensitivities to chemicals in a population-based survey, *Am. J. Epidemiol.*, 150, 1–12, 1999.

110. Wolfe, F., Ross, K., Anderson, J., Russell, I.J., Hebert, L., The prevalence and characteristics of fibromyalgia in the general population, *Arthritis Rheum.*, 38, 19–28, 1995.

111. Wesseley, S., Chalder, T., Hirsch, S. et al., The prevalence and morbidity of chronic fatigue and chronic fatigue syndrome: a prospective primary care study, *Am. J. Public Health*, 87, 1449–1455, 1997.

112. Pankow, D., Carbon monoxide formation due to metabolism of xenobiotics, in *Carbon Monoxide*, Penney, D., Ed., CRC Press, Boca Raton, FL, 1996, 25–44.

113. Stewart, R.D. and Hake, C.L., Paint-remover hazard, *J. Am. Med. Assoc.*, 235, 398–401, 1976.

114. Beard, G., *A Practical Treatise on Nervous Exhaustion (Neurasthenia): Its Symptoms, Nature, Sequences, Treatment*, William Wood, New York, 1880.

115. Jewell, J., Influence of our present civilization in the production of nervous and mental disease, *J. Nerv. Ment. Dis.*, 8, 1–24, 1881.

116. Johnson, G., Lectures on some nervous diseases that result from overwork and anxiety, *Lancet*, 85–87, 1875.

117. Leubuscher, P. and Bibrowicz, W., Die neurasthenie in arbeitkreisen, *Dtsch. Med. Wochenshcr.*, 31, 820–824, 1905.

118. Hallock, F., The sanatorium treatment of neurasthenia and the need of a colony sanatorium for the nervous poor, *Boston Med. Surg. J.*, 44, 73–77, 1911.

119. Goulon, M., Barois, A., Rapin, M., Nouailhat, F., Grosbuis, S., and Labrousse, J., Intoxication oxycarbonee et anoxie aigue par inhalation de gaz de charbon et d'hydrocarbures, *Ann. Med. Intern.* (Paris), 120, 355, 1969.

120. Guss, D.A. and Neuman, T.S., Carbon monoxide poisoning: how to detect — and what to do, *J. Respir. Dis.*, 11, 773–785, 1990.

121. Jain, K.K., *Carbon Monoxide Poisoning*, Warren H. Green, St. Louis, MO, 1990.

122. Choi, I.S., Delayed neurologic sequelae in carbon monoxide intoxication, *Arch. Neurol.*, 40, 433, 1983.

123. Hart, I.K., Kennedy, P.G., Adams, J.H. et al., Neurological manifestations of carbon monoxide poisoning, *Postgrad. Med. J.*, 64, 213, 1988.

124. Kindwall, E.P., Delayed sequelae in carbon monoxide poisoning and the possible mechanisms, in *Carbon Monoxide*, Penney, D., Ed., CRC Press, Boca Raton, FL, 1996, 239–252.

125. Shillito, F.H., Dringer, C.K., and Shaughnessy, T.J., The problem of nervous and mental sequellae in carbon monoxide poisoning, *J. Am. Med. Assoc.*, 106, 669, 1936.

126. Lasater, S.R., Carbon monoxide poisoning, *Can. Med. Assoc. J.*, 134, 991, 1986.

127. Min, J.K., A brain syndrome associated with delayed neuropsychiatric sequelae following acute carbon monoxide intoxication, *Acta Psychiatr. Scand.*, 73, 80, 1986.

128. Thom, S.R., Carbon monoxide-mediated brain lipid peroxidation in the rat, *J. Appl. Physiol.*, 68, 997, 1990.

129. Killick, E.M., The nature of acclimatization occurring during repeated exposure of the human subject to atmospheres containing low concentrations of carbon monoxide, *J. Physiol.*, 87, 41, 1936.

130. Collier, C.R. and Goldsmith, J.R., Interactions of carbon monoxide and hemoglobin at high altitude, *Atmos. Environ.*, 1, 723–728, 1983.

131. Abraham, N.G., Drummond, G.S., Lutton, J.D., and Kappas, A., The biological significance and physiological role of heme oxygenase, *Cell Physiol. Biochem.*, 6, 129–168, 1996.

132. Stewart, R.D., Stewart, R.S., Stamm, W., and Seelen, R.P., Rapid estimation of carboxyhemoglobin level in fire fighters, *J. Am. Med. Assoc.*, 235, 390–392, 1976.

133. Smith, J.R. and Landau, S., Smokers' polycthemia, *N. Engl. J. Med.*, 298, 6, 1978.

134. Astrup, P., Intraerythrocytic 2,3-diphosphoglycerate and carbon monoxide exposure, *J. Appl. Physiol.*, 41, 893, 1976.

135. Penney, D.G., Helfman, C.C, Hull, J.C., Dunbar, J.C., and Verma, K., Elevated blood glucose is associated with poor outcome in the carbon monoxide poisoned rat, *Toxicol. Lett.*, 54, 287, 1990.

136. Leikin, J.B., Goldenberg, R.M., Edwards, D., Zell-Kantor, M., Metabolic predictors of carbon monoxide poisoning, *Vet. Hum. Toxicol.*, 30, 40–42, 1988.

137. Lawrence, J.H. and Berlin, N.I., Relative polycythemia — the polycythemia of stress, *Yale J. Biol. Med.*, 24, 498, 1952.

138. Helfaer, M.A. and Traystman, R.J., Cerebrovascular effects of carbon monoxide, in *Carbon Monoxide*, Penney, D., Ed., CRC Press, Boca Raton, FL, 69–86, 1996.

139. Allred, E.N., Bleecker, E.R., Chaitman, B.R. et al., Effects of carbon monoxide on myocardial ischemia, *Environ. Health Perspect.*, 91, 89–132, 1991.

140. Hiramatsu, M., Kameyama, T., and Nabeshima, T., Carbon monoxide-induced impairment of learning, memory, and neuronal dysfunction, in *Carbon Monoxide*, Penney, D., Ed., CRC Press, Boca Raton, FL, 1996, 187–210.

140a. Beck, H.G., Slow carbon monoxide asphyxiation: a neglected problem, *J. Am. Med. Assoc.*, 107, 1025–1029, 1936.

141. Randolph, T., Allergic type reactions to indoor utility gas and oil fumes, *J. Lab. Clin. Med.*, 44, 913, 1954.

142. Randolph, T.G., The specific adaptation syndrome, *J. Lab. Clin. Med.*, 48, 934–941, 1956.

143. Haldane, J., The relation of the action of carbonic oxide to oxygen tension, *J. Physiol.* (London), 18, 201, 1895.

144. Thom, S.R., Taber, R.L., Mendiguer, I.I., Clark, J.M., Hardy, K.R., and Fisher, A.B., Delayed neuropsychological sequelae following carbon monoxide poisoning and its prophylaxis by treatment with hyperbaric oxygen, *Ann. Emerg. Med.*, 25, 474, 1995.

145. Engen, T., The combined effect of carbon monoxide and alcohol on odor sensitivity, *Environ. Int.*, 12, 207–210, 1986.

146. Wald, N.J., Idle, M., Boreham, J., and Bailey, A., Carbon monoxide in breath in relation to smoking and carboxyhaemoglobin levels, *Thorax*, 36, 366–369, 1981.

11 Update on the Clinical Treatment of Carbon Monoxide Poisoning

Suzanne R. White

CONTENTS

11.1 INTRODUCTION

Carbon monoxide (CO) poisoning is the leading cause of toxicologic death in the United States, with 5600 fatalities reported annually.[1] Worldwide, CO remains the most lethal toxin in every community in which it has been studied.[2] In addition to the high mortality rates associated with acute exposure to CO, significant long-term morbidity exists as well. Most notably, numerous studies have documented delayed neuropsychiatric sequelae in a significant percentage of CO survivors.[3-8]

This chapter focuses primarily on the various treatment aspects of acute CO poisoning. It should be kept in mind that the present knowledge regarding therapy for CO poisoning is limited for several reasons. First, effective medical treatment is ideally guided by predictors for either positive or negative outcomes following exposure to toxic substances. For example, blood or urine levels of toxins, combined with characteristic signs or symptoms of toxicity, often aid in the institution of appropriate therapy. Unfortunately, symptoms relating to CO exposure are notoriously vague, and some studies estimate that the diagnosis is missed in 30% of cases presenting to the emergency department.[9] With regard to inpatients, Balzan et al.[10] screened all neurological admissions over a 5-month period and found that 3 out of 29 patients admitted with impaired consciousness and no lateralizing neurological signs had serious CO intoxication.[10] Furthermore, carboxyhemoglobin (COHb) levels neither correlate with toxicity nor predict the risk for development of long-term effects.[11,12] Although other predictors of long-term neuropsychiatric sequelae are proposed (i.e., loss of consciousness,[13] cerebral edema on brain computed tomography (CT),[14] elevated blood glucose,[15] or a history of a "soaking"-type exposure[16]), their sensitivity and specificity are unproved. As a result, how best to treat patients with such warning signs remains controversial. Second, appropriate therapy for poisoned patients is ideally guided by an understanding of the toxic mechanisms of that poison. Unfortunately, even though CO has most likely been present since the beginning of time, and has been studied clinically for over 100 years, an adequate understanding of its toxic mechanisms eludes us. Finally, treatment guidelines should ideally be based on prospective, well-controlled, peer-reviewed studies. There is, however, a dearth of such studies as they relate to CO-poisoning treatment in the literature.

Despite these limitations, a general approach to treatment will be described. As an overview, treatment is based on the cessation of tissue hypoxia, the removal of CO from the body, the consideration of potential neuroprotective interventions, and the management of the long-term sequelae of CO poisoning. First, a review of historical, often-failed treatments for CO poisoning will be presented, followed by a discussion of promising neuroprotective agents. Finally, a clinical approach to the CO-poisoned patient will be outlined.

11.2 HISTORICAL PERSPECTIVE

Oxygen therapy has been the mainstay of treatment for CO poisoning since it was first used therapeutically by Linas and Limousin[17] in 1868. Haldane[18] subsequently was able to demonstrate experimentally that mice exposed to "carbonic oxide" were unaffected if oxygen was provided during the exposure. In this seminal work, Haldane concluded that "the higher the oxygen tension the less dependent an animal is on its red corpuscles as oxygen carriers, since the oxygen simply dissolved in the blood becomes considerable when the oxygen tension is high."[18] Indeed, 100% oxygen at sea-level pressure decreases the half-life of CO from 320 to 80 min. Unfortunately, 100% at sea level alone has not been entirely effective in the treatment of CO poisoning, particularly with regard to the prevention of delayed neuropsychiatric sequelae. This realization has prompted researchers and physicians to search for yet other treatment modalities.

The use of resuscitative gases other than oxygen has been proposed. Studies by Killick and Marchant[19,20] demonstrated more rapid clearance of COHb with 5% carbogen (5% carbon dioxide, plus 95% oxygen), which was thought to be related to increased ventilatory drive.[19,20] Schwerma et al.[21] exposed dogs to 0.3% CO until near respiratory arrest occurred. Upon removal from exposure, 36% survived with fresh air alone. The survival rate increased to 50% when mechanical ventilation was used, 69% when ventilation with 100% oxygen was applied, and 66% when mechanical ventilation was combined with 7% carbogen. There was no clinical advantage to the use of carbogen in terms of improved survival, normalization of pH or lactate, or decreased incidence of neurological sequelae in animals relative to breathing 100% oxygen alone. Thus, this method of treatment has subsequently fallen out of favor.

Numerous fascinating therapies for CO poisoning have not proved to be effective, and are mentioned here only for historical interest. Methylene blue, succinic acid, persantine, iron and cobalt preparations, and ascorbic acid have all been tried, without benefit.[22] In animals, cytochrome c, theorized to activate cytochrome oxidase upon supplementation has not been associated with clinical improvement.[23] Hydrogen peroxide infusions do reduce COHb content in experimental animals, but the absence of human experience with this chemical and the danger of air embolism preclude its clinical use.[24] While ultraviolet radiation was proposed to facilitate the dissociation of COHb from erythrocytes during transit through skin capillaries and to decrease mortality in animals,[25] these results were not able to be duplicated in a subsequent animal trial.[26] Intravenous procaine hydrochloride does not improve the anoxia of CO poisoning in humans,[27] and intravenous lidocaine, advocated based on its facilitation of neuronal recovery after cerebral ischemia in experimental animals, has not yet been employed in humans.[28]

Dipyridamole pretreatment in rats with inhalational CO toxicity was associated with protective effects, in that it inhibited ultrastructural changes in capillary endothelial cells, myocardial mitochondria, and myocardial myofilament arrangement.[29] Further follow-up studies with dipyridamole have not been performed.

Exchange transfusion has been reported to improve survival following CO poisoning in the animal model.[30] Despite the fact that this method has been utilized in only a single patient,[31] it is still promoted by some clinicians as an alternative to hyperbaric oxygen therapy.[32] While exchange transfusion does in fact lower COHb levels, given the complex mechanisms for CO toxicity, which extend well beyond the toxicity of COHb, this technique is not likely to be effective as sole therapy. Furthermore, given the potential for exchange transfusions to deplete valuable blood product resources and place the patient at risk for blood-borne pathogen infections, this treatment modality can no longer be recommended.

Perfluorochemical infusions have been used in animal models as treatment for CO toxicity.[33,34] Recently, pyridoxalated hemoglobin–polyoxyethylene conjugates (PHPs) have been developed. These agents act as blood substitutes capable of transporting oxygen through chemical modification of hemoglobin derived from human erythrocytes whose shelf-life has expired. The affinity of PHP for oxygen is almost identical to that of whole blood. PHP use in rabbits poisoned by CO was associated with prolonged survival time, temporary recovery of PO_2 and PCO_2, and elevations in pH and blood pressure in comparison with animals treated with saline.[35] Beyond the fact that human use of this product has not yet been reported, its efficacy as sole therapy is unlikely, for the same reasons as discussed above with exchange transfusion.

Hyperbaric oxygen (HBO) therapy was first suggested as treatment for CO poisoning in 1901 by Mosso.[36] The first clinical use of HBO in the treatment of human CO poisoning, however, did not occur until 1960.[37,38] This modality has subsequently become the mainstay of therapy in severe CO poisoning and is discussed in more detail below.

11.3 MECHANISMS OF CARBON MONOXIDE TOXICITY

To gain an understanding of the available methods for treatment of CO poisoning, a review of what is known about mechanisms for CO toxicity, albeit an incomplete comprehension of the problem, is presented here.

Hypoxic ischemia plays a significant role in the neurotoxicity of CO, which binds slowly to hemoglobin, but with extremely high affinity (240 times that of oxygen). Oxygen-binding sites are occupied by CO at very low partial pressures of the gas, decreasing the oxygen-carrying capacity of the blood and subsequently decreasing the usual facilitation phenomenon for further unloading of oxygen at the tissue level. The net result is an abnormal hyperbolic oxygen-dissociation curve that is shifted to the left. Those tissues most susceptible to the hypoxic effects of CO are those that are the most metabolically active. Oxygen delivery may be further impaired through the alteration of erythrocyte diphosphoglycerate concentration.[39] In adults, COHb half-life is dependent upon the concentration of inspired oxygen, and is most commonly reported to be approximately 4.5 h on room air, 90 min on normobaric 100% oxygen, and 20 min with oxygen applied at 2.5 to 3.0 atmospheres absolute (atm abs). It should be noted that reported half-lives are extremely varied in the literature. In children, the half-life of COHb has not been well studied, but is reported by one author to be 44 min on 100% oxygen at normobaric pressure, based on measurements performed on 26 school-aged children.[40] The half-life of COHb in the fetus is approximately 7 h.[38] In addition to hypoxia, CO induces ischemia secondary to hypotension. The degree of central nervous system (CNS) damage correlates well with degree of hypotension.[41] Hypotension may be mediated through increased cyclic guanosine monophosphate (cGMP), resulting in vasculature smooth muscle dilatation or myocardial suppression.

Additional mechanisms of toxicity have been sought following observations that (1) COHb levels did not correlate with toxicity, (2) COHb formed by noninhalational routes did not produce the same lethal consequences as inhalational exposure to CO, and (3) that delayed neuropsychiatric sequelae were common after apparent complete recovery from the initial CO insult. Early researchers such as Haldane,[42] Drabkin et al.,[43] and Goldbaum[44–46] suggested cellular uptake of the gas as a possible mechanism for toxicity. In competition with oxygen, CO will bind iron- or copper-containing proteins such as myoglobin, mixed-function oxidases, and cytochrome c oxidase *in vitro*. The binding to the cytochrome c oxidase (a,a_3) has been proposed as the mechanism for intracellular CO toxicity and has been demonstrated in animals.[47] It has also been demonstrated that during recovery, the ultimate restoration of mitochondrial function lags behind clearance of COHb.[48] However, the Warburg constant for cytochrome oxidase is unfavorable for CO binding relative to the other heme-proteins.[49] Furthermore, only reduced cytochrome a,a_3 binds CO. It is likely, then, that other hemeproteins act as "CO buffers," thus preventing significant binding to cytochrome c oxidase at COHb levels of less than 50%. At high levels of COHb, depletion of high-energy stores and intracellular neuronal acidosis occurs, which may favor CO–cytochrome binding.

On the other hand, *in vivo* data from Smithline[50] supports hemoglobin binding with impaired oxygen delivery, rather than mitochondrial poisoning as the etiology of the metabolic acidosis in CO poisoning. In this model, even at extremely high COHb levels, dogs were able to extract and utilize oxygen fully, indicating a lack of mito-chondrial effect. Additional work by Ward et al.[51] demonstrated that expression of the heat shock proteins 72 and 32 (sensitive markers of acute neuronal stress) did not occur following CO poisoning in rats that were maintained normotensive throughout the exposure. This caused the authors to question the role of CO as a direct neurotoxin, and to suggest that neuronal injury results from hypotension-induced ischemia.

The role of iron as a promoter or attenuator of CO toxicity is not clear. Iron deficiency results in lowered hemoglobin, cytochrome, and myoglobin levels in the animal model.[52] These combined effects could potentially predispose to CO toxicity. Conversely, neuronal tissues high in iron content, such as the basal ganglia, seem particularly vulnerable to the effects of CO.

Myoglobin binding may play a role in CO-mediated toxic effects. Like hemo-globin, myoglobin is a hemeprotein with similar three-dimensional configuration that can bind CO reversibly. Myoglobin binds CO more slowly and with greater affinity than does hemoglobin *in vivo*. Normally, myoglobin is an O_2 carrier protein that facilitates oxygen diffusion into skeletal or cardiac muscle cells and serves to place oxygen stores in close proximity to mitochondria. The clinical significance of high carboxymyoglobin levels is not yet clear but may in part explain the cardiac and skeletal muscle toxicity seen after CO poisoning. These effects likely come into play at COHb levels of 20 to 40%.[29,53]

The mechanism for delayed effects of CO poisoning have been a medical conun-drum. An increasing body of research suggests that brain ischemia/reperfusion injury, lipid peroxidation, vascular oxidative stress, neuronal excitotoxicity, and apoptosis play significant roles. After removal from the CO environment, animal models dem-onstrate marked changes in neutrophil structure and function. Abnormal adherence to brain endothelial cell receptors quickly occurs, possibly as a result of endothelial damage. Upregulation of endothelial intercellular adhesion molecules (ICAM) is demonstrated on endothelial cells as a result of activation by inflammatory mediators. ICAMS bind beta-2 integrins located on polymorphonuclear cells (PMNs), resulting in aggregation of PMNs onto endothelial cell surfaces. Subsequent degranulation of PMNs results in release of destructive proteases, which cause further oxidative injury.

Thom's[54-56] work has been instrumental in demonstrating these CO-induced perivascular oxidative changes that occur during recovery from CO poisoning and ultimately lead to superoxide formation, prolonged lipid peroxidation reactions, reactive oxygen species (ROS) generation, vascular injury, and neuronal death. Even low-level CO can produce vascular oxidative stress as evidenced by platelet-mediated deposition of peroxynitrate, a highly oxidative substance.[57] Peroxynitrite, which forms from NO released from platelets and endothelial cells, can further inactivate mitochondrial enzymes and damage vascular endothelium of the brain.[58,59] ROS generation can be attributed to several other sources including mitochondria and cycloxygenase. ROS production increases notably during CO hypoxia, with the highest oxidative stress occurring in the most vulnerable brain regions.[60] This stress may result from lower antioxidant capacity or higher tissue concentrations of iron. In mitochondria, decreased ratios of reduced to oxidized glutathione are seen following CO poisoning, and may reflect decreased ability to detoxify ROS.[61]

Neuronal excitation may also play a role in the development of delayed toxicity following CO poisoning. These effects have been extensively reviewed by Piantadosi,[66] and are paraphrased here. Excitatory amino acids (EAA), such as glutamate, accumulate in synaptic clefts during neuronal depolarization due to both excessive presynaptic release and failure in ATP-dependent reuptake mechanisms.[62] Interstitial glutamate concentration increases in the hippocampus during and after CO exposure.[51] Postsynaptic binding to at least three glutamate receptors including N-methyl-D-aspartate (NMDA), kainic acid (KA), and (S)-alpha-amino-3-hydroxy-5-methylisoxazole-4-propionate (AMPA) occurs with a secondarily increased influx of calcium into postsynaptic neurons. This hypercalcemia is associated with neuronal death. ROS production follows increased EAA release as well. NMDA receptor antagonism attenuates delayed neuronal degeneration in the hippocampus after CO poisoning, and is discussed further below. Another modulating factor may be local NO production, however, the exact role of NO as an attenuator of CO toxicity remains unclear.[63]

Catecholamine excess may also be detrimental following CO exposure. EAA release causes excessive surges of norepinephrine and dopamine release with synaptic accumulation. This effect appears to be related to NO production, as both events can be prevented by nitric acid synthase (NOS) inhibition. Therefore, EAA release, catecholamine release, and NO production are closely regulated in vivo and can be influenced by CO hypoxia. Furthermore, auto-oxidation or oxidative deamination of catecholamines occurs during ischemia and reperfusion by type B monoamine oxidase.[64] ROS production after CO exposure can be inhibited by partially blocking type B monoamine oxidase, located predominantly in glia.[65] Gliosis, a known neuronal response to injury, and a condition found in Alzheimer's disease, may play a role in the delayed toxicity seen after CO hypoxia.

ROS are capable of triggering programmed cell death or apoptosis. Apoptotic cell death requires activation and/or expression of specific cellular processes, some of which may act through oxidant pathways. In animals, CO-induced neuronal loss was slight at Day 3, increased at Day 7, and persistent at Day 21 following exposure. Neuronal apoptosis was observed at histopathological examination in this model.[66] Others have not demonstrated apoptosis in rats, despite moderately severe poisoning,[67] but gliosis was observed.[68]

As eloquently summarized by Piantadosi,[69] "impaired mitochondrial energy provision in CO hypoxia/ischemia leads to neuronal depolarization, EAA and catecholamine release, and failure of re-uptake until energy metabolism is restored during re-oxygenation. These processes, normally modulated by NO production, could contribute to degeneration of neurons in vulnerable regions, possibly by enhancing mitochondrial ROS generation which can initiate apoptosis" (Figure 11.1).

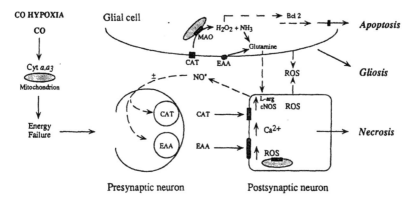

FIGURE 11.1   Hypothesis diagram of possible mechanisms of neuronal injury after CO hypoxia. Dashed lines indicate untested possibilities. EAA = excitatory amino acids, CAT = catecholamines, Cyt a,a₃ = cytochrome oxidase, MAO = monoamine oxidase, L-arg = L-arginine, cNOS = constitutive nitric oxide synthase, ROS = reactive oxygen species. Chemical abbreviations given in the text.

11.4 CURRENTLY AVAILABLE NEUROPROTECTIVE TREATMENTS

11.4.1 NORMOBARIC OXYGEN

In one study, 33 patients with acute CO poisoning (mean COHb 29.4%, 10 patients with levels above 40%, 7 comatose on arrival) were treated with 100% normobaric oxygen. Recovery was reportedly rapid, with no neurological deficits at discharge. Formal neuropsychiatric (NP) testing and follow-up were not performed.[70] Similarly, four patients presenting comatose from CO poisoning who did not receive HBO therapy were identified retrospectively, but then evaluated with formal neuropsychological testing at 6 and 12 months after exposure. All had normal NP examinations. These data suggest that normal NP outcome is possible in a subset of patients presenting comatose, but, given the small number of patients studied, this does not add to understanding of predictors for good or adverse outcome.[71]

Meert et al.[72] looked retrospectively at the outcome of children treated with normobaric oxygen, and concluded that acute neurological manifestations resolve rapidly without HBO therapy. However, neurological outcome was assessed from nursing and physician records, physical and occupational therapy evaluations, and unspecified neuropsychological testing in an unspecified number of patients. Therefore, the major limitation of this study is the lack of detailed neurological assessment both at presentation and at follow-up that would allow detection of those specific NP changes known to result from CO poisoning. Nonetheless, the authors noted "gross" neurological abnormalities in nine (8.5%) survivors, with seven of those persistent at various stages of follow-up (2 months to 3.3 years). Three patients developed delayed neurological syndromes including tremors, hallucinations, seizures, occipital lobe infarctions, and defects in cognitive and interpersonal skills. The presence of serious comorbidities, such as smoke inhalation, burns, need for mechanical ventilation, and need for surgical procedures certainly confounds the outcomes reported by these investigators, which are not reflective of pure CO exposure.

11.4.2 Hyperbaric Oxygen Therapy

In addition to the obvious effects of increasing both the amount of dissolved oxygen in the blood and the rate of displacement of CO from hemoglobin, HBO may have other beneficial effects. HBO at 2.5 to 3.0 atm abs reversibly inhibits PMN CD18 beta-2 integrin activation and therefore decreases adherence of PMNs to endothelial cells.[73,74] *In vivo*, HBO at 3.0 atm abs also prevents functional neurological impairment in rats.[75] Hyperbaric oxygen also regenerates inactivated cytochrome oxidase, and may thereby restore mitochondrial function.[76] Other proposed beneficial effects of HBO include decreased production of reactive oxygen species,[77] protection against cerebral edema and increased CSF pressure,[78,79] induction of protective stress proteins (SP-72), and antagonism of NMDA excitotoxic neuronal injury. Conversely, HBO was not shown to be effective in animal models as a treatment modality for non-CO-mediated acute cerebral ischemia with reperfusion.[80]

Anecdotal case reports of clinical improvement from CO poisoning during HBO therapy exist.[31,81–87] Additionally, numerous other case series report beneficial effects of HBO therapy, but are either retrospective in nature or prospectively performed without the use of randomization, double-blinding, or controlled methodology.[88–94]

Three different research groups have gone one step farther toward answering questions of potential benefit from HBO therapy, through randomization of CO-poisoned patients into HBO and normobaric (NBO) treatment groups. Raphael[107] studied 343 mildly poisoned CO patients (those who had not lost consciousness) and found no difference at 1-month follow-up in neurologic outcome between HBO- and NBO-treated groups, and no benefit to multiple HBO treatments in a more severely poisoned group (those who had lost consciousness), randomized to either one or two HBO treatments. Criticisms of this study regarding the method of neurological evaluation used, the application of inadequate doses of oxygen therapy, and potential delays in HBO treatment have been raised.[95] Ducasse et al.[96] randomized noncomatose patients to HBO and NBO groups and found significantly improved differences in quantitative EEG and cerebral vascular responsiveness to acetazolamide in the HBO group during the first 24 h, but differences at 3-week follow-up are not reported consistently. Early improvements in clinical signs and symptoms, such as headache, reflex impairment, and asthenia, were significant in the HBO group.[96] Thom randomized patients with mild to moderate CO poisoning into NBO vs. HBO groups.[173] The incidence of delayed neuropsychiatric sequelae (DNS) was 23% after treatment with oxygen at ambient pressures and 0% in the HBO group. DNS persisted for an average of 41 days following exposure. The author concluded that HBO therapy decreased the incidence of DNS. Limitations of this study are the lack of double-blinding, inconsistent NP testing methods used, and exclusion of severely poisoned patients.[97,98]

Very recently, three even more rigorous, elegantly designed trials have been reported. All are prospective, randomized, double-blinded, controlled clinical trials assessing the benefit of HBO vs. NBO in the treatment of CO-poisoned patients.

One has been completed and two have performed interim analyses of their data, which are reviewed below.

Scheinkestel et al.[99] recently completed the first prospective, randomized, double-blinded, controlled clinical trial investigating the neurological sequelae in 191 patients with all grades of CO poisoning after treatment with HBO and NBO. Sham HBO treatments were given to those randomized to the NBO group. Pregnant women, children, and burn victims were excluded. Higher doses of oxygen were utilized than those reported in most previous studies, averaging approximately 37 COHb-dissociation half-lives in the HBO group and 28.5 in the NBO group (up to three daily treatments in those not improving). NP evaluations were performed at completion of treatment and at discharge. No benefit and possible adverse effects of HBO were found. Overall mortality was 3%, with persistent neurological deficits in 71% at hospital discharge and 62% at 1-month follow-up. All five patients with delayed onset of neurological deficit occurred in the HBO group. Limitations of the study are that 44% of victims had ingested other drugs, there was a mean treatment delay of 7.1 h, 56% of patients were lost to 1-month follow-up, and 76% were suicidal, which could impact neuropsychological testing. These results led the investigators to conclude that HBO therapy can no longer be recommended for CO-poisoning treatment.

Similarly, an ongoing, longitudinal follow-up study[100] conducted in Utah demonstrated that approximately 30% of the patients with acute CO poisoning have neurocognitive problems 1 year after poisoning. Of these patients, approximately one third have the DNS and two thirds have persistent neurocognitive problems, primarily difficulties with memory and executive function.[100] Blinded interim analysis of treatment of these patients showed no difference in outcome between the HBO and NBO groups at 50 and 100 patients.[101] A third randomized clinical trial is under way by Mathieu et al.[102] on noncomatose CO-poisoned patients. Interim results showed no benefit between HBO and NBO at 1 year postexposure, but final results are pending.[102]

Some investigators routinely perform additional treatments if lack of improvement is noted after the first HBO treatment.[103–105] Turner et al.[104] recently proposed the use of the initial hydrogen ion concentration (degree of metabolic acidosis) as a marker for *repetitive* HBO treatment requirement, based on a retrospective analysis of 48 patients. McNulty et al.[106] found impairment of short-term memory for verbal material to be predictive of the number of HBO treatments needed. Others report no benefit to multiple HBO treatments.[107]

Risks of HBO include complications of transport to an HBO facility, barotrauma, oxygen toxicity with resultant seizures in 1 to 3% of CO-poisoned patients,[108] and fire/explosion hazard. Historically, because of the lack of results from randomized clinical trials to guide therapy, the indications for HBO therapy in CO poisoning have been arbitrary, tremendously varied, and primarily based upon history, presenting neurological and cardiovascular signs and symptoms, and laboratory data such as glucose, lactate, arterial blood gases, ECG, and COHb levels. Unfortunately, clinical and laboratory findings at presentation are not entirely predictive of long-term outcome following CO exposure. Furthermore, it appears that HBO therapy may not prevent persistent neurologic sequelae (PNS) or DNS. Therefore, while

HBO will likely continue to be used, clinicians should attempt to be much more selective in prescribing this treatment modality. Greater future research emphasis will likely be placed on non-HBO methods of therapy.

11.4.3 ALLOPURINOL AND N-ACETYLCYSTEINE

There is considerable evidence that reactive oxygen metabolites mediate neurological injury in models of CO poisoning. Lipid peroxidation is documented in rats as a result of exposure to CO at a concentration sufficient to cause unconsciousness. Products of lipid peroxidation are increased by 75% over baseline values 90 min after CO exposure. Unconsciousness is associated with a brief period of hypotension, so brief that in itself it causes no apparent insult. Lipid peroxidation occurs only after the animals are returned to CO-free air, and there is no direct correlation with the COHb level.[54] Xanthine oxidase has a central role in this toxicity. During the above-described CO-induced PMN degranulation, the released proteases convert xanthine dehydrogenase to xanthine oxidase. Xanthine oxidase generates superoxide, free radicals, and lipid peroxidation occurs.[56] The xanthine oxidase enzyme is an NAD-dependent dehydrogenase that under ischemic conditions converts to an oxidase, utilizing molecular oxygen rather than NAD as an energy source and generates the superoxide radical and hydrogen peroxide. These products in turn cause tissue injury, the brain being particularly susceptible with its low content of catalase and glutathione peroxidase.[109] The restoration of xanthine dehydrogenase functional activity is accomplished through the use of xanthine oxidase inhibitors (allopurinol)[110] and sulfhydryl donors (NAC)[111] in non-CO-mediated neuronal injury. Fechter et al.[112] noted that acute CO poisoning produces preferential high-frequency hearing impairment, also noted to be a consequence of other types of anoxic exposure. These investigators also discovered that either allopurinol or phenyl-n-tert-butyl-nitrone (PBN), a free-radical scavenger, blocked the formation of characteristic compound action potential threshold elevation and cochlear microphonic amplitude. Therefore, both agents were effective in blocking loss of CO-mediated auditory threshold, if given prophylactically in the guinea pig.[112]

Similarly, the above two agents have been used in the treatment of CO-induced neuronal injury. Thom[56] demonstrated decreased conversion of xanthine dehydrogenase to xanthine oxidase with decreased lipid peroxidation in rats pretreated with allopurinol. Only one human case report demonstrating effectiveness of such combined therapy for the treatment of CO poisoning is reported. A 26-year-old male with a COHb level of 25%, 40 h postexposure, who was comatose for 4 days with cerebral edema on CT, was treated with both a xanthine oxidase inhibitor (allopurinol) and a sulfhydryl donor, N-acetylcysteine (NAC). NAC was given intravenously over a 20-h period and allupurinol was given orally for 2 weeks. The patient became responsive 8 h after the completion of this regimen and gradually improved over the next 3 weeks. "Neurological and mental examination at 6 weeks follow-up were normal."[113] Although no formal NP testing was reported, this type of therapy may perhaps provide a basis for further study. Such agents may one day serve as useful adjuncts with HBO in limiting free-radical-mediated injury.

11.4.4 INSULIN

Numerous studies in humans and animals have shown that elevated blood glucose is associated with worsened neurological outcome after brain ischemia caused by stroke or cardiac arrest. Similarly, acute severe CO poisoning is characterized by hyperglycemia and this elevation has been linked to increased severity of brain dysfunction in the rat.[114] Indeed, animal studies also show that CO exposure raises blood glucose in a dose-dependent manner, and is an independent predictor of neurological outcome. A few similar observations have been made in CO-poisoned patients. Penney[115] observed that elevated admission blood glucose was associated with worse neurological outcome after CO poisoning in patients. Leikin et al.[116] found elevated blood glucose in most patients presenting with COHb saturation above 25%. Furthermore, anecdotal evidence in the literature on humans suggests that the neurological outcome in people with diabetes poisoned with CO is generally worse than in those without diabetes.[117]

Considerable basic research has been directed at identifying molecular mechanisms of tissue injury and potential interventions to allow the preservation or rescue of neurons after stroke and cardiac arrest. For example, because of the aforementioned association between elevated blood glucose and poor outcome, insulin has recently been investigated as a potential therapeutic agent in various models of brain and spinal cord ischemia, and indeed appears to ameliorate substantially neuronal death induced by ischemia in these studies. Surprisingly, this neuron-sparing effect is known to be independent of insulin-induced reductions in blood glucose, and is hypothesized to be mediated through cell signal transduction mechanisms, in common with other growth factors.

To date, very few studies have approached the problem of neurological damage induced by CO by utilizing molecular-based concepts similar to those now being applied to the problems of brain ischemic anoxia. Hence, the question of whether insulin ameliorates neuronal injury secondary to CO toxicity was investigated in rats that were exposed to a CO LD_{50} of 2400 ppm for 90 min. Survivors were treated for 4 h with (1) normal saline infusion, (2) continuous infusion of glucose to clamp blood levels at 250 to 300 mg/dl, and (3) continuous infusion of glucose to maintain blood levels at 250 to 300 mg/dl with intraperitoneal (IP) injection of 4 units/kg regular insulin. Neurological scoring was performed at time 0, 5.5, 24, 48, 72, and 96 h later using a standardized system. It was noted that significant neurological deficit occurred in all groups after the CO exposure and treatment period. Induced hyperglycemia after CO exposure was associated with significantly worsened neurological scores as compared with saline-treated controls. Insulin therapy simultaneous with induced hyperglycemia significantly improved neurological scores at all times despite maintenance of comparable hyperglycemia with respect to the group treated only with glucose. No significant difference in mortality was found between treatment groups.[118]

Several theories have evolved regarding the postreceptor binding protective effect of insulin on the neuron. Insulin has been shown to provide neuromodulatory inhibition of synaptic transmission *in vivo* and *in vitro*. As an inhibitor of glial uptake of gamma aminobutyric acid (GABA), insulin may increase the availability of this inhibitory neurotransmitter, which may decrease neuronal firing, beneficially reducing cell metabolism.[119] The additional effect of sodium extrusion from the cell which affords subsequent protection against water accumulation may prevent neuronal swelling.[120] Furthermore, it has been suggested that an insulin-induced elevation of brain cate-cholamines through both inhibition of catecholamine uptake and stimulation of release might be a contributory neuroprotective mechanism since catecholamines have been found to attenuate ischemic brain damage.

Of these theories regarding the neuroprotective activity of insulin, however, the most recent highlights its role in stimulating second messengers, and emphasizes its potential genomic effects, i.e., the regulation of protein synthesis, enzymatic activity, and the signaling of cell proliferation. It is well established that the neonatal brain is rich in insulin-like growth factor receptors. Indeed, insulin is similar in structure to other growth factors such as platelet-derived growth factor (PDGF), epidermal growth factor (EGF), insulin-like growth factor-1 (IGF-1). Such peptides are involved in basic neuron development and differentiation. Once bound to its receptor, like other growth factors, insulin triggers signal transduction by internal autophosphorylation of tyrosine on the insulin receptor, which subsequently enhances further phosphorylation reactions of other tyrosine-containing substrates by tyrosine kinase, also located on the insulin receptor.

This type of tyrosine phosphorylation is important in signaling pathways for growth factors and products of proto-oncogenes. Insulin is a progression growth factor in replication G0G1 phase and works synergistically with other growth factors to generate both competence and progression of cells. Through tyrosine phosphorylation of phos-phokinase C, other second messengers are formed, such as diacyglycerol, which causes increased intracellular calcium, activation of the sodium–hydrogen pump, and increased intracellular pH. This pH change in turn activates the sodium–potassium ATPase pump which signals cell proliferation.[121] Insulin also regulates specific mRNA levels through diacylglycerol[122] and may increase messenger-ribonucleic acid (mRNA) efflux from the nucleus via nuclear triphosphatase activation.[123–125] Of even greater importance is the fact that insulin also stimulates lipid neogenesis. It appears, therefore, that the effects of insulin are fundamental with regard to cell signaling, proliferation, replication, and repair following injury. These processes are crucial to cells such as neurons which normally are terminally differentiated and contain little if any capacity to replicate or to synthesize repair lipids.

Certainly, anatomic correlates to the above-proposed mechanisms are in place. For example, it has been demonstrated that the location of insulin and IGF-1 receptors correlates with phosphotyrosine products in the brain.[125] Moreover, the basal ganglia areas typically found to be damaged by CO possess low levels of insulin receptors. Although the initial results of animal studies such as those above may provide building blocks for clinical work, any recommendations regarding the use of insulin in humans as treatment for CO poisoning will of course await further studies.

11.4.5 NMDA RECEPTOR ANTAGONISTS

Recent evidence implicates the endogenous excitatory amino acids such as NMDA in ischemic neurodegeneration.[126–128] Moreover, the NMDA receptor antagonist, MK801, prevents non-CO-induced ischemic neurodegeneration in Mongolian gerbils.[129]

Successive CO exposures induce a consistent pattern of degeneration of hippocampal CA1 pyramidal neurons, a selective neuronal death that resembles that seen with other models of cerebral ischemia. This observation has prompted the study of NMDA receptor antagonists in CO poisoning in mice. Ishimaru et al.[130] pretreated animals with a competitive NMDA antagonist, CPP; a noncompetitive NMDA antagonist, MK-801; a glycine-binding site antagonist, 7-CK; a polyamine-binding site antagonist, ifenprodil; glycine and saline. The number of hippocampal CA1 pyramidal cells was quantified 7 days postexposure using an image analyzer. A decrement of 20% in the number of hippocampal CA1 pyramidal cells was noted relative to the control group. Those animals receiving high doses of MK-801, 7-CK, and CPP had significant reduction in neuronal damage. No clear protective effect was obtained with ifenprodil. Interestingly, glycine, a facilitory neurotransmitter at the NMDA receptor complex, did not exaggerate the CO-induced neuronal damage as might be expected.[130] Although no neurological outcome correlates or survival data are reported, this work may provide valuable mechanistic and possibly future therapeutic insights. Similar work by Lui et al.[131] revealed beneficial effects of MK801 when administered either systemically or directly to the cochlea in protecting against CO-induced ototoxicity.[131]

Ketamine, widely used a dissociative anesthetic agent, is known to have NMDA receptor blocking properties.[132] It has been shown to be neuroprotective in various animal models of ischemic and anoxic neuronal injury. It has also been observed to blunt hypotension, a condition known to worsen CO-mediated neuropathologic changes.[14] Promising work by Penney and Chen[133] demonstrated significantly reduced cerebral edema, more rapid recovery from hypotension, and suppressed lactate formation following CO poisoning when 40 mg/kg ketamine was administered to rats before and during the exposure period. This same study did not yield positive results with the use of verapamil, which could theoretically block NMDA-mediated postsynaptic calcium uptake in neurons.

11.4.6 Brain-Derived Neuropeptides

Cerebrolysin (CL), a drug produced by enzymatic breakdown of lipid free proteins of porcine brain, is a putative neuroprotective agent of unknown mechanism. A proposed neurotrophic effect was supported by reports that CL-treated rats had increased brain protein synthesis,[134] prevention of neuronal degeneration,[135] and enhanced neuronal growth in tissue culture.[136] Effects on the blood–brain barrier have also been noted. Interestingly, CL increases expression of the blood–brain barrier glucose–transported gene in brain endothelial cell cultures. It is hypothesized that CL may accelerate repair of the blood–brain barrier in regions compromised by hypoxia.[137]

Recently, a model of acute CO poisoning combined with spreading depression-induced metabolic stress was used to examine the protective effects of CL on the development of electrophysiological, behavioral, and morphological signs of hypoxic damage in rats. Spreading depression waves reflect the recovery of cerebral cortex in the peri-ischemic areas, or penumbra zone. After a 90-min exposure to 0.8 to 5% CO, microinjections of 5% KCl into the cortical and hippocampal areas were performed, and the duration of spreading depression was noted. At 9- and 18-day follow-up, repeat spreading depression measurements were taken, and a decrease in amplitude was used as an index of brain damage. Postexposure CL-treated rats significantly improved hippocampal recovery. Better performance was also noted on behavioral testing, and no apparent histological damage was apparent in the hippocampus as compared with controls.[138] This very promising neuroprotective agent, which appears to be effective even if given postexposure, certainly deserves further study to elucidate any possible beneficial role in humans.

11.4.7 HYPOTHERMIA

Lowered body temperature was found to be beneficial in the management of CO poisoning by Sluijter,[139] an effect that was thought to be secondary to the increased amounts of dissolved oxygen found in the bloodstream at lower temperatures. Peirce et al.,[140] however, was unable to demonstrate any synergistic effect when hypothermia was used in conjunction with hyperbaric oxygen in a dog model. An interesting report of the use of mechanical ventilation and hypothermia in patients with abnormal motor activity or coma to treat CO toxicity noted complete reversal of these manifestations in three patients when therapy was initiated within the first 24 h. No beneficial effects were noted in a fourth patient who did not receive hypothermic treatment until 5 days after exposure. HBO was not available to these patients.[141]

11.5 APPROACH TO THE PATIENT WITH CARBON MONOXIDE POISONING

11.5.1 GENERAL

The most critical step in managing the patient poisoned with CO is the cessation of tissue hypoxia. This involves supplementation with 100% oxygen, delivered either by a tight-fitting continuous positive airway pressure (CPAP) mask or by endotracheal intubation. Intubation may be necessary in the patient with chronic obstructive pulmonary disease (COPD), to avoid CO_2 retention secondary to high concentrations of oxygen. NBO oxygen should be initiated as soon as the diagnosis is entertained, and should not be delayed for confirmatory COHb levels. As discussed above, the use of 5% CO_2 mixed with 95% oxygen (carbogen) has been proposed by some to facilitate the release of CO from hemoglobin by increasing the ventilatory response. This therapy is of questionable value, and has fallen out of favor. Furthermore, even though not demonstrated in animals,[19,142] the potentially life-threatening possibility of carbogen-induced CO_2 retention and subsequent worsening of already existing acidosis would contraindicate its use in the patient with COPD, concurrent poisoning with respiratory depressants, or altered mental status. The duration of oxygen therapy is guided by a knowledge of the half-life of CO and allows for a margin of safety. Generally, this would involve at least 6 h of therapy on 100% oxygen, longer if the patient is gravid or an infant (see Section 11.5.2). An early chest radiograph is mandatory to assess for evidence of pulmonary edema resulting from CO or other inhaled toxins.

As airway control and oxygenation proceed, attention should be directed toward the cardiovascular system. Continuous cardiac monitoring is advisable and a 12-lead

ECG should be obtained to assess for subclinical cardiac ischemia. Myocardial enzymatic changes without ECG changes has recently been noted in adults with CO poisoning.[143] It is not clear whether this "occult" evidence of myocardial injury is clinically significant. Should arrhythmias, ischemia, or hemodynamic instability occur despite therapy with 100% oxygen, the patient could be considered a candidate for HBO therapy. Myocardial depression and arrhythmias may occur at extremely low arterial pH, which is generally secondary to severe lactic acidosis. Correction of mild acidosis with sodium bicarbonate, however, is not advisable, as this could result in a further shift of the oxyhemoglobin dissociation curve to the left, and thus impair the unloading of oxygen to hypoxic tissues.

In those patients with altered mental status, treatment of rapidly reversible causes of coma should be carried out through the bedside assessment of a finger-stick glucose and the administration of thiamine and naloxone, a narcotic antagonist. Supplemental glucose may be needed to correct hypoglycemia, but every attempt should be made to maintain euglycemia and avoid iatrogenic hyperglycemia. A thorough physical assessment for burns, odors, toxidromes, skin findings, and signs of smoke inhalation, trauma, or abuse is indicated. A careful history regarding the circumstances surrounding the exposure must be obtained once the patient is stabilized. Considerations should be made to gastric or skin decontamination and activated charcoal administration in the setting of suspected intentional drug abuse, suicide attempt, or dermal chemical exposure (methylene chloride).

Removal of CO from the body is best accomplished through displacement by oxygen, either normobaric or hyperbaric. Other less conventional therapies have been used anecdotally with favorable outcomes, and include both exchange transfusion[31] and extracorporeal oxygenation.[144] These invasive approaches, however, would be recommended only in unusual circumstances, for example, if HBO therapy was not available in a deteriorating or moribound patient.

Should the patient have been CO poisoned during the inhalation of smoke from a fire, numerous other products of combustion may be contributing to the metabolic and pulmonary derangements seen. Of particular interest is cyanide, a lethal combustion product which is commonly elaborated when plastics or synthetic materials burn. The patient with CO poisoning who remains significantly acidotic despite treatment with oxygen after smoke inhalation should be suspected of having cyanide toxicity. Some authors advocate empiric treatment of fire victims with the sodium thiosulfate component of the cyanide antidote kit. The methemoglobin-forming agents such as amyl nitrite and sodium nitrite have been traditionally withheld if CO poisoning is suspected to avoid further hemoglobinopathy and worsened hypoxia. Moore et al.[145] demonstrated a 25% increase in mortality in sodium nitrite-treated animals with CO poisoning compared with the untreated controls. Despite this, a recent human study demonstrated that five of seven patients with CO poisoning were safely treated with the antidote kit in its entirety.[146] These patients, however, had only moderate COHb levels with a mean level of 26%. This coupled with the fact that only a small number of patients were studied, indicated that this form of empiric treatment for cyanide poisoning in victims of smoke inhalation cannot yet be widely recommended. Other proposed cyanide antidotes such as hydroxycobalamin require that large volumes of fluid be given (2.4 l per average dose), are costly, and are not readily available at this time in the United States. Therefore, given the safety of the sodium thiosulfate component of the cyanide kit when given alone, this is the only antidote advisable at this time for the treatment of the patient dually poisoned with CO and cyanide.

Once the patient has been stabilized, consideration of the use of possible neuro-protective agents, including HBO, must ensue. If the patient is awake, a mental status examination should be performed. Abbreviated neuropsychologic tests have been developed specifically for the CO-poisoned patient, but often are impractical to perform in the emergency department in those with moderate or severe acute intox-ication. Examples of tasks performed by the patient include placing pegs in a board, complete rapid finger tapping, memorization, construction, number processing, and subjective stress response.[147] Should the patient perform abnormally on the CO neuropsychiatric screening battery (CONSB), have a history of a soaking-type expo-sure or loss of consciousness, have any abnormal neurological findings, exhibit cerebral edema on CaT scan, or have evidence of cardiovascular involvement, HBO could be considered, although predictors of poor neurological outcome are as yet not proved.

The use of COHb levels to guide therapy is controversial. A survey of medical directors of U.S. and Canandian HBO facilities[148] indicated that 62% would use a specific COHb level as the sole criterion for HBO in an asymptomatic patient. The same survey found that when a specific level was used as an indication for HBO therapy, 25% was the most common level chosen. Others suggest that HBO therapy is prescribed based on COHb level in 40% of patients and in 60% based on CNS or cardiac dysfunction.[149] Other patients who could be considered candidates for HBO would include the pregnant patient with a COHb level greater than 15%, the patient with a history of coronary artery disease and a COHb level greater than 20%, the asymptomatic patient with a level greater than 25 to 40% COHb, or the patient with recurrent or persistent symptoms despite 6 h of therapy with NBO (Figure 11.2). If HBO therapy is to be administered, treatment within 6 h is desirable.[150] Patients should undergo full NP evaluation prior to discharge. Close follow-up is necessary with repeat NP examinations at 6 weeks, 6 months, and 12 months.

11.5.2 PREGNANCY

The effects of CO on the fetus are extensively reviewed by Penney.[151] The fetus is particularly vulnerable to the effects of CO, which readily crosses the placenta and is even more tightly bound to fetal hemoglobin than adult hemoglobin. The fetus also reaches higher peak COHb levels than does the mother. Fetuses that survive a sig-nificant CO poisoning may be left with limb malformation, hypotonia, areflexia, persistent seizures, mental and motor disability, and microcephaly.[152,153] The only prospective, multicenter study of acute CO poisoning in pregnancy recently reported adverse outcomes in 60% of children whose mothers suffered severe CO toxicity. Of those babies born to mothers with mild to moderate CO exposure, normal physical exams and neurobehavioral development were reported.[154] Since CO elimination from the fetus is prolonged (7 to 10 h), it is generally accepted that HBO therapy is indicated at lower maternal COHb levels than would be acted upon in the nongravid patient. In addition, NBO therapy should be extended to four to five times the normal duration.

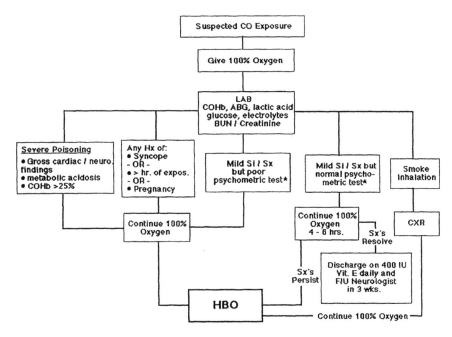

FIGURE 11.2   Suggested carbon monoxide diagnosis/therapeutic decision tree. Hx = history; ABG = arterial blood gases; BUN = blood urea nitrogen; Si = sign; Sx = symptom; F/U = follow-up; CoHb = carboxy-hemoglobin.
*Option: check orientation, phone number, address, date of birth, serial 7s, digit span, forward and backward spelling of three- and four-letter words, and short-term memory. (Modified from Emerson, T.S., *Oxygen Currents*, Vol. 1(2), p. 5, Fall 1993. With permission.)

TABLE 11.1
Proposed Indications for Hyperbaric Oxygen Therapy in Pregnancy

1. Maternal COHb level > 15 to 20% at any time during the exposure
2. Any neurological sign or symptom other than headache
3. Evidence of fetal distress (fetal tachycardia, decreased beat-to-beat variability, late decelerations)
4. If maternal neurological symptoms or fetal distress persist 12 h after initial therapy, additional HBO treatments may be necessary

Although controversial, HBO has been reported to be safe in pregnancy,[155] despite theoretical dangers of fetal hyperoxia in animal models.[156–158] (Such animal models exceeded the time and pressure routinely used in clinical therapy.) A recent report of 44 women undergoing HBO during pregnancy for CO exposure suggests that HBO is safe and should be considered, although miscarriages did occur, and 6 patients were lost to follow-up.[159] It should be noted that HBO was implicated in the induction of labor in one pregnant patient; the pregnancy, however, was near term when the CO exposure occurred.[160] Proposed indications for HBO therapy in the pregnant patient are listed in Table 11.1, although these are not well studied.

11.5.3 CHILDREN

Younger children have traditionally been thought to be more susceptible to CO poisoning based on more their rapid metabolic rate and higher oxygen demands (See Chapter 21, by White for a complete review of pediatric CO poisoning). They may also present more atypically than adults. Both persistent and delayed sequelae are described in children; however, formal NP testing is generally difficult and not well documented in such case series. The use of HBO therapy in the treatment of pediatric CO poisoning is controversial, and recommendations vary, even among pediatric toxicology experts. Until further knowledge and experience are gained in this area, children will likely be treated as aggressively as adults who are CO poisoned.

11.6 MANAGEMENT OF THE SEQUELAE OF CARBON MONOXIDE POISONING

Delayed sequelae from CO poisoning are devastating and occur in 10 to 43% of persons recovering from acute exposure. Parkinsonism, the most dramatic long-term neurological complication, has a grim prognosis. Unfortunately, conventional therapy with agents such as dopa has been disappointing. Another centrally acting dopaminergic agonist, bromocriptine, may be more promising. Nine patients (mean age 61 years) suffering from CO-induced parkinsonism who were given bromocriptine (5 to 30 mg/day) displayed improvement in Webster's scores while under treatment.[161] Clearly, no definitive conclusions regarding bromocriptine therapy can be made based on this small study, but perhaps this study will provide a basis for future investigations.

One reported case involving hyperpyrexia and muscle rigidity as sequelae of CO poisoning was treated successfully with a prolonged course of dantrolene sodium, a peripheral skeletal muscle relaxant.[162] Given that the patient manifested signs characteristic of severe hypoxic/ischemic encephalopathy, this therapy was symptomatic for that condition, and not specific to CO poisoning. Dantrolene would not likely provide any benefit beyond other safer sedatives, such as benzodiazepines, in treating such complications.

Another common sequelae from CO poisoning is memory impairment. Recent work by Hiramatsu et al.[163] focused on treating delayed amnesia in mice. The investigators treated mice with documented amnesia 5-days postexposure with dynorphin A (1-13). They found this treatment regimen to be effective in reversing CO-induced memory impairment. Nor-binaltrophimine (kappa opioid receptor antagonist) blocked the effect of dynorphin A (1-13), suggesting that kappa receptors mediated the reversal of impairment in memory seen from CO in this animal model. The authors reported similar findings with a second kappa receptor agonist, U-50488H, which appeared additionally to activate the cholinergic neuronal system, known also to play an important role in cognitive deficits associated with other conditions such as aging and neurodegenerative diseases.[164] These agents may hold promise for the future in treating the persistent or delayed detrimental effects of CO on acquisition and consolidation of memory.

Delayed, sometimes repetitive, HBO therapy has been advocated by some to improve the long-term neurological deficits from CO toxicity, even if instituted weeks after the initial CO exposure.[86,108,165–171] Such practice, which is advocated by several treatment centers in the United States, lacks validation by well-controlled, blinded clinical studies that utilize NP testing data. Interestingly, behavioral treatment has been successful when guided by formal NP testing. In certain patients, indirect measures of learning are better predictors of treatment efficacy.[172]

Patients who present to health-care facilities late or have suffered recurrent or chronic low-level exposures provide particular treatment challenges for the clinician. Any proposed therapeutic approach to such patients should be considered carefully given the fact that few definitive clinical or animal studies in this area exist.

11.7   CONCLUSIONS

There is much yet to be learned regarding the mechanisms of CO toxicity, predictors of outcome, and treatment of CO poisoning. Necessary areas for further research include the delineation of clear-cut clinical indications for the use of potentially neuroprotective agents (i.e., insulin, sulfhydryl donors, allopurinol, ketamine, brain-derived peptides, kappa receptor agonists, and hypothermia). In addition, how best to use these agents synergistically remains to be seen. Other areas that deserve more study is the further clarification of risk factors for adverse fetal outcome following CO exposure during pregnancy, and delineation of the true incidence of PNS and DNS in children, along with how best to test for these sequelae. The future for HBO therapy in CO poisoning remains to be seen. Recent data from randomized trials using HBO therapy now compel even more carefully selecting adult patients for this therapy and the study of subsets of patients such as children and pregnant women who may potentially benefit.

REFERENCES

1. Cobb, N. and Etzel, R.A., Unintentional carbon monoxide-related deaths in the United States, 1979 through 1988, *J. Am. Med. Assoc.*, 266, 659, 1995.
2. Runciman, W.W. and Gomman, D.F., Carbon monoxide poisoning: from old dogma to new uncertainties, *Med. J. Aust.*, 158, 439–440, 1993.
3. Smith, J.S. and Brandon, S., Morbidity from acute carbon monoxide poisoning at three-year follow-up, *Br. Med. J.*, 1, 279–281, 1983.
4. Lugaresi, A., Montagna, P., Morreale, A. et al., "Psychic akinesia" following carbon monoxide poisoning, *Eur. Neurol.*, 30, 167–169, 1990.
5. Klawans, H.L., Stein, B.W., Tanner, C.M. et al., A pure parkinsonian syndrome following acute carbon monoxide intoxication, *Arch. Neurol.*, 39, 302–304, 1982.
6. Wemer, B., Back, W., Akerblom, H. et al., Two cases of acute carbon monoxide poisoning with delayed neurologic sequelae after a "free" interval, *Clin. Toxicol.*, 23, 249–265, 1985.
7. Hart, I.K., Kennedy, P.G., Adams, J.H. et al., Neurological manifestations of carbon monoxide poisoning, *Postgrad. Med. J.*, 64, 213–216, 1988.
8. Sauk, G.M., Watson C.P., Tebragge, K. et al., Delayed encephalopathy following carbon intoxication, *Can. J. Neurol. Sci.*, 8, 77–79, 1981.

9. Barret, L., Danel, V., and Faure, J., Carbon monoxide poisoning, a diagnosis frequently overlooked, *Clin. Toxicol.*, 23, 309–313, 1985.

10. Balzan, M.A., Agius, G., and Debono A.G., Carbon monoxide poisoning, easy to treat but difficult to recognise, *Postgrad. Med. J.*, 72, 470, 1996.

11. Sokal, J.A., Majka, J., and Palus, J., The content of carbon monoxide in the tissues of rats intoxicated with carbon monoxide in various conditions of acute exposure, *Arch. Toxicol.*, 56: 106–108, 1984.

12. Lasater, S.R., Carbon monoxide poisoning, *Can. Med. Assoc. J.*, 134, 991–992, 1986.

13. Olson, K.R., Carbon monoxide poisoning: mechanisms, presentation, and controversies in management, *J. Emerg. Med.*, 1, 233–243, 1984.

14. Ikeda, T., Kondo, T., Mogami, H., Miura, T., Mitomo, M., Shimazaki, S., and Sugimoto, T., Computerized tomography in cases of acute carbon monoxide poisoning, *Med. J. Osaka Univ.*, 29, 253–262, 1978.

15. Penney, D.G., Hyperglycemia exacerbates brain damage in acute severe carbon monoxide poisoning, *Med. Hypotheses*, 27, 241–244, 1988.

16. Bogusz, M., Cholewa, L., Pach, J., and Mlodkowska, K., A comparison of two types of acute carbon monoxide poisoning, *Arch. Toxicol.*, 33, 141–149, 1975.

17. Linas, A.J. and Limousin, S., Asphyxie lente et graduelle par l'oxyde de carbone, traitement et guerison par les inspirations d'oxygene, *Bul. Mem. Soc. Ther.*, 2, 32, 1868.

18. Haldane, J., The relation of the action of carbonic oxide to oxygen tension, *J. Physiol.*, 18, 201–217, 1895.

19. Killick, E.M. and Marchant, J.V., The effect of barbiturates on the resuscitation of dogs from severe acute CO poisoning, *J. Physiol.*, 180, 80, 1965.

20. Killick, E.M. and Marchant, J.V., Resuscitation of dogs from severe, acute carbon monoxide poisoning, *J. Physiol.*, 147, 274, 1959.

21. Schwerma, H., Ivy, A.C., Friedman, H., and Brosse, E.L., A study of resuscitation from the juxtalethal effects of exposure to carbon monoxide, *Occup. Med.*, 5, 24, 1948.

22. Jain, K.K., *Carbon Monoxide Poisoning*, Warren H. Green, St. Louis, MO, 1990, 140.

23. Gros, J.F. and Leandri, P., Traitment de l'intoxication oxycarbonee par le cytochrome, *Presse Med.*, 64, 1356–1357, 1956.

24. Bentolila, P., Tran, G., and Olive G., Essai de traitment de l'intoxication oxycarbonee par perfusion de solutions diluees de paroxyde d'hysrogene, resultats obtnus chez le lapin, *Therapie*, 28, 1043–1049, 1973.

25. Koza, F., Die kohlenmonoxidevergiftungund deren neurartige Therapie mit Bestrahlung, *Med. Klin.*, 26, 422–425, 1930.

26. Estler, W., Experimentelle Untersuchungen uber die Anwendung der Bestrahlung mit ultravoilettem Licht zur Behandlung der Kohlenoxidvergiftung, *Arch. Hyg. Bakteriol.*, 115, 152–257, 1935.

27. Amyes, E.W., Ray, J.W., and Brockman, N.W., Carbon monoxide anoxia; intravenous administration of procaine hydrochloride in the treatment of acute and chronic effects, *J. Am. Med. Assoc.*, 142, 1054–1058, 1950.

28. Evans, D.E., Catron, P.W., McDermott, J.J. et al., Effect of lidocaine after experimental cerebral ischemia induced by air embolism, *J. Neurosurg.*, 70, 97–102, 1989.

29. Jeda-Sahagun, J.L., Accion protectora del dipiridamol frente a la intoxicacion por gas del alumbrado, *Rev. Esp. Fisiol.*, 27, 305, 1971.

30. Agostini, J.C., Ramirez, R.G., Albert, S.N., Goldbaum, L.R., and Absolon, K.B., Successful reversal of lethal carbon monoxide intoxication by total body asanguineous hypothermic infusion, *Surgery*, 75, 213, 1974.

31. Yee, L.M. and Brandon, G.K., Successful reversal of presumed carbon monoxide-induced semicoma, *Aviat. Space Environ. Med.*, 54, 641, 1983.

32. Shumate, M.J., Carbon monoxide poisoning, *Chest*, 107, 1474, 1995.

33. Geyer, R.P. and Haggard, H.W., Review of perfluorochemical-type blood substitutes, in *Proc. 10th Int. Cong. Nutrition — Symposium on Perfluorochemical Artificial Blood*, Igakushobe, Osaka, Japan, 1976, 3.

34. Levine, E.M. and Tremper, K.K., Perfluorochemical emulsions: potential clinical uses and new developments, *Int. Anesthesiol. Clin.*, 23, 211, 1985.

35. Uchiyama, T. and Harafuji, K., Life-saving effect of pyridoxalated hemo-globin–polyoxyethylene conjugate on carbon monoxide intoxication of rabbits, *Artif. Organs*, 18:576, 1994.

36. Mosso, A., La mort apparente du coer secours l'empoisonneement par l'oxyd de carbonee, *Arch. Ital. Biol.*, 35, 75–89, 1901.

37. Smith, G. and Sharp, G.R., Treatment of carbon monoxide poisoning with oxygen under pressure, *Lancet*, 2, 905, 1960.

38. Sluitjer, M.E., The treatment of carbon monoxide poisoning by administration of oxygen at high pressure, *Proc. R. Soc. Med.*, 56, 123, 1963.

39. Dinman, B., Eaton, J., and Brewer, G., Effects of carbon monoxide on DPG concentrations in the erythrocyte, *Ann. N.Y. Acad. Sci.*, 1974, 246, 1970.

40. KIasner, A.E., Smith, S.R., Thompson, M.W., and Scalzo, A.J., Carbon monoxide mass exposure in a pediatric population, *Acad. Emerg. Med.*, 5, 992, 1998.

41. Okeda, R., Funata, N., Song, S.J. et al., Comparative study of selective cerebral lesions in carbon monoxide poisoning and nitrogen hypoxia in cats, *Acta Neuropathol.*, 56, 265, 1982.

42. Haldane, J.B.S., Carbon monoxide as a tissue poison, *Biochem. J.*, 21, 1086, 1927.

43. Drabkin, D.L., Lewey, F.H., Bellet, S. et al., The effect of replacement of normal blood by erythrocytes saturated with carbon monoxide, *Am. J. Med. Sci.*, 205, 755, 1943.

44. Goldbaum, L.R., Orellano, T., and Dergal, E., Studies on the relationship between carboxyhemoglobin concentration and toxicity, *Aviat. Space Environ. Med.*, 48, 969, 1977.

45. Goldbaum, L.R., Orellano, T., and Dergal, E., Mechanism of the toxic action of carbon monoxide, *Ann. Clin. Lab. Sci.*, 6, 372, 1976.

46. Goldbaum, L.R., Ramirez, R.G., and Absalon, K.B., What is the mechanism of carbon monoxide toxicity? *Aviat. Space Environ. Med.*, 46, 1289, 1975.

47. Brown, S. and Piantadosi, C., *In vivo* binding of CO to cytochrome oxidase in rat brain, *J. Appl. Physiol.*, 68, 604, 1990.

48. Brown, S. and Piantadosi, C.A., Recovery of energy metabolism in rat brain after CO hypoxia, *J. Clin. Invest.*, 89, 666, 1992.

49. Cobum, R.F. and Forman, H.J., Carbon monoxide toxicity, in *Handbook of Physiology*, Sect. 3: *The Respiratory System*, The American Physiological Society, Washington, D.C., Vol. IV, 1987, chap. 21, 439.

50. Smithline, H.A., Rivers, E.P., Chiulli, D.A., Rady, M.Y., Baltarowich, L.L., Blake, H.C. et al., Systemic hemodynamic and oxygen transport response to graded carbon monoxide poisoning, *Ann. Emerg. Med.*, 5, 203, 1993.

51. Ward, K.R., Lin, J., Zhang, L., and Chopp, M., Carbon monoxide poisoning does not cause neuronal heat shock protein 72 or 32 expression, *Acad. Emerg. Med.*, 5, 487, 1998.

52. Dallman, P. and Schwartz, H., Distribution of cytochrome C and myoglobin in rats with dietary iron deficiency, *Pediatrics*, 35, 677, 1965.

53. Wittenberg, B.A. and Wittenberg, J.B., Effects of carbon monoxide on isolated heart muscle cells, *Res. Rep. Health Effects Inst.*, 62, 1, 1993.

54. Thom, S.R., Carbon monoxide-mediated brain lipid peroxidation in the rat, *J. Appl. Physiol.*, 68, 997, 1990.

55. Thom, S.R., Leukocytes in carbon monoxide-mediated brain oxidative injury, *Toxicol. Appl. Pharmacol.*, 123, 234, 1993.

56. Thom, S.R., Dehydrogenase conversion to oxidase and lipid peroxidation in brain after carbon monoxide poisoning, *J. Appl. Physiol.*, 73, 1584, 1992.

57. Thom, S.R., Gamer, S., and Fisher, D., Vascular oxidative stress from carbon monoxide exposure, *Undersea Hyperb. Med.*, 25 (Suppl.), 47, 1998.

58. Thom, S.R., Xu, Y.A., and Ischiropoulos, H., Vascular endothelial cells generate peroxynitrite in response to carbon monoxide exposure, *Chem. Res. Toxicol.*, 10, 1023, 1997.

59. Ischiropoulos, H., Beers, M.F., Ohnishi, S.T., Fisher, D., Gamer, S.E., and Thom, S.R., Nitric oxide production and perivascular nitration in brain after carbon monoxide poisoning in the rat, *J. Clin. Invest.*, 97, 2260, 1996.

60. Hall, E.D., Andrus, P.K., Althaus, J.S. et al., Hydroxyl radical production and lipid peroxidation parallels selective post-ischemic vulnerability in gerbil brain, *J. Neurosci. Res.*, 34, 107, 1993.

61. Zhang, J. and Piantadosi, C.A., Mitochondrial oxidative stress after carbon monoxide hypoxia in the rat, *J. Clin. Invest.*, 90, 1193, 1992.

62. Amara, S.G., Neurotransmitter transporters, *Nature*, 360, 420, 1992.

63. Shinomura, T., Nakao, S., and Mori, K., Reduction of depolarization-induced glutamate release by hemoxygenase inhibitor, possible role of carbon monoxide in synaptic transmission, *Neurosci. Lett.*, 166, 131, 1994.

64. Zhang, J. and Piantadosi, C.A., Prevention of H_2O_2 generation by monoamine oxidase protects against CNS O_2 toxicity, *J. Appl. Physiol.*, 71, 1057, 1991.

65. Piantadosi, C.A., Tatro, L., and Zhang, J., Hydroxyl radical production in the brain after CO hypoxia in rats, *Free Radical Biol. Med.*, 18, 603, 1995.

66. Piantadosi, C.A., Zhang, J., Levin, E.D. et al., Apoptosis and delayed neuronal damage in acute carbon monoxide poisoning in the rat, *Exp. Neurol.*, 147, 103, 1997.

67. Ward, K.R., Junmin, L., Zhang, L., and Chopp, M., Moderately severe carbon monoxide poisoning does not cause neuronal apoptosis [abstract], *Acad. Emerg. Med.*, 5, 487, 1998.

68. Ward, K., Lin, J., and Chopp, M., Carbon monoxide poisoning does not cause significant cerebral heat shock protein 72 expression [abstract], *Crit. Care Med.*, 26 (Suppl.), A39, 1998.

69. Piantadosi, C.A., Toxicity of carbon monoxide, hemoglobin vs. histotoxic mechanisms, in Penney, D.G., Ed., *Carbon Monoxlde*, CRC Press, Boca Raton, FL, 1996, 163.

70. Van Der Hoeven, J.G., Compier, E.A., and Meinders, A.E., Goede resultaten van behandeling met 100% zuurstop wegens acute koolmonoxide-intoxicatie; voorlopig geen indicatie, voor hyperbare zuurstoftoediening, *Ned. Tijdschr. Geneeskd.*, 137, 864, 1993.

71. Weaver, L.K., Hopkins, R.O., and Larson-Lohr, V., Neuropsychologic and functional recovery from severe carbon monoxide poisoning without hyperbaric oxygen therapy, *Ann. Emerg. Med.*, 27, 736, 1996.

72. Meert, K.L., Heidemann, S.M., and Sarnaik, A.P., Outcome of children with carbon monoxide poisoning treated with normobaric oxygen, *J. Trauma*, 44, 149, 1998.

73. Thom, S., Functional inhibition of neutrophil beta-2 integrins by HBO in CO-mediated brain injury, *Toxicol. Appl. Pharmacol.*, 123, 248–256, 1993.

74. Thom, S.R., Mendiguren, I., Hardy, K. et al., Inhibition of human neutrophil beta-2 integrin dependent adherence by hyperbaric oxygen, *Am. J. Physiol.*, 272, C770, 1997.

75. Tomaszewski, C., Rudy, J., Wathen, J., Brent, J., Rosenberg, N., and Kulig, K., Prevention of neurologic sequelae from carbon monoxide by hyperbaric oxygen in rats, *Ann. Emerg. Med.*, 21, 631–632, 1992.

76. Brown, S.D. and Piantadosi, C.A., Reversal of carbon monoxide-cytochrome C oxidase binding by hyperbaric oxygen *in vivo*, in *Advances in Experimental Medicine and Biology*, Rakusan, K., Biro, G.D., and Goldstick, T.K., Eds., Vol. 248, Plenum Press, New York, 1989, 747–754.

77. Thom, S.R., Antagonism of carbon monoxide-mediated brain lipid peroxidation by hyperbaric oxygen, *Toxicol. Appl. Pharmacol.*, 105, 340, 1990.

78. Sukoff, M.H. and Ragatz, R.E., Hyperbaric oxygen for the treatment of acute cerebral oedema, *Neurosurgery*, 10, 29, 1982.

79. Jiang, J. and Tyssebotn, I., Cerebrospinal fluid pressure changes after acute carbon monoxide poisoning and therapeutic effects of normobaric and hyperbaric oxygen in conscious rats, *Undersea Hyperb. Med.*, 24, 245, 1997.

80. Roos, J.A., Jackson-Friedman, C., and Lyden, P., Effects of hyperbaric oxygen on neurologic outcome for cerebral ischemia in rats, *Acad. Emerg. Med.*, 5, 18, 1998.

81. Kokame, G.M. and Shuler, S.E., Carbon monoxide poisoning, *Arch. Surg.*, 96, 211, 1968.

82. Myers, R.A.M., Snyder, S.K., Linberg, S., and Cowley R.A., Value of hyperbaric oxygen in suspected carbon monoxide poisoning, *J. Am. Med. Assoc.*, 246, 2478, 1981.

83. Winter, A. and Shatin, L., Hyperbaric oxygen in reversing carbon monoxide coma, *N.Y. State J. Med.*, 1, 880, 1970.

84. Thomson, L.F., Mardel, S.N., Jack, A., and Shields, T.G., Management of the moribund carbon monoxide victim, *Arch. Emerg. Med.*, 9, 208, 1992.

85. Welsh, F., Matos, L., and DeTreville, R.T.P., Medical hyperbaric oxygen therapy: 22 cases, *Aviat. Space Environ. Med.*, 51, 611, 1980.

86. Gibson, A.J., Davis, F.M., and Ewer, T., Delayed hyperbaric oxygen therapy for carbon monoxide intoxication — two case reports, *N. Z. Med. J.*, 104, 64, 1991.

87. Dean, B.S., Verdile, V.P., and Krenzelok, E.P., Coma reversal with cerebral dysfunction recovery after repetitive hyperbaric oxygen therapy for severe carbon monoxide poisoning, *Am. J. Emerg. Med.*, 11: 61, 1993.

88. Roche, L., Bertoye, A., Vincent, P. et al., Comparison de deux groupes de vingt intoxications oxycarbonees traitees par oxygene normobare et hyperbare, *Lyon Med.*, 49, 1483, 1968.

89. Mathieu, D., Nolf, M., Durocher, A., Saulnier, F., Frimat, P., Furon, D., and Wattel, F., Acute carbon monoxide poisoning risk of late sequelae and treatment by hyperbaric oxygen, *Clin. Toxicol.*, 23, 315, 1985.

90. Gorman, D.F., Clayton, D., Gilligan, J.E., and Webb, R.K., A longitudinal study of 100 consecutive admissions for carbon monoxide poisoning to the Royal Adelaide Hospital, *Anaesth. Intensive Care*, 20, 311, 1992.

91. Myers, R.A.M., Snyder, S.K., and Emhoff, T.A., Subacute sequelae of carbon monoxide poisoning, *Ann. Emerg. Med.*, 14, 1163, 1985.

92. Hsu, L.H. and Wang, J.H., Treatment of carbon monoxide poisoning with hyperbaric oxygen, *Chin. Med. J.*, 58, 407, 1996.

93. Gozal, D., Ziser, A., Shupak, A., and Melamed, Y., Accidental carbon monoxide poisoning, *Clin. Pediatr.*, 24, 132, 1985.

94. Norkool, D.M. and Kirkpatrick, J.N., Treatment of acute carbon monoxide poisoning with hyperbaric oxygen: a review of 115 cases, *Ann. Emerg. Med.*, 14, 1168, 1985.

95. Weaver, L.K., Carbon monoxide poisoning, *Crit. Care Clin.*, 15, 297, 1999.

96. Ducasse, J.L., Izard, P.H., Celsis, P. et al., Moderate carbon monoxide poisoning: hyperbaric or normobaric oxygenation? in *Proceedings of the Joint Meeting of the 2nd European Conf. and the 2nd Swiss Symp. on Hyperbaric Medicine*, Basel, Switzerland, 1990, 289–297.

97. Olson, K.R. and Seger, D., Hyperbaric oxygen for carbon monoxide poisoning: does it really work? *Ann. Emerg. Med.*, 25, 535, 1995.

98. Weaver, L.K., Hopkins, R.O., and Larson-Lohr, V., Hyperbaric oxygen and carbon monoxide poisoning, *Ann. Emerg. Med.*, 2693, 390, 1995.

99. Scheinkestel, C.D., Bailey, M., Myles, P.S., Jones, K., Cooper, D.J., Millar, I.L., and Tuxen, D.V., Hyperbaric or normobaric oxygen for acute carbon monoxide poisoning: a randomised controled clinical trial, *Med. J. Aust.*, 170, 203, 1999.

100. Weaver, L.K., Hopkins, R.O., Howe, S., Larson-Lohr, V., and Churchill, S., Outcome at 6 and 12 months following acute CO poisoning, *Undersea Hyperb. Med.*, 23S, 9, 1996.

101. Weaver, L.K., Hopkins, R.O., Larson-Lohr, V., Howe, S., and Haberstoek, D., Double-blind, controlled, prospective, randomized clinical trial (RCT) in patients with acute carbon monoxide CO poisoning: outcome of patients treated with normobaric oxygen or hyperbaric oxygen — an interim report, *Undersea Hyperb. Med.*, 22S, 14, 1995.

102. Mathieu, D., Wattel, F., Mathiew-Nolf, M. et al, Randomized prospective study comparing the effect of HBO versus 12 hours NBO in non-comatose CO poisoned patients: results of the interim analysis, *Undersea Hyperb. Med.*, 23(S), 7, 1996.

103. Gorman, D.F., Clayton., D., and Gilligan, J.E., A longitudinal study of 100 consecutive admissions for carbon monoxide poisoning to the Royal Adelaide Hospital, *Anesth. Intensive Care*, 20, 311, 1992.

104. Turner, M., Esaw, M., and Clark, R.J., Carbon monoxide poisoning treated with hyperbaric oxygen: metabolic acidosis as a predictor of treatment requirements, *J. Acid. Emerg. Med.*, 16, 96, 1999.

105. Tibbles, P.M. and Edelsberg, J.S., Hyperbaric oxygen therapy, *N. Engl. J. Med.*, 334, 1642, 1996.

106. McNulty, J.A., Maher, B.A., Chu, M., and Sitnikova, T., Relationship of short-term verbal memory to the need for hyperbaric oxygen treatment after carbon monoxide poisoning, *Neuropsychiatr. Neuropsychol. Behav. Neurol.*, 10, 174, 1997.

107. Raphael, J.C., Elkharrat, D., Jars-Guincestre, M.C. et al., Trial of normobaric and hyperbaric oxygen for acute carbon monoxide intoxication, *Lancet*, 1, 414, 1989.

108. Hampson, N.B., Simonson, S.G., and Kramer, C.C., Central nervous system oxygen toxicity during hyperbaric treatment of patients with carbon monoxide poisoning, *Undersea Hyperb. Med.*, 23, 215, 1996.

109. Marklund, S.L., Westman, G., Lundgren, E., and Roos, G., CuZn superoxide dismutase, Mn superoxide dismutase, catalase and glutathione peroxidase in normal and neoplastic human cell lines and normal human tissues, *Cancer Res.*, 42, 1955–1961, 1982.

110. Toledo-Pereyra, L.H., Simmons, R.L., and Najarian, J.S., Effect of allopurinol on the preservation of ischemic kidneys perfused with plasma or plasma substitutes, *Ann. Surg.*, 180, 780–782, 1974.

111. Stewart, J.R., Blackwell, W.H., Crute, S.L., Loughlin, V., Hess, M.L., and Greenfield, L.J., Prevention of myocardial ischemia/reperfusion injury with oxygen-free radical scavenger, *Surg. Forum*, 33, 317–320, 1982.

112. Fechter, L.D., Liu, Y., and Pearce, T.A., Cochlear protection from carbon monoxide exposure by free radical blocks in the guinea pig, *Toxicol. Appl. Pharmacol.*, 142, 47, 1997.

113. Howard, R.J., Blake, D.R., Pall, H., Williams, A., and Green, I.D., Allopurinol/ *N*-acetylcysteine for carbon monoxide poisoning, *Lancet*, 2, 628–629, 1987.

114. Penney, D.G., Helfman, C.C., Hull, J.C., Dunbar, J.C., and Verma, K., Elevated blood glucose is associated with poor outcome in the carbon monoxide poisoned rat, *Toxicol. Lett.*, 54, 287, 1990.

115. Penney, D.G., Hyperglycemia exacerbates brain damage in acute severe carbon monoxide poisoning, *Med. Hypotheses*, 27, 241–244, 1988.

116. Leikin, J.B., Goldenberg, R.R., Edwards, D., and Zell-Kantor, M., Metabolic predictors of carbon monoxide poisoning, *Vet. Hum. Toxicol.*, 30, 40-42, 1988.

117. Pulsinelli, W., Waldman, S., Sigsbee, B., Rawlinson, D., Scherer, P., and Plum, F., Experimental hyperglycemia and diabetes mellitus worsen stroke outcome, *Trans. Am. Neurol. Assoc.*, 105, 21–24, 1980.

118. White, S.R. and Penney, D.G., Effects of insulin and glucose treatment on neurologic outcome after carbon monoxide poisoning, *Ann. Emerg. Med.*, 23, 606, 1994.

119. Bouhaddi, K., Thomopoulos, P., Fages, C., Khelil, M., and Tardy, M., Insulin effect on GABA uptake in astroglial primary cultures, *Neurochem. Res.*, 13, 1119–1124, 1988.

120. Stahl, W., The Na-K-ATPase of nervous tissue, *Neurol. Int.*, 8, 449–476, 1988.

121. MaCara, I.G., Oncogenes, ions, and phospholipids, *Am. J. Physiol.*, 248, C3–C11, 1985.

122. Standaert, M.L. and Pollet, R.J., Insulin-glycerolipid mediators and gene expression, *FASEB J.*, 2: 2453–2461, 1988.

123. Goldfine, I.D., The insulin receptor: molecular biology and transmembrane signaling, *Endocrinology*, 8, 235–255, 1987.

124. Goldfine, I. D., Purrello, F., Vigneri, R., and Clawson, G.A., Direct regulation of nuclear functions by insulin: relationship to mRNA metabolism, in *Molecular Basis of Insulin Action*, Czech, M.P., Ed., Plenum, New York, 1985, 329–345.

125. Moss, A.M., Unger, J.W., Moxley, R.T., and Livingston, J.N., Location of phospho-tyrosine containing proteins by immunocytochemistry in the rat forebrain corresponds to the distribution of the insulin receptor, *Proc. Natl. Acad. Sci. U.S.A.*, 87, 4453–4457, 1990.

126. Beneviste, H., Drejer, J., Schousboe, A., and Diemer, N.H., Elevation of extracellular concentrations of glutamate and aspartate in rat hippocampus during transient cerebral ischaemia monitored by intracerebral microdialysis, *J. Neurochem.*, 43, 1369–1374, 1984.

127. Jorgensen, M.B. and Diemer, N.H., Selective neuron loss after cerebral ischaemia in the rat: possible role of transmitter glutamate, *Acta Neurol. Scand.*, 66, 536–546, 1982.

128. Simon, R.P., Swan, J.H., Griffith, T., and Meldrum, B.S., Blockade of *N*-methyl-D-aspartate receptors may protect against ischaemic damage in the brain, *Science*, 226, 850–852, 1984.

129. Gill, R., Foster, A.C., and Woodruff, G.N., Systemic administration of MK-801 protects against ischemia-induced hippocampal neurodegeneration in the gerbil, *J. Neurosci.*, 7, 3343–3349, 1987.

130. Ishimaru, H., Katoh, A., Suzuki, H., Fukuta, T., Kameyama, T., and Nabeshima, T., Effects of *N*-methyl-D-aspartate receptor antagonists on carbon monoxide induced brain damage in mice, *J. Pharmacol. Exp. Ther.*, 261, 349–352, 1991.

131. Liu, Y. and Fechter, L.D., MK-801 protects against carbon monoxide induced hearing loss, *Toxicol. Appl. Pharmacol.*, 132, 196, 1995.

132. Anis, N.A., Berry, S.C., Burton, N.R., and Lodge, D., The dissociative anaesthetics, ketamine and phencyclidine, selectively reduce excitation of the central mammalian neurons by *N*-methylasparate, *Br. J. Pharmacol.*, 79, 565, 1983.

133. Penney, D.G. and Chen, K., NMDA receptor-blocker ketamine protects during acute carbon monoxide poisoning, while calcium channel-blocker verapamil does not, *J. Appl. Toxicol.*, 16, 297, 1996.

134. Piswanger, A., Paier, B., and Windisch, M., Modulation of protein synthesis in a cell free system from rat brain by cerebrolysin during development and aging, in *Amino Acids*, G. Lubec, G.A. Rosenthal, Eds., ESCOM Science Publishers, 1990, 651–657.

135. Akai, F., Himura, S., Sato, T., Iwamoto, N., Fujimoto, M., Ioku, M., and Hashimoto, S., Neurotrophic factor-like effect of FPF 1070 on septal cholinergic neurons after transec-tions of fimbria-fornix in the rat, *Histol. Histopathol.*, 7, 213, 1992.

136. Sato, T., Imano, F., Akai, S., Himura, S., Hashimoto, T., Itoh, T., and Fujimoto, M., Morphological observation of effects of cerebrolysin on cultured neural cells, in *Alzheimer's Disease and Related Disorders*, M. Nicolini, P.F. Zatta, and B. Coraine, Eds., Pergamon Press, Oxford, 1993, 195–196.

137. Boado, R.J., Brain-derived peptides increase the expression of a blood–brain barrier GLUT1 glucose transporter reporter, *Neurosci. Lett.*, 220, 53, 1996.

138. Koroleva, V.I., Korolev, O.S., Mares, V., Pastalkova, E., and Bures, J., Hippocampal damage induced by carbon monoxide poisoning and spreading depression is alleviated by chronic treatment with brain derived polypeptides, *Brain Res.*, 816, 618, 1999.

139. Sluitjer, M.E., The treatment of carbon monoxide poisoning by administration of oxygen at high pressure, *Proc. R. Soc. Med.*, 56, 1002–1008, 1963.

140. Peirce, E.C., Zacharias, A., Alday, J.M., Hoffman, B.A., and Jacobson, J.H., Carbon monoxide poisoning: experimental hypothermic and hyperbaric studies, *Surgery*, 72, 229–237, 1972.

141. Boutros, A.R. and Hoyt, J.L., Management of carbon monoxide poisoning in the absence of hyperbaric oxygenation chamber, *Crit. Care Med.*, 4, 144–147, 1976.

142. Norman, J.N. and Ledingham, I. McA., Carbon monoxide poisoning: investigations and treatment, *Proc. Roy. Soc. Med.*, 56, 101–102, 1963.

143. Aurora, T., Chung, W., Dunne, R., Martin, G., Ward, K., Rivers, E., Knoblich, B., Nguyen, H.B., and Tomlanovich, M.C., Occult myocardial injury in severe carbon monoxide poisoning [abstract], *Acad. Emerg. Med.*, 6, 394, 1999.

144. Radushevich, V.P. and Koroteeva, E.L., Parallel blood circulation with oxygenation of blood in severe poisoning with carbon monoxide fumes, *Vestn. Khir.*, 116, 131–134, 1976.

145. Moore, S.J., Norris, J.C., Walsh, D.A. et al., Antidotal use of methemoglobin-forming cyanide antagonists in concurrent carbon monoxide/cyanide intoxication, *J. Pharmacol. Exp. Ther.*, 242, 70–73, 1987.

146. Kirk, M.A., Gerace, R., and Kulig, K.W., Cyanide and methemoglobin kinetics in smoke inhalation victims treated with the cyanide antidote kit, *Ann. Emerg. Med.*, 22, 1413–1418, 1993.

147. Seger, D. and Welch, L., Carbon monoxide controversies: neuropsychologic testing, mechanism of toxicity, and hyperbaric oxygen, *Ann. Emerg. Med.*, 24, 242–248, 1994.

148. Hampson, N.B., Dunfored, R.G., Kramer, C.C. et al., Selection criteria utilized for hyperbaric oxygen treatment of carbon monoxide poisoning, *J. Emerg. Med.*, 13, 227, 1995.

149. Sloan, E.P., Murphy, D.G., Hart, G., Cooper, M.A., Turnbull, T., Berreca, R.S., and Ellerson, B., Complications and protocol considerations in carbon monoxide-poisoned patients who require hyperbaric oxygen therapy: report from a ten year experience, *Ann. Emerg. Med.*, 18, 629, 1989.

150. Goulon, M., Barois, A., Rapin, M. et al., Carbon monoxide poisoning and acute anoxia due to breathing coal gas and hydrocarbons, *J. Hyperb. Med.*, 1, 23, 1986.

151. Penney, D.G., Toxicity of carbon monoxide, hemoglobin vs. histotoxic mechanisms, in Penney, D.G., Ed., *Carbon Monoxide*, CRC Press, Boca Raton, FL, 1996, chap. 6.

152. Ginsberg, M.D. and Myers, R.E., Fetal brain injury after maternal carbon monoxide intoxication, *Neurology*, 26, 15–23, 1976.

153. Ginsberg, M.D. and Myers, R.E., Fetal brain damage following maternal carbon monoxide intoxication, *Acta Obstet. Gynecol.*, 53, 309–317, 1974.

154. Koren, G., Sharav, T., Pastuszak A., Garrettson, L.K., Hill, K. Samson, I., Rorem, M., King, A., and Dolgin, J.E., A multicenter, prospective study of fetal outcome following accidental carbon monoxide poisoing in pregnancy, *Reprod. Toxiol.*, 5, 397–403, 1991.

155. Brown, D.B., Mueller, G.L., and Golich, F.C., Hyperbaric oxygen treatment for carbon monoxide poisoning in pregnancy: a case report, *Aviat. Space Environ. Med.*, 63, 1011–1014, 1992.

156. Femm, V.H., Teratogenic effects of hyperbaric oxygen, *Proc. Soc. Exp. Biol. Med.*, 116, 975–976, 1964.

157. Fujikura, T., Retrolental fibroplasia and prematurity in newborn rabbits induced by maternal hyperoxia, *Am. J. Obstet. Gynecol.*, 90, 854–858, 1964.

158. Miller, P.D., Telfored, I.D., and Haas, G.R., Effects of hyperbaric oxygen on cardiogenesis in the rat, *Biol. Neonate*, 17, 44–52, 1971.

159. Elkaharrat, D., Raphael, J.D., Korach, J.M, Jars-Guincestre, M.C., Chastang, C., Harbom, C., and Gajdos, P., Acute carbon monoxide intoxication and hyperbaric oxygen in pregnancy, *Intensive Care Med.*, 17, 289–292, 1991.

160. Farrow, J.R., Davis, G.J., Roy, T.M. et al., Acute carbon monoxide intoxication and hyperbaric oxygen in pregnancy, *Intensive Care Med.*, 17, 289–292, 1991.

161. De Pooter, M.C., Leys, D., Godefroy, O., DeReuck, J., and Peter, H., Parkinsonian syndrome caused by carbon monoxide poisoning. Preliminary results of the treatment with bromocriptine, *Rev. Neurol.*, 147, 399–403, 1991.

162. Ten Holter, J.B.M. and Schellens, R.L.L.A.M., Dantrolene sodium for treatment of carbon monoxide poisoning, *Br. Med. J.*, 296, 1772, 1988.

163. Hiramatsu, M., Sasaki, M., Nabeshima, T., and Kaemyama, T., Effects of dynorphin A (1-13) on carbon monoxide-induced delayed amnesia in mice. *Pharmacol. Biohem. Behav.*, 56, 73, 1997.

164. Hiramatsu, M., Hyodo, T., and Kameyama T., U-50488H a selective k-opioid receptor agonist, improves carbon monoxide-induced delayed amnesia in mice, *Eur. J. Pharmacol.*, 315, 119, 1996.

165. Myers, R.A., Snyder, S.K., and Emhoff, T.A., Subacute sequelae of carbon monoxide poisoning, *Ann. Emerg. Med.*, 14, 1163–1167, 1985.

166. Smith, J.S. and Brandon, S., Morbidity from acute carbon monoxide poisoning at three-year followup, *Br. Med. J.*, 1, 318–321, 1973.

167. Myers, R.A., Mitchell, J.T., and Cowley, R.A., Psychometric testing and carbon monoxide poisoning, *Disaster Med.*, 1, 279–281, 1983.

168. Myers, R.A., Snyder, S.K., Linberg, S. et al., Value of hyperbaric oxygen in suspected carbon monoxide poisoning, *J. Am. Med. Assoc.*, 246, 248, 1981.

169. Coric, V., Oren, D.A., Wolkenberg, A., and Kravitz, R.E., Carbon monoxie poisoning and treatment with hyperbaric oxygen in the subacute phase, *J. Neurol. Neurosurg. Psychiatr.*, 65, 245, 1998.

170. Maeda, Y., Kawasaki, Y., Jibiki, I., Yamaguchi, N., Matsuda, H., and Hisada, K., Effect of therapy with oxygen under high pressure on regional cerebral blood flow in the interval form of carbon monoxide poisoning: observation from subtraction of Technetium-99m HMPAO SPECT brain imaging, *Eur. Neurol.*, 31, 380–383, 1991.

171. Samuels, A.H., Vamos, M.J., and Taikato, M.R., Carbon monoxide, amnesia and hyperbaric oxygen therapy, *Aust. N. Z. J. Psychiatr.*, 26, 316, 1992.

172. Heinrichs, R.W., Relationship between neuropsychological data and response to behavioral treatment in a case of carbon monoxide toxicity and dementia, *Brain Cognition*, 14, 213–219, 1990.
173. Thom, S.R., Taber, R.L., Mendiguren, I., Clark, J.M., Hardy, K.R., and Fisher, A.B., Delayed neuropsychologic sequelae after carbon monoxide poisoning: prevention by treatment with hyperbaric oxygen, Ann. Emerg. Med., 25, 474–480, 1995.

12 Treatment of Carbon Monoxide Poisoning in France

Monique Mathieu-Nolf and Daniel Mathieu

CONTENTS

12.1 INTRODUCTION

Carbon monoxide (CO) poisoning has long been a matter of interest in France. Hartman gave the first clinical description of coal gas poisoning in 1775. In 1842, Leblanc identified CO as the toxic substance in coal gas. Claude Bernard described the hypoxic effects of CO in 1865.

Probably sporadic before 1750, the frequency of CO poisoning dramatically increased during the industrial revolution of the 19th century, because of rapid urbanization and the great use of charcoal for steam production and heating. This led to a heightened awareness on the part of both physicians and the general public about the toxicity of coal gas. Perhaps the most famous victim of CO poisoning during this period was Émile Zola, the well-known French novelist, who died in 1902.

During the late 1960s there was a decline in the coal mining industry and a changeover from the use of coal gas to natural gas for heating houses. These developments changed the pattern of CO poisoning and progressively decreased the public's and physicians' awareness of CO. Yet, despite recent efforts in prevention, public information, and medical education, CO intoxication is still a public health problem in France. It remains frequent, severe, and too often overlooked.[1]

12.2 EPIDEMIOLOGY

CO is a colorless, odorless gas produced during the combustion of carbon materials in an oxygen-deficient atmosphere. CO is ubiquitous in industrialized society and still induces large numbers of poisonings.

12.2.1 THE LILLE POISON CENTRE SURVEILLANCE SYSTEM FOR CARBON MONOXIDE POISONING

Faced with large numbers of CO poisonings in its region (northern part of France, 7.6 million inhabitants), a continuous surveillance program was set up in 1986 by the Lille Poison Centre (PC) in which every case of CO poisoning in the region involving hospitalization is recorded. This toxicovigilance program is based on a network involving the Critical Care Unit and Hyperbaric Oxygen Centre of Lille University Hospital, every emergency department and critical care unit of the regional and general hospitals, medical rescue services, and firefighters and environmental health engineers (Regional and Departmental Public Health Service).

All medical and technical records are reviewed by the PC staff and duplicate records are eliminated. Validation of poisoning and the relationship to CO is assessed according to a decision tree (Figure 12.1). It is classified as certain, probable, suspected, not related, or unknown.

For every case a systematic follow-up is done by PC medical staff by phone or visit at 1, 3, 6, and 12 months after poisoning. The same is done by an obstetrician during pregnancy and a pediatrician after the delivery in every pregnant woman. Data (certain, probable, suspected cases) are analyzed continuously and annually at a local level with regard to frequency, severity, mortality, identification of high-risk

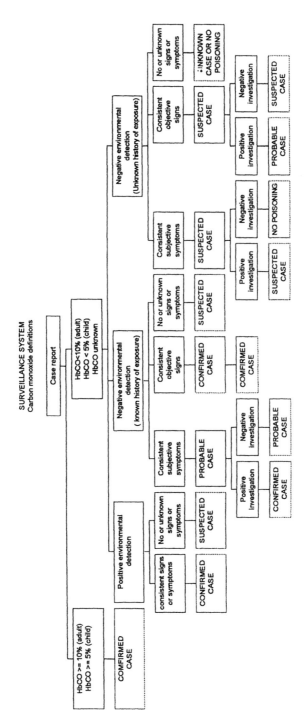

FIGURE 12.1 Decision tree for carbon monoxide poisoning definition.

circumstances or populations. If necessary, an alert is sent to the authorities (health, environment, consumers) and/or to the general public via the media.

12.2.2 DEATH FROM CARBON MONOXIDE POISONING

Until the 1970s, CO poisoning by inhalation of domestic gas was the major means of suicide in France as well as in other European countries. Now natural gas that contains no CO has replaced coal gas in household use in France. This change sharply decreased the incidence of intentional CO poisoning from 36% in 1968 to 7% in 1981.[2] A similar measure in Switzerland reportedly also dramatically reduced CO poisoning suicide there from 1971 to 1990.[3]

Despite the reduction of domestic gas toxicity, CO poisoning remains a serious health concern. In France, many other countries of Europe, and throughout the world, CO remains the single most common cause of death by poisoning.[4-7] In 1981 in France CO killed an estimated 470 people annually (estimated mortality incidence rate: 8 per million inhabitants),[8] with no decrease apparent. In 1991 there were approximately 400 deaths by CO poisoning (7 per million inhabitants; 5% of CO poisonings).[9]

Based on this continuous surveillance program, it was stated in the 1997/98 regional epidemiological report, that CO remains the single most common cause of toxic death. Among the 119 people who died from poisonings in 1998, CO was involved in 35 (29.4%). This represents a regional mortality incidence rate of 4.6 per million inhabitants. Most of the victims (29/35; 82%) did not reach a hospital alive; they were found dead at home or died before hospitalization. The fact that most people died before or without being admitted to a hospital explains why hospital mortality data do not represent a true incidence rate for CO-induced mortality. A specific survey has to be undertaken to take this phenomenon into account.

Among the 35 CO-induced deaths, 83.5% were adults and 16.5% children. Mean age was 47 years (minimum 2; maximum 85). Most of the deaths were unintentional (34/35; 97%); 21 (60%) were caused by CO produced by a malfunctioning coal or gas heating device or a gas water heater, 13 (37%) were related to fire, and only 1 (3%) was a suicide by inhalation of car exhaust. These results are similar to other studies in France, where domestic gas heaters represented 52% of the CO sources involved in deaths,[9] and in other European countries.[10]

These results bring out the important point that CO poisoning deaths are related to domestic air pollution, caused by defective stoves or water heaters, and insufficient ventilation of houses. This pattern is very different from North American data where more deaths are related to suicide (46.1%), followed by fire, (27.6%), and then unintentional poisoning, mostly related to motor-vehicle exhaust.[11] These different patterns have to be considered when interpreting clinical studies and prevention programs from both sides of the Atlantic Ocean.

12.2.3 NONFATAL CARBON MONOXIDE POISONING

Nonfatal poisonings are much more frequent than fatal poisonings. Even with a lack of precise information at the national level, it is commonly estimated that 5000 to 8000 intoxications occur each year in France. The distribution of CO intoxication may vary throughout the country, from one region to another, depending on local

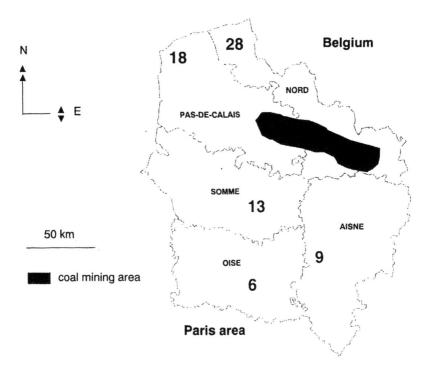

FIGURE 12.2   Carbon monoxide poisoning incidence rate in northern part of France (number of cases per 100,000 inhabitants per year).

heating or cooking habits, socioeconomic situation, and climate. For example, morbidity incidence rate per 100,000 inhabitants varies from 9 in Picardie (northwest of Paris), to 17.5 in the Paris area, to 24 in the Nord-Pas de Calais region (north of Paris, near Belgium) (Figure 12.2). This high incidence rate in the last region is probably explained on the basis of historical (coal mining industry until 1980), socioeconomic (high rate of unemployment), and climatic (fog) reasons.

In the 5-year period 1993 to 1997, the surveillance system discussed here recorded 4902 CO poisonings. Poisonings occur mostly during the cold season (75% from October to February) (Figure 12.3) and involve predominantly young families (40% of CO-poisoned people were young adults between 20 and 39 years old) in collective intoxication (1.7 patients per episode) (Figure 12.4). This explains the important proportion (23%) of CO-poisoned children (<15 years) (Table 12.1). The sex ratio (female/male) is 1.2 and a substantial fraction of the intoxicated women (159/2393; 6%) were pregnant at the time of poisoning, placing the fetus at special risk.

As stated above, most of the circumstances can be classified as "unintentional" (98.5%), related to domestic condition (89%), fire (8%), or occupation (1.5%). "Intentional" circumstances (1.5%) are far less frequent, related to attempted suicide (1.1%) or intentional fire (0.4%). Suicide attempts using automobile exhaust inhalation are rare, but are a progressively increasing cause of CO poisoning (0 in 1993; 20 in 1997).

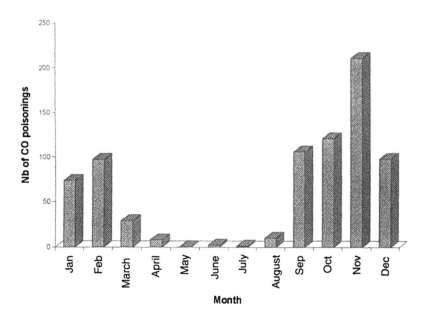

FIGURE 12.3 Numbers of carbon monoxide poisonings (note the two peaks in February and in November).

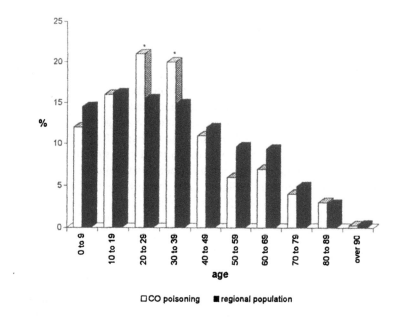

FIGURE 12.4 Age distribution of carbon monoxide poisonings compared with regional population (note the significantly higher rate in the 20- to 39-year-old range).

TABLE 12.1
Carbon Monoxide Poisoning Incidence Rate by Age Group

Age (years)	CO Incidence Rate (per 100,000 inhabitants)
0–9	24
10–19	25
20–29	32
30–39	35
40–49	22
50–59	14
60–69	15
70–79	19
80–89	22
>90	18
Whole population	24

TABLE 12.2
Origin of Carbon Monoxide Poisoning (4902 cases; 5 years: 1993 to 1997)

Unintentional	4828 (98.5%)
Domestic pollution	4365 (89%)
Coal stove	1910 (39%)
Gas water heater	1088 (22%)
Gas furnace	1076 (22%)
Kerosene space heater	195 (4%)
Other	96 (2%)
Fire	390 (8%)
Occupational	73 (1.5%)
Nonautomobile gasoline-powered	34 (0.7%)
Others (steel industry, etc.)	39 (0.8%)
Intentional	74 (1.5%)
Car exhaust	56 (1.1%)
Fire (immolation, etc.)	18 (0.4%)

Most of the CO poisonings occur in homes (98%), and domestic sources are mainly heating appliances, predominantly coal stoves, gas furnaces, and kerosene space heaters. Gas water heaters are secondary (Table 12.2).

The predominance of coal stoves as CO sources is a peculiarity of the northern region and can be explained by the local history of coal mining. In contrast, gas water heaters are the most common source in other regions (57%), where coal is much less important (9%).[9]

12.2.4 Specific Aspects of Carbon Monoxide Poisoning in the North of France

In addition to the high incidence rate of CO intoxication, the north of France is characterized by some epidemiologic specificities. These include the occurrence of sudden outbreaks where the number of CO-poisoned victims in a brief period of time may overwhelm the capacity of emergency departments and the recurrence of CO poisoning in the same victim or the family of a victim because of the lack of preventive measures.

12.2.4.1 Carbon Monoxide Poisoning Outbreaks

One of the peculiarities of the north of France is the occurrence of sudden outbreaks of CO poisonings. During such outbreaks, the number of CO-poisoned patients admitted to the hospitals is such that rescue service and emergency department capacities are overwhelmed and the whole region has to be considered a medical disaster area. Public authorities are alerted, all hospitals are mobilized, and the public is warned by messages in newspapers and on radio and TV.

During the past 10 years, the authors have experienced two such outbreaks. One occurred in September 1988, when 136 people were hospitalized in 48 h for CO poisoning. A second incident occurred in November 1993, when 365 people were hospitalized in 72 h in 26 hospitals of the region. The second outbreak also affected the south and central parts of Belgium, where more than 50 people were hospitalized during 3 days.[12] This outbreak was related to climatic conditions and involved a very typical pattern — fog and lack of wind associated with milder temperatures following a cold period. Nevertheless, these climatic conditions must be viewed as being only a revealing factor. The true source of CO was heating appliances using coal as the fuel. Analysis of the furnaces pointed out that it was an accumulation of defects that led to CO poisoning — clogging, lack of chimney sweeping, and absence of home ventilation. A mean of 3.5 defects was found associated with each CO source (Figure 12.5).

12.2.4.2 Carbon Monoxide Poisoning Recurrence

When faced with a CO-poisoned patient, preventive measures concerning the CO source are mandatory to prevent recurrence of the poisoning in the same patient or family. Prevention is usually introduced before hospital discharge by the physician talking with the patient or the patients family, followed by a visit to the home by a health engineer.

Despite of these measures, 152 patients out of 4902 (3%) experienced recurrence of CO poisoning in this region. Even though the rate of recurrences has decreased from 4.5% in 1993 to 1.9% in 1997, some people do not follow professional advice and fail to undertake preventive measures. In such cases, information is insufficient and other actions (social support, financial help, regulatory measures, and so forth) are probably needed to reduce the risk.

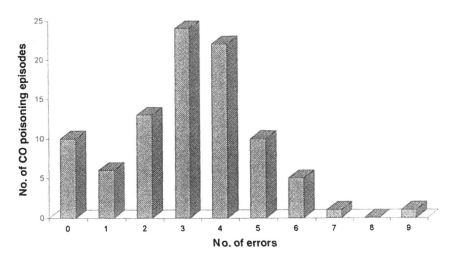

FIGURE 12.5 Number of defects potentially involved in carbon monoxide poisonings which were detected by health engineers during visits to homes.

12.3 TREATMENT

Thanks to Haldane's work,[13,14] oxygen has been long recognized as the major treatment for CO poisoning. Following Smith's[15] paper in 1962, the most commonly used treatment in France for CO poisoning is hyperbaric oxygen (HBO).[1,2,16]

12.3.1 PATHOPHYSIOLOGY OF CARBON MONOXIDE POISONING AND RATIONALE FOR HYPERBARIC OXYGEN TREATMENT

The toxic effects of CO lie in its ability to bind to heme proteins, whose functions are then blocked.

12.3.1.1 Effect on Oxygen Transport

CO crosses the alveolocapillary membrane where it binds to hemoglobin, making hemoglobin unavailable for oxygen transport. CO has about 250 times greater affinity for hemoglobin than it has for oxygen. Because of alterations in the structure of carboxyhemoglobin (COHb), the association–dissociation curve is shifted to the left. Erythrocyte 2,3-diphosphoglycerate is reduced, accentuating the left shift further.[17]

The effect of CO exposure depends on the concentration of CO in inspired air, the alveolar ventilation, and the duration of exposure.[18] Exposure to high concentrations for a short period is less harmful than exposure to a lower concentration but over a longer period of time.

The partial pressure of CO relative to oxygen also determines its clinical effects. CO uptake is inversely proportional to the partial pressure of oxygen. This explains why confinement significantly increases the severity of CO poisoning.

The binding of CO to hemoglobin leads to a nonfunctional form of hemoglobin. Consequently, arterial blood oxygen content decreases, and because the peripheral oxygen delivery in ambient air mostly relies on this form, CO poisoning induces an hypoxemic peripheral hypoxia.

12.3.1.2   Effect on Tissues

Decrease in peripheral oxygen delivery is not the sole mechanism of CO toxicity. Extravascular uptake of CO has been estimated at 10 to 50% of the total-body CO.[19] Goldbaum et al.[20] showed that, with the same blood level of COHb, dogs inhaling CO died, whereas dogs transfused with erythrocytes binding CO did not. It was concluded that the decrease in oxygen delivery induced by the interaction of CO with hemoglobin was not the most important component of CO toxicity, but that extravascular CO had a toxic action probably mediated by interfering with oxygen utilization. Yokoyama[21] questioned this finding by showing that infusion of perfluorochemical prior to CO exposure prevents CO toxicity. However, these studies are compatible with the conclusion that if tissue hypoxia is needed to allow the development of CO toxicity, it is the tissue CO that leads to clinical toxicity.

CO reacts with a number of other heme compounds besides hemoglobin: myoglobin, hydroperoxidase, cytochrome oxidase, and P-450.[22-24] Although the affinity of CO for these hemelike compounds is lower than its affinity for hemoglobin, as blood oxygen tension falls, a PO_2 level is reached at which CO avidly binds to myoglobin and cytochrome oxidase.

12.3.1.2.1   Carbon monoxide binding to myoglobin

Myogolobin is an O_2 carrier protein that acts to facilitate oxygen diffusion in skeletal and cardiac muscle cells. The binding of CO to myoglobin leads to a nonfunctional form of myoglobin, the carboxymyoglobin (MbCO), with a ratio MbCO/COHb of approximately 1.[24] The decrease in facilitated oxygen diffusion in muscle cells combined with the decrease in oxygen delivery to the muscles may play a role in the limitation of maximal oxygen consumption (VO_2 max.),[25,26] and in the decrease in cardiac output that appears in patients even with a mild CO poisoning.[27,28]

12.3.1.2.2   Effect on carbon monoxide binding to cytochrome a_3

Cytochrome a,a_3 is the terminal member of intramitochondrial electron transport chain. It catalyzes the reduction of molecular dioxygen to water in a four-electron process. The enzyme complex accounts for perhaps 90% of the total O_2 uptake of the body. Inhibition of cytochrome a,a_3 by CO binding blocks the flow of electrons from substrate to O_2, which normally provides the source of conservation energy by oxidative phosphorylation.[29,30]

The binding of CO to cytochrome a_3 is a well-known biochemical fact. Nevertheless, an argument often advanced to deny the role of the CO inhibition of cytochrome a_3 in clinical intoxication is that intracerebral PO_2 does not decrease enough during CO poisoning to allow CO to bind to cytochrome a_3. However, as COHb increases, oxygen content in jugular venous blood decreases to a very low level where cerebral blood flow becomes heterogeneous. Thus, depending on local O_2 demand, blood flow, and capillary density, areas of severe hypoxia may exist, allowing CO to bind heavily to cytochrome a_3.[31] Moreover, such a failure to adapt microcirculatory O_2 delivery to local O_2 demand has also been reported to be a potential direct toxic effect of CO on vascular smooth muscle,[32] which could contribute to observed regional differences in CO sensitivity in the central nervous system by enhancing intracellular CO binding in the most hypoxic areas.

Last, direct evidence of CO binding to cytochrome a_3 has been shown by Brown and Piantadosi,[33] who demonstrated, using *in vivo* reflectance spectrophotometry, that as COHb level rises and hypotension occurs, cytochrome a_3 inhibition is detected. This is followed by a decrease in intracellular energetic compounds and pH.[34] The impaired mitochondrial energy level induced by CO leads to neuronal depolarization, excitatory amino acid (glutamate) and catecholamine release, and failure of reuptake. These processes, normally modulated by nitric oxide production, could contribute to degeneration of neurons in vulnerable regions, possibly by enhancing mitochondrial reactive O_2 species generation that initiate apoptosis.[35]

In summary, in addition to hypoxemic hypoxia, CO poisoning induces a histotoxic hypoxia, and this process is self-worsening. This is in good agreement with the clinical experience.

12.3.1.3 Effect of Reoxygenation

12.3.1.3.1 Dissociation of carbon monoxide hemoproteins
Dissociation of CO–hemoprotein complexes follows the law of mass action. Thus, dissociation rate depends only on the proportion of O_2, CO, and proteins, and their relative affinity. Thus, removal of a patient from a toxic atmosphere, eliminating CO pressure and giving O_2 represent two fundamental therapeutic measures.

COHb dissociation begins as soon as the patient is removed from the CO atmosphere. It follows an exponential law with a half-life of 230 to 320 min in room air. Dissociation rate is much increased when O_2 pressure is raised, with a half-life of 90 min in pure normobaric O_2, 35 min in 2 atm abs pure O_2, and 22 min in 3 atm abs pure O_2.[36]

Dissociation rates of other CO–hemoprotein complexes are less well known, but are much slower because they rely on the amount of O_2 delivered to tissues, which depends itself on the remaining COHb level. Thus, dissociation of other CO–hemoprotein complexes may only occur when tissue oxygen pressure normalizes, which means that COHb dissociation has to be far advanced.

12.3.3.1.2 Evidence for reoxygenation injury in carbon monoxide poisoning
The fact that cytochrome a_3 may remain inhibited when a patient is reoxygenated leads to consideration of the possibility of free oxygen radical formation and then occurrence of reoxygenation injury.

It has long been recognized that, with regard to pathological aspects of brain lesions, similarities are striking between CO poisoning and postischemic reperfusion injury; a common mechanism has been hypothesized for these different forms of brain injury.[37–39]

Recently, Thom[40] provided evidence for the occurrence of lipid peroxidation in the brain of CO-poisoned rats. An increase of conjugated diene and malonyldialde- . hyde concentrations appears only after a 90-min period of normal air breathing following CO exposure. He was able to demonstrate that blocking xanthine oxidase by allopurinol or depleting animals of this enzyme by feeding them a tungsten-supplemented diet decreases the magnitude of brain lipid peroxidation.[41] This offers further evidence that at least in part a common mechanism exists between CO poisoning and reperfusion injury.

Brown and Piantadosi[42] offered further evidence for oxygen free-radical generation during the reoxygenation phase after CO exposure. They showed a decrease in brain catalase activity that demonstrates hydrogen peroxide production and a decrease in the ratio of reduced to oxidized glutathione, with an increase in salicylate hydroxylation products that demonstrate hydroxyl radical production.[42] Moreover, they were able to demonstrate that decreases in intracellular pH and energetic compounds were rapidly corrected by HBO as opposed to normobaric O_2 where it continues to decrease during the first 45 min.

In summary, it can be said that during CO exposure, cerebral O_2 pressure may decrease to a level allowing CO to bind to cytochrome a_3, which leads to a decrease in intracellular pH and energetic compounds. Exposure to an inspired O_2 pressure not sufficient to induce an increase in tissue O_2 pressure delays CO–cytochrome a_3 dissociation and may promote oxygen free-radical generation and tissue injury. On the other hand, exposure to a sufficiently high inspired O_2 pressure to induce a tissue O_2 pressure increase, such as in HBO, allows immediate CO–cytochrome a_3 dissociation and normalization of mitochondrial oxygen metabolism.

Thom[43] offered another explanation for the role of HBO in preventing reoxygenation injury in CO poisoning. He showed that under HBO there was a dramatic decrease in leukocyte adhesion to endothelial cells when compared with reoxygenation with normal O_2 pressure. This decrease in leukocyte adhesion was due to a decrease in beta-2 integrin expression by leukocytes allowing leukocytes to move away from endothelial cells, thus decreasing vascular wall injury induced by biochemical agents released by activated leukocytes.[44]

Thus, it may be possible that the beneficial effect of HBO in preventing reoxygenation injury in CO poisoning is due to prevention of microvascular injury more than to cytochrome a_3 reactivation. Further studies are needed to confirm this point.

12.3.2 MANAGEMENT OF CARBON MONOXIDE POISONING

Management of CO-poisoned patients consists first of removing the patient from toxic exposure and supplying pure O_2. Respiratory and circulatory conditions should be assessed rapidly and resuscitative measures initiated if needed.

Evaluation includes assessment of neurological status with consciousness level, motor responses and reflexes, and a complete physical examination looking for complications, associated trauma or intoxication, and previous disease.

Laboratory exams should include a blood gas analysis to look for acidosis and a COHb measurement. A COHb level over 5% in a nonsmoker and over 10% in a smoker confirms the diagnosis.[45]

Unfortunately, CO poisoning is difficult to recognize. Faced with a patient with headache and dizziness, questions concerning the heating apparatus and cohabitants with similar symptoms can identify those cases with occult CO poisoning. If the patient is comatose, circumstances of discovery are often suggestive. Measurement of CO concentration in the room air is of major interest for the diagnosis. In situations with a large number of potentially intoxicated victims, such as in building fires, measurement of CO concentration in expired air may be a good triage test.[46]

Treatment consists of removing the patient from exposure to CO and supplying pure O_2 to accelerate the elimination of CO and improve tissue oxygenation. At the accident scene the administration of O_2 can best be done by using a high O_2 flow rate with a tight-fitting facial mask. Oxygen has to be given as soon as possible. Endotracheal intubation and mechanical ventilation are needed if there is any risk regarding permeability of the airways or adequacy of the ventilation.

Oxygen administration has to be continued in the hospital. HBO provides O_2 at a pressure greater than 1 atm and increases the COHb dissociation rate. Moreover, HBO immediately improves tissue oxygenation, increases the dissociation of the other CO–hemoprotein complexes, and decreases cerebral edema.[47]

The indications for HBO are still a controversial point but, in the authors' view, it appears to be useful. Every patient remaining comatose, showing any neurological abnormality, or with a history of loss of consciousness should be treated with HBO. The COHb level is not, in itself, a good criterion on which to rely. Myers et al.[48] have proposed to improve the neurological examination by using psychometric tests to detect subtle psychic perturbations.

After each HBO session, a new clinical examination should be done to detect residual manifestations. In those cases where there are persisting neurological abnormalities, a new HBO session is done with a 6-h interval. The number of repeated HBO sessions is still unsettled, but it seems of little use to give more than five HBO sessions.

In the case of normobaric O_2 treatment, length of O_2 administration is also controversial, but it appears that it must be long enough to ensure total CO detoxication. Proposed durations are often between 12 and 48 h.

12.3.3 Specific Problems

12.3.3.1 Carbon Monoxide Poisoning and Pulmonary Edema

Pulmonary edema is a fairly common feature of CO poisoning. The appearance of pulmonary edema aggravates tissue hypoxia in adding a respiratory hypoxia to the CO-induced oxygen delivery lack. It makes O_2 administration less active because of the intrapulmonary arteriovenous shunting and decreased rate of CO elimination.

The mechanism of pulmonary edema is controversial. In a personal series of 1850 CO-poisoning patients, 120 had pulmonary edema. Two groups may be distinguished. The first group consisted of 92 cases where pulmonary edema was of cardiac origin. In this group, patients were older and previous cardiac diseases were more often present. The important point is that this type of pulmonary edema improves during HBO. This correlates well with the echocardiographic and angioscintigraphic studies, which show that left ventricle ejection fraction is decreased significantly in CO poisoning as soon as the COHb level reaches 30%.[27,28] The second group includes 28 cases, younger than the first group. Pulmonary edema is recognized as being of noncardiac origin, but in 20 of these patients vomiting has been reported and aspiration was the most probable cause of pulmonary edema.

12.3.3.2   Late Neurological Sequelae

Smith and Brandon[49] published a study in 1973 where they found 10% of patients had immediate gross neurological sequelae; 33% had delayed personality deterioration and 43% had memory disturbances. In a literature review, Ginsberg and Romano[50] found between 15 and 40% of patients developed late neurological sequelae.

The explanation may lie in misdiagnosis (30% in a French Poison Control Center study),[51] inadequate therapy (40% of Smith and Brandon's patients received no O_2 in emergency treatment),[49] or delayed therapy. Complete recovery is obtained in more cases if HBO is applied in less than 6 h.[16]

In their unit, the authors undertook a study of 774 patients, who were divided into five groups: group 0 were patients who suffered only from headache or nausea; group I were patients with an abnormality on neurological examination; group II were patients who had lost consciousness regardless of their clinical state on admission; group III were patients who were comatose (Glasgow coma scale over 6); and group IV were patients who were deeply comatose. Group 0 patients received only normobaric pure O_2 while groups I, II, III, and IV received HBO.

At 1 year, only 4.4% of the patients suffered from persistent manifestations, and only 1.6% had major functional impairments. However, persistent neurological manifestations occurred only in group I through IV patients. Thus, the authors advocate the use of HBO in every CO-poisoned patient who had suffered loss of consciousness during CO exposure, or who has a neurological abnormality upon clinical examination, or who remains comatose upon admission.[52]

These results are in accordance with those obtained by others.[47,48] A French study has added some controversy about the use of HBO in minor forms of CO poisoning.[53] This study has been heavily criticized, and two multicentered randomized studies are in development, one in the United States and one in France. An interim analysis of the latter study has shown a significant improvement at 3 and 6 months in the rate of persistent neurological manifestations in the group receiving HBO as compared with the group receiving 12 h of normobaric O_2. This study continues to allow more precise definition of the criteria for HBO referral.[54]

12.3.3.3   Carbon Monoxide Poisoning and Pregnancy

Fetal death frequently occurs in CO poisoning. Fetal and maternal intoxications differ because the placenta delays fetal intoxication. However, the placenta also delays fetal detoxification.[55] Fetal hemoglobin has more affinity for CO than adult hemoglobin; thus CO dissociation is decreased. Fetal hypoxia is more pronounced than maternal tissue hypoxia. All these facts have important clinical implications. The severity of fetal intoxication cannot be assessed on the maternal state. A difference exists between the evolution of the COHb level in the mother and in the fetus with an important delay in CO elimination.

During a 7-year study period, 90 pregnant women who had been referred to the authors' center for CO poisoning were enrolled. Follow-up data were obtained from 86 of the women. When compared with the general CO-poisoned nonpregnant female population, no difference could be observed in CO sources, clinical severity, COHb, or plasma bicarbonate. When compared with the matched group, short-term complications were more frequent in the pregnant women, but long-term outcome did not differ.

A total of 77 women (89.5%) had a successful pregnancy. CO intoxication led to fetal death (a fourfold increase in relative risk) in five cases. However, neither the prematurity rate, the rate of fetal hypotrophy, nor the malformation rate was increased when compared with the general population.

Thus, CO intoxication during pregnancy induces an increase in short-term maternal complication rate and in fetal death rate. However, in this population where every pregnant woman has been treated with HBO, the long-term outcome of both the mother and the child is not different from that of a control population. Thus, the recommendation is to use HBO in each CO-poisoned pregnant woman whatever her clinical state.

In summary, HBO is the most commonly used treatment modality for CO poisoning and its widely accepted indications in France are as follows:

1. Comatose patient;
2. Patient who loses consciousness during exposure;
3. Patient with abnormal objective neuropsychological manifestations;
4. A pregnant woman.

12.4   PREVENTION

The high incidence of CO poisoning in France and its consequences in mortality, morbidity, and socioeconomic cost has led health authorities to implement prevention programs at the national level. As for other occupational or domestic hazards, prevention is implemented through both technical measures and public information.

Technical measures involve, first, safety standards for space and water heating devices. These standards are actually established at the European Union (EU) level and they regulate every device marketed in the EU Recommendations concerning installation and maintenance remain the responsibility of the national level but, except for professionals, these recommendations are only incentives and no enforcement measures exist. The public is informed through national campaigns on TV or in the news media, but results are often disappointing.

Considering the high incidence rate of CO poisoning in the north of France, the authors adopted a systemic approach to act on the multiple factors involved in the occurrence of CO poisoning. Objectives and programs are decided on the basis of information obtained from a surveillance system, which is also used to monitor program implementation and to evaluate results. This approach considers each of the three levels recognized by the World Health Organization.

Primary prevention consists of preventing the occurrence of poisoning. This is accomplished through information campaigns directed to the general public using multiple media: newspaper, radio, TV, leaflets, posters, Web sites, and so forth. Considering the seasonal occurrence of CO poisoning, these campaigns take place twice a year, in the autumn and in February. They are launched when the surveillance system detects a sudden increase in the number of CO-poisoned people. Cycling prevention campaigns in accord with the occurrence of numerous CO poisonings obtains a better compliance by the media for the prevention actions and a better efficacy may be expected.

In addition to this action directed toward the general public, certain prevention programs are focused more specifically on populations at risk, for example, students, who form a group at high risk of CO poisoning as pointed out by the surveillance system. In such cases, every person or organization able to act as a relay for prevention is involved and is used to pass information especially designed for the target population (for students: universities, student associations, sport clubs).

Technical measures are also part of primary prevention. The surveillance system is used to detect the occurrence of new circumstances or a sudden change in frequency. An illustrative example is the action directed to CO poisoning caused by kerosene space heaters. Although space heaters using kerosene as fuel are a well-known source of CO poisoning in the United States, this means of heating was nearly unknown in the north of France and began to be used in 1987/88. The surveillance system detected the occurrence and the rapid increase of cases due to kerosene space heaters (0 in 1987, 2 in 1988, 14 in 1989) and health authorities and the Consumer Council were alerted (1989). Technical preventive measures were taken and manufacturers were obliged to equip those apparatus with CO detectors and other safety devices. During the next 4 years, only one case of CO poisoning due to a kerosene space heater was recorded, but in 1994 six new cases occurred and this number increases progressively each year. Health authorities were again alerted and new technical measures are currently under investigation. This points out the role of a surveillance system to monitor the toxic events and to evaluate preventive measures.

Secondary prevention consists of preventing complications and sequelae of poisoning. This is done through initial training and continuous education of both physicians and paramedics. Actions have been taken to include information on CO poisoning in the syllabi of medical students, nurses, firefighters, as well as in continuous medical education in emergency medicine, neurology, pneumology, and family practice programs. The main points are to detect CO poisoning when a patient complains only of unremarkable signs, to identify high-risk members of the population (pregnant women, children, aged people), and to improve patient management (oxygen, transportation, indications for HBO).

Tertiary prevention consists of preventing recurrence. CO intoxication recurrence may frequently happen if no preventive measures are taken. Faced with a high recurrence rate, a program was implemented in 1993 in emergency departments and in the hyperbaric center of the Lille University hospital. Just before patient discharge, information on the origin of CO poisoning and preventive measures are given to patients during an interview with the medical staff. A visit to the patient's home by a health engineer is proposed in order to help patients identify CO sources and important factors. Since the implementation of this program, the surveillance system has recorded a progressively decreasing recurrence rate: 4.5% in 1993, 3.2% in 1994, 2.8% in 1995, 1.8% in 1996. This decreasing trend is encouraging; however, efforts must be maintained.

12.5 CONCLUSIONS

CO poisoning remains a serious public health problem in France. Even if its therapeutic management is actually well known and applied, too many cases remain overlooked, and this leads to severe consequences in mortality, morbidity, and socioeconomic costs.

Prevention must be the cornerstone of public health actions and the implementation of a surveillance system is, in the authors' view, the key factor for prevention program efficacy.

REFERENCES

1. Mathieu, D., Mathieu-Nolf, M., and Wattel, F., Intoxication par le monoxyde de carbone: aspects actuels, *Bull. Acad. Natl. Med.*, 180, 965, 1996.
2. Larcan, A. and Lambert, H., Aspects epidemiologiques, clinico-biologiques et therapeutiques actuels de l'intoxication oxycarbonee aigue, *Bull. Acad. Natl. Med.*, 165, 471, 1981.
3. La Harpe, R., Selbsttotungen in kanton Genf (1971–1990). Eine Analyse des rechtsmedizinischen Sektionsgutes, *Arch. Kriminol.*, 195, 65, 1995.
4. Faure, J., Arsac, P., and Chalandre, P., Prospective study in carbon monoxide intoxication by gas water heaters, *Hum. Toxicol.*, 2, 422, 1983.
5. Cobb, N. and Etziel, R., Unintentional carbon monoxide-related deaths in the United States, 1979 through 1988, *J. Am. Med. Assoc.*, 266, 659, 1991.
6. Investigation Committee of the Medico-Legal Society of Japan, Reports on medico-legal data from the massive investigation performed by the Medico-Legal Society of Japan, A statistical study of death by poisoning, *Nippon Hoigaku Zasshi*, 45, 258, 1991.
7. Dukes, P., Robinson, G., Thompson, K., and Robinson, B., Wellington coroner autopsy cases 1970–89: acute deaths due to drug, alcohol and poisons, *N. Z. Med. J.*, 105, 25, 1992.
8. Anonymous, Les intoxications oxycarbonées aigues, *Bull. Epidèmiologique Hebdomadaire*, 18, 1984.
9. Gajdos, P., Conso, F., Korach, J., Chevret, S., Raphael, J., Pasteyer, J., Elkharrat, D., Lanata, E., Geronimi, J., and Chastang, C., Incidence and causes of carbon monoxide poisoning. Results of an epidemiologic survey in a French department, *Arch. Environ. Health*, 46, 373, 1991.
10. Risser, D. and Schneider, B., Carbon monoxide-related deaths from 1984 to 1993 in Vienna, Austria, *J. Forensic Med.*, 40, 368, 1995.
11. Hampson, N., Incidence of carbon monoxide poisoning in the United States, presented at II Corso di Aggiornamento in Tossicologia Clinica, Milano, March 12, 1999.

12. Guermonpre, P., Jeuneau, C., and Van Renterghem, A., Carbon monoxide intoxication. A three year experience of the first multiplace hyperbaric chamber in Belgium, in *Proceedings of the XXIst Annual Meeting of European Underwater and Baromedical Society*, Sipinen, S. and Leinio, M., Eds., Finnish Society of Diving and Hyperbaric Medicine, Helsinki, 1995, 90.

13. Haldane, J., The relation of the action of carbonic oxide to oxygen tension, *J. Physiol.*, 18, 201, 1895.

14. Haldane, J., Carbon monoxide as a tissue poison, *Biochem. J.*, 21, 1068, 1927.

15. Smith, G., The treatment of carbon monoxide poisoning with oxygen at two atmosphere absolute, *Ann. Occup. Hyg.*, 5, 259, 1962.

16. Barois, A., Grosbuis, S., and Goulon, M., Les intoxications aigues par l'oxyde de carbone et les gaz de chauffage, *Rev. Prat.*, 29, 1211, 1979.

17. Roughton, F. and Darling, R., The effect of carbon monoxide on the oxyhemoglobin dissociation curve, *Am. J. Physiol.*, 141, 17, 1944.

18. Pace, N., Consolazio, W., White, W., Formulation of the principal factors affecting the rate of uptake of carbon monoxide by normal man, *Am. J. Physiol.*, 174, 352, 1946.

19. Luomanmaki, K. and Coburn, R., Effects of metabolism and distribution of carbon monoxide on blood and body stores, *Am. J. Physiol.*, 217, 354, 1969.

20. Goldbaum, L., Orenallo, T., and Dergal, E., Mechanism of the toxic action of carbon monoxide, *Ann. Clin. Lab. Sci.*, 6, 372, 1976.

21. Yokoyama, K., Effect of perfluorochemical emulsion in acute carbon monoxide poisoning in rats, *Jpn. J. Surg.*, 4, 342, 1978.

22. Coburn, R. and Mayers, L., Myoglobin O_2 tension determined from measurements of carbomyoglobin in skeletal muscle, *Am. J. Physiol.*, 220, 66, 1971.

23. Keilin, D. and Hartree, E., Cytochrome and cytochrome oxidase, *Proc. R. Soc. London Ser B*, 127, 167, 1939.

24. Coburn, R. and Forman, H., Carbon monoxide toxicity, in *Handbook of Physiology of the Respiratory System*, Vol. 4: *Gas Exchange*, Fishman, A., Ed., American Physiological Society, Bethesda, 1987, 439.

25. King, C., Dodd, S., and Cain, S., O_2 delivery to contracting muscle during hypoxic or CO hypoxia, *J. Appl. Physiol.*, 63, 726, 1987.

26. Hogan, M., Bebout, D., Gray, P., Wagner, J., West, J., and Haab, P., Muscle maximal O_2 uptake at constant O_2 delivery with and without CO in the blood, *J. Appl. Physiol.*, 69, 830, 1990.

27. Corya, B., Black, M., and McHenry, P., Echocardiographic findings after acute carbon monoxide poisoning, *Br. Heart J.*, 38, 712, 1976.

28. Elkharrat, D., Raphael, J.C., Tainturier, C., Brunel, D., De Truchis, P., and Goulon, M., Angioscintigraphie cardiaque au cours de l'intoxication argue par l'oxyde de carbone, *Rean. Soins Intens. Med. Urg.*, 2, 61, 1986.

29. Piantadosi, C., Carbon monoxide, oxygen transport and oxygen metabolism, *J. Hyperb. Med.*, 2, 27, 1987.

30. Sylvia, A., Piantadosi, C., and Jobsis-Vandervliet, F., Energy metabolism and *in vivo* cytochrome a oxidase redox relationships in hypoxic rat brain, *Neurol. Res.*, 7, 81, 1985.

31. Brown, S. and Piantadosi, C., Reversal of carbon monoxide–cytochrome c oxidase binding by hyperbaric oxygen *in vivo*, *Adv. Exp. Med. Biol.*, 248, 747, 1989.

32. Graser, T., Vedernikiv, Y., and Li, D., Study on the mechanism of carbon monoxide induced endothelium-independent relaxation in porcine coronary artery and vein, *Biochem. Acta*, 4, 293, 1990.

33. Brown, S. and Piantadosi, C., *In vivo* binding of carbon monoxide to cytochrome c oxidase in rat brain, *J. Appl. Physiol.*, 62, 604, 1990.
34. Zhang, J. and Piantadosi, C., Mitochondrial oxidative stress after carbon monoxide hypoxia in rat brain, *J. Clin. Invest.*, 90, 1193, 1992.
35. Piantadosi, C.A., Zhang, J., Levin, E.D. et al., Apoptoris and delayed neuronal damage after carbon monoxide poisoning in the rat, *Exp. Neurol.*, 147, 103, 1997.
36. Pace, N., Strajman, E., and Walker, E.L., Acceleration of carbon monoxide elimination in man by high pressure oxygen, *Science*, 11, 652, 1950.
37. Ginsberg, M., Hedeley White, E., and Richardson, E., Hypoxic-ischemic leukoencephalopathy in man, *Arch. Neurol.*, 33, 5, 1976.
38. Okeda, R., Funata, N., Takano, T., Miyazaki, Y., Yokoyama, K., and Manabe, M., The pathogenesis of carbon monoxide encephalopathy in the acute phase physiological morphological correlation, *Acta Neuropathol.*, 54, 1, 1981.
39. Okeda, R., Funata, N., Song, S.J., Higashino, F., Takano, T., and Yokoyama, K., Comparative study on pathogenesis of selective cerebral lesions in carbon monoxide poisoning and nitrogen hypoxia in cats, *Acta Neuropathol.*, 56, 265, 1982.
40. Thom, S., Carbon monoxide-mediated brain lipid peroxidation in the rat, *J. Appl. Physiol.*, 68, 997, 1990.
41. Thom, S., Dehydrogenase conversion to oxidase and lipid peroxidation in brain after carbon monoxide poisoning, *J. Appl. Physiol.*, 73, 1584, 1992.
42. Brown, S. and Piantadosi, C., Recovery of energy metabolism in rat brain after carbon monoxide hypoxia, *J. Clin. Invest.*, 89, 666, 1992.
43. Thom, S., Leucocytes in carbon monoxide mediated brain oxidative injury, *Toxicol. Appl. Pharmacol.*, 123, 234, 1993.
44. Thom, S., Functional inhibition of leukocyte B_2 integrins by hyperbaric oxygen in carbon monoxide–mediated brain injury in rats, *Toxicol. Appl. Pharmacol.*, 123, 248, 1993.
45. Ilano, A. and Raffin, T., Management of carbon monoxide poisoning, *Chest*, 97, 165, 1990.
46. Mathieu, D., Mathieu-Nolf, M., Bocquillon, N., Linke, J.C., and Wattel, F., Accuracy of carbon monoxide measurement in expired air for the diagnosis of CO poisoning, *Clin. Toxicol.*, (in press)
47. Ducasse, J.L., Izard, P., and Celcis, P., Intoxication modérée l'oxyde de carbone: oxygenotherapie normobare ou hyperbare. Etude randomisée avec mesure du debit sanguin cerebral, *Rean. Soins Intens. Med. Urg.*, 4, 364, 1988.
48. Myers, R., Snyder, S., Linberg, S., and Cowley, A., Value of hyperbaric oxygen in suspected carbon monoxide poisoning, *J. Am. Med. Assoc.*, 246, 2478, 1981.
49. Smith, J. and Brandon, S., Morbidity from acute carbon monoxide poisoning at three years follow-up, *Br. Med. J.*, 1, 318, 1973.
50. Ginsberg, R. and Romano, J., Carbon monoxide encephalopathy. Need for appropriate treatment, *Am. J. Psychol.*, 133, 317, 1976.
51. Barret, L., Danel, V., and Faur, J., Carbon monoxide poisoning, a diagnosis frequently overlooked, *Clin. Toxicol.*, 23, 309, 1985.
52. Mathieu, D., Nolf, M., Durocher, A., Saulnier, F., Frimat, P., Furon, D., and Wattel, F., Acute carbon monoxide intoxication. Treatment by hyperbaric oxygen and risk of late sequellae, *Clin. Toxicol.*, 2, 315, 1985.
53. Raphael, J.C., Elkharrat, D., Jars-Guincestre, M.C., Chastang, C., Chasles, V., Vercken, J.B., and Gajdos, Ph., Trial of normobaric and hyperbaric oxygen for acute carbon monoxide intoxication, *Lancet*, 2, 414, 1989.

54. Mathieu, D., Wattel, F., Mathieu-Nolf, M., Durak, C., Tempe, J.P., Bouachour, G., and Sainty, J.M., Randomized prospective study comparing the effects of HBO versus 12 hours NBO in non comatose CO poisoned patients: result of the interim analysis, *Undersea Hyperb. Med.*, 23(5), 7, 1996.

55. Longo, L., The biological effects of carbon monoxide on pregnant woman, fetus and newborn infant, *Am. J. Obstet. Gynecol.*, 129, 69, 1977.

13 Acute Carbon Monoxide Poisonings in Poland — Research and Clinical Experience

Jerzy A. Sokal and Janusz Pach

CONTENTS

13.1 ACUTE CARBON MONOXIDE POISONINGS IN POLAND

Acute carbon monoxide (CO) poisonings are still an epidemiological problem in Poland due to the frequency of occurrence and the relatively high fatality rate.

The system of treatment of acute poisonings in Poland that was organized in 1967 is still operating today. Following the regulation of the Minister of Health, nine regional acute poisoning centers were set up, which in addition to treatment also offer toxicological information. The treatment of poisoning is supervised by the national specialist in clinical toxicology appointed by the Minister of Health. Information on acute poisonings is collected at the National Poison Information Center at the Institute of Occupational Medicine in Lodz. The center does not collect data on poisoning cases treated in other hospitals or fatal cases among unhospitalized patients. Regional acute poisoning centers treat 7000 to 11,000 cases annually.

TABLE 13.1
Acute Carbon Monoxide Poisoning in Poland[a]

Year	No. of Individual CO Poisoning Cases	No. of Fatalities Due to CO Poisoning	Fatality Rate
1980	632	19	3.0
1981	476	25	5.3
1982	463	22	4.8
1983	442	32	7.2
1984	420	21	5.0
1985	435	21	4.8
1986	322	34	10.6
1987	431	14	3.25
1988	450	24	5.3
1989	820	19	2.3
1990	817	19	2.3
1991	678	14	2.0
1997	664	4	0.6
1998	523	2	0.4

[a] The data are for patients hospitalized at regional acute poisoning centers.

Suicide attempts are the most frequent cause of poisonings, followed by accidental poisonings, drug overdosages at home, chemical and fire accidents, and poisonings from other causes.

CO poisonings make up about 10% of all poisoning cases and, considering the number of patients, are the third most frequent cause of poisoning, following drugs (over 50%) and alcohol (10 to 25%). The most frequent causes of CO poisonings include domestic gas installations, coal stoves, and fires.[1,2]

Table 13.1 shows the number of patients hospitalized in acute poisoning centers in the years 1980 to 1997 because of CO poisoning. The numbers do not cover patients hospitalized in other hospitals or unhospitalized fatal cases.

The number of patients treated in the toxicological centers for CO poisoning has not changed significantly since the beginning of the 1980s, while the fatality rate has decreased. This figure increases considerably, however, if fatal cases at the sites of accidents are taken into account.

The most recent data for the population of the city of Krakow[3] indicate that in 1997 there were 217 hospitalized cases of CO poisoning. Two patients died in the hospital (fatality rate 0.92%), while 23 individuals died at the site of accident, which increases the fatality rate to 10.4%.

13.2   RESEARCH ON ACUTE CARBON MONOXIDE POISONING IN POLAND

Research on acute CO poisoning in Poland during the past 20 to 30 years has been concentrated mainly in scientific institutions operating acute poisoning clinical centers.

TABLE 13.2

Relation Between Duration of Exposure and the Severity of Carbon Monoxide Poisoning

Exposure Time, h	No. of Patients	COHb (%)	Severe and Very Severe Poisonings (%)[a]	Poisonings with Complications (%)
<8	25	29 ± 13	12	8
≥8	14	30 ± 11	64	36

[a] Criteria of the severity of CO poisoning by Sokal and Kralkowska.[6]

Source: Sokal, J.A., *Stud. Mat. Monogr.*, IMP Lodz, 26, 47–56, 1987. With permission.

The centers located in Krakow at the Collegium Medicum of the Jagiellonian University and in Lodz at the Nofer's Institute of Occupational Medicine have been the most active. The Krakow center has focused mainly on clinical studies, whereas the Lodz center has been involved both in experimental and clinical studies. A few other research centers have also been involved in CO research in Poland. Basic metabolic and morphologic research on CO intoxications and the other hypoxemic and ischemic conditions were extensively performed in the 1970s at the Medical Research Center of the Polish Academy of Sciences.

This section focuses on the results of studies on the pathology and mechanisms of acute CO poisoning.

13.2.1 DURATION OF CARBON MONOXIDE EXPOSURE AND THE SEVERITY OF POISONING

Clinical observations show that the severity of acute CO poisoning does not correlate well with blood carboxyhemoglobin (COHb) levels, while the duration of exposure is an important factor affecting the degree of intoxication.[4–6]

Observations illustrating the extent of the problem are presented in Table 13.2. It has been shown in patients with similar COHb levels (mean 29 to 30%) and transportation times (poisoning site to clinic) that the percentage of severe and very severe poisonings with complications is four to five times higher in those with exposure times exceeding 8 h, as compared with those with exposure times shorter than 8 h.

An explanation for this phenomenon is important both for the diagnosis and the treatment of CO poisoning. In experimental studies in rats and in clinical studies two hypotheses have been tested, assuming certain facts about toxicokinetics and toxicodynamics:

1. CO accumulates in the organism during long-term acute exposure.
2. Long-term exposure results in more extensive and prolonged biological effects, related to the hypoxemic action of CO.

The results of studies have not shown CO accumulation in the organism under long-term acute exposure. CO concentration in the extravascular compartment of skeletal and heart muscles depends on blood COHb level, but not on the exposure duration.[8] These results provide direct evidence for the lack of CO accumulation in the extravascular compartment of heart and skeletal muscles during long-term acute exposure to CO.

The time course of CO elimination from the blood and brain of rats and from the blood of patients poisoned with CO after the termination of exposure has been found not to be dependent on the duration of exposure.[9,10] Although extravascular CO could not be determined in brain tissue, a similar time course of CO elimination from brain and blood during restitution suggests that there is no significant retention of CO in the brain and that this process is independent of exposure conditions.

However, CO accumulation in the tissues is still considered by some authors as an important factor in the severity of CO poisoning after prolonged exposure,[11] despite the lack of evidence of accumulation, and despite evidence to the contrary.

On the other hand, there is adequate evidence of substantial differences in the biological effect of short vs. long single exposures to CO at the same COHb level.

In the authors' comparative studies of CO intoxication in rats, three different durations of exposure and CO concentration were used: 4 min at about 1% CO, 40 min at about 0.5% CO, or 12 h at about 0.15% CO. Blood COHb at the end of the 4-min or 40-min exposures was usually 60 to 70%, and after 12 h of exposure, 50 to 60%. After the 4-min and 40-min exposures, rats were unconscious and did not survive higher CO concentrations. The rats exposed for 12 h were in a better clinical state, but they could not survive even a slight increase in CO concentration in the chamber.

In the extrapolation of the biochemical effects of short- and long-duration CO exposures to CO poisonings in humans, models of 4-min and 40-min exposure seem to be more appropriate than those of 40-min and many hours exposure. CO intoxication that develops in human from long exposure is associated with extensive metabolic acidosis, which is not the case after 12 h of exposure in rats. The increase in blood lactate after 4 or 40 min of CO exposure in rats is comparable with that observed in CO poisonings in humans during short (≤ 1.5 h) or long (10 to 14 h) exposure. This phenomenon is most probably related to the much slower CO uptake and carbohydrate metabolism in humans as compared with rats.

Disturbances of motor coordination in rats have been observed for much longer periods of time after prolonged exposure at the same blood COHb level[12] (Figure 13.1).

Disturbances in carbohydrate metabolism in terms of elevations in pyruvate and lactate concentrations in blood and brain are much higher after prolonged exposure,[6,10,13] indicating apparently that tissue hypoxia is more profound under the conditions of longer exposure, which could not be detected by the measurement of COHb alone. Data on blood lactate are presented in Figure 13.2..

High-energy phosphates in the brain were substantially reduced only following short high-CO exposure.[8] After 4-min exposure to 1.3% CO, the brain content of ATP and PC was substantially reduced. After 40-min exposure to 0.5% CO, the cerebral ATP level was slightly increased, whereas the content of both ATP and PC in the brain of rats exposed to CO for 12 h was significantly higher than in the

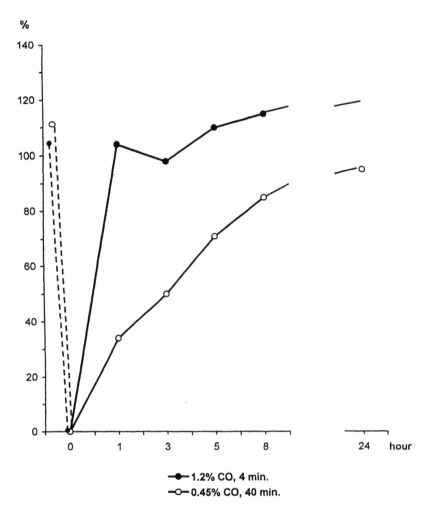

FIGURE 13.1   Motor coordination in rats after different conditions of acute exposure to CO, expressed in percentage of control values (assumed 100%). Consists of mean values of 9 to 15 rats in experimental groups and of 7 to 9 rats in the controls. (From Sokal, J.A., DSc. dissertation, Medical Academy, Lodz, 1985.)

controls. Similar time dependence of brain ATP level in rats exposed to CO was observed in another laboratory.[14]

The mechanism of the observed change in high-energy phosphates with different conditions of acute CO exposure is not clear. It probably reflects the balance between metabolic physiological compensation of tissue hypoxia and the energy-consuming metabolic activity of the brain. Several mechanisms have been postulated, but none has been demonstrated. Clearly, it is not specifically an effect of CO. Similar effects have been observed in rats subjected to experimental hypoxic hypoxemia,[8,15] when chamber oxygen content is gradually reduced (Figure 13.3). As with the CO exposure, hypoxemic rats were in a critical state after the 4-min or 40-min experiments.

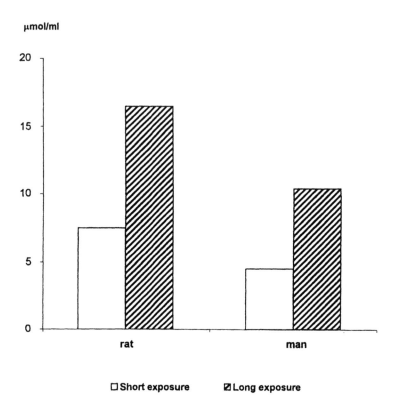

FIGURE 13.2   Blood lactate in CO poisonings after short and long exposure in rat and human. Short exposure: rat, 4 min (1–1.2% CO), mean COHb = 58%; human, <1.5 h, mean COHb = 42%. Long exposure: rat, 40 min (0.4–0.5%), mean COHb = 57%; human, 10–14 h, mean COHb = 39%. (Based on results from Sokal.[12,13])

No differences in the levels of cerebral high-energy phosphates were evident between the rats intoxicated with CO and those subjected to experimental hypoxic hypoxemia. Changes in the oxidation–reduction state of nicotinamide adenine dinucleotide in the brains of rats exposed to CO or subjected to experimental hypoxemia were in good agreement with changes in the high-energy phosphates (Figure 13.3).

The decrease in high-energy phosphates in the rat brains was accompanied by ultrastructural changes in the hippocampal mitochondria. Both CO exposure and experimental hypoxic hypoxemia caused mitochondria to swell and show damaged cristae. Swollen mitochondria were not seen if the level of high-energy compounds in the brain was unchanged or increased.[8] The biochemical and morphological results of these studies suggest that severe clinical effects of CO exposure may be due, at least in part, to factors other than reduction in energy reserves of the brain. Such an additional factor could be excessive metabolic acidosis, which is much more pronounced after prolonged exposure to lower levels of CO than after short exposures to high CO concentrations. It correlates well with the severity of CO poisonings in humans.[5,6,16]

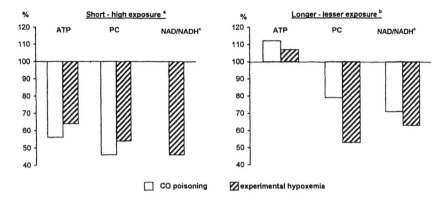

FIGURE 13.3 Cerebral high-energy phosphates (ATP, PC) and the oxidation–reduction state of nicotinamide adenine dinucleotide (NAD/NADH) in the brain of rats intoxicated with CO and in rats subjected to experimental hypoxic hypoxemia. Expressed as percent of control (assumed 100%) (a) 4 min to 1.3% CO or 2 to 4 min to nitrogen with exponentially decreased oxygen content. (b) 15–45 min to 15% CO or 24 to 40 min to nitrogen with exponentially decreased oxygen content. (c) Based on the determination of total oxidized and total reduced nicotinamide adenine dinucleotide. (Based on results from Sokal et al.[8] and Sokal.[15])

13.2.2 IN VIVO BINDING OF CARBON MONOXIDE TO MYOGLOBIN AT REST AND DURING INCREASED WORKLOAD

The authors have demonstrated that CO is bound *in vivo* to myoglobin (Mb) in heart and skeletal muscle of rats intoxicated by CO,[9,17,18] which is compatible with data obtained by Coburn et al.[19,20]

The carboxymyoglobin (COMb) content in heart and skeletal muscles was determined using modification of a method[9] based on that of Coburn and Mayers.[19] COMb level was calculated after independent determinations of

1. Total CO content in muscles by gas chromatography,
2. CO concentration in circulating blood by gas chromatography,
3. Blood content in muscles with ^{51}Cr, and
4. Mb content in muscles by spectrophotometric methods.

It can be assumed that practically all extravascular CO in muscle is bound to Mb.[21] Thus, COMb saturation can be estimated from the concentrations of extravascular CO and tissue Mb.

In this experiment, COMb content was investigated in heart and skeletal muscles of rats, under various exposure conditions: 1% CO for 4 min, 0.4% CO for 40 min, and 0.12% CO for 12 h. After CO exposure, COMb saturation was higher in heart than in skeletal muscle. In both types of muscles, saturation of Mb with CO depended on blood COHb level and not on the duration of exposure[9] (Table 13.3).

The above experiment has proved that in rats severely intoxicated with CO, a substantial proportion of the Mb is saturated with CO. The biological significance

TABLE 13.3
The Content of COMb in the Heart and Skeletal Muscles of
Rat after Varous Conditions of CO Exposure[a]

Exposure Conditions	COHb (%)	COMb (%)	COHb/COMb Ratio
Heart Muscle			
1% CO, 4 min	72 ± 5	65 ± 9	1.13 ± 0.19
0.4% CO, 40 min	76 ± 3	61 ± 9	1.29 ± 0.23
0.12% CO, 12 h	48 ± 4	49 ± 15	1.07 ± 0.3
Skeletal Muscle			
1% CO, 4 min	73 ± 4	33 ± 13	2.63 ± 1.33
0.4% CO, 40 min	79 ± 5	44 ± 11	1.97 ± 0.8
0.12% CO, 12 h	48 ± 4	21 ± 8	2.49 ± 0.88

[a] The determinations were performed immediately after termination of exposure. Values represent means (± SD for 5 to 10 animals).

Source: Adapted from Sokal, J.A. et al., Arch. Toxicol., 56, 106–108, 1984.

of the formation of COMb is not clear. Evidence that Mb can serve as an oxygen store and is involved in facilitating oxygen delivery to the mitochondria makes it reasonable to believe that CO binding to Mb contributes to overall CO toxicity, especially cardioxicity.

Increased binding of CO to Mb during physical exercise was inferred by Clark and Coburn[22] from a decrease in blood COHb in men exercising in a rebreathing system. In rat experiments the authors have provided direct evidence that the binding of CO to Mb is increased during physical exercise in both skeletal and heart muscles.[17]

In this experiment, rats were exposed to CO at concentrations of 130, 296, and 1030 mg/m^3 for 40 min. During the last 20 min of the experiment, the rats were subjected to increased workload in the form of a forced run on a treadmill moving at a rate of 400 m/h. Blood COHb and lactic acid and heart and skeletal muscle COMb were determined immediately after termination of exposure.

With CO exposure of rats at rest, the concentration of COMb in the heart and skeletal muscle increased with increasing concentration of blood COHb. Data for heart muscle are presented in Figure 13.4. COMb concentration in heart muscle at the same COHb saturation was higher than that in skeletal muscle. In the conditions of this experiment, workload did not affect the concentration of COHb in the blood, but it considerably enhanced the concentration of COMb in both skeletal and heart muscle (Figure 13.5). The concentration of lactic acid in blood increased only in the group of rats exposed to CO at the highest concentration (1030 mg/m^3), with COHb at 50% and the subject being simultaneously subjected to increased workload.

It seems clear that the simultaneous effect of CO exposure and increased workload induces greater tissue hypoxia than would be expected on the basis of the blood COHb concentration alone.

An interplay of factors may be operating under conditions of increased workload: a fall in intracellular pO$_2$ during exercise or the binding of CO to hemoglobin would,

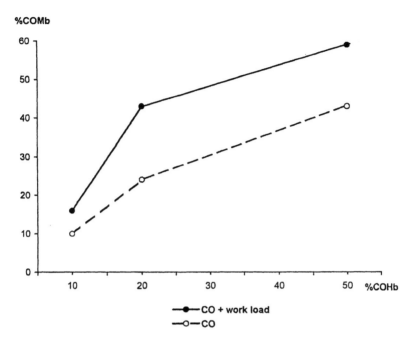

FIGURE 13.4 Carboxymyoglobin in the heart of rats exposed to CO at rest and during work. (Based on Sokal, J.A. and Majka, J., *J. Hyg. Epidemiol. Microbiol. Immunol.*, 30, 57–62, 1986.)

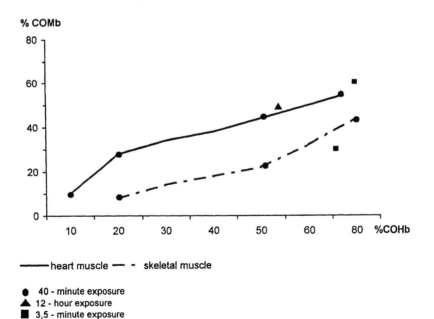

FIGURE 13.5 Relationship between COHb in blood and COMb in the muscles of rats: 40-min exposure; 12-h exposure; 3.5-min exposure.

in turn, cause an increase in COMb, further limiting O_2 supply to muscle mitochondria. As a consequence, the effect of exposure to CO with simultaneous increased workload will be more harmful than the effect of CO alone on persons at rest.

Some evidence of the increased cardiotoxicity of combined CO exposure and increased workload was provided in the authors' rat experiments.[18] Rats were exposed to CO at a concentration of 876 (56 mg/m³) for 40 min, twice daily, for 9 consecutive days. Increased workload consisted of running on a treadmill (400 m/h) for the last 20 min of CO exposure. Rats were sacrificed immediately after exposure and their heart muscle was examined by light and electron microscopy.

Histopathological examination in the light microscope did not reveal any significant changes. No ultrastructural changes were seen in the heart muscles of rats subjected to physical exercise (not exposed to CO). Exposure to CO at rest induced injuries in mitochondria. Groups of mitochondria were enlarged with low electron density of the matrix (edematous mitochondria), and the cristae were broken in some edematous mitochondria.

Similar changes but of greater degree were seen in the myocardium of rats exposed to CO also subjected to increased workload. Apart from the edematous mitochondria, there were giant mitochondria with completely destroyed structures. Although the evidence is not direct, there is the suggestion of a close linkage between increased binding of CO to Mb and heart muscle injury.

13.2.3 OTHER FINDINGS

Increased alkaline phosphatase activity of neutrophilic granulocytes and increased activity of beta-glucuronidase of lymphocytes in blood occurred in a case of acute CO poisoning.[23,24] Increased Mb was noted in the blood serum of patients intoxicated with CO,[25,26] suggestive of myolysis.

Nine males poisoned with CO were examined for spermatoxic effects. It was reported that sperm cell kinetic activity was injured.[27]

In a histochemical study, significant reduction of nucleoside phosphatase activity was observed in the brains of rats intoxicated by CO. During the first period of recovery the activity of this enzyme increased.[28]

Increased glycogen content and uridine diphosphate (UDP) glucose: glycogen (4-glucosyltransferase) activity was reported in rats intoxicated with CO.[29] In these CO studies with rats there were no significant changes of brain glycogen. However, changes in liver glycogen and in blood and brain glucose were observed.[10] The direction and magnitude of these changes were dependent on the conditions of exposure (Figure 13.6).

High doses of ethanol (4 to 6 g/kg body weight) shortened the survival time of CO-intoxicated rats.[30] On the other hand, carbon dioxide at a concentration of 5% did not influence blood COHb, glucose, lactate, or pH in rats intoxicated with CO.[31]

13.3   DIAGNOSIS AND TREATMENT OF ACUTE
CARBON MONOXIDE POISONING IN POLAND

The principles and methods of diagnosis and treatment of acute CO poisoning in clinical practice in Poland do not differ from those used in toxicological centers in

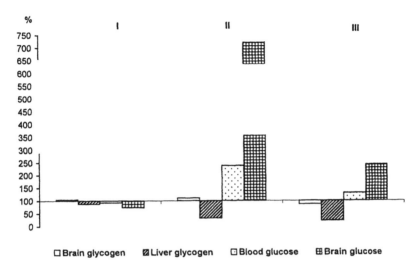

FIGURE 13.6   The content of glycogen and glucose in the tissues of rats intoxicated with CO after various conditions of CO exposure. Expressed as percentage of control (assumed 100%). CO exposures: I — 4 min to 1.2% CO, COHb = 65%; II — 40 min to 0.4 to 0.5% CO, COHb = 63%; III — 12 h to 0.13 to 0.17% CO, COHb = 55%.

other countries. Detailed diagnostic procedures and criteria as well as treatment methods may depend on the experience and research carried out by a center and its equipment.

For example, at the Krakow Poison Center severity of CO poisoning is estimated on the basis of age, duration of exposure, COHb level, and patient state on admission. This is the product of its clinical experience and studies[4,16,32,33] (Table 13.4). Patient state is evaluated according to neurological symptoms (Table 13.5).

The mechanism of acute CO poisoning and its clinical picture have for many years remained a major focus at the Krakow center. Its staff has been very successful in applying methods making it possible to assess changes in the central nervous system (CNS) in patients. One of them is brain computed tomography (CT). Evaluation by CT is made after considering ventricular system and cranial vault dimensions. This permits the determination of some basic dimensions and the values of such indexes as anterior horn index, ventricular index (VI), lateral ventricle index, and Huckman's number (HN), which shows changes in the inner cerebrospinal fluid (CSF) space. Measurement of the external CSF space is performed, measuring the width of the longitudinal fissure, and the width of insular cisterns and cerebral sulci. The values obtained were compared with norms for each age group in healthy people according to Meese et al.[34] and modified by Kuśmiderski and colleagues[35] (Table 13.6). A special computer program was developed to obtain final results more easily and faster. After data input, this program provides the results together with a data report on the kind of the cerebral atrophy (cortical, subcortical, generalized), the degree of intensification (small, medium, large), and the localization. Incorrect CT results were found in over 74.8% of examined patients. The most common lesion was cortical atrophy and subcortical atrophy of the brain. More rarely, isolated cortex

TABLE 13.4

Point Scale of the Severity of Carbon Monoxide Poisoning Used at the Center of Acute Poisoning in Krakow

Parameters	Point Scale			
	0	1	2	3
Age (years)	≤29	30–39	40–49	>50
Time of exposure (min)	≤30	31–60	61–120	>120
Patient condition	Good	Medium	Serious	Very serious
COHb (%)	—	<15	15–30	>30
Blood lactate (μmol/ml)	1.0–1.7	1.8–3.6	3.7–5.4	>5.5

Light poisoning: 1 to 4 points; medium poisoning: 5 to 8 points; serious poisoning: >9 points.

Source: Adapted from Reference 33.

TABLE 13.5

Objective Evaluation of a Patient's Condition Based on Neurological Syndromes

Degree	Specification of Neurological Syndromes
I Minor	No disturbances of consciousness; possible muscular twitching; tonic or clonic convulsions
II Moderate	Disturbances of consciousness, and possible simultaneous excessively intensified tendon reflexes, Babinski reflex, tonic or clonic convulsions, and intensified muscular tonus
III Severe	Total loss of consciousness without other neurological symptoms
IV Very severe	Complete loss of consciousness with simultaneous excessively intensified tendon reflexes, Babinski reflex, tonic convulsions, clonic convulsions, intensified muscular tonus, partial or total elimination of tendon reflexes, pupillary, corneal, and swallowing reflexes with a diffusely diminished muscular tonus

Source: Adapted from Reference 33.

and subcortical atrophy with simultaneous areas of low density in the subcortical nuclei was found. Unilateral or bilateral areas of low density in the globus pallidus were not observed. Incorrect images using brain CT were noted in 95% of the patients stated as being severely poisoned at the time of clinic admission. The risk of occurrence of pathological changes increased when the age of the patients was under 40 years, when the duration of CO exposure was longer than 30 min, and when blood lactate was higher than 1.8 mmol/l.[35,36]

Evaluation of CO neurotoxicity at the Krakow Poison Center includes complex psychological and psychiatric examination and electroencephalography. Results of these examinations are evaluated using the point scale in Table 13.7. CNS damage is diagnosed when the score is a minimum of 5 points. A significant correlation

TABLE 13.6
Range of Numerical Values (in mm) of Measurements of the Inner and External CSF Spaces in Each Age Group Considering the Degree of Change Intensification

	Degree of Atrophy											
	Small				Medium				Large			
	0–20	21–40	41–60	>60	0–20	21–40	41–60	>60	0–20	21–40	41–60	>60
Inner CSF spaces												
III Ventricle	3.90–4.55	4.56–6.95	6.96–8.95	8.96–11.55	4.56–6.95	6.96–8.95	8.96–11.55	11.56–14.95	6.96–8.95	8.96–11.55	11.56–14.95	14.96
F/A	3.70–3.56	3.55–3.36	3.35–3.16	3.15–2.96	4.56–6.95	3.35–3.16	3.15–2.96	2.95–2.76	3.35–3.16	3.15–2.96	2.95–2.76	2.75
D/A	1.69–1.58	1.57–1.48	1.47–1.38	1.37–1.28	1.57–1.48	1.47–1.38	1.37–1.28	1.27–1.18	1.47–1.38	1.37–1.28	1.27–1.18	1.17
H/E	4.50–4.28	4.27–3.98	3.97–3.68	3.67–3.38	4.27–3.98	3.97–3.68	3.67–3.38	3.37–3.08	3.97–3.68	3.67–3.38	3.37–3.08	3.07
A+B	40.00–47.5	47.51–55.50	55.51–63.50	63.51–71.50	47.51–55.0	55.51–63.50	63.51–71.50	71.51–79.50	55.51–63.50	63.51–71.50	71.51–79.50	79.51
External CSF spaces												
FI, IC, SW	3.0–4.15	4.16–5.35	5.36–6.55	6.56–7.7	4.16–5.35	5.36–6.55	6.56–7.75	7.76–8.95	5.36–6.65	6.56–7.75	7.76–8.95	8.96

Source: Based on Kroch, S. et al., *Przegl. Lek.*, 52, 267–270, 1995.

TABLE 13.7
The Point Scale to Evaluation of Complex Psychiatric Examination

Psychiatric Examination		Psychological Examination		EEG	
Result	Points	Results[a]	Points	Record	Points
Normal results without changes	0	Normal results	0	Normal record	0
Features of quasi-neurological syndrome	1	Suspicion of CNS damage	1	Flat without paroxysmal and focal features record	1
Quasi-neurological syndrome	2	Slight CNS damage	2	Mild pathology with dispersed, slow, low power waves record	2
Psycho-organic syndrome	3	Pronounced CNS damage	3	Marked paroxysmal and focal pathological wave record	3

[a] Results of Benton, Graham–Kendall, and Bender tests.
Note: The CNS damage is recognized when the point score in psychiatric, psychological and EEG examinations amounted at least 5 points.

was noted between the CT and the complex psychiatric examination in CO-poisoned patients.

The brain CT combined with the complex psychiatric examination allows an appropriate assessment of the morphological and functional state of the CNS in the course of acute CO poisoning.[36,37]

Magnetic resonance techniques have been used for 2 years to evaluate neurological sequelae.[38] Because proton magnetic resonance spectroscopy (MRS) allows an evaluation of cerebral metabolic changes, it show previously unrecognized neuronal activity in CO poisoning and reflects the severity of symptoms. The superiority of proton MRS over conventional radiological examinations in CO poisoning has been stressed.

The ECG and echocardiography are used to evaluate pathological changes in the myocardium. Cardiac enzyme activity (CK, CK-MB, ALT, AST) is measured on a daily basis by clinical toxicologists. Immediately after intoxication, however, the harmful effects of CO are not always reflected in enzyme activity and the ECG. Thus, the rest and stress single-photon emission CT (SPECT) is used to evaluate myocardial perfusion in acute CO-poisoned patient at the Krakow center.[39,40] It has the advantages of sensitivity, specificity, and noninvasiveness, using technetium-99m-labeled myocardial tracers (e.g., Tc 99 dimethyloxyisobuthylisonitril, MIBI). The use of SPECT, ECG, and measurement of the activity of chosen enzymes performed simultaneously are methods well suited for evaluation of the morphological and functional status of the myocardium after acute CO intoxication. Tc 99m MIBI SPECT scintigraphy is a more sensitive method than ECG and enzyme activity measurements for the evaluation of CO cardiotoxicity in acute poisoning. This method provides information on localization and extent of disease.

The types of respiratory disorders caused by CO poisoning and their dynamics after treatment were analyzed using computer spirometer. The obturation of central and small bronchi was the most frequent disturbance caused by CO poisoning and was dependent on poisoning severity.[41] An analysis of the respiratory pattern to assess the central and peripheral mechanisms of respiratory control in CO-intoxicated patients showed that the parameters were also dependent on poisoning severity.[42]

CO hepatotoxicity was revealed in longitudinal clinical observations using liver scintigraphy and enzyme activity measurements.[27] Widespread morphological liver damage was manifested by variously intensified changes in the liver scintigraphy as well as by changes in serum indicator enzymes (AST, LDH). Liver morphological changes were accompanied by impaired metabolic function of hepatocytes, with resultant decreases in secretory enzyme activity (CHE and prothrombin). The state of the liver in CO-poisoned patients should be evaluated directly after intoxication and during follow-up examinations.

The basic method for treatment of acute CO poisoning in Polish centers is oxygen therapy as well as the elimination of metabolic acidosis. Especially good results are obtained through the use of hyperbaric oxygen (HBO) treatment. Many years of experience using HBO in the treatment of CO poisoning as well as comparative research carried out by the Poisoning Center in Sosnowiec have indicated the advantages of this method as well as revealed its limitations. The treatment of acute poisoning by use of HBO accelerates normalization of the bioelectrical function of the brain, shortens the time of coma, diminishes the frequency of occurrence of severe neurological syndromes, and helps to overcome brain swelling and circulatory collapse. In cases involving damage to brain structures, the results of HBO treatment are unfavorable.[43,44] Unfortunately, in Poland HBO treatment is not commonly available. Therefore, 100% oxygen therapy is generally applied. Acidosis is usually treated with the intravenous administration of sodium bicarbonate.

Table 13.8 shows the most common complications of acute CO poisoning and their treatment.

13.4   PREVENTION OF CARBON MONOXIDE POISONING

Regional acute poisoning centers disseminate information on the risk of poisoning due to exposure to various chemicals, including CO. They also provide information on the measures to be followed at the site of poisoning and at the health-care unit when such poisonings occur. Such information is also made available by round-the-clock telephone toxicology services from the regional poisoning centers. The centers may also organize training courses on acute poisonings for medical professionals. The National Poison Information Center publishes and disseminates a review periodical for physicians which summarizes information from important articles on acute poisonings published in foreign journals.

The experience of the Krakow center indicates that special preventive actions developed for local conditions can be effective and can lead to the reduction of poisoning incidents.[36]

In the mid-1980s a special report was prepared on the risk of CO exposure within the population of Krakow. According to this report, the main cause of CO poisoning

TABLE 13.8
Complications and Their Treatment in Carbon Monoxide Poisoning

Organs and Systems Affected	Complications	Treatment
Circulatory system	Arrhythmia	Treatment dependent on type of arrhythmia
	Toxic damage to the cardiac muscle	Rest and pharmacological treatment if changes in ECG and AspAT, CPK, LDH activities as well as clinical symptoms indicate myocardial ischemia
Respiratory system	Pulmonary edema, bleeding	Oxygen, diuretics, PEEP
Muscular system	Myonecrosis and swelling of soft tissues	Tension of fascias in case of pressure on arterial vessels or if neuropathy due to pressure occurs
Kidneys	Myoglobinuria and acute renal failure	Early: intensive diuresis and alkalization of urine; limited supply of liquids and/or peritoneal dialysis, forced hemofiltration. Constant monitoring of electrolytes.
Nervous system	Convulsions	Anti-convulsive medicines, phenobarbital
	Cerebral edema	Oxygen, steroids, diuretic medicines and medicines stimulating hyperventilation; possibility of monitoring of intracranial pressure

Source: Based on Reference 33.

that was avoidable was the inefficient ventilation of bathrooms and kitchens, as well as the inefficiency of heaters. The report included a map of Krakow showing districts with the highest CO-poisoning rate. The report was presented to the administrative authorities of the city as well as the inhabitants through the mass media. Soon afterward, city services began to overhaul heaters and ventilation installations in apartments, thus helping to improve the condition of the equipment and to draw the residents' attention to the safe operation of these units for their own safety. This resulted in a threefold increase in the number of individuals reporting to the poisoning center with minor symptoms of CO poisoning in the period just after the media campaign. Only in 5% of the cases did residents report to the center unnecessarily.

The preventive activities undertaken in Krakow were facilitated by the sensitivity to health and safety of both the authorities and the residents of Krakow, as well as the actions and experience of the local center for acute poisoning. Various preventive actions were also undertaken on a smaller scale in other areas of the country.

REFERENCES

1. Czerczak, S. and Jaraczewska, W., Acute poisonings in Poland, *Clin. Toxicol.*, 33(6), 669–675, 1995.
2. Jaraczewska, W. and Czerczak, S., The pattern of acute poisonings in Poland, *Vet. Hum. Toxicol.*, 36(3), 228–233, 1994.

3. Sancewicz-Pach, K., Kamenczak, A., Klag, E., and Klus, M., Acute poisonings with chemical compounds among adolescent and adult inhabitants of Krakow in the year 1997 [in Polish], *Przegl. Lek.*, 56, 409–414, 1999.

4. Bogusz, M., Cholewa, L., Mlodkowska, K., and Pach, J., Biochemical criteria of hypoxia in acute carbon monoxide poisoning, *J. Eur. Toxicol.*, 5, 306–309, 1972.

5. Burmeister, H. and Neuhaus, G.A., Die Behandlung der schweren subakuten Leuchtgasvergiftung beim Menschen, *Arch. Toxicol.*, 26, 277–292, 1970.

6. Sokal, J.A. and Kralkowska, E., The relationship between exposure duration, carboxyhemoglobin, blood glucose, pyruvate and lactate and the severity of intoxication on 39 cases of acute carbon monoxide poisoning in man, *Arch. Toxicol.*, 57, 196, 1985.

7. Sokal, J.A., Problems of biochemical diagnostics in acute carbon monoxide poisoning related to different time of exposure [in Polish], *Stud. Mat. Monogr.*, IMP Lodz, 26, 47–56, 1987.

8. Sokal, J.A., Opacka, J., Gorny, R., and Kolakowski, J., Effect of different conditions of acute exposure to carbon monoxide on the cerebral high-energy phosphates and ultrastructure of brain mitochondria in rats, *Toxicol. Lett.*, 11, 213–219, 1982.

9. Sokal, J.A., Majka, J., and Palus, J., The content of carbon monoxide in the tissues of rats intoxicated with carbon monoxide in various conditions of acute exposure, *Arch. Toxicol.*, 56, 106–108, 1984.

10. Sokal, J.A., The effect of exposure duration on the blood level of glucose, pyruvate and lactate in acute carbon monoxide intoxication in man, *J. Appl. Toxicol.*, 5, 395–397, 1985.

11. Broome, J.R., Pearson, R.R., and Skrine, H., Carbon monoxide poisoning: forgotten not gone, *Br. J. Hosp. Med.*, 39, 298, 300, 302, 304–305, 1988.

12. Sokal, J.A., Disturbances of Motor Coordination in Rats after Carbon Monoxide Poisoning in Different Experimental Conditions, The Significance of Duration of Exposure in Acute Carbon Monoxide Poisonings [in Polish], DSc. dissertation, Medical Academy, Lodz, 1985.

13. Sokal, J.A., Lack of correlation between biochemical effects on rats and blood carboxyhemoglobin concentrations in various conditions of single acute exposure to carbon monoxide, *Arch. Toxicol.*, 34, 331–336, 1975.

14. Sikorska, M., Bicz, W., Smialek, M., and Mossakowski, M.J., ATP, ADP, AMP concentrations in rat brain following carbon monoxide intoxication and experimental ischema [in Polish], *Neuropatol. Pol.*, 12, 327–395, 1974.

15. Sokal, J.A. and Klyszejko-Stefanowicz, L., Nicotinamide adenine dinucleotides in acute poisoning with some toxic agents, *Soc. Sci. Lodz.*, 112, 1–104, 1972.

16. Bogusz, M., Cholewa, L., Pach, J., and Mlodkowska, K., A comparison of two types of acute carbon monoxide poisoning, *Arch. Toxicol.*, 33, 141–149, 1975.

17. Sokal, J.A. and Majka, J., Effect of work load on the content of carboxymyglobin in the heart and skeletal muscles of rats exposed to carbon monoxide, *J. Hyg. Epidemiol. Microbiol. Immunol.*, 30, 57–62, 1986.

18. Sokal, J.A., Majka, J., and Palus, J., Carbon Monoxide and Workload — An Experimental Study, Health Effects of Combined Exposures to Chemicals in Work and Community Environments (Proceeding of a Course, Lodz, Poland, October 18–22, 1982), WHO Regional Office for Europe, Copenhagen, 1983, 356–370.

19. Coburn, R.F. and Mayers, L.B., Myoglobin O_2 tension determined from measurements of carboxymyoglobin in skeletal muscle, *Am. J. Physiol.*, 220, 66–74, 1971.

20. Coburn, R.F., Ploegmakers, F., Gondrie, P., and Abboud, R., Myocardial myoglobin oxygen tension, *Am. J. Physiol.*, 224, 870–876, 1973.

21. Coburn, R.F., Mechanism of carbon monoxide toxicity, *Prev. Med.*, 8, 310–322, 1979.

22. Clark, B.J. and Coburn, R.F., Mean myoglobin oxygen tension during exercise at maximal oxygen uptake, *J. Appl. Physiol.*, 39, 135–144, 1975.

23. Moszczynski, P. and Wiernikowski, A., Alkaline phosphatase activity of neutrophilic granulocytes in acute carbon monoxide poisoning [in Polish], *Pol. Tyg. Lek.*, 32, 53–56, 1977.

24. Moszczynski, P., Wiernikowski, A., and Slowinski, S., The activity of three-glucoramydase of lymphocytes in acute carbon monoxide poisoning [in Polish], *Przegl. Lek.*, 34, 763–765, 1977.

25. Kochanska-Dziurowicz, A., Klopotowski, J., and Smolicha, W., Assessment of determination of myoglobin concentration in acute carbon monoxide poisoning [in Polish], *Post. Med. Nukl.*, 9, 55–59, 1991.

26. Kochanska-Dziurowicz, A., Klopotowski, J., Smolicha, W., and Jablonska, E., Intensity of changes of myoglobin concentration in serum in patients with acute carbon monoxide poisoning [in Polish], *Probl. Ter. Mon.*, 3(1–2), 28–33, 1992.

27. Pach, J., Malolepszy, A., Dudek, J., Feret, J., and Filipek, M., Attempt at the evaluation of the spermatotoxic activity of carbon monoxide in acute poisonings [in Polish], *Stud. Mat. Monogr.*, IMP, Lodz, 26, 209–213, 1987.

28. Szumanska, G., Ostenda, M., and Mossakowski, M.J., Activity of nucleoside phosphatases in rat brain following carbon monoxide intoxication [in Polish], *Neuropatol. Pol.*, 14(2), 197–207, 1976.

29. Smialek, M., Sikorska, M., Korthals, J., Bicz, W., and Mossakowski, M.J., The glycogen content and its topography and UDP glucose: glycogen (-4 glucosyltransferase (EC 2.4.1.11) activity in rat brain after experimental carbon monoxide intoxication, *Acta Neuropathol.*, (Berlin), 24, 222–231, 1973.

30. Molenda, R., Effect of ethanol on carbon monoxide toxicity [in Polish], *Arch. Med. Sad.*, 41, 178-184, 1991.

31. Sokal, J.A., Biochemical changes in the blood of rats intoxicated with carbon monoxide or the mixture of carbon monoxide and carbon dioxide [in Polish], *Bromatol. Chem. Toksykol.*, 16, 48–51, 1986.

32. Pach, J., Prognosis and guidelines in treatment of acute carbon monoxide poisoning [in Polish], *Folia Med. Cracov.*, 17, 211, 1975.

33. Pach, J., Cases of acute poisoning with chemically asphyxiating gases — carbon monoxide, cyanogen compounds [in Polish], Chemiczne Substancje Toksyczne w Srodowisku, Komisja Nauk Medycznych-Oddzial PAN w Krakowie, Krakow, 83, 102, 1990.

34. Meese, W., Kluge, W., Grumme, T., and Hopfenmuller, W., CT Evaluation of the CSF spaces of healthy persons, *Neuroradiology*, 19, 131, 1980.

35. Kroch, S., Kuśmiderski, J., and Urbanik, A., Progress in the CT evaluation of the cerebral atrophy in acute poisoning with carbon monoxide [in Polish], *Przegl. Lek.*, 52, 267–270, 1995.

36. Pach, J., Neuropsychiatric complications of carbon monoxide poisoning [in Polish], *Przegl. Lek.*, 52, 221–222, 1995.

37. Pach, J., Mitka, A., and Billewicz, O., Usefulness of brain computer tomography, electroencephalographic, psychological and psychiatric tests in evaluating the CNS state in severe poisonings by carbon monoxide, *J. Toxicol. Clin. Toxicol.*, 23, 430, 1985.

38. Pach, J., Groszek, B., Urbanik, A., Chojnacka, I., Herman-Sucharska, I., and Szczepanska, L., Neurological sequelae in severe carbon monoxide poisoning — case report [in Polish], *Przegl. Lek.*, 55, 554–557, 1998.

39. Pach, J., Hubalewska-Hola, A., Pach, D., Targosz, D., and Szybinski, Z., Usefulness of rest and forced perfusion scintigraphy SPECT to evaluation of carbon monoxide cardiotoxicity in acute poisoning, in *Abstract Book*, EAPCCT XVIII International Congress, Zurich, 1998, 102.

40. Pach, J., Pach, D., Hubalewska-Hola, A., Krach, S., and Targosz, D., The assessment of cardiotoxic activity of carbon monoxide in group poisoning [in Polish], *Przegl. Lek.*, 55, 505–507, 1988.

41. Kolarzyk, E., The effect of acute carbon monoxide poisoning on the respiratory system efficiency. II: Types of ventilatory disorders and dynamics of changes according to the severity of carbon monoxide poisoning, *Intern. J. Occup. Med. Environ. Health*, 7, 237–243, 1994.

42. Kolarzyk, E., Regulation of breathing in cases of acute carbon monoxide poisoning, *Intern. J. Occup. Med. Environ. Health*, 8, 89–101, 1995.

43. Klopotowski, J., Langauer-Lewowicka, H., and Zajac-Nedza, M., Oxygen hyperbaric treatment of acute carbon monoxide poisoning [in Polish], *Med. Pr.*, 26, 315–322, 1975.

44. Klopotowski, J., Treatment of carbon monoxide poisoning using hyperbaric oxygen hyperbarism [in Polish], Scientific Conference Materials, Oxygen hyperbarism treatment, Lancut, October 6, 1988, Rzeszow, Przemysl, 1989, 63–70.

14 Treatment of Carbon Monoxide Poisoning in the United Kingdom

Martin R. Hamilton-Farrell and John Henry

CONTENTS

14.1 CARBON MONOXIDE — FACTS AND PERCEPTIONS

It has been widely perceived in the United Kingdom that carbon monoxide (CO) poisoning is rare. Yet it is the most common cause of death by poisoning in this country. Mortality is approximately 1500 deaths per year in the United Kingdom, the majority of which are successful suicides and victims of fires. Therefore, when "town gas" (containing CO in the fresh gas supplied to households) was replaced by North Sea gas in the 1970s, there was little change in mortality from CO poisoning.

CO poisoning is sometimes misdiagnosed, especially where the source has been accidental or domestic. Chronic CO poisoning has only recently been acknowledged as a source of serious morbidity as well as mortality. The late complications of CO poisoning are not widely recognized in the literature or in clinical practice.

The treatment of CO poisoning, once diagnosed, involves oxygen in most cases. Yet, until the last 10 years, there has been no universally agreed standard for the clinical management of this disease. The place of hyperbaric oxygen (HBO) has recently been defined by the British Hyperbaric Association (BHA). However, not all the sources of clinical advice on poisoning support the use of HBO.

In the last 10 years, there have been hopeful signs of increased awareness and concern about this problem. Physicians, engineers, public health experts, government officials and politicians, and also those who have suffered from CO poisoning have begun to talk with each other. The result may be a radical change, and improvement of services provided for those affected by this complex disease.

14.2 CONVENTIONAL MANAGEMENT OF CARBON MONOXIDE POISONING

Clinical practice in hospital emergency departments and in-patient wards usually involves the measurement of blood carboxyhemoglobin (COHb) levels. This is essential in establishing the diagnosis, especially in domestic exposure, and where the symptoms are chronic and ill-defined. Oxygen is given to most patients by nonsealed face mask, yielding a maximum of 85% FIO_2. This continues until either the patient is recovered or until the COHb level has fallen to <5%.

The treatment of smoke inhalation from fires often forms a part of the broader management of burn injuries in a dedicated burn unit. In any event, intensive care and artificial ventilation with a high FIO_2 are generally available for the unconscious patient.

It is in the prevention of late complications of CO poisoning that HBO is increasingly considered.

14.3 ESTABLISHMENT OF HYPERBARIC OXYGEN FACILITIES

In Britain, CO poisoning has been treated with HBO since hyperbaric facilities became available for nondiving injuries in the 1960s. Early work by McDowall et al.[1] pioneered the clinical use of HBO. However, this work did not persist as a clinical service. Until 1990, referrals were mainly by direct contact between hospitals, and based on request rather than standard criteria.

HBO chambers were installed mainly in coastal towns and cities, and many were dedicated to the treatment of diving injuries. Some HBO chambers were installed in the 1960s and 1970s, but were then closed for lack of dedicated medical supervision. A national network of multiplace HBO chambers grew in the 1970s and 1980s, initially for clinical trials into multiple sclerosis therapy, and subsequently for the treatment of multiple sclerosis; however, these tended to operate independently from hospitals, and mostly did not take CO-poisoning cases. ·

In the 1970s and 1980s, several new HBO chambers were developed, initially for the treatment of diving injuries. However, the use of HBO for nondiving illnesses began to increase. In 1990, all British HBO chamber personnel began meeting as the British Isles Group of Hyperbaric Therapists, which later became the BHA. With this, communication and service development became much more rapid.

FIGURE 14.1 Distribution of HBO chambers in the United Kingdom. (From *J. Accident Emergency Med.*, 1999, 16:98–103. With permission from the BMJ Publishing Group.)

Following increased awareness of the applications of HBO, demand for this service began to grow; and over the first few years of the 1990s more new HBO chambers were established. Some existing HBO facilities were provided with new equipment. These new chambers required negotiations with National Health Service local health authorities, who generally agreed to pay for the treatment provided, as long as it was clinically justifiable. Much use was made of the publications of the Undersea and Hyperbaric Medical Society in these discussions.

Since the publication of the First European Consensus Conference report in 1994,[2] increasing emphasis has been placed upon European coordination.

14.4 RECENT DEVELOPMENTS IN HYPERBARIC OXYGEN THERAPY

The distribution of HBO chambers is still mainly coastal (Figure 14.1). However, although the treatment of CO poisoning has formed a large part of the emergency caseload of some chambers, others have received very few such cases.

To assess the activity of British HBO chambers in CO poisoning, a data set was developed at Whipps Cross Hospital in 1992, as a tool of clinical audit. This was based on clinical symptoms and signs commonly observed in patients attending that facility. In 1992, the BHA agreed to adopt this data set nationally, and members agreed to send completed anonymized forms in medical confidence to Whipps Cross Hospital for each patient treated. As a result, from April 1993 until March 1996, a national database of treated cases was built. The exercise resulted in a view of the management of CO poisoning with HBO that reflected practice at the time.

The activity in HBO chambers between 1993 and 1996 is shown in Figure 14.2. However, this may be an underestimate, as there was underreporting from some centers.

There were 575 cases of CO poisoning reported as being treated in HBO chambers in the 3-year period. Of these, 292 cases (51.0%) were accidental; 280 cases (49.0%) were nonaccidental; in 3 cases, the cause was not recorded. The ratio of accidental to nonaccidental cases was therefore 1.05:1. Of the accidental cases, 81 were exposed continuously or intermittently for more than 24 h. Of the accidental cases, 150 (51.9%) were male and 139 (48.1%) were female. Of the nonaccidental cases, 237 (85.3%) were male and 41 (14.7%) were female.

Sources of accidental exposure were varied, although the majority were in connection with domestic central heating (Figure 14.3). Central heating sources were entirely from blocked or leaking flue systems (leading to exposure to CO as a product of combustion).

While the presentation of nonaccidental cases was continuous throughout the year, accidental cases predominated in the winter months (November through February). Peaks occurred during individual winter months where there was a period of severe cold (Figure 14.4).

The mean delay from time of removal from exposure to CO to time of arrival at a hyperbaric facility was in excess of 6 h in all 3 years. The mean for the whole study period was in excess of 9 h.

Each HBO facility used its own treatment schedules, resulting in great variability. 100% oxygen was generally used. Schedules included the U.S. Air Force CO table, Royal Navy Decompression Table 60, U.S. Navy Decompression Table 5, and others. Patients tended to be retreated in case of failure to resolve symptoms and signs after the first exposure to HBO, until signs and symptoms resolved or stabilized.

Since 1996, British HBO facilities have become well established, partly due to developing contractual relationships with health service commissioning agencies such as local health authorities. The restriction of clinical activity mainly to those conditions that form part of international consensus statements has strengthened the case for the public funding of HBO facilities.

Nevertheless, in England, only three HBO facilities are funded directly by the National Health Service, the others being funded and managed by other agencies. In Scotland and Northern Ireland, HBO facilities have been funded directly by the departments of health of the relevant governments.

The BHA has led developments in the field, with publications on fire and electrical safety,[3,4] as well as a core curriculum for the training and education of staff.[5] The BHA worked with the Faculty of Occupational Medicine of the Royal College of Physicians of London to produce, in 1994, guidelines for the operation of HBO facilities,[6] including descriptions of staffing, equipment, and operating and

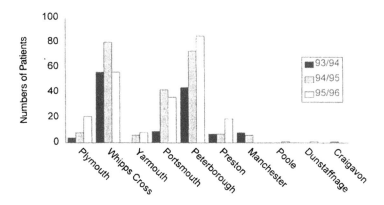

FIGURE 14.2 The activity of HBO chambers in the treatment of CO poisoning, 1993-1996. (From *J. Accident Emergency Med.*, 1999, 16:98–103. With permission from the BMJ Publishing Group.)

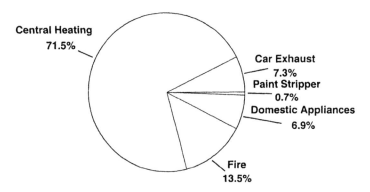

FIGURE 14.3 Sources of Accidental Exposure to CO. *Note:* Based on a study of 292 patients. (From *J. Accident Emergency Med.*, 1999, 16:98–103. With permission from the BMJ Publishing Group.)

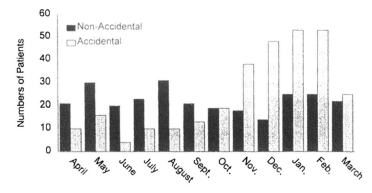

FIGURE 14.4 Presentation for CO poisoning in the United Kingdon by Month. (From *J. Accident Emergency Med.*, 1999, 16:98–103. With permission from the BMJ Publishing Group.)

emergency procedures. These have been superseded, but not markedly altered, by the outcome of the First European Consensus Conference on Hyperbaric Medicine, published in late 1994.[2] The BHA has taken a national and international lead on health and safety matters in the operation of HBO facilities, and it is working with the Health and Safety Executive to define good practice. Finally, the BHA Gazetteer of British HBO facilities,[7] published every 3 years, has drawn attention to the availability and accessibility of this form of treatment to medical referral agencies, such as hospital accident and emergency departments. A national referral system for severe CO poisoning has been developed at the Diving Diseases Research Centre in Plymouth, thus providing an additional resource for inquirers.

As a result of these initiatives, the use of HBO in the treatment of CO poisoning has become progressively safer and better organized.

The National Health Service has to date consistently funded the use of HBO in the management of severe CO poisoning; and this has been facilitated by good communication with poisons information services.

14.5 COMMUNICATION: POISONS ADVICE SERVICES AND HYPERBARIC OXYGEN FACILITIES

The need for a clear referral system led in 1988 to initial contact being made between the HBO facility at Whipps Cross Hospital, London, and the National Poisons Information Service Centre, also in London. As a result, criteria for the definition and referral of cases of severe of CO poisoning were agreed to. These were precisely the same as those internationally quoted in the medical literature. This led directly to an increase in caseload of the Whipps Cross Hospital facility.

Inquiries from throughout the south of England were thus given consistent advice, and some cases were referred on this basis by the Guy's Hospital Poisons Unit to HBO facilities outside London.

In the British Isles, although the majority of inquiries about the diagnosis and management of CO poisoning were dealt with by the London center of the National Poisons Information Service, several other centers were operational from the late 1960s onward. These were located in Birmingham, Cardiff, Dublin, Edinburgh, Leeds, and Newcastle. They tended to operate from a common database initially provided by the London center and later made available in a teletext form by the Edinburgh center. However, the centers frequently differed in the advice provided on major clinical problems, and the management of CO poisoning was no exception. This led to uncertainty among the callers, who were aware that different advice was available from different centers. By the late 1990s, the centers were providing much more consistent data, and liaison with the medical profession and the hyperbaric facilities improved markedly.

The inconsistency of clinical opinion, and of the advice given by poisons information services, was expressed in the concentration of referrals from different hospitals, with some hospitals sending cases regularly for treatment, while others referred very few, if any, patients. The weighting of referrals during the BHA 1993 to 1996 study (Figure 14.2) toward referrals from the south and east of England may also be explained by the consistent advice given by the London-based National

Poisons Information Service. A further analysis of trends since 1996 is awaited. The effect of publication of the BHA Gazetteer will need to be studied.

Changes since 1996 will also be subject to the operation of the Carbon Monoxide Advice Line operated by the Plymouth-based Diving Diseases Research Centre, since January 1996. The advice given through this agency has been designed to match patients' clinical condition with the capability of the respective HBO facilities. The categorization of HBO facilities in 1994[5] has helped to prevent the inappropriate referral of critically ill patients to HBO facilities unable to cope with them. The definitions of these categories are as follows:

1. Comprehensive hyperbaric facilities capable of supporting the treatment of patients who are critically ill from any cause, and who may require hyperbaric intensive therapy — multiplace.
2. Hyperbaric facilities capable of receiving elective or emergency referrals, excluding patients who are critically ill at the time of referral or are considered likely to become so — multiplace.
3. Facilities without some of the capabilities of categories 1 and 2, sited specifically for the treatment of diving emergencies — multiplace.
4. Monoplace hyperbaric chambers operating at relatively low pressure and without an air lock.

However, despite the simplification of the referral process, delay continues to be incurred by the perceived need to obtain COHb levels before referral. While it is clearly stated in the medical literature that COHb levels are inconsistently related to the severity of the symptoms of poisoning, and also to its prognosis, most referral agencies insist on this investigation before referral for HBO. As many hospitals do not have the facilities to do this investigation, and in others considerable time is taken in obtaining the result, delay is inevitable. The diagnosis may not always be clear, particularly in chronic exposure or domestic heating accidents. However, after parasuicide or smoke inhalation, there is often little doubt that CO poisoning is at least a factor in the patient's illness. In such cases, there may be the coincidental effects of drug and/or alcohol ingestion; even so, the effects of CO cannot be ignored in such cases, and treatment should be promptly instituted.

It can be envisaged that with further progress, it may be possible to establish a system of direct transport of patients by the emergency services to an HBO facility, without the delay of prior passage through a hospital accident and emergency department. The same result could be achieved by having a hyperbaric facility located close to, or within, selected major accident and emergency departments.

14.6 CURRENT ISSUES

Chronic CO poisoning is only now emerging as a recognized clinical entity. In the BHA 1993 to 1996 study, approximately 27% of accidental cases of severe CO poisoning were documented as chronic (i.e., of greater than 24-h duration). Experience at a follow-up clinic at one HBO facility (Whipps Cross Hospital) suggests that retrospective analysis of symptoms reveals further cases of chronic exposure.

This is not surprising, in view of the high proportion of cases of accidental exposure (approximately 71%) due to domestic or workplace heating systems. These cases may have had relatively mild poisoning for some time before the index episode which led to their assessment and treatment in a hyperbaric facility. This history may not be documented for a variety of reasons. It has yet to be seen whether the long-term outcome from the use of HBO in such cases is any different from that of acute cases. Yet, as many acute cases are nonaccidental, the two groups may be difficult to compare using neuropsychological testing.

More challenging to HBO treatment providers is the demand for randomized controlled trial data, comparing the use of HBO and normobaric oxygen for severe CO poisoning. The provision of this service may soon be questioned, in a national climate of increasingly critical analysis of outcomes from many medical treatments. As this treatment is sometimes expensive, analysis of cost-effectiveness will also become essential.

With this in mind, members of the BHA have entered into discussions with the National Poisons Information Service (London), with a view to designing an appropriate trial. Data from trials conducted abroad may not provide information relevant to a British population, as sources and durations of exposure to CO may not be comparable.

Obstacles to such research may prove difficult to surmount. Randomization of some patients not to receive HBO treatment may be ethically questioned. In addition, well-validated neuropsychological testing must be used, with a follow-up of perhaps as long as 2 years.

14.7   THE BRITISH HYPERBARIC ASSOCIATION: CONVERGENT STATEMENT

To carry out any useful analysis of the clinical effectiveness of HBO treatment for severe CO poisoning in Britain, it will be necessary to standardize practice as far as possible.

While the referral patterns for these cases are variable throughout the country, there is some consistency provided by the National Poisons Information Service (London) and some of its counterparts elsewhere. Furthermore, the national advice line provided by the Diving Diseases Research Centre in Plymouth has reinforced the criteria now in use in most of the poisons information centers.

Consequently, the members of the BHA have arrived at a statement of convergence in the management of these cases, based on a draft prepared by Dr. Andrew Pitkin of the Institute of Naval Medicine in Portsmouth. The following is a summary of this agreement.

Selection — The following criteria should be used to select which patients would be offered HBO treatment:

1. *Neurological* — The presence of any neurological or neuropsychiatric sign or symptom (other than mild headache or nausea) at any stage during or after definite exposure to CO. Particular reference should be made to elements of physical examination commonly omitted, such as hearing, balance, and gait.

2. *Neurological* — Cognitive impairment, especially arithmetic skills and memory, with or without an abnormal mini-mental score.
3. *Neurological* — History of unequivocal loss of consciousness.
4. *Cardiovascular* — Evidence of myocardial infarction or ischemia (chest pain, abnormal electrocardiogram or cardiac enzymes).
5. Pregnancy of any gestation.
6. Inability to assess adequately (e.g., concurrent drug overdose).

COHb levels are useful to confirm the diagnosis, particularly where there was a possibility of more than one diagnosis. They were commonly used as a guide to the level of severity of poisoning; they had been shown to precipitate discussion with hyperbaric facilities by which advice could be offered to referring hospitals; and they were commonly used as an end point to determine the duration of face mask oxygen for those patients not receiving HBO. However, COHb levels had consistently been shown not to relate well to the prognosis from CO poisoning, and should not be used as the sole selection criterion for HBO treatment.

If none of these criteria were initially found positive, the patient should receive 100% oxygen by face mask and be reassessed after 2 h. If any of these criteria were then positive, the patient should be referred for HBO.

If, however, there was still no abnormal finding, the patient should receive 100% oxygen by face mask for at least 10 h in total (see below).

Contraindications to HBO treatment include:

1. State of patient unfit for interhospital transfer, or nonavailability of suitable hyperbaric facility.
2. Patient dependency unsuitable for the individual hyperbaric unit.
3. Patient refusal.

It is accepted that clinical judgment would be necessary in every case.

14.8 ASSESSMENT

Assessment in the hyperbaric facility before treatment should be carried out by or in the name of a physician qualified in undersea and hyperbaric medicine.

The unconscious patient should be assessed according to vital signs and any other observations appropriate to the critically ill state, taking into account any available medical history. The conscious patient should be asked for details of history, including previous medical problems. Examination should include standard assessment of respiratory and cardiovascular systems. Neurological assessment should include:

1. Gait and balance;
2. Limb reflexes;
3. Clonus;
4. Nystagmus;
5. Coordination of upper and lower limbs;
6. Cranial nerves, including smell and gag reflex;

7. Romberg's test;
8. Sensory testing, where possible;
9. Motor capacity;
10. Glasgow Coma Score;
11. Mini-mental score.

For a patient with impaired consciousness, it may not be possible to evaluate the above signs. Neuropsychiatric testing may not be appropriate before the first treatment with HBO. Investigations should include:

1. Cardiac enzymes;
2. Electrocardiogram, repeated later in case of any abnormality;
3. Chest radiograph;
4. Arterial blood gases in artificially ventilated patients, or when metabolic acidosis is suspected;
5. Drug screen for paracetamol, salicylate, opiates, and other incidental central nervous system or cardiovascular depressants;
6. COHb level, if not taken at referring hospital.

Records of these assessments would best be based on a standard computer-based form, which could be used by all hyperbaric facilities.

14.9 TREATMENT ALGORITHMS AND SCHEDULES

14.9.1 NORMOBARIC OXYGEN ADMINISTRATION

Oxygen administered by face mask prior to or instead of HBO should be through a mask with at least a reservoir bag to prevent rebreathing, with an oxygen flow rate of at least 10 l/min. It was realistic to acknowledge that interruption of face mask oxygen would be necessary during clinical assessment.

However, the preferred method of oxygen administration would be by tight-fitting oronasal mask, such as an anesthetic mask, CPAP mask, or aviator's mask.

If no significant symptoms or signs were found, and HBO was not given or arranged immediately, the patient should be reassessed 2 h after the first assessment, to check on any missed significant symptoms or signs. If these were again not found, oxygen should be given by face mask for at least a further 10 h in total. Further oxygen would only be given in the case of residual symptoms.

14.9.2 HYPERBARIC OXYGEN ADMINISTRATION

If significant symptoms or signs were found, HBO would be arranged through a local hyperbaric unit, or through one of the following two agencies:

1. Duty Diving Medical Officer, Institute of Naval Medicine, Portsmouth. Phone: 0831-151523.
2. Diving Diseases Research Centre, Plymouth. Phone: 01752-209999.

HBO therapy should consist of a minimum of 2.8 atm abs 100% oxygen for at least 60 min, decompressing according to a schedule applied for the individual patient's requirements and safety.

Further treatments of at least 2.4 atm abs 100% oxygen for at least 60 min will almost always be required, based on whether there is recurrence or persistence of symptoms or signs. These treatments should be between 2 and 12 h after the first treatment. Treatments should be repeated until symptoms and/or signs fail to show sustained improvement. It is not necessary to give oxygen by face mask between HBO treatments.

Admission to the hospital would be desirable, but not essential. However, if this was not done, a telephone assessment of the patient should be carried out between 2 and 12 h after the first treatment.

Intensive Care Algorithm — Patients with any of the following should receive intensive care support:

1. Glasgow Coma Score ≤ 10;
2. Compromised airway;
3. Known or suspected aspiration;
4. $P_aCO_2 > 8$ kPa, or $P_aO_2 < 12$ kPa;
5. Inability to deliver 100% oxygen by other methods.

Any patient with cardiovascular instability or requiring tracheal intubation, according to the criteria outlined below, should proceed to intensive care management. At the referring hospital, intensive care should be initiated, with appropriate hemodynamic and respiratory monitoring.

A place in a Category 1 HBO unit should be sought, using the advice agencies listed above. In a Category 1 HBO facility, initial HBO treatment should be at least 2.8 atm abs 100% oxygen for at least 60 min. A further treatment of at least 2.4 atm abs 100% oxygen for at least 60 min should be considered in every case. Thereafter, further HBO may carry the risks of oxygen toxicity and/or barotrauma, such that it should only be considered in the case of documented clinical neurological abnormality, for which an MRI scan may prove useful. There is no fixed maximum on the number of HBO treatments that should be given.

After HBO treatment is completed, conventional intensive care and medical management should be continued as dictated by the individual case.

14.9.3 Discharge

A final clinical assessment should be carried out before discharge, with particular attention to clinical symptoms and signs initially found to be abnormal. A follow-up appointment should be arranged. Account should be taken of the possibility of late sequelae arising up to 3 weeks after exposure.

14.9.4 Follow-up

A standardized questionnaire should be used to assess patients, probably by telephone.

14.10   CURRENT OUTCOMES

During the BHA 1993 to 1996 study, patients attending the Whipps Cross Hospital Hyperbaric Unit were invited to return 8 to 12 weeks after their treatment, for follow-up assessment. Of the 50 patients (26%) who did so during the period of the study, 37 complained of further symptoms after discharge, including memory loss, head-aches, and fatigue. While 44 were normal on neurological examination, 6 were found to have persistent short-term memory loss. Post-treatment symptomatology was not significantly different between acutely and chronically poisoned patients.

While attendance at such a follow-up clinic might be self-selecting for those with continuing problems, the existence of such continuing ill-health should no longer be doubted. Studies are in progress to follow up patients who have been treated with HBO, at up to 1 year after discharge, using neuropsychological testing. Identification of retrospective control patients, treated with normobaric oxygen, is difficult.

14.11   CONCLUSIONS

Perceptions of CO poisoning by the public and by physicians are not clear or accurate. The conventional management of this condition includes oxygen; yet this may be inconsistently applied.

The rapid development of safe and well-organized HBO treatment for severe CO poisoning in Britain has been achieved as a result of cooperation within the BHA. The BHA has drawn up an algorithm for the management of CO poisoning, which is consistent with the advice given by the National Poisons Information Service (London).

There is no doubt that a prospective randomized clinical trial is necessary in a British context, to assess the clinical effectiveness of this form of treatment for severe CO poisoning.

Most HBO treatment providers are content to take part in this work, expecting it to define better those patients for whom this treatment is appropriate.

REFERENCES

1. McDowall, D.G., Ledingham, I.M., Jacobson, I., and Norman, J.N., Oxygen admin-istration by mask in a pressure chamber, *Anesthesiology*, 26, 720–726, 1965.
2. First European Consensus Conference on Hyperbaric Medicine, Reports and Recom-mendations, Wattel, F. and Mathieu, D., Eds., Lille, September, 1994. Available from ASPEPS, Batiment Pierre Swynghedauw, Rue du 8 Mai 1945, 59037 Lille CEDEX, France.
3. Working Party of the British Hyperbaric Association, Gough-Allen, R., Chairman, Fire Safety, British Hyperbaric Association, London, 1994.
4. Working Party of the British Hyperbaric Asociation, Ross, J., Chairman, Electrical Safety, British Hyperbaric Association, London, 1995.
5. Working Party of the British Hyperbaric Association, Ross, J., Chairman, A Core Curriculum for Training and Education in Hyperbaric Medicine, 1999.

6. Working Party of the Faculty of Occupational Medicine, Cox, R., Chairman, A Code of Good Working Practice for the Operation and Staffing of Hyperbaric Chambers for Therapeutic Purposes, Faculty of Occupational Medicine, Royal College of Physicians, London, 1994.
7. Working Party of the British Hyperbaric Association, Colvin, A., Chairman, Gazetteer of Members, British Hyperbaric Association, London, 1998.

15 Carbon Monoxide Air Pollution and Its Health Impact on the Major Cities of China

Qing Chen and Lihua Wang

CONTENTS

15.1 CARBON MONOXIDE IN AMBIENT AIR IN CHINA'S MAJOR CITIES

Before 1980, the traditional method to make use of energy was the burning of coal directly. About 80% of fuel is used in this way in the whole country. A great number of industrial furnaces have rather short chimneys. The exhaust air is released into the ordinary air. Hence, ashes and substances that have not been burned completely spread in large measure into the living environment. Vast territories in north China have a long heating season of about 4 to 6 months per year. Small stoves are very common utensils for cooking and keeping warm in a household. When coal is burned in a stove, a great number of ash particles are exhausted with the gases. Among the exhaust gas, sulfur dioxide (SO_2), carbon monoxide (CO), and other compounds are pollutants harmful to human health. The annual average concentration of total suspended particulate (TSP) in urban districts is 220 to 500 mg/m³ and the annual average concentration of SO_2 is 70 to 150 mg/m³. Thus, the environment of the city is normally highly polluted. This is the so-called smoke-type pollution.[1]

TABLE 15.1

Concentrations of CO in the Air in Some Cities of China (mg/m³) (1979 to 1980)

District	Type of District			
	Busy Traffic	Center of Urban	Industrial	Clean
Beijing	12.3	5.5	5.1	1.3
	(highest value = 68)	(highest value = 35.6)	(highest value = 113)	
Shenyang	8.08	7.72	5.99	4.97
Lanzhou	12.20	10.2	4.2	—
Nanjing	2.39	3.03	1.14	1.15
Average values of 30 Chinese cities	2.40–12.20	2.10–6.90	1.10–6.30	1.5–2.30

Source: Ye, Z., *J. Ind. Health Environ. Med.* [in Chinese], 4, 174–177, 1981. At heating/winter season (in Nov., Jan., March).

. CO concentration in the ambient air varies considerably, not only in suburban areas, but also within cities. The daily concentrations of CO in urban districts of 30 cities in China in the 1980s are shown in Table 15.1.[2] Average daily CO levels of busy traffic areas, centers of urban districts, industrial districts, and "clean" districts were about 2.4 to 12.20, 2.10 to 6.90, 1.10 to 6.30, and 1.5 to 2.30 mg/m³, respectively. Much higher concentrations occurred in north China (e.g., Beijing, Shenyang, Lanzhou). The highest CO level of 113 mg/m³ was observed in the industrial district of Beijing city.

From 1981 to 1991, the Beijing Municipal Center for Hygiene and Epidemic Control was responsible for the affairs of the air-monitoring system in Beijing. It cooperated with the World Health Organization (WHO) in the program of Global Air Monitoring. TSP and SO_2 were the major parameters monitored. Since 1992, the Environmental Protection Agency and the Environmental Monitoring Center of Beijing have had this responsibility. More than 50 fixed-site monitoring stations were constructed in urban and suburban areas. Eight automobile-monitoring stations were established with monitoring indexes for TSP, SO_2, CO, NO_x, O_3, benz(a)pyrene (Bap), and Pb, while 27 monitoring stations were located in busy traffic areas monitoring NO_x and CO. The annual average concentrations of CO in Beijing (1984 to 1997) are seen in Figures 15.1 and 15.2.[3]

Figure 15.1 shows the trend of the annual average daily concentrations of CO in both urban areas and the traffic environment in Beijing. The concentrations of CO for the whole year are mostly below 4 mg/m³. The concentrations of CO during the heating season (approximately November 7 to March 20) are generally higher than for the nonheating season. The average levels of CO in busy traffic areas are the highest, ranging from 4 to 7 mg/m³ and showing an increasing trend from 1984 to 1997.

Figure 15.2 shows the annual average concentrations of CO at the automobile-monitoring stations in 1997, including a station for control (1), a station for urban areas (2, 3, 4, 5), and stations for the suburban area (6, 7, 9). It is very clear that the levels of CO in the center of urban areas are much higher than for suburban areas, especially during the heating season.

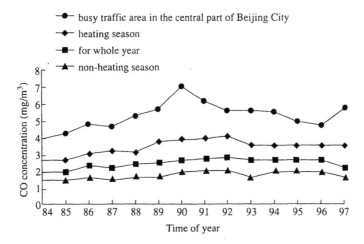

FIGURE 15.1   Annual average concentration of ambient CO in urban and suburban areas, and in the busy traffic area of Beijing city. (From Government Report, 1990 to 1997.)

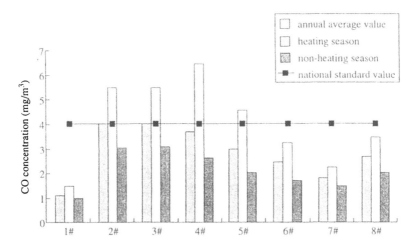

FIGURE 15.2   Annual average concentration of CO at the automobile monitoring stations of Beijing city. (From Government Report, 1997.)

Associated with the rapid development of cities, the ownership of automobiles has increased by 13% annually since the 1980s (Figures 15.3 and 15.4). In 1997 there were 11,000,000 automobiles or mobile vehicles in the whole country. There were about 1,200,000 automobiles in each of four cities, Beijing, Tianjin, Shanghai, and Canton. This has resulted in increased total emissions from automobiles (Figure 15.5). This has been due to improper operation, inefficient practice or combustion techniques, and unsatisfactory controls on vehicle emissions. Pollution due to automobile use is becoming serious.

Emissions from automobiles are increasing year by year, including CO as one component. An investigation showed that 3000 kg of CO was discharged from every

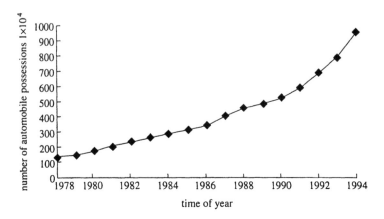

FIGURE 15.3   Increase in the number of automobiles in China. (From He, K., *J. Environ. Sci.* [in Chinese], 17, 80–83, 1996. With permission.)

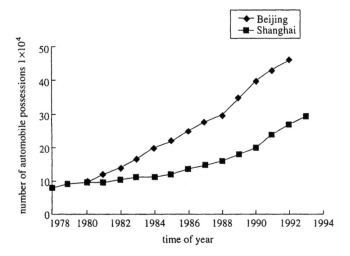

FIGURE 15.4   Increase in the number of automobiles in two Chinese cities. (From He, K., *J. Environ. Sci.* [in Chinese], 17, 80–83, 1996. With permission.)

1000 automobiles each day. In 1994 this was four times what it was in 1980. Vehicle CO emissions in Beijing, Shanghai, and Canton accounted for about 48 to 70% of the total emissions. The concentration of ozone also appears to be increasing. The number of automobiles in China in the year 2000 is 18,000,000 to 21,000,000. The emissions of NO_x and CO would reach 1,190,000 and 14,120,000 tons, respectively, in 2000 unless pollution control is undertaken.[4] It is likely that the urban air pollution might change from a coal-smoke type of pollution to a photochemical-smog type. Because of the rapid increase in the total amount of CO that is produced chiefly from coal combustion (both industrial and civil purposes) and the great emission from automobile exhaust gas, it will certainly cause harm to human health. No matter what type of pollution develops, this is a considerable problem to which attention should be directed.

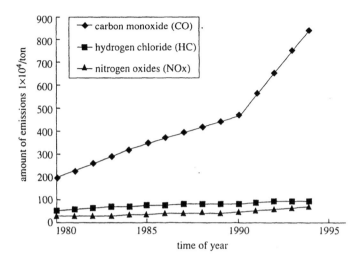

FIGURE 15.5 Quantity of emissions from automobiles in China. (From He, K., *J. Environ. Sci.* [in Chinese], 17, 80–83, 1996. With permission.)

15.2 HEALTH EFFECTS OF AUTOMOBILE EXHAUST ON TRAFFIC POLICE

Traffic police are maximally vulnerable to automobile exhaust. They have long working hours — usually 9 to 10 h/day. They are required to be alert when at work and have only short rests. Traffic police who are on duty at busy intersections are exposed to more than one kind of harmful pollutant coming from automobiles. The emissions contain CO, NO_x, hydrogen chloride (HCl), lead aerosol, particulate, and organic volatiles. The CO concentration in the environment of these police is closely related to traffic flow rate (i.e., number of automobiles per hour), controls on automobile exhaust pollutants, and so forth. Differences in CO levels depend on location (urban, suburban), what is happening (rush hour, nonrush hour), the season (summer, winter), etc. In Fuzhou, capital of Fujian Province,[5] CO concentrations at intersections were 3.52 to 45.6 mg/m^3. Carboxyhemoglobin saturation (COHb%) of the traffic police was 2.3%. These individuals often suffered headache, dizziness, insomnia, amnesia, tiredness, anxiety, etc.

In Beijing,[6] the ownership of automobiles is high. Traffic flow is heavy. CO concentrations at the intersections were 35.9 (21.0 to 44.5) mg/m,^{3} which is about six- to sevenfold what it was in the 1980s. The COHb saturations of 550 traffic police were measured. The average value was 6.8%, which is about 1.7 times higher than that in 1983 (i.e., 2.5%). This shows that the concentration of CO at Beijing traffic crossings has increased significantly.

In recent years, traffic in Shanghai city has improved along with the reconstruction of the urban districts. Nonetheless, air pollution due to traffic is in some places substantial.[7,8] The Dapu Highway Tunnel beneath the Huangpu River is one example. This tunnel was built 20 years ago. Present-day traffic volume has already surpassed designed capacity. CO concentration in the tunnel is usually around 50 mg/m³. The COHb saturation of traffic police after work was found to be 4.0 to 8.8%.

When blood COHb reaches about 4%, longer reaction times, vigilance decrements, and changes in visual sensitivity occur.[2,22] When blood COHb reaches 6 to 8%, decreases in both maximal work time and maximal oxygen uptake take place. This suggests that sensory function and judgment will be impaired, and people will not be able to work to their best for as long as usual. Hence, municipal improvements related to CO pollution at traffic crossings and highway tunnels should be considered.

Besides CO, there are more than 100 other organic components in automobile exhaust. Shunhua et al.[9] systematically studied the harmful effects of motor-vehicle exhaust gases on human health. They used a series of assays representing different end points, such as gene, DNA, chromosome, cells, and animals to assess the risk of genetic toxicity and the harmful effects of various (gasoline-powered and diesel-powered) vehicle exhaust particulate extracts, especially that of mutagenicity and potential carcinogenicity. At the same time, they studied the health of the exposed workers. The results were as follows:

1. They all showed positive effects in the Ames assay, UDS, micronucleus assay and SHE cell malignantly transformed assay.
2. The extracts caused transformed foci in KMB-13 cells, in which oncogene proteins such as c-myc, p21, bcl-2 were elevated.
3. The inhaled extracts harmed experimental animals' pulmonary and immune function.
4. Respiratory symptoms (cough, nasal, obstruction, sore throat and short-ness of breath) occurred more often, and the risk of chronic obstructive pulmonary disease rose in occupationally exposed workers.

These results suggested the extracts had carcinogenic potential.

Song et al.[10] used the scrape-loading and dye transfer technique (SLDT) to study the effects of diesel exhaust particulates (DEP) on gap-junction intercellular communication (GJIC). This study demonstrated that DEP extract caused inhibition of GJIC in Balb/c3t3 cells. It was suggested that inhibition of GJIC may be a cause of cell transformation.

Yang et al.[11] compared the composition and mutagenicity of engine exhaust produced by leaded and lead-free gasoline. The composition and mutagenicity of the organic exhaust particulate extract was examined with gas chromatography, mass spectrometry (GC/MS), and an *in vitro* CHL cell micronucleus assay. There was no significant difference in mutagenicity of the organic extracts from the engine exhaust gases produced by the two kinds of gasoline. The lead-free gasoline-fueled engine produced less CO, HCl, and particulate in the exhaust gas.

In summary, many measures should be considered to protect populations at risk. These might include control of the urban population and the magnitude of traffic volume, improvement of city construction projects and the functioning of automobiles, and promotion of the use of lead-free vs. leaded gasoline.

TABLE 15.2
CO Level (mg/m3) in Residential Rooms in Four Cities (geometric value)
(1987 to 1988)

		Chengdu		Shenyang		Shanghai		Wuhan	
Season	Fuel	Kitchen	Bedroom	Kitchen	Bedroom	Kitchen	Bedroom	Kitchen	Bedroom
Winter	Coal	7.39	7.11	7.46	6.25	14.07	13.7	13.46	9.91
	Gas	5.20	4.91	4.69	4.32	3.45	3.58	4.83	4.38
Summer	Coal	2.41	1.82	3.43	2.92	13.8	7.00	9.67	6.07
	Gas	2.24	1.56	2.65	2.16	2.26	2.03	1.59	1.37

Source: Qin, Y., *J. Environ. Health* [in Chinese], 8, 100–102, 1991.

15.3 LEVELS OF INDOOR AIR CARBON MONOXIDE IN THE RESIDENTIAL DISTRICTS

An investigation was carried out in 1987 to 1988 to study indoor air pollution in summer and winter in Shanghai, Wuhan, Chengde, and Shenyang.[12] Shanghai and Wuhan are south China cities while Chengde and Shenyang are north China cities. The 30 households examined were classified by the fuel used into a "coal-as-fuel" group and a "gas-as-fuel" (natural gas or liquified gas) group. The pollutants studied were respirable particulates, SO_2, NO_x, and CO. The results read as follows:

1. *Seasons*: Winter lasts for about 4 to 6 months in north China. Prior to the 1980s, most urban households had one stove for both cooking and keeping warm. The fuel mostly used was coal, resulting in a great quantity of coal fumes being produced. Because the residents kept rooms with poor ventilation warm during the long winter, indoor air pollutants were higher than in summer. Generally, it was found that the levels of indoor CO (average value in kitchen and other rooms) in winter were about two to five times higher than in summer (Table 15.2). This is the general situation in the above-mentioned four cities.
2. *Fuel*: No matter whether in summer or in winter, no matter whether it be in the north or in the south, the levels of indoor CO are about three to five times as high when coal is used as fuel as when gas is used as the fuel.
3. The CO concentration in the kitchen air is high. It is the source of pollution of indoor air. The highest average CO concentration in kitchen air in winter may reach 13 to 14 mg/m³.

15.3.1 METHODS OF HEATING

In the 1980s[13] in China there were three major ways for urban residents to keep the indoors warm:

TABLE 15.3
The Concentration of CO in Indoor Air When Three Kinds of
Fuel Are Used

		Conc. of CO	
Kind of Fuel	Sampling Spots	In Summer	In Winter
Natural gas	Kitchen	7.1 (3.8–12.5)	10.8 (1.9–16.3)
	Dwelling rooms	3.5 (1.4–10.0)	5.2 (1.9–12.5)
Coal gas	Kitchen	6.5 (4.4–10.6)	10.9 (6.3–32.5)
	Dwelling rooms	2.7 (1.9–5.0)	6.8 (3.8–15.0)
Liquefied petroleum gas	Kitchen	20.3 (5.0–56.3)	21.8 (8.8–48.8)
	Dwelling rooms	5.4 (2.5–0.3)	10.3 (5.0–31.3)

Source: Wang, J. et al., J. Environ. Sci. [in Chinese], 16, 44–46, 1995.

1. The use of central steam-heating systems.
2. The use of small furnaces operated by government offices, institutions, or enterprises.
3. The use of small stoves in every household used both for cooking and keeping warm.

The average CO concentrations found were 3.72 (0.6 to 18.6) mg/m^3, 8.86 (2.22 to 18.8) mg/m^3, and 15.42 (3.2 to 62.5) mg/m^3, respectively. The highest level of CO pollutants came from heating stoves fueled by coal. Fuels for daily living are of more than one kind. In addition to coal, there is natural gas, coal gas, and liquified petroleum gas.

A study by Wang et al.[14] described the differences in indoor air pollution when these three kinds of fuel were used (Table 15.3). They analyzed the organic volatile substances, NO_x, CO_2, and CO in the air when the three kinds of fuel are burning, finding that natural gas is the most effective fuel. It produces the least quantity of particulate and CO, together with a moderate quantity of NO_x and CO_2. Liquified petroleum gas is the least effective fuel of the three, as it produces the greatest amount of CO. Obviously, for the future, clean fuels should be used. In Beijing in 1993, 240,000 households used natural gas as fuel, a volume of 1.5×10^9 m^3. In the year 2000, as the number of customer households increases, the gas supply also increases to 8.5×10^9 m^3 per year.

15.3.2 TYPES OF RESIDENTIAL QUARTERS

In the 1980s there were three major types of urban residential quarters in Beijing city.

Type 1 consists of "dwelling-units" like apartments. In such units the living rooms are separate from the kitchen. There is a stove in the kitchen fueled by natural gas. The trend in constructing residential quarters in urban districts is such dwelling units in multistory buildings (10 stories or more). The units are supplied with central steam-heat in winter.

Type 2 consists of single-story houses in a row. This is a rearrangement of the original Beijing dwelling house, the *Si-he-yuanr*. *Si-he-yuanr* is a compound with houses around a square courtyard, or quadrangle. Thus there are usually three or four rows of houses in the compound. Each row of houses may contain from three to seven rooms. One compound usually connects with the smaller or larger compound next door. Before the founding of the People's Republic in 1949, this was the conventional style of residential quarters in Beijing. Each family usually owned one *Si-he-yuanr*. Recently, as the population in the cities has outstripped construction, the *Si-he-yuanr* has been modified to house more than one family. In such a household, the kitchen and living rooms can be separate. If the household has only one room, the kitchen and livingroom are the same. As a whole, stoves are scattered about in such households for cooking and keeping warm. Before the 1980s this was the most common type of dwelling arrangement for the vast number of urban residents in Beijing and other cities.

Type 3 are three- to five-story dwelling buildings. They have two rows of equal-sized rooms with a public-owned passage between. This is similar to the dormitories of colleges and universities. The passage is used by each family as a place to put the cooking stove. While these buildings have central heating, ventilation is relatively poor. They house mainly young people, who later marry and gradually move on to new dwelling units.

A comparison of the CO concentrations in these three types of housing units showed the following.[15] The CO concentration in Type 1 was mostly below 4 mg/m^3, and COHb% of the residents was below 2%. The CO level in Type 2 was mostly above 4 mg/m^3, the highest value (in winter) occasionally reaching 52 mg/m^3 (original value was 47 ppm). The COHb% of the residents was about 6.5% (in winter) and 2.0% (in summer). The CO concentration in Type 3 was 4.5 to 14.3 mg/m^3 (original value wase 4 to 13 ppm), for 5 to 7 months of the year, while the COHb% of the residents was 1.3 to 4.4%. Residing in Type 3, with the longer duration of CO exposure, could result in more harm to the residents.

During the 1990s there was a gradual implementation of urban reconstruction, resulting in many new buildings and houses in urban and rural areas of China. Based on Beijing government data in 1990 and 1997, the average living space per person increased from 7.7 m^2 to 9.66 m^2. The percent of households with central heating increased from 21.2 to 34.5%. The percent of gas-burning stoves for cooking in one family in a residential district increased from 84 to 94%. People living in Type 2 and Type 3 dwelling buildings were able to leave their old dwelling and move into new ones.

There is more than one kind of dwelling unit in the apartment buildings: units of one room and one entrance corridor; units of two rooms and one entrance corridor; units of three rooms and two corridors; and so forth. In each unit there are bedrooms, a kitchen, a toilet, a livingroom, corridors, etc. Living conditions and the average living space have markedly improved since the 1980s. Based on a study of two residential communities in the Western District of Beijing proper,[16] we know that the average living space there was 6 m^2 to 15 m^2 per person. The building designs were good, including adequate ventilation and natural lighting. CO concentrations in the livingrooms and bedrooms, both in the heating season and in the nonheating in all the major cities of China is similar to those in Beijing. As living quarters are being improved, CO pollution is gradually being controlled.

15.3.3 THE KITCHEN

Usually the kitchen in a house is the main source of pollution. Although the design of kitchens in apartment houses has improved, the limited space (3 to 4 m²) at present still causes inconvenience. In winter, especially when the doors and the windows are tightly closed, or when cooking takes longer than usual, or due to lack of a ventilator or ventilating fans, the CO concentration might increase rapidly to 10, 20, or above 48 mg/m³. The water heater ("geyser" or "gayster")[17] is usually situated in the kitchen. Cases of acute CO poisoning sometimes occur, as a result of the limited kitchen space, the presence of the geyser, and especially a cooking fire that is burning incompletely.

The types of houses, the living quarters, the methods of heating, and the composites of the fuel used are very different in the countryside and villages of various geological regions of China. There is not enough systematic data at the present stage for a study in this field. There is a report[18] about the condition of five counties in the Yangtze Gorge region. This is in the border region of Hubei and Sichuan provinces, at an altitude of 800 to 1200 m, in the southwestern region of China. People there arrange their living quarters as follows. They build an open fire in the center part of the room, and cook their meals and warm the dwelling house with this earth oven, or "fire-pool." No chimney is provided for the fire. The fuel used is mostly coal, rich in fluoride and sulfide. The weather in this region is frigid and damp, so the fire-pool usually is kept burning all year round. All activities, including the eating of meals, gatherings, and reunions of family members are held around the fire. Without question, the air in the room is very much polluted. It contains large amounts of fluoride, SO_2, particulates, compounds of carbon and hydrogen, some metallic elements, and others. Studies show that the daily average CO concentration is 31 to 84 mg/m³, with fluctuations of 11.3 to 310.0 mg/m³. This CO concentration is the highest recorded in the whole country. To correct this serious indoor air pollution, the local residents changed their traditional open earth oven to an oven with a chimney tunnel. In doing so,[19] indoor air pollution was markedly reduced. All the pollutants, including TSP, Bap, SO_2, and CO were reduced to about 10% of those before adding the chimney. Although the example described here is unique in that region, it is still a matter needing attention.

15.4 DERIVATION OF HYGIENIC STANDARDS FOR CARBON MONOXIDE IN AMBIENT AIR

From 1949 to 1982, the Chinese "Hygienic Standards for Industrial Enterprise Design" (TJ36-79) was based on standards of the former Soviet Union. According to these standards, the highest permissible daily average CO concentration in ambient air was 1 mg/m³ over 24 h. The maximal CO concentration for 1 h was 3 mg/m³. In 1982 the Environmental Protection Leading Division of the State Council (formerly the National Environmental Protection Agency) developed new standards for ambient

TABLE 15.4
Standards of CO in Ambient Air by Country

Country	Standard for Longer Time			Standard for Shorter Time		
	mg/m³	ppm	Duration, h	mg/m³	ppm	Duration, s
Former Soviet Union	1.0	0.9	24	3.0	2.7	30
Former East Germany	1.0	0.9	24	3.0	2.7	30
Czechoslovakia	1.0	0.9	24	6.0	5.4	30
Bulgaria	1.0	0.9	24	3.0	2.7	3.0
Poland	0.5	0.45	24	3.0	2.7	3.0
Romania	2.0	1.8	24	6.0	5.4	3.0
Former Yugoslavia	1.0	0.9	24	3.0	2.7	30
Hungary	1.0	0.9	24	3.0	2.7	30
People's Republic of China	1.0	0.9	24	3.0	2.7	30
United States	10	9	8	40	35	1 h
Canada (accepted level)	15	13	8	35	30	60
Canada (prospective level)	6.0	5.0	8	15	13	60
Former West Germany	10	8.6	8	40	35	60
Japan	11.5	10	24	23	20	8 h
Italy	23	20	8	57.7	50	30
Finland	10	9	8	40	35	60

the National Environmental Protection Agency) developed new standards for ambient air quality (GB3095-82), recommending more flexibility for air exposure limits. That is, ambient air quality control could be met at different CO levels in different areas. Three different area types, each with its CO exposure limit, were established by the Environmental Protection Agency of China. The Type 1 area contains nature preservation zones, landscape and scenic spots, and locations that need to be protected. They observe the first grade standard. The Type 2 area contains areas of residence, commerce, traffic, cultural activities, general industries, as well as rural areas. They observe the second grade standard. The Type 3 area involves special industrial areas, which observe the third grade standard.

Based on the ambient CO concentrations in major Chinese cities at that time, combined with the state of the national economic development, and the expertise and experience of the regulatory agencies, Wang's group (Wang Guanqun, Song Huaqin, Wen Tianyou, Wang Zhengang, Song Hongpeng, LiuJunzhuo, et al.) recognized that the CO exposure limits in the standard (TJ36-79) were too strict to be instituted at that time. They proposed 3 mg/m³ and 9 mg/m³ as the average daily value and the maximal value rather than 1 and 3 mg/m³, respectively.[21] The bases for their conclusions are as follows:

1. Having reviewed information on environmental health standards in various nations (Table 15.4), they recognized that there was a great difference in the methods and principles of developing CO standards and critical indicators. Table 15.4 shows that the limit values are so different that they could be divided into two groups: one group of nations included the former Soviet Union, and the other included the United

lower than those of the latter. The questions could be posed, "What methods and principles should be used in the development of standards? Which critical indicators should be selected for an evaluation of CO exposure? All of these should be taken into consideration for further study.

2. Blood COHb level is accepted as a biological indicator of CO exposure.[22] As is well known, CO is absorbed from the lungs into the blood vessels. The final step in the process involves the competitive binding between CO and O_2 to hemoglobin (Hb) in erythrocytes to form COHb and O_2Hb, respectively. The binding of CO to Hb produces COHb, which decreases the O_2-carrying capacity of the blood. This is probably the principal mechanism behind the toxic effect of low-level CO exposure. The toxic effects produced by COHb formation are likely to include the induction of a hypoxic state in various tissues and organs, and metabolic dysfunction. Wang's group decided to use COHb% as the critical end point of CO exposure instead of the original inhibition of metabolism of phyrin and reversible morphological change in rat brain which had been used for development of standards in the former Soviet Union.

3. It was important to establish the maximal level of COHb that would still protect the general population.

In Wang Zhengang's report,[23] the COHb levels of 192 college students (non-smoking, healthy adults, age 17 to 27, male 86, female 106, no occupational exposure history) were measured in Beijing during the nonheating season. The mean of the COHb level was 0.6%, with an upper limit at 95% confidence of 1.2%. It might be noted that COHb levels of residents in urban districts have usually been below 1%, with maximal levels seldom exceeding 2%.

An epidemiological study of the correlation between COHb and visual reaction time was carried out.[21] Students 12 to 13 years of age from five primary schools were examined for COHb saturation and visual reaction time. Three of the five schools were situated in a heavily polluted district, while the other two were in a lightly polluted district. A high correlation ($r = 0.83$) between COHb level and delay in time of visual reaction was obtained. When COHb reached 2.02%, the visual reaction time of pupils from the heavily polluted district was significantly delayed.

According to Guan Ren's report (cited in Ye Zheng[2]), when the COHb saturation reached 2% in volunteers (after an exposure to 12 ppm CO), a decrease in memory function was observed. In the Environmental Health Criteria,[22] maximal values of 2.5 to 3.0% COHb are recommended for the protection of the general population, including those with impaired health.

Based on the above findings, Wang's group suggested that a level of <2% COHb be one of the criteria in the ambient CO standard.

4. It is believed[22] that normal blood is in equilibrium when ambient CO is at a concentration of approximately 5 mg/m³ (4.3 ppm). When the concentration of CO is below 5.0 mg/m³, elimination of CO from the blood will occur. The CO concentration in ambient air should not exceed this level.

5. Although an ideal dose–response relationship involving CO is not yet available, it is recognized that there is an exposure–effect relationship between the concentration of air CO and COHb saturation. In Wang's study,[21] 253 air samples for measuring CO were collected in four locations in urban districts of Beijing:

TABLE 15.5
**Ambient Air Quality Standard for CO in the National Standard of the People's
Republic of China (GB3095 — 1996)**

	Concentration Limits[a]		
Area/Grade	Average Daily Conc.[b] (mg/m³)	Maximal Conc. for 1 h[c] (mg/m³)	Interpretation
Type 1 area (first grade)	4.00	10.00	Areas of nature preservation, landscape, scenic and historic spots
Type 2 area (second grade)	4.00	10.00	Areas of residence, traffic, commerce, cultural activities, and general industries
Type 3 area (third grade)	6.00	20.00	Special industrial areas

[a] Limits that should not be exceeded.
[b] Average concentration for a day at any time.
[c] Maximal concentration for 1 h at any time.

Source: Chinese Environmental Protection Agency (1996).

a. Facilities with central heating;
b. Facilities with small heating boilers;
c. Facilities with small coal-burning stoves;
d. Control location.

The average concentrations of air CO in these four locations were 2.8, 4.1, 15.8, and 0.7 mg/m³. The study collected 1217 blood samples from exposed persons for the measurement COHb level in these four locations. They were students from primary and middle schools and colleges, residents, and soldiers, all nonsmokers. A clear exposure–effect relationship was noted between ambient CO and the COHb level, according to the equation:

$$Y = 0.065\ X + 0.99,$$

where Y = COHb(%), and X = CO concentration in mg/m³.
According to this relationship, when the air CO concentration is 15 mg/m³, COHb will be about 2%, whereas, when COHb attains 1.2%, ambient CO is 3 mg/m³.
Based on a comprehensive evaluation of reports and the information discussed above, together with the WHO guidelines for CO (1979 and 1987), Wang's group suggested that 3 mg/m³ should be the average maximum CO concentration allowed over 8 to 24 h, while 9 mg/m³ should be the maximal CO concentration allowed over 1 h. These studies provided China with background values and scientific criteria for developing national CO standards. They were accepted by the Chinese Environmental

Protection Agency. These national standards were once again approved in 1996. They became part of the "Ambient Air Quality Standard" in the "National Standard of the People's Republic of China" (GB 3095-1996). This standard for ambient air CO level was issued and went into effect October 1, 1996 (Table 15.5).

15.5   RELATED STUDIES

Song and Wang[24] reported on the relationship between CO concentration and brain monoamine neurotransmitter. Many studies of this type have been carried out at high CO levels (above 1500 ppm). Based on the relatively lower levels of CO in ambient air to which humans are normally exposed, Song and Wang selected 11.5, 57, and 286 mg/m^3 of CO as the doses. Rats 21 days of age were used and exposure occurred over 28 days.

The amounts of norepinephrine (NE), dopamine (DA), and 5-hydroxytryptamine (5-HT) were measured on Days 1, 14, and 28 using fluorospectrophotometry. The amounts of NE, DA, and 5-HT decreased in the exposed groups at the beginning compared with controls, then increased gradually with duration of exposure. It was found that CO at 57 mg/m^3 is the critical dose for changing levels of NE, DA, and 5-HT. What is the significance of these results? Conceivably, NE, DA, and 5-HT might be used as biochemical indicators of CO neurotoxicity.

Gong and Liu[25] studied the effects of CO on the content of malanic dialdehyde (MDA) and glutathione peroxidase (GSH-Px) activity. Rats were exposed to CO for 2 h/day for 30 days. The exposure doses used were 25, 50, and 500 mg/m^3. Blood or blood plasma and heart and brain tissues of rats were obtained and assayed for MDA and GSH-Px, 10 days each. Plasma lipid peroxide increased significantly with 20 days exposure to CO at 500 mg/m^3; also in heart after 30-day exposure (Tables 15.6 and 15.7). These changes were not observed in the 25 and 50 mg/m^3 groups. Changes in blood and brain GSH-Px activity took place in rats at 25, 50, and 500 mg/m^3. The results indicate that CO induced free-radical formation, promoted lipid peroxidation, and affected the formation of MDA. It was suggested that GSH-Px might be a sensitive, early indicator of CO effects.

Gu and Lu[26] studied the relationship between delayed encephalopathy after CO poisoning (DEACMP) and free radical formation in humans. Blood serum samples were collected from 31 DEACMP patients and 30 healthy volunteer controls. The levels of vitamin E (VE), glutathione peroxidase (GSP-Px), a water-soluble fluorescense substance (WSFS), and lipid peroxide (LPO) were determined. As compared with the controls, the levels of LPO and WSFS in the patients were elevated, while the levels of GSH-Px and VE were low ($p < 0.001$). After 1 month of treatment, the measurements were repeated in the DEACMP patients. The levels of LPO, WSFS, GSH-Px, and VE had returned to near normal values. This study suggests that enhancement of lipid peroxidation induced by free radicals may be correlated with brain leisons in patients with DEACMP. Free radicals might therefore play a role in the initiation of this disease.

TABLE 15.6
Effect of Different Levels of CO Exposure on the MDA Content of Rat Tissues

Tissues	CO Conc. (mg/m³)	Exposure Duration (days) 10	20	30
Plasma	0	11.7 ± 0.5	6.2 ± 1.4	10.5 ± 1.2
(n mol /ml)	25	10.8 ± 1.7	7.2 ± 1.7	10.4 ± 1.5
	50	11.1 ± 0.9	8.1 ± 1.3	12.1 ± 3.1
	500	10.7 ± 1.2	13.9 ± 1.2**	11.5 ± 4.0
Heart	0	9.6 ± 1.2	7.8 ± 0.7	8.5 ± 0.8
(n mol/100mg wet weight)	25	10.0 ± 1.2	7.2 ± 0.6	8.4 ± 0.9
	50	9.4 ± 1.1	7.4 ± 0.3	9.7 ± 0.9
	500	10.1 ± 2.3	7.7 ± 0.7	10.8 ± 1.0**
Brain	0	18.0 ± 2.7	11.7 ± 1.6	12.6 ± 2.1
(n mol/100mg wet weight)	25	16.2 ± 3.7	11.3 ± 1.2	12.5 ± 2.2
	50	16.8 ± 1.5	12.7 ± 1.2	12.2 ± 0.8
	500	15.0 ± 2.0	12.4 ± 2.1	13.2 ± 3.1

** $P < 0.01$.

Source: Gong, C.H. et al., *J. Environ. Health* [in Chinese], 12, 111–114, 1995.

TABLE 15.7
Effect of Different Levels of CO Exposure on the GSH-Px Activity of Rat Tissues

Tissues	CO Conc. (mg/m³)	Exposure Duration (days) 10	20	30
Whole blood	0	20.2 ± 2.0	21.7 ± 2.3	10.7 ± 3.0
(u/ml·min)	25	18.1 ± 1.8	15.5 ± 1.1**	20.9 ± 2.7
	50	15.1 ± 1.5**	15.2 ± 0.9**	18.7 ± 3.3
	500	15.3 ± 1.9**	16.8 ± 2.4*	20.3 ± 1.4
Heart tissues	0	52.9 ± 1.1	49.4 ± 7.5	41.3 ± 2.1
(u/mg Pr·min)	25	50.9 ± 1.3	61.3 ± 5.5*	45.1 ± 4.3
	50	50.4 ± 1.2	66.7 ± 6.8**	45.9 ± 5.4
	500	55.3 ± 1.6	61.2 ± 4.0*	44.0 ± 5.0
Brain tissues	0	27.2 ± 2.8	10.5 ± 1.2	9.2 ± 0.9
(u/mg Pr·min)	25	22.3 ± 1.7**	9.7 ± 1.2	2.7 ± 2.1**
	50	17.9 ± 3.9**	16.1 ± 1.1**	11.1 ± 1.3*
	500	15.2 ± 3.7**	13.1 ± 1.8*	10.6 ± 2.3

* $P < 0.05$.
** $P < 0.01$.

Source: Gong, C.H. et al., *J. Environ. Health* [in Chinese], 12, 111–114, 1995.

REFERENCES

1. Song, R. and Cui J., An investigation of situation of ambient air pollution in the five cities in China, *J. Hyg. Res.*, [in Chinese], 35, 338, 1996.
2. Ye, Z., Discussion on the grades of health criteria of carbon monoxide produced from coal-smoke pollution, *J. Ind. Health Environ. Med.*, [in Chinese], 4, 174–177, 1981.
3. Editorial Group of Environmental Quality Report, Environmental Quality Reports of Beijing City, Reports from Beijing Government, 1995.
4. He, K., The current situation and its prospect of automobile exhaust emission in our country, *J. Environ. Sci.*, [in Chinese], 17, 80–83, 1996.
5. Luo, J., Health impact of traffic environment on traffic police *J. Occup. Med.*, [in Chinese], 18, 144, 1991.
6. Li, P., et al., An occupational investigation to traffic police in Beijing, *J. Hyg. Res.*, [in Chinese], 24, 198–200, 1995.
7. Lu, H., An approach of the factors on the current situation of vehicle exhaust in the highway tunnel and its influence on human health, *J. Environ. and Health*, [in Chinese], 14, 263–265, 1997.
8. Lu, H., Health effect of carbon monoxide pollution in the under-river-bed highway tunnel, *J. China Public Health*, [in Chinese], 13, 608–609, 1997.
9. Ye, S. et al., Harmful effects of vehicle exhaust on human health, *J. Shanghai Environ. Sci.*, [in Chinese], 17, 16–18, 1989.
10. Song, J., Ye, S. et al., Effects of diesel exhaust particulates on gap junction intercellular communication, *J. Hygiene Res.*, [in Chinese], 26, 145–147, 1997.
11. Yang, D., Zhou, W. et al., The composition and mutagenicity of engine exhaust induced by leaded gasoline and lead-free gasoline, *J. Environ. Health*, [in Chinese], 16, 125–127, 1999.
12. Qin, Y. et al., Study on the indoor air pollution *J. Environ. Health*, [in Chinese], 8, 100–102, 1991.
13. Wang, J. et al., Evaluation on natural gas to indoor air pollution, *J. Environ. Sci.*, [in Chinese], 16, 44–46, 1995.
14. Wang, J. et al., Health evaluation of indoor air pollution of natural gas, *J. Environ. Sci.*, [in Chinese], 16, 44–63, 1995.
15. Wang, J. et al., Analysis of indoor CO, particulates and other metallic elements in the three types of Beijing residential quarters, *J. Environ. Sci.*, [in Chinese], 11, 28–31, 1990.
16. Working Group of Center for Hygiene and Epidemic Control in Western District of Beijing City, Evaluation of living environmental quality in Beijing Proper [in Chinese], Beijing Government Report, 1988.
17. Tong, R. et al., How the hot water heater (gayster or geyser) influences the quality of indoor air when operating, *J. Environ. Health*, [in Chinese], 12, 281, 1995.
18. Zhao, B. et al., Analysis of indoor air pollution in endemic area of coal-smoke-fluorosis, *J. Hyg. Res.*, [in Chinese], 20, 16–19, 1991.
19. He, X. and Yang, R., *Air Pollution by Indoor Coal-Burning and Lung Cancer*, [in Chinese], Yunnan Sci. and Tech. Publishing House, 1994, 56.
20. National Environmental Agency of China, National Standards of People's Republic of China (GB3095-1996) — Ambient Air Quality Standard, Government Report [in Chinese], 1996.
21. Wang, L. et al., Revision of Standard of Carbon Monoxide in Ambient Air [in Chinese], Restricted Publication of Ministry of Health, 1989.

22. World Health Organization, *Environmental Health Criteria 13 — Carbon Monoxide*, World Health Organization, Gèneva, 1979.
23. Wang, Z. et al., Normal values of carboxyhemoglobin of college students in Haidian District of Beijing, *J. Chin. Environ. Sci.*, [in Chinese], 5, 71–73, 1985.
24. Song, H., Wang, Z. et al., Study of the relationship between the concentration of carbon monoxide and the monoamines neurotransmitter in young rat's brain, *J. Chinese Environ. Sci.*, [in Chinese], 10, 127–130, 1990.
25. Gong, C.H., Liu, J.Z. et al., The effects of carbon monoxide at low concentration on the lipid peroxidation process and the activity of glutathione peroxidase in rat's tissues, *J. Environ. Health*, [in Chinese], 12, 111–114, 1995.
26. Gu, R., Lu X. et al., The primary study of the correlation of free radicals and patients with delayed encephalopathy after acute carbon monoxide poisoning (DEACMP), *J. Chin. Neuroimmunol. Neurol.*, [in Chinese], 3, 172–175, 1996.

16 Use of Scanning Techniques in the Diagnosis of Damage from Carbon Monoxide

I.S. Saing Choi

CONTENTS

16.1   INTRODUCTION

In Korea, carbon monoxide (CO) poisoning is still one of the most important conditions causing brain damage as a result of cellular oxygen lack; however, the incidence of CO poisoning is decreasing annually.[1]

CO affects nearly all the organs and tissues including the brain, heart, kidney, skeletal muscle, skin, peripheral nerves, etc. Clinical manifestations involve a wide range of abnormalities. Various systemic complications and neurological sequelae develop after CO poisoning.[2-6]

Of 2759 patients with acute CO poisoning examined clinically between 1979 and 1982, 654 were admitted to Severance Hospital, Yonsei University Medical Center, Seoul, Korea. This included 315 men and 339 women. Mean age was 42.5 years (range, 8 months to 89 years). In 243 patients (37.2% of those admitted and 8.8% of the total group), 132 systemic complications and 154 neurological sequelae developed. Of the 132 systemic complications, there were 42 skin lesions, 35 pneumoniae, 20 local swellings, 10 pulmonary edemas, 8 atrial fibrillations, and others, in order frequency. Of the 154 neurological sequelae, there were 75 delayed neurological sequelae, 35 prolonged comae, 23 peripheral neuropathies, 5 hemiplegiae, and others, in order frequency (Table 16.1).[5]

The clinical symptomatology of CO poisoning may take one of two forms: the monophasic form, in which survival may range from hours to years without remission, and the biphasic form, in which a period of unconciousness is followed by an interval

0-8493-2065-8/00/$0 00+$ 50

TABLE 16.1
Systemic Complications and Neurological Sequelae in
654 patients with Acute CO Poisoning

Systemic Complication	No. of Cases	Neurological Sequelae	No. of Cases
Skin lesions	42	DNS	75
Pneumonia	35	Prolonged coma	35
Local swelling	20	Peripheral neuropathy	23
Pulmonary edema	10	Hemiplegia	5
Atrial fibrillation	8	Speech disturbance	3
Acute renal failure	7	Amnesia	3
Fetal death	3	Cortical blindness	3
Gastrointestinal-I bleeding	2	Hearing disturbance	2
Volkman's contracture	2	Parkinsonism	2
Gastric ulcer	1	Mental retardation	1
Osteomyelitis	1	Epilepsy	1
Shock	1	Facial palsy	1
ARDS	1		

ARDS = adult respiratory distress syndrome, DNS = delayed neurological sequelae.

Source: Choi, I.S., *Muscle Nerve*, 9, 965, 1986. With permission.

The clinical symptomatology of CO poisoning may take one of two forms: the monophasic form, in which survival may range from hours to years without remission, and the biphasic form, in which a period of unconciousness is followed by an interval of apparent normality (lucid interval) lasting 1 week to 1 month.[7-9] The systemic manifestations and complications after CO poisoning are seen in Table 16.2.[2-5,10-24] Among various neurological sequelae (Table 16.3),[7-9,25-36] delayed neurological sequelae are more characteristic of anoxic encephalopathy after CO poisoning than of other types of anoxia, and they have the characteristic symptom triad of mental deterioration, urinary incontinence, and gait disturbance.[9]

The diagnosis of acute CO poisoning is usually suggested by the circumstances under which the patient is found and is confirmed by the subsequent demonstration of carboxyhemoglobin in the blood. Consequently, it is easy to make a diagnosis of acute CO poisoning, but it is not easy to determine the prognosis of CO poisoning. The prediction of outcome during the acute insult is difficult in most cases, because of variations in age, duration and severity of exposure, and individual susceptibility.[9,37,38] Initial laboratory findings do not provide any prognostic clues of CO poisoning.

Recent improvement in neuroimaging techniques has made it possible to demonstrate the morphological and functional changes in the brain during life. Attempts at predicting the damage from CO poisoning and the subsequent prognosis by these means have been tried, but remain unsatisfactory.[39,40]

The brain computed tomography (CT) scan was introduced in the Yonsei Medical Center, Seoul, Korea in 1979. Magnetic resonance imaging (MRI) and single-photon

TABLE 16.2
Systemic Manifestations in CO Poisoning

System	Clinical Findings
Cardiovascular	ECG changes (T wave and ST segment), angina pectoris, myocardial infarct, atrial fibrillation, tachycardia, bradycardia, A-V block, premature venticular contraction, shock
Respiratory	Pneumonia, adult respiratory distress syndrome, pulmonary edema
Genitourinary	Glycosuria, proteinuria, hematuria, myoglobulinuria, acute renal failure, abortion, still-birth, menstrual disturbance
Gastrointestinal	Gastrointestinal disturbance, Gastrointestinal bleeding, gastric ulcer, hepatomegaly
Hematological	Leucocytosis, erythrocytosis, anemia, pernicious anemia, thrombotic thrombocytopenic purpura
Metabolic and endocrinological	Hyperglycemia, decreased T_3, acute Basedowís disease, reduction in weight of testes and in number of spermatozoa
Dermatological	Bulla, erythema, swelling, ulcer, gangrene, alopecia
Skeletomuscular	Muscle necrosis, Volkman's contracture, secondary osteomyelitis
Otologic	Disturbances of hearing and vestibular functions
Ophthalmological	Retinal hemorrhage, retinopathy, optic atrophy, amblyopia, scotoma, hemianopsia, blindness

TABLE 16.3
Neuropsychiatric Sequelae in Carbon Monoxide Poisoning

Condition	Sequelae
Psychosis	Dementia, mental retardation, hallucination, catatonia, manic depressive state, Korsakoff's syndrome, Kluver–Bucy syndrome
Psychoneurosis	Depression, anxiety, insomnia, melancholia, personality and judgment changes, amnesia
Striatal syndrome	Parkinsonism, chorea, athetosis, ballism, myoclonus, tremor, dystonia, Gille de la Tourette's disease
Motor deficit	Hemiplegia, apraxia, hyperkinetic state
Sensory deficit	Hemianopsia, cortical blindness, agnosia, anosmia, hearing disturbance
Speech deficit	Motor or sensory aphasia, anomia, agraphia
Convulsive disorder	Epilepsy
Spinal cord deficit	Syringomyelia
Peripheral nerve deficit	Polyneuropathy, mononeuropathy
Prolonged coma	Vegetative state, akinetic mutism
Delayed deficit	Delayed encephalopathy with/without basal ganglia signs

emission computed tomography (SPECT) scanning may also be used for patients with CO poisoning.

16.2 NEUROIMAGING FINDINGS

16.2.1 Brain Computed Tomography

The literature contains descriptions of various CT findings in CO poisoning, but most of them were reports predicting the clinical outcome of acute CO poisoning by CT scan.[41-48]

Lapresle and Fardeau[49] reviewed 22 patients with CO poisoning and classified them into four main types pathologically:

1. 16 had globus pallidus lesions, consisting of varying degrees of necrosis depending on the duration of survival;
2. 16 had white matter lesions containing many scattered or focal necrotic areas or confluent areas of demyelination;
3. 12 had cerebral cortex lesions, consisting of spongy changes, intense capillary proliferation, degeneration, and reduction of neurons;
4. 10 had hippocampal lesions, consisting of clearly delimited coagulation necrosis.

Of these four types, only examples of the first two have been documented in the radiologic literature. Radiologic demonstration of lesions of the hippocampal area and the cortex are less common, possibly because it is difficult to detect lesions near bony structures with low-resolution CT.

The most characteristic features of CO poisoning in the CT scan are symmetrical low-density lesions in the globus pallidus and/or cerebral white matter. An initial CT scan may fail to detect such damage.[41-48] Of the series of Miura et al.[46] and Choi et al.,[40] abnormal CT findings were found in 23 of 60 patients (38.5%) and 62 of 129 patients (48%), respectively.

The most common abnormal finding is symmetrically diffuse low density in the cerebral white matter, particularly in the frontal areas (Figure 16.1a). According to an experimental study of acute CO poisoning in rhesus monkeys, the extent of cerebral white matter damage correlated with the degree of metabolic acidosis and arterial hypotension sustained during CO exposure.[50] CO-related white matter abnormalities have been categorized into three groups, although there is much overlap among the groups. The first category consists of multiple small necrotic foci in the centrum semiovale and inter hemispheric commissures. The second category of the lesions consists of extensive, confluent areas of necrosis throughout the deep periventricular white matter. The third category consists of demyelination with relative preservation of axons in the deep white matter. The lesions may be small and discrete, extensive, or even confluent. The lesions are most prominent in the frontal lobes, with sparing in the subcortical arcuate U fibers. This third category is the type that is seen most often in patients with delayed encephalopathy or the so-called biphasic myelinopathy of Grinker.[49,50] This was found in 21 of the 23 patients (91.3%) in the Miura et al.[46] study and in 42 of the 62 patients (67.7%) in the Choi et al.[40] study.

A second not pathognomonic characteristic is a finding of low density in the globus pallidus, bilaterally (Figure 16.1b). These lesions result from the combined action of hypoxia and hypotension in areas of poor anastomotic blood supply.[51] This

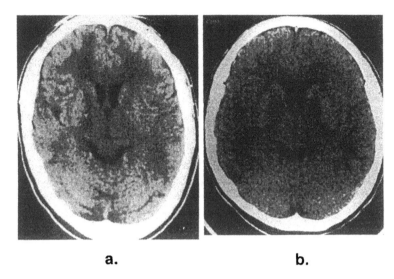

a.   **b.**

FIGURE 16.1   (a) CT Scan 7 days after CO poisoning shows low density in the cerebral white matter. (b) CT Scan 1 day after CO poisoning reveals low density in the globus pallidus bilaterally. (From Choi, I.S., *Eur. Neurol.*, 33, 462, 1993. With permission.)

was found in 18 of the 23 patients (78.3%) in the Miura et al.[46] study, and in 33 of the 62 patients (53.2%) in the Choi et al.[40] study. In addition, the CT scan occasionally shows low-density lesions in the frontal and occipital cortex, thalamus, or caudate nucleus, and cortical atrophy presumably not related to CO poisoning.[40]

Low density on CT scan is observed in various states, which produce edema, necrosis, and demyelination, and reflect the fatty or watery content of the affected tissue.[42,45] In general, the edema of hypoxia is most evident during the first week and begins to recede by 2 weeks. Within 3 weeks, necrosis is evident, and at times neovascularization or dilatation of normally unperfused vessels is evident on contrast infusion in so-called luxury perfusion. Enhancing is commonly found with CT scans in stroke, tumor, and inflammatory disease, but rarely in CO poisoning,[40,42,46] possibly because there is little neovascularization or alteration of the vascular permeability of the necrotic central nervous system (CNS) tissue during the anoxic process.[45,49] Nardizzi[41] reported a case of CO poisoning with enhancing lesions of the globus pallidus on CT scan 8 days after anoxia, and Miura et al.[46] reviewed CT brain scans of 60 cases with CO poisoning and found only one case with subtle enhancement in the center of the globus pallidus on CT scan 24 days after anoxia. Of 129 brain CT scans in a CO-poisoning series, only two had enhanced lesions (Figures 16.2 and 16.3).[40]

Initial CT lesions may have disappeared, diminished, or remained unchanged at follow-up. All the initial CT lesions in this study did not disappear at follow-up (Figure 16.4), even up to 26 months after anoxia; moreover, some revealed more aggravating findings, with cortical atrophy. This suggests that the majority of low-density lesions on CT scans of acute CO poisoning consists essentially of areas of necrosis rather than area of edema, as Lapresle and Fardeau[49] demonstrated pathologically.

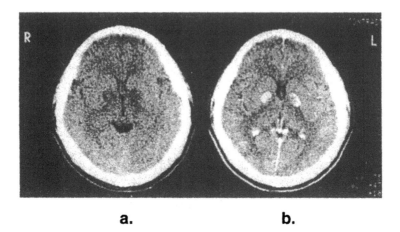

a.   **b.**

FIGURE 16.2   (a) CT scan 2 days after CO poisoning shows low density in the globus pallidus bilaterally. (b) CT scan with contrast clearly demonstrates increased uptake in the globus pallidus bilaterally.

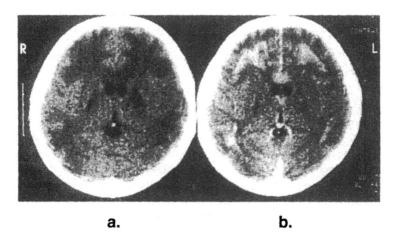

a.   **b.**

FIGURE 16.3   (a) CT scan 15 days after anoxia shows low-density in the globus pallidus and cerebral white matter bilaterally. (b) CT scan with contrast demonstrates increased uptake in the subcortical white matter, particularly frontal areas.

Prediction of outcome following acute CO poisoning is in most cases difficult. Sawada et al.[42] described the outcome of 21 patients with acute CO poisoning. They reported that the outcome of 11 patients with globus pallidus lesions was relatively poor, while that of 10 patients with no abnormal findings was good. They mentioned that the outcome following acute CO poisoning is correlated with the pallidal low-density areas in the CT scan. In contrast, Miura et al.[46] and Choi et al.[40] believe that the prognosis of acute CO poisoning depends on the cerebral white matter lesions and not on the pallidal low-density lesions. Of 29 patients with pure white matter lesions on CT during acute stage in the Choi et al. study, 17 had delayed sequelae,

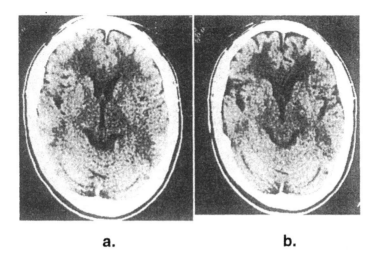

a.   b.

FIGURE 16.4   (a) CT scan 7 days after CO poisoning shows low density of the cerebral white matter. (b) CT scan 1 year later has remained unchanged.

6 remained in a vegetative state, 1 died, and only 5 recovered completely. All recovered patients except one were below 40 years of age. Of 20 patients with pure globus pallidus lesions on CT, 3 had delayed sequelae, 4 remained in a vegetative state, 1 died, and 11 recovered completely. There is a significant correlation between the cerebral white matter lesions on acute stage CT scans and the development of delayed CO sequelae. In the author's experience, delayed CO sequelae occur within 1 month after acute insult in more than one half of patients with cerebral white matter changes on acute stage CT scans, particularly for those in middle age or older (Table 16.4).

The outcome of delayed CO sequelae is relatively good. About one half to three fourths of patients with delayed neurological sequelae recover within 1 year,[7,9] but some have persisting late sequelae, including memory disturbance and parkinsonism. In the author's experience, of 16 delayed sequelae patients with initial abnormal CT findings, 10 recovered completely, 3 were disabled, and 3 remained in a vegetative state. Of 12 patients with initial normal CT finding, 6 recovered completely, 3 were disabled, and 3 died (Table 16.5). This suggests there is no correlation between the initial CT findings and the outcome of delayed CO sequelae.

Some patients with delayed CO sequelae and with the clinical improvement showed a more aggravating pattern with cortical atrophy seen on follow-up CT scan.[48] Of 25 patients with delayed sequelae, 9 showed cortical atrophy with an aggravating pattern on follow-up CT scans (Figure 16.5). Of 13 patients with cerebral white matter changes on initial CT scans, 6 revealed a more aggravating pattern with cortical atrophy seen on follow-up CT scans, but none of 4 patients with pallidal low density on initial CT scans showed cortical atrophy. This suggests that there is a good correlation between cerebral white matter low-density lesions seen on CT scan and the development of cortical atrophy in patients with delayed CO encephalopathy. The exact mechanism producing such cortical atrophic changes on CT

TABLE 16.4
Correlation between the Initial CT Findings and the Outcome of Acute CO Poisoning

	Number of Patients					
Initial CT findings	Full Recovery	Delayed Sequelae	Disability	Vegetative State	Death	Total
White matter lesion	5[a]	17*	0	6	1	29
Pallidal lesion	11**	3	1	4	1	20
White matter and pallidal lesions	1	3	1	6	2	13
Normal and others	46	14	2	5	0	67
Total	63	37	4	21	4	129

[a] All except one were below 40 years of age.

**P < 0.006; *P < 0.002, chi square test.

TABLE 16.5
Correlation between the Initial CT Findings and the Outcome of Delayed CO Encephalopathy

	Number of Patients				
Initial CT findings	Full Recovery	Disability	Vegetative State	Death	Total
White matter lesion	7	3	2	0	12
Pallidal lesion	2	0	1	0	3
White matter and pallidal lesions	1	0	0	0	1
Normal and others	6	3	0	3	12
Total	16	6	3	3	28

scan in patients with delayed CO sequelae is unknown. It is possible that the pallidal and hippocampal lesions do not affect the presence of cortical atrophy on the CT scan because of their smaller volumes as compared with the large volumes of the cerebral white matter and the cerebral cortex. The invisible lesions of the cerebral cortex on the CT scan and the cerebral white matter lesions may gradually progress during the clinical course, and these may hasten the development of the atrophic changes of the brain in delayed CO encephalopathy.

16.2.2 BRAIN MAGNETIC RESONANCE IMAGING

With its invention as an imaging tool, MRI has been shown superior in sensitivity to CT scanning in studies of the brain. MRI findings in CO poisoning have been published in several reports.[52-56]

In general, MRI findings in CO poisoning are similar to those using the CT scan. The author recently obtained five MRIs in five patients with acute CO poison-

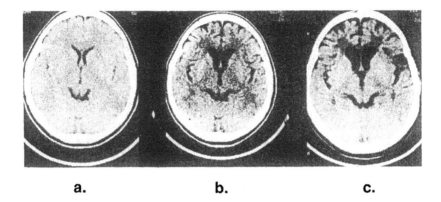

a.   b.   c.

FIGURE 16.5   (a) CT scan 1 month after CO poisoning shows no abnormality. (b and c) CT scans 5 and 20 months after anoxia show a gradual aggravating cortical atrophy with low-density lesions of the cerebral white matter.

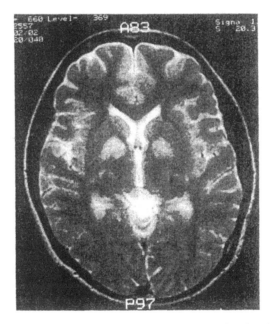

FIGURE 16.6 T_2-weighted axial MR image 6 days after CO poisoning shows high intensity in the globus pallidus bilaterally.

ing. Four had abnormal findings — three with bilateral hyperintensities of the globus pallidus on T_2-weighted images (Figure 16.6) and two with symmetric hyperintensity lesions of the cerebral white matter on T_2-weighted images (Figure 16.7).

Chang et al.[55] in Korea, reviewed MRIs in 15 patients with delayed encephalopathy after CO poisoning. They described three characteristic findings. Seen in all patients was bilateral symmetric hyperintensity of the cerebral white matter on proton

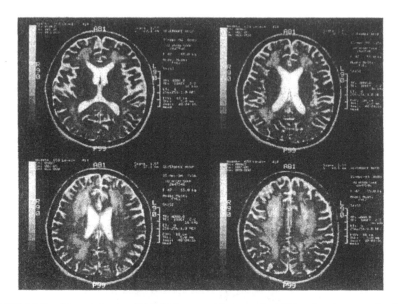

FIGURE 16.7   T$_2$-weighted axial MR images 10 days after CO poisoning reveals diffuse high-intensity lesions of the cerebral white matter.

and T$_2$-weighted images. The hyperintensity varied in degree from slight to severe. The abnormality appeared as slight hypodensity or isodensity on T$_1$-weighted images. The white matter lesions were extended into the corpus callosum, subcortical U fibers, and external and internal capsules. The second finding was diffuse hypointensity of the thalamus on T$_2$-weighted images, suggesting iron deposition. This was seen in 10 patients. Among 10 patients, hypointensity also was seen in the putamen in 6 and in the caudate nucleus in 2. This finding has not been described in the literature on CO poisoning. Dietrich et al.[53] reported increased iron deposition in the basal ganglia and thalamus in children with hypoxic-anoxic leukoencephalopathy. They explained that interruption of normal axonal transportation of nonheme iron caused by white matter abnormality might lead to increase the accumulation of iron at the basal ganglia and thalamus.

The third finding was bilateral hyperintensity lesions of the globus pallidus on T$_2$-weighted images, which was noted in 9 patients. MRI occasionally showed hyperintensity lesions in the thalamus, putamen, caudate nucleus, occipital cortex, and pons, as well as cortical atrophy. They mentioned that a decrease in the extent and signal intensity of the white matter lesions on follow-up MRI study usually accompanied a lessening of the clinical symptoms, and suggested that the main pathological feature of delayed CO sequelae might be a reversible demyelinating process of the cerebral white matter.

Recently, there have been a few reports concerned with cerebral metabolic changes in delayed CO sequelae, studied by proton MR spectroscopy (MRS).[28,29] The author also experienced MRS in two patients with delayed CO encephalopathy, revealing a lowered NAA (*N*-acetyl-asparatate)/Cr (creatine and phosphocreatine) ratio and an increased Cho (choline containing compounds)/Cr ratio (Figure 16.8).

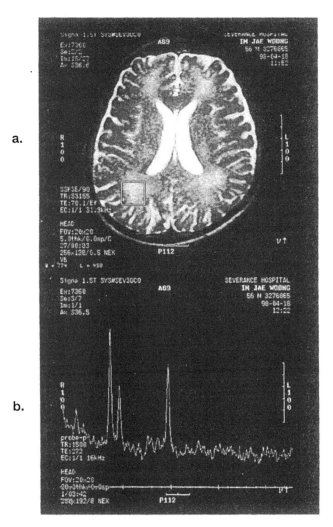

FIGURE 16.8   (a) T_2-weighted MR image 1 month after delayed CO sequelae shows bilateral high intensity in the cerebral white matter. (b) Proton MRS shows relatively lowered NAA/Cr ratio, compared with high Cho/Cr ratio. Peak assignments are as follows: first peak, Cho, second peak, Cr, and third peak, NAA.

Theoretically, every metabolite that contains visible protons can be demonstrated by MRS. In practice, however, only four signals are constantly detected, namely, NAA, Cho, Cr, and lactate. All but Cr are thought to be of clinical significance. The role of NAA, which predominates in neurons, is still unknown. It is thought that NAA enhances the transport of high-energy phosphates by facilitating the diffusion of phosphocreatine and adenosine phosphate (ATP). A decrease in NAA results in impaired transport of high-energy phosphates which reflects loss or degeneration of neurons. Increased Cho may reflect active membrane metabolism associated with demyelination. Lactate is produced by anaerobic glycolysis. Proton MRS shows previously unrecognized normal activity in CO poisoning and may be a useful predictor to determine neuron viability and prognosis early in the course of delayed CO encephalopathy.

16.2.3 BRAIN SINGLE-PHOTON EMISSION COMPUTED TOMOGRAPHY

Brain imaging that reflects regional cerebral blood flow (rCBF) using ^{133}Xe inhalation, SPECT, and positron emission computed tomography (PET) has recently come to be used for the topographical identification of brain lesions. Brain SPECT imaging, in particular, has the advantage of providing three-dimensional information on lesions. In addition, the cost of the equipment is low. Compared with PET, SPECT is an indirect and less accurate method for measuring rCBF. In spite of more than 3000 bibliographic references to CO poisoning, there have been only a few studies where tomographic CBF techniques were used in patients with CO poisoning.[30-35]

The author obtained SPECT scans of five patients with CO poisoning during the acute insult (5 to 26 days). Three (patients 1, 4, and 5) showed normal scans, while the others (patients 2 and 3) had focal hypoperfusion at the left cerebral cortex which appeared several days prior to the onset of delayed CO sequelae (Figure 16.9). Of three patients with normal SPECTs, two recovered completely and one had a delayed CO sequelae (Table 16.6). The exact mechanism producing such early change in SPECT is unknown, but there be two possibilities. First, this early abnormality may be part of the process leading to the delayed sequelae. Second, it may represent a slow recovery toward normal CBF pattern following acute illness, during which CBF abnormalities were almost certainly present. Of the two possibilities, the author thinks the second is less likely by reason of the following. First, the initial SPECT findings were obtained 10 and 30 days after full recovery clinically, when CSF might be normal, even with a slow recovery. Second, the second SPECTs 10 and 13 days after delayed sequelae also showed less perfusion in the left frontal areas, thus the same findings as the initial SPECTs.

Lee et al.[59] reported a study of rCBF measured using ^{133}Xe inhalation. They found a marked decrease in global rCBF in the white matter of the cerebral cortex in four patients with delayed CO encephalopathy. Jibiki et al.[60] reported diffuse hypoperfusion of N-isopropyl-(^{131}I)-P-iodoamphetamine with SPECT imagings performed repeatedly in a patient with the interval form of CO poisoning. rCBF increased in both the gray and the white matter in follow-up SPECT scans which paralleled the clinical improvement.

Maeda et al.[61] described changes in SPECT images with ^{99}m Tc HM-PAO before and after therapy with oxygen under high pressure (HBO) in a patient with delayed CO encephalopathy. They observed that the pre-HBO SPECT image showed frontal dominant hypoperfusion and the post-HBO image showed increased perfusion throughout the cortex, particularly in the frontal areas. All initial SPECTs in 13 patients with delayed CO sequelae in this study[63] showed diffuse hypoperfusion in the cerebral cortex, particularly in the frontal areas (Figures 16.9b and 16.10a). rCBF in follow-up SPECTs in six patients with delayed encephalopathy increased in the cerebral cortex, parallel to the clinical improvement (Figure 16.9), but the follow-up SPECT in a patient in a vegetative state remained unchanged (Figure 16.11). A review of the literature and these studies suggests there is a good correlation between the SPECT

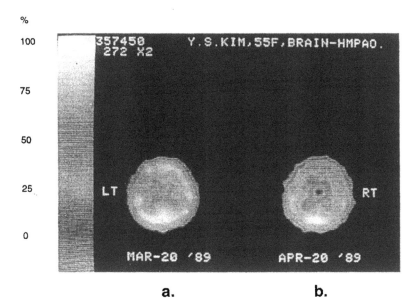

%
100
75
50
25
0

a.   b.

FIGURE 16.9   (a) SPECT with [99]m Tc HM-PAO 10 days after acute CO poisoning shows hypoperfusion in the left frontal area. (b) SPECT 10 days after delayed CO encephalopathy shows diffuse hypoperfusion throughout the cerebral cortex. Regional tracer distribution is expressed as a percentage of maximum according to the color scale alongside. (From Choi, I.S., *Eur. Neurol.*, 33, 462, 1993. With permission.)

TABLE 16.6
Clinical Summary and SPECT Findings in Five Patients with Acute CO Poisoning

		Duration of unconsciousness, h	Lucid Interval, days	CT finding	Initial (on day)	Follow-Up (on day)	Outcome
No.	Age/Sex						
1	70/F	8	20	Low density of white matter	Normal (5)	D.H. (23)	DNS
2	55/F	24	33	Low density of globus pallidus	F.H. (10)	D.H. (43)	DNS
3	57/F	3	45	Normal	F.H. (30)	D.H. (58)	DNS
4	57/M	24	—	Normal	Normal (10)	Normal (40)	Normal
5	46/F	16	—	Normal	Normal (14)	Normal (54)	Post CO depression

DNS = delayed neurological sequelae; D.H. = diffuse hypoperfusion; F.H = focal hypoperfusion.

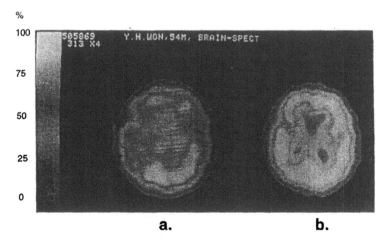

FIGURE 16.10   (a) SPECT with ^{99}m Tc HM-PAO 10 days after delayed sequelae shows diffuse hypoperfusion throughout the cerebral cortex. (b) SPECT 3½ months afterward shows markedly increased perfusion with clinical improvement.

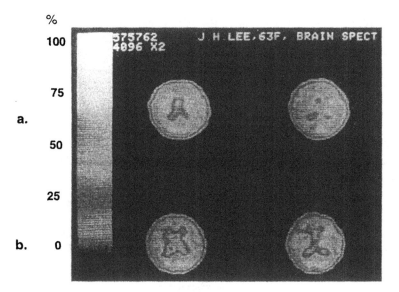

FIGURE 16.11   (a) SPECT with ^{99}m TC HM-PAO 11 days after delayed sequelae shows diffuse hypoperfusion throughout the cerebral cortex. (b) SPECT 6 months afterward has remained unchanged in a patient in a vegetative state. (From Choi, I.S., *Eur. Neurol.*, 35, 141, 1995. With permission.)

findings and the clinical outcome of delayed CO sequelae. Brain SPECT scanning may be useful in evaluating the outcome of delayed sequelae after CO poisoning.

De Reuck et al.[64] recently reported on seven patients with acute CO poisoning who were treated with HBO and underwent a PET study 2 to 5 days post-anoxia. Their PET findings showed that ischemia was prolonged after CO poisoning, even after normalization of clinical status. They suggest that PET scanning can show the regions at risk for development of late complications after CO poisoning, although PET cannot predict the final outcome.

16.3   SUMMARY

Various systemic clinical manifestations and neurological sequelae develop after CO poisoning, and the development of delayed neurological sequelae is more characteristic of anoxic encephalopathy after CO poisoning than of other types of anoxia.

Initial abnormal CT findings related to CO poisoning are found in about one third to one half of patients with acute CO poisoning. The most common CT finding is symmetric low density in the cerebral white matter. The second most common characteristic is bilateral pallidal low-density lesions.

The prognosis of acute CO poisoning depends on the cerebral white matter changes rather than those of the globus pallidus. There also seems to be a significant correlation between the cerebral white matter low-density areas on initial CT scan and the development of delayed CO sequelae, but there is no correlation between the abnormal CT findings and the clinical outcome of delayed CO sequelae.

The most common MRI finding is symmetric hyperintensity of the cerebral white matter. The second most common MRI finding is hypointensity of the thalamus, while the third is hyperintensity of the globus pallidus bilaterally on T_2-weighted images. In contrast with the CT scan, the follow-up MRI of patients with delayed CO sequelae will show a decrease in extent and signal intensity of the white matter lesions accompanied by lessening of the clinical symptoms.

Proton MRS may show previously unrecognized neuronal activity in delayed CO sequelae and reflect the severity of symptoms.

Some SPECTs in patients with acute CO poisoning will show early focal hypoperfusion in the cerebral cortex which will appear several days prior to the onset of delayed CO sequelae. All SPECTs in patients with delayed CO sequelae show diffuse hypoperfusion throughout the cerebral cortex, and follow-up SPECTs of delayed CO sequelae patients reveal increased cerebral perfusion with simultaneous clinical improvement.

At present, brain MRI and SPECT are the more sensitive in detection and evaluation of brain damage from CO. They are also more useful in evaluating the clinical outcome of CO poisoning, as compared with brain CT scan.

REFERENCES

1. Hwang, S.H. and Choi, I.S., Clinical and laboratory analysis in acute carbon monoxide intoxication, *J. Kor. Med. Assoc.*, 33, 997, 1990.
2. Raskin, N. and Mullaney, O.C., The mental and neurological sequelae of carbon monoxide asphyxia in a case observed for fifteen years, *J. Nerv. Ment.*, 92, 640, 1940.
3. Finck, P.A., Exposure to carbon monoxide: review of the literature and 567 autopsies, *Mil. Med.*, 131, 1513, 1966.
4. Lilienthal, J.L., Jr., Carbon monoxide, *Pharmacol. Rev.*, 2, 324, 1966.
5. Choi, I.S., Peripheral neuropathy following acute carbon monoxide poisoning, *Muscle Nerve*, 9, 965, 1986.

6. Choi, S.A. and Choi, I.S., Clinical manifestations and complications in carbon monoxide intoxication, *J. Kor. Neurol. Assoc.*, 16, 500, 1998.

7. Shillito, F.M., Drinker, C.K., and Shaughnessy, T.J., The problem of nervous and mental sequelae in carbon monoxide poisoning, *J. Am. Med. Assoc.*, 106, 669, 1936.

8. Plum, F., Posner, J.B., and Hain, R.F., Delayed neurological deterioration after anoxia, *Arch. Intern. Med.*, 110, 56, 1962.

9. Choi, I.S., Delayed neurologic sequelae in carbon monoxide intoxication, *Arch. Neurol.*, 40, 433, 1983.

10. Anderson, R.F., Allenworth, D.C., and deGroot, W.J., Myocardial toxicity from carbon monoxide poisoning, *Ann. Intern. Med.*, 67, 1172, 1967.

11. Lee, M.S., Choi, I.S., and Choi, K.G., Electrocardiographic and serum enzymes changes in acute carbon monoxide poisoning, *J. Kor. Med. Assoc.*, 32, 86, 1989.

12. Ellenhorn, M.J., Respiratory toxicology, in *Ellenhorn's Medical Toxicology*, Cooke, D.B., Ed., Williams & Wilkins, Baltimore, 1997, 1465.

13. Longhridge, L.W., Acute renal failure due to muscle necrosis in carbon monoxide poisoning, *Lancet*, 1, 349, 1958.

14. Lee, S.S., Choi, I.S., and Song, K.S., Hematologic changes in acute carbon monoxide intoxication, *Yonsei Med. J.*, 35, 245, 1994.

15. Stonesifer, L.D., Bon, R.C., and Hiller, F.C., Thrombotic thrombocytopenic purpura in carbon monoxide poisoning, *Arch. Intern. Med.*, 140, 104, 1980.

16. Kim, O.J., Choi, I.S., and Kim, K.W., Blood levels of thyroid hormones and sugar in acute carbon monoxide poisoning, *J. Kor. Neurol. Assoc.*, 13, 67, 1995.

17. Long, P.I., Dermal changes associated with carbon monoxide intoxication, *J. Am. Med. Assoc.*, 205, 50, 1962.

18. Nagy, R., Greer, K.E., and Harman, L.E., Cutaneous manifestations of acute carbon monoxide poisoning, *Cutis*, 24, 381, 1975.

19. Lee, J.B., Chang, K.H., and Choi, I.S., Cutaneous manifestations of carbon monoxide poisoning, *Kor. J. Dermatol.*, 21, 279, 1983.

20. Lee, S.A., Choi, I.S., and Kim, J.S., A serum enzyme study in acute carbon monoxide intoxication, *J. Kor. Med. Assoc.*, 31, 70, 1988.

21. Baker, S.R. and Lilly, D.J., Hearing loss from acute carbon monoxide intoxication, *Ann. Otorhinolaryngol.*, 86, 323, 1977.

22. Choi, I.S., Brainstem auditory evoked potentials in acute carbon monoxide poisoning, *Yonsei Med. J.*, 26, 29, 1985.

23. Dempsey, L.S., O'Donell, J.J., and Hoft, J.J., Carbon monoxide retinopathy, *Am. J. Ophthalmol.*, 82, 692, 1976.

24. Kelley, J.S. and Sophocleus, G.J., Retinal hemorrhage in subacute carbon monoxide poisoning. Exposures in homes with blocked furnace flues, *J. Am. Med. Assoc.*, 239, 1515, 1978.

25. Garland, H. and Pearce, J., Neurological complications of carbon monoxide poisoning, *Q. J. Med.*, 36, 445, 1967.

26. Choi, I.S., A study of neurologic sequelae in carbon monoxide intoxication, *J. Kor. Med. Assoc.*, 25, 341, 1982.

27. Sandson, T.A., Lilly, R.B., and Sodkol, M., Kluver-Bucy syndrome associated with delayed postanoxic lencoencephalopathy following carbon monoxide, *J. Neurol. Neurosurg. Psychiatr.*, 51, 156, 1988.

28. Klawans, H.L., Stein, R.W., Tanner, C.M., and Goetz, C.G., A pure parkinsonian syndrome following acute carbon monoxide intoxication, *Arch. Neurol.*, 39, 302, 1982.

29. Choi, I.S., Hwang, Y.M., and Kim, K.W., Chorea as a clinical manifestation of delayed neurologic sequelae in carbon monoxide poisoning, *J. Kor. Neurol. Assoc.*, 2, 91, 1984.

30. Schwartz, A., Hennerici, M., and Wegener, O.H., Delayed choreoathetosis following acute carbon monoxide poisoning, *Neurology*, 35, 98, 1985.

31. Pulst, S.M., Walshe, T.M., and Romero, J.A., Carbon monoxide poisoning with features of Gille de la Tourette's syndrome, *Arch. Neurol.*, 40, 443, 1983.

32. Scott, B.L. and Jankovic, J., Delayed onset progressive movement disorders after static brain lesions, *Neurology*, 46, 68, 1996.

33. Snyder, R.D., Carbon monoxide intoxication with peripheral neuropathy, *Neurology*, 20, 177, 1970.

34. Meigs, J.W. and Hughes, J.P.W., Acute carbon monoxide poisoning. An analysis of 105 cases, *Arch. Ind. Hyg.*, 5, 344, 1954.

35. Walton, J.N., *Brain's Disease of The Nervous System*, 8th ed., Oxford University Press, Oxford, 1997, 810.

36. Lee, M.H., Clinical studies on delayed sequelae of carbon monoxide intoxication, *J. Kor. Neuropsychiatr. Assoc.*, 15, 374, 1978.

37. Jennett, B. and Bond, M., Assessment of outcome after severe brain damage: a practical scale, *Lancet*, 1, 480, 1975.

38. Bang, O.Y., Choi, B.O., Choi, I.S., Jung, J.H., and Rho, J.H., Predicting factors in prognosis of acute carbon monoxide intoxication, *J. Kor. Neurol. Assoc.*, 14, 229, 1996.

39. Vieregge, P., Klostermann, W., Blumm, R.G., and Borgis, K.J., Carbon monoxide poisoning: clinical, neurophysiological, and brain imaging obervations in acute disease and follow-up, *J. Neurol.*, 236, 478, 1989.

40. Choi, I.S., Kim, S.K., Choi, Y.C., Lee, S.S., and Lee, M.S., Evaluation of outcome after acute carbon monoxide poisoning by brain CT, *J. Kor. Neurol. Assoc.*, 8, 78, 1993.

41. Nazdizzi, L.R., Computerized tomographic correlate of carbon monoxide poisoning, *Arch. Neurol.*, 36, 38, 1979.

42. Sawada, Y., Takahashi, M., Ohashi, N., Fusamoto, H., Maemura, K., Kobayashi, H., Yoshioka, T., and Sugimoto, T., Computerized tomography as an indication of long-term outcome after acute carbon monoxide poisoning, *Lancet*, 2, 783, 1980.

43. Kim, K.S., Weinberg, P.E., Suh, J.H., and Ho, S.U., Acute carbon monoxide poisoning: computed tomography of the brain, *AJNR*, 1, 399, 1980.

44. Kono, E., Kono, R., and Shida, K., Computerized tomographies of 34 patients at the chronic stage of acute carbon monoxide poisoning, *Arch. Psychiatr. Nervenkr.*, 233, 271, 1983.

45. Kodayashi, K., Isaki, K., Fukutani, Y., Kurachi, J.M., Eboshida, A., Matsubara, R., and Yamaguchi, N., CT findings of the interval form of carbon monoxide poisoning compared with neuropathological findings, *Eur. Neurol.*, 23, 34, 1984.

46. Miura, T., Mitomo, M., Kawai, R., and Harada, K., CT of the brain in acute carbon monoxide intoxication: characteristic features and prognosis, *AJNR*, 6, 739, 1985.

47. Choi, I.S., Choi, Y.C., Kim, S.K., and Lee, M.S., Two cases of carbon monoxide poisoning showing low-density lesions with unusual enhancement on computed tomographic brain scan, *J. Kor. Neurol. Assoc.*, 10, 358, 1992.

48. Choi, I.S., Kim, J.H., and Jung, W.Y., Cortical atrophy following delayed encephalopathy after carbon monoxide poisoning, *J. Kor. Neurol. Assoc.*, 14, 560, 1996.

49. Lapresle, J. and Fardeau, M., The central nervous system and carbon monoxide poisoning. II. Anatomical study of brain lesion following intoxication with carbon monoxide (22 cases), *Prog. Brain Res.*, 24, 31, 1967.

50. Ginsberg, M.D., Myers, R.E., and McDonagh, B.F., Experimental carbon monoxide encephalopathy in the primate: clinical aspects, neuropathology, and physiologic correlation, *Arch. Neurol.*, 30, 209, 1974.

51. Okeda, R., Funata, N., Takano, J. et al., The pathogenesis of carbon monoxide encephalopathy in the acute phase — physiological and morphological correlation, *Acta Neuropathol.* (Berlin), 54, 1, 1981.

52. Horowitz, A.L., Kaplan, R., and Sarpel, G., Carbon monoxide toxicity: MR imaging in the brain, *Radiology*, 162, 787, 1987.

53. Dietrich, R.B. and Bradley, W.G., Jr., Iron accumulation in the basal ganglia following severe ischemic-anoxic insults in children, *Radiology*, 168, 203, 1988.

54. Tuchman, R.F., Moser, F.G., and Moshe, S.l., Carbon monoxide poisoning: bilateral lesions in the thalamus on MR imaging of the brain, *Pediatr. Radiol.*, 20, 478, 1990.

55. Chang, K.H., Han, M.H., Kim, H.S., Wie, B.A., and Han, M.C., Delayed encephalopathy after acute carbon monoxide intoxication: MR imaging features and distribution of cerebral white matter lesions, *Radiology*, 184, 117, 1992.

56. Murata, T., Itoh, S., Koshino, Y., Sakamoto, K., Nishio, M., Maeda, M., Yamada, H., Ishii, Y., and Isaki, K., Serial cerebral MRI with FLAIR sequences in acute carbon monoxide poisoning, *J. Comput. Assist. Tomogr.*, 19, 631, 1995.

57. Kamada, K., Houkin, K., Aoki, T., Koiwa, M., Kashiwaba, T., Iwasaki, Y., and Abe, H., Cerebral metabolic changes in delayed carbon monoxide sequelae studied by proton MR spectroscopy, *Neuroradiology*, 36, 104, 1994.

58. Murata, T., Itoh, S., Koshino, Y., Omori, M., Murata, I., Sakamoto, K., Isaki, K., Kimura, H., and Ishii, Y., Serial proton magnetic resonance spectroscopy in a patient with the interval form of carbon monoxide poisoning, *J. Neurol. Neurosurg. Psychiatr.*, 58, 100, 1995.

59. Lee, M.S., Kim, J.S., Chung, T.S., and Suh, J.H., Measurement of cerebral blood flow in delayed carbon monoxide sequelae using xenon inhalation CT scan, *Yonsei Med. J.*, 29, 185, 1988.

60. Jibiki, I., Kurokawa, K., and Yamaguchi, N., ^{123}I-IMP brain SPECT imaging in a patient with the interval form of CO poisoning, *Eur. Neurol.*, 31, 149, 1991.

61. Maeda, Y., Kawasaki, Y., Jibiki, I., Yamaguchi, N., Matsuda, H., and Hisada, K., Effect of therapy with oxygen under high pressure on regional cerebral blood flow in the interval form of carbon monoxide poisoning: observation from subtraction of technetium-99m HM-PAO SPECT brain imaging, *Eur. Neurol.*, 31, 380, 1991.

62. Choi, I.S. and Lee, M.S., Early hypoperfusion of technetium-99m hexamethylpropylene amine oxime brain single photon emission computed tomography in a patient with carbon monoxide poisoning, *Eur. Neurol.*, 33, 461, 1993.

63. Choi, I.S., Kim, S.K., Lee, S.S., and Choi, Y.C., Evaluation of outcome of delayed neurologic sequelae after carbon monoxide poisoning by technetium-99m hexamethylpropylene amine oxime brain single photon emission computed tomography, *Eur. Neurol.*, 35, 137, 1995.

64. De Reuck, J., Decoo, D., Lemanhieu, I., Strijckmans, K., Boon, P., Van Maele, G., Buylaert, W., Leys, D., and Petit, H., A positron emission tomography study of patients with acute carbon monoxide poisoning treated by hyperbaric oxygen, *J. Neurol.*, 240, 430, 1993.

17 Low-Level Carbon Monoxide and Human Health

Robert D. Morris

CONTENTS

17.1 INTRODUCTION

Evaluating the public health impact of a particular toxic exposure requires integrating knowledge of the dose–response relationship for the toxicant with estimates of the probability of human exposure to that chemical. Understanding of the toxic effects of most chemicals decreases with decreasing level of exposure. However, for almost all toxic exposures, the proportion of the population with high levels of exposure is far less than the proportion of the population with low exposures. In other words, the greater the number of people exposed to a contaminant at or above a particular concentration, the less likely one is to understand the health effects of that exposure.

This pattern certainly holds for carbon monoxide (CO). The levels of exposure discussed in much of this book are anomalous, occurring as the result of accidents, equipment malfunction, or acts of self-poisoning, and affect relatively small numbers of people. Far greater numbers of people are exposed to CO at much lower levels due to ambient air pollution. This chapter considers the possible health effects associated with exposure to CO at these levels.

Researchers investigating the health effects of low levels of CO have two general options available to them, chamber studies or epidemiological studies. Both designs have significant limitations when seeking to estimate the public health consequences of particular concentrations of ambient CO.

Chamber studies of animals or humans allow for the carefully controlled experimental study of exposure to specific concentrations of ambient CO. These studies suffer from two important limitations. First, by isolating a particular exposure, they do not include the effect of additional environmental stressors such as temperature, other air pollutants, or physical exertion. A limited number of studies have attempted to consider the impact of a single additional stressor, but none has considered multiple stressors. Second, chamber studies, with a few important exceptions to be discussed below, are usually conducted on healthy subjects. A large portion of the general population suffers from some form of chronic disease that can diminish the ability to respond to environmental stress. Ambient air pollution will have its greatest effect on this segment of the population. Both of these limitations will cause underestimation of the effects of ambient CO on public health.

Epidemiological studies can provide a more direct measurement of the health effects of ambient CO, but they pose a different set of challenges to researchers seeking to estimate the impact of CO on public health. The most daunting challenge of any study in environmental epidemiology is the measurement of exposure. Ambient concentrations of CO are highly heterogeneous and, as a consequence, the exposure of individual members of the population can vary widely depending upon the activity patterns of those individuals. In other words, levels of CO at monitoring stations for ambient CO do not accurately predict the personal exposure of individuals in the population. Rather, levels at these monitors correspond to a distribution of exposure with a central tendency that is, in some way, related to the measurements at the monitors. Estimating personal exposure from that monitoring information will result in substantial misclassification of exposure. This exposure misclassification will tend to be random and random misclassification generally results in underestimation of risk to the individual. One can reduce misclassification through the use of personal monitors, but studies relying on this design are generally costly to conduct, include small numbers of subjects, and, as a consequence, have limitations for estimating health impacts in the general population. In considering the total impact of ambient air pollution on public health, detailed exposure measurement may not be necessary. An alternative approach is based on the use of ambient monitors of air pollution as indicators of exposure for the population of a particular urban area during a specific time period to generate a time series of CO levels. These data can then be integrated with measurements of other time variant exposures such as weather conditions and other air pollutants. These exposure data can then be combined with time series data reflecting the daily rates of disease and mortality in the area of the monitors. Multivariate statistical models that have been adapted for use with time series data are then used to evaluate the relationship between air pollutant levels, in this case CO levels, and morbidity and mortality in the community.

These methods do not give precise information on the relationship between air pollution and health for any particular individual. However, they give valuable information on the overall health impact of air pollution as measured at fixed-site monitoring stations. Since most interventions to reduce air pollution take the form of regulations intended to reduce levels at these monitoring sites, these studies have an interpretation that is directly relevant to measures that might be taken to protect public health. Most of the recent evidence of adverse effects of low-level CO derives from epidemiological time series studies. This chapter will focus on these studies.

17.2 RECENT EPIDEMIOLOGICAL STUDIES OF CARDIOVASCULAR MORBIDITY

In a 1995 study, Morris et al.[1] investigated the possibility that even these lower levels of CO might have significant health effects. This study used nonlinear, multivariate time series models to demonstrate an association between ambient CO and hospital admissions for congestive heart failure (CHF) among the elderly in seven, large U.S. cities. This association persisted after adjustment for seasonal effects, temperature, and other gaseous pollutants, including NO_2, SO_2, and O_3. Of these gaseous pollutants, only CO showed consistent, significant associations with CHF admissions. These relationships are shown in Figure 17.1. (Note that this figure includes eight cities with Denver added to the initial group of seven.) A 1997 study by Burnett et al.[2] demonstrated a similar investigation of 10 Canadian cities. The combined results for all 10 cities are shown in Figure 17.2. Subsequent studies discussed below have expanded on these results by accounting for the possible confounding effect of airborne particulate matter, by considering additional health outcomes, and by including the possible synergistic effect of additional environmental stressors.

A substantial body of epidemiological evidence suggests that current ambient levels of airborne particulate matter (PM) can have significant adverse health effects. Although the initial study by Morris et al.[1] did not include PM, subsequent studies in Detroit,[3] Chicago,[4] and Tucson[5] have demonstrated that including PM in the model does not substantially diminish the effect of CO on CHF. In the Detroit study,[3] CO was associated with admissions for ischemic heart disease (IHD) as well as CHF, but the risk associated with the interquartile range decreased from 1.010 (95% confidence interval or CI: 1.001, 1.018) to 1.006 (95% CI: 0.996, 1.018) after adjustment for PM.

Two European studies have also found associations between hospital admissions for cardiovascular disease and CO levels. Pantazopoulou et al.[6] found that emergency admissions in Athens for cardiac disease were associated with CO in the winter, but not in the summer. Poloniecki7 found that combined admissions for cardiovascular disease in London were significantly associated with CO, particularly myocardial infarction (MI), angina, and nonspecified circulatory disease. Additional analyses found that CO was only associated with MI in the winter, and the data related to MI showed that this effect was limited to the cold season. Unlike the American and Canadian studies, the British study did not find a significant association between CO and CHF admissions. In general, these studies suggest that ambient CO, even at relatively low levels, is consistently associated with hospital admissions for cardiovascular disease, particularly CHF. The specific outcomes involved show some variation from one city to the next. These discrepancies may reflect regional differences in pollutant mix, climate, or medical practice patterns.

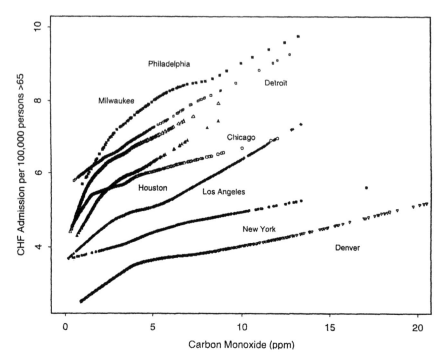

FIGURE 17.1   Locally smoothed curve of the association between ambient levels of CO and hospital admissions for CHF among the elderly after adjustment for temperature, month, day of week, and year for the period 1986 to 1989 in eight U.S. cities.

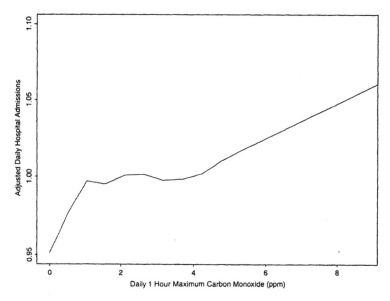

FIGURE 17.2   Nonparametric smoothed curve of the adjusted hospitalization rates for CHF in the elderly plotted against daily 1-h maximum CO levels based on 10 Canadian cities. (From Burnett, R.T. et al., *Epidemiology*, 8, 162–167, 1997. With permission.)

17.3 EPIDEMIOLOGICAL STUDIES OF CARDIOVASCULAR MORTALITY

In their seminal 1968 study, Goldsmith and Landaw[8] demonstrated that daily variation in CO levels was associated with daily variation in mortality in Los Angeles County. In particular, they demonstrated that the likelihood of death among persons admitted to Los Angeles–area hospitals with an acute MI increased when CO levels were elevated.[9] The effect of CO appeared to be far greater than could be attributed to measures of the combined oxidants.[10] Two aspects must be considered in using these studies to estimate health effects of CO. First, although these investigators used the most sophisticated tools available to them, the computational power and environmental data available in 1969 limited the analytical methods used in these studies. Second, the daily average CO levels in Los Angeles in the period from 1962 through 1965 when these studies were conducted ranged from 7 to 20 ppm, with concentrations in excess of 10 ppm for most of the period. These levels far exceed those currently found in the United States, but do have relevance for many areas of the developing world. Since 1969, air ambient levels of CO have dropped steadily in most U.S. cities. Today, the maximum CO levels in most cities rarely exceed the minimum level observed by Hexter and Goldsmith[10] in their Los Angeles study.

More recent studies in areas with lower CO concentrations have tended to find weaker associations between CO and mortality.[11-13] The fact that the epidemiological studies of CO were less likely to demonstrate an association with mortality than with morbidity (as indicated by hospitalization) may reflect both the pathophysiology of CO toxicity and the subsequent care of affected individuals. Persons who experience an MI or acute CHF will usually be hospitalized. Mortality, if it occurs, is likely to occur at some time after hospitalization. In the hospital or even in the ambulance, they will receive oxygen. Since the detrimental effects of CO are most likely to be related to impaired delivery of oxygen, this counteracts the persistent effect of ambient CO directly while counteracting other pollutants only indirectly. Hence, once an individual is hospitalized, the effect of ambient CO levels may be more attenuated than the effect of other pollutants. Two findings from the CO mortality studies tend to support this assertion. First, studies that examined the effect of CO levels at a time lag of 1 to 3 days found stronger effects for the lagged values.[13,14] Second, studies in cities with higher CO levels continued to find an effect on mortality.[6,15]

17.4 PLAUSIBILITY OF HEALTH EFFECTS RELATED TO LOW LEVELS OF CARBON MONOXIDE AT MONITORING STATIONS

The consistency of these results suggests that ambient CO levels or factors closely correlated with CO levels either precipitate or exacerbate CHF and IHD. However, any assertion that CO could play a causal role must include an explanation of the fact that the observed association with CHF occurs at levels of CO below current federal standards and well below levels at which effects have been observed in laboratory studies. Several factors could help to explain this apparent inconsistency.

First, levels at ambient monitors poorly represent individual exposures. It is well documented that individual exposures can be far higher than those measured at ambient monitors,[16] especially those of persons in traffic.[17] Heavy traffic and unfavorable meteorological conditions will increase ambient levels of CO, but heavy traffic will also dramatically increase the level and duration of exposure experienced by drivers and passengers in that traffic. In other words, elevated levels at ambient monitors translate into a shift in the distribution of individual exposure and a greater probability of individual exposure to elevated levels.

Second, persons with underlying heart disease, particularly those prone to CHF, may be uniquely susceptible to CO. Most studies of the effects of CO have been conducted among healthy young and middle-aged adults. Those studies that have considered persons with heart disease have focused on subjects with coronary artery disease rather than CHF. A review of the literature failed to identify any chamber studies of the effects of CO on persons with CHF.

Third, the presence of additional environmental stressors not directly related to ambient air quality may modify the effect of CO. These stressors could include any of the factors known to exacerbate cardiovascular disease such as dietary intake, exertion, or infection. Two stressors that deserve particular attention are temperature-related stress and CO from sources unrelated to ambient CO levels.

17.4.1 COMBINED EFFECTS OF CO AND TEMPERATURE

The findings of the study by Morris and Naumova[4] investigating the interaction between temperature and CO in relation to CHF admissions in Chicago support the third assertion. They found that most of the effect of CO occurred at low temperatures. There was little effect of CO in temperatures above 75°. Figure 17.3 shows the relative risk associated with the exposure to percentiles of CO at specific temperature strata for both the single-pollutant and multipollutant model.

Essentially all of the chamber studies of the effect of CO have been conducted at room temperature. This raises questions about the applicability of dose–response data from laboratory studies to cold weather conditions.

Several other studies have found results consistent with a synergistic effect of temperature and CO. In a study of hospital admissions for cardiovascular disease in Toronto, Burnett et al.[18] found that the effect of CO was weak when analyses were limited to summer, consistent with the finding of diminished effect at warmer temperatures. In a study of cardiovascular mortality in Los Angeles, Shumway et al.[19] concluded that the rate of increase in mortality associated with CO at low temperatures was greater than the rate at higher temperatures. As noted above, Pantazopoulou et al.[6] found that the CO had an effect on hospital admissions in the winter, but not in the summer.

The possibility that CO and cold temperatures act synergistically to exacerbate underlying cardiovascular disease appears biologically plausible. Exposure to cold air results in increased heart rate, increased systolic and diastolic blood pressure, and increased cardiac output in young adults.[20-22] This probably results from sharp increases in sympathetic activity and peripheral vasoconstriction,[23] a response also seen in subjects with coronary artery disease[24] and CHF.[25] The available research indicates that this increase in blood pressure is greater among older persons.[22,26]

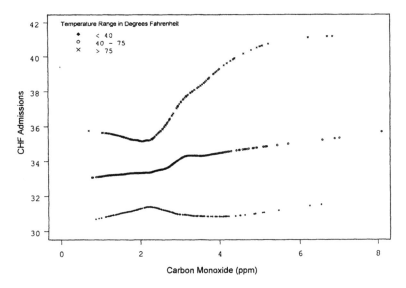

FIGURE 17.3 The results of the generalized additive model of hospital admissions for heart failure among the elderly as a function of CO in Chicago after stratification by temperature. (From Morris, R.D. and Naumova, E.N., *Environ. Health Perspect.*, 10, 643–647, 1998.)

In addition, exposures to temperatures as high as 12°C in still air for as little as 1 h can induce decreases in rectal temperature[22,26] with greater decreases in older men.[26] Finally, hypoxic conditions will increase the rise in diastolic blood pressure and the degree of shivering associated with cooling.[27]

Each of these changes will place an increased load on the failing heart. Increased vascular resistance will require increased cardiac work to maintain the same output. Patients using vasodilators may experience blunting of this response, but will then be more susceptible to cooling, which, itself, can place greater demand on the heart. Shivering, in particular, will require increased metabolic activity and increased oxygen consumption.[28] Overall, laboratory studies demonstrate that exposure to cold can increase the workload on the heart and cause oxygen demand to rise substantially.[29] It seems plausible that cold exposure would amplify the adverse effects of elevations in carboxyhemoglobin (COHb).

It has also been suggested that extremely high temperatures may also increase the adverse effects of air pollution.[30] The limited data from laboratory studies on the combined effects of air pollution with either heat stress or cold stress seem to support the possibility of enhanced toxicity of CO associated with extreme temperatures.[31]

Studies in Chicago and Tucson failed to demonstrate an increase in the effect of CO at high temperatures. It is possible that the cut point for the high temperature range was not high enough in the Chicago paper, while the summer CO levels were not high enough in Tucson. Further studies in areas with higher temperatures will be required to determine if the interaction between temperature and CO is "u-shaped."

The impact of CO on CHF patients has not been well studied, but in coronary artery disease patients, this effect has been clearly shown to decrease the time to angina at COHb levels as low as 3.0%,[32] and perhaps even lower.[33] It is possible that this threshold for adverse effects may be even lower among persons with CHF than persons with coronary artery disease. The studies described above indicate that cold exposure may reduce this threshold still further. In this way, susceptibility, thermal stress, and relatively low levels of ambient CO may combine to induce acute cardiovascular disease at an individual level.

The findings could also be explained by an unmeasured covariate that is correlated with temperature. These might include seasonal changes in diet, temporal variation in rates of respiratory diseases, particularly infection, or increases in physical stress during the winter months. Without a meaningful surrogate for these exposures, one cannot exclude these covariates as an explanation for the findings of this study.

17.5   CARBON MONOXIDE FROM HIGHLY LOCALIZED SOURCES

In considering the potential public health impact of ambient CO, it is important to understand that raising CO levels over broad geographic areas for large segments of the population has fundamentally different implications than exposure due to small, localized sources. Smoking and other combustion sources, particularly indoors, may contaminate microenvironments with levels of CO far above those caused by pollution of the ambient environment. The most extreme example of this is the cigarette, which exposes the smoker to levels of CO far in excess of those found in ambient air.

It is important that one not view exposures from these local sources as distinct from ambient CO simply because the concentrations are markedly higher. If these sharp peaks are considered as large waves that can cause damage when they hit shore, then ambient CO represents the sea level. To continue the analogy, increasing ambient CO is equivalent to raising sea level. A small increase in sea level can markedly increase the destructive effect of these waves as they hit the shore. In other words, exposure to a localized source producing CO concentrations of 19 ppm results in exposure to 25 ppm when ambient CO levels are 6 ppm. In this way, exposures of large numbers of people to CO (or any other pollutant for that matter) can have effects that are disproportionate to the relatively low concentrations produced in the ambient air.

This effect could also provide an alternative explanation for the greater effect of CO during cold temperatures. Decreased indoor air exchange and increased use of furnaces and other indoor sources of combustion products during the winter could increase the impact of increases in CO based on the mechanism described above. If the sea level metaphor is extended, increases in the number and size of waves will increase the effect of a change in sea level.

17.6   LIMITATIONS OF TIME SERIES STUDIES

The above discussion relies heavily on the results of epidemiological time series studies. It is conceivable that the observed effects reflect the fact that exposure and potential confounders were assessed at an ecological level. Nonrandom misclassification of exposure or time variant confounding variables could explain the observed temporal associations.

The use of ambient monitors to describe exposure to air pollutants represents a relatively crude measure of individual exposure. In particular, outdoor monitors do not accurately reflect indoor exposures, which account for the major portion of exposures for most people. The errors in the measurement of individual exposure will tend to reduce the magnitude of the observed effect. This source of error is not likely to explain the observed associations.

One of the major concerns in any epidemiological study is the possibility that an observed association between a particular exposure and disease could be the result of a confounding factor. A confounder is a factor that is responsible for the observed health outcome and occurs in correlation with the exposure of interest, in this case CO. This can create an association that is not truly causal. Time series methods limit potential confounders to those factors that covary in time with levels of air pollutants. Factors known to exacerbate heart disease such as smoking and diet are not likely to be confounders in these studies because they are not likely to covary with CO levels. Time variant factors with the potential to confound the observed association include temperature, seasonal weather patterns, and outdoor activity patterns. Inclusion of temperature and dummy variables for the month in the analysis should control for the potential effects of weather and other factors with major seasonal fluctuation. Exposure to other pollutants may also confound the association between a single pollutant and CHF. Since the published studies indicate that the criteria pollutants including PM do not act as confounders, the potential confounders are limited to pollutants that are not routinely measured, but are closely correlated with CO. Although the existence of such a pollutant is conceivable, it seems more parsimonious to posit a causal role for CO.

Even if CO does contribute to hospital admissions for cardiovascular disease, the impact of removing that CO is unclear. A person who is hospitalized for cardiovascular disease generally has substantial underlying disease that has simply reached a level of severity that necessitates medical intervention. The public health implications are very different depending on whether CO has simply accelerated the disease process by a few days or caused admissions to occur that would not have occurred otherwise. The extent to which the CO-related admissions simply reflect the acceleration of admissions by a few days among an extremely ill subpopulation (an effect often referred to as "harvesting") is unclear and needs to be explored.

17.7 CONCLUSIONS

There is sufficient evidence to suggest that ambient CO levels have a significant adverse effect on persons with underlying heart disease. This probably results from the combined effect of increases in background CO and exposure to other environmental stressors. Cold weather, in particular, appears to increase the effect of CO. These findings indicate that members of sensitive subpopulations who are exposed concurrently to other stressors may be affected by CO at concentrations previously thought to be safe. The exact level of the individual exposures is unclear, since these exposures are measured at fixed monitoring sites. However, these exposures are occurring when the fixed-site monitors are usually well below EPA standards.

REFERENCES

1. Morris, R.D., Naumova, E.N., and Munasinghe, R.L., Ambient air pollution and hospitalization for congestive heart failure among elderly people in seven large U.S. cities, *Am. J. Public Health*, 85, 1361–1365, 1995.
2. Burnett, R.T., Dales, R.E., Brook, J.R., Raizenne, M.E., and Krewski, D., Association between ambient carbon monoxide levels and hospitalizations for congestive heart failure in the elderly in 10 Canadian cities, *Epidemiology*, 8, 162–167, 1997.
3. Schwartz, J. and Morris, R.D., Cardiovascular disease and airborne particulate levels in Detroit, Michigan, *Am. J. Epidemiol.*, 142, 23–35, 1995.
4. Morris, R.D. and Naumova, E.N., Congestive heart failure among the elderly in Chicago and ambient carbon monoxide, temperature and particulates, *Environ. Health Perspect.*, 10, 643–647, 1998.
5. Schwartz, J., Air pollution and hospital admissions for cardiovascular disease in Tucson, *Epidemiology*, 8, 371–377, 1997.
6. Pantazopoulou, A., Katsouyanni, K., Kourea-Kremastinou, J., and Trichopolous, D., Short–term effects of air pollution on hospital emergency outpatient visits and admissions in the greater Athens, Greece area. *Environ. Res.*, 69, 31–36, 1995.
7. Poloniecki, J.D., Atkinson, R.W., Ponce de Leon, A., and Anderson, H.R., Daily time series for cardiovascular hospital admissions and previous day's air pollution in London, U.K., *Occup. Environ. Med.*, 54, 535–540, 1997.
8. Goldsmith, J.R. and Landaw, S.A., Carbon monoxide and human health, *Science*, 162, 1352–1359, 1968.
9. Cohen, S.I., Deane, M., and Goldsmith, J.R., Carbon monoxide and survival from myocardial infarction, *Arch. Environ. Health*, 19, 510–517, 1969.
10. Hexter, A.C. and Goldsmith, J.R., Carbon monoxide: association of community air pollution with mortality, *Science*, 172, 265–266, 1971.
11. Verhoeff, A.P., Hoek, G., Schwartz, J., and Van Wijnen, J.H., Air pollution and daily mortality in Amsterdam, *Epidemiology*, 7, 225–230, 1996.
12. Burnett, R.T., Cakmak, S., and Brook, J.R., The effect of the urban ambient air pollution mix on daily mortality rates in 11 Canadian cities, *Can. J. Public Health*, 89, 152–156, 1998.
13. Wietlisbach, V., Pope, C.A., and Ackermann-Liebrich, U., Air pollution and daily morality in three Swiss urban areas, *Soz. Praeventivmed.*, 41, 107–115, 1996.
14. Kelsall, J.E., Samet, J.M., Zeger, S.L., and Xu, J., Air pollution and mortality in Philadelphia, 1974–1988, *Am. J. Epidemiol.*, 146, 750–762, 1997.
15. Saldiva, P.H.N., Pope, C.A., Schwartz, J., Dockery, D.W., Lichtenfels, A.J., Salge, J.M., Barone, I., and Bohm, G.M., Air pollution and morality in elderly people: a time-series study in Sao Paulo, Brazil, *Arch. Environ. Health*, 50, 159–163, 1995.
16. Akland, G.G., Hartwell, T.D., Johnson, T.R., and Whitmore, R.W., Measuring human exposure to carbon monoxide in Washington, D.C. and Denver, CO, during the winter of 1982–1983, *Environ. Sci. Technol.*, 19, 911–918, 1985.
17. Flachsbart, P.G., Mack, G.A., Howes, J.E., and Rodes, C.E., Carbon monoxide exposures of Washington commuters, *J. Air Pollut. Control Assoc.*, 37, 135–142, 1987.
18. Burnett, R.T., Cakmak, S., Brook, J.R., and Krewski, D., The role of particulate size and chemistry in the association between summertime ambient air pollution and hospitalizations for cardiorespiratory diseases, *Environ. Health Perspect.*, 105, 614–620, 1997.

19. Shumway, R.H., Azari, A.S., and Pawitan, Y., Modeling mortality fluctuations in Los Angeles as functions of pollution and weather effects, *Environ. Res.*, 45, 224–241, 1988.

20. Walsh, J.T., Andrews, R., Batin, P.D., and Cowley, A.J., Haemodynamic and hormonal response to a stream of cooled air, *Eur. J. Appl. Physiol. Occup. Physiol.*, 72, 76–80, 1995.

21. Vogelaere, P., Deklunder, G., Lecroart, J., Savourey, G., and Bittel, J., Factors enhancing cardiac output in resting subjects during cold exposure in air environment, *J. Sports Med. Phys. Fitness*, 32, 378–386, 1992.

22. Inoue, Y., Nakao, M., Araki, T., and Ueda, H., Thermoregulatory responses of young and older men to cold exposure. *Eur. J. Appl. Physiol. Occup. Physiol.*, 65, 492–498, 1992.

23. Postolache, T., Gautier, S., Laloux, B., Safar, M., and Benetos, A., Positive correlation between the blood pressure and heart rate response to the cold pressor test and the environmental temperature in older hypertensives, *Am. J. Hypertens.*, 6, 376–381, 1993.

24. DuBois-Rande, J.L., Dupouy, P., Aptecar, E., Bhatia, A., Teiger, E., Hitting, L., Berdeaux, A., Castaigne, A., and Geschwind, H., Comparison of the effects of exercise and cold pressor test on the vasomotor response of normal atherosclerotic coronary arteries and their relation to the flow-mediated mechanism, *Am. J. Cardiol.*, 76, 467–473, 1995.

25. Grassi, G., Seravelle, G., Catteneo, B.M., Lanfranchi, A., Vailati, S., Gianattasio, C., Del Bo, A., Sala, C., Bolla, G., Pozzi, M., and Mancia, G., Sympathetic activation and loss of reflex sympathetic control in mild congestive heart failure, *Circulation*, 92, 3206–3211, 1995.

26. Collins, K.J., Abdel-Rahman, T.A., Easton, J.C., Sacco, P., Ison, J., and Dore, C.J., Effects of facial cooling on elderly and young subjects: interactions with breath-holding and lower body negative pressure, *Clin. Sci.*, 90, 485–492, 1996.

27. Robinson, K.A. and Haymes, E.M., Metabolic effects of exposure to hypoxia plus cold at rest and during exercise in humans, *J. Appl. Physiol.*, 68, 720–725, 1990.

28. McArdle, W.D., Magel, J.R., Lesmes, G.R., and Pechar, G.S., Metabolic and cardiovascular adjustment to work in air and water at 18, 25, and 33 degrees C, *J. Appl. Physiol.*, 40, 85–90, 1976.

29. Frank, S.M., Higgins, M.S., Fleicher, L.A., Sitxmann, J.V., Raff, H., and Breslow, M.J., Adrenergic, respiratory and cardiovascular effects of core cooling in humans, *Am. J. Physiol.*, 272, R557–R562, 1997.

30. Katsouyanni, K., Pantazopoulou, A., Touloumi, G., Tselepidaki, I., Moustris, K., Asimakopoulos, D., Poulopoulou, G., and Trichopoulos, D., Evidence for interaction between air pollution and high temperature in the causation of excess mortality, *Arch. Environ. Health*, 48, 235–242, 1993.

31. Yang, L., Zhang, W., He, H.Z., and Zhang, G.G., Experimental studies of combined effects of high temperature and carbon monoxide, *J. Tongji. Med. Univ.*, 8, 60–65, 1988.

32. Allred, E.N., Bleecker, E.R., Chaitman, B.R., Dahms, T.E., Gottlieb, S.O., Hackney, J.D., Pagano, M., Selvester, R.H., Walden, S.M., and Warren, J., Short-term effects of carbon monoxide exposure on the exercise performance of subjects with coronary heart disease, *N. Engl. J. Med.*, 321, 1426–1432, 1989.

33. Aronow, W.S. and Isbell, M.W., Carbon monoxide effect on exercise induced angina pectorix, *Ann. Intern. Med.*, 79, 392–395, 1973.

18 Chronic Carbon Monoxide Poisoning

David G. Penney

CONTENTS

18.1 INTRODUCTION

18.1.1 DEFINITIONS

The purpose of this chapter is to bring into sharper focus what is meant by chronic carbon monoxide (CO) poisoning, to review some of the literature on the subject, and to present data from recently completed studies.

0-8493-2065-8/00/$0.00+$.50
© 2000 by CRC Press LLC

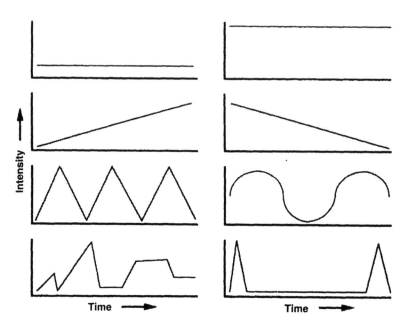

FIGURE 18.1   Possible scenarios of chronic carbon monoxide exposure.

As a working definition, chronic CO poisoning involves an exposure to CO that occurs more than once and lasts longer than 24 h. Accordingly, acute CO poisoning involves an exposure to CO that occurs only once and lasts no longer than 24 h. Chronic CO poisoning usually involves lower levels of the gas in the air and lower blood COHb saturations; higher CO concentrations and COHb often end in death, thus never becoming chronic. Exposure usually continues for many days, months, and even years. The boundary limit between acute and chronic exposure is indistinct.

According to *Webster's New College Dictionary*,[1] the word *chronic* derives from the Greek word *khronos*, or time. In Latin the word is *chronicus*, and in French *chronique*. Its meaning is "1) of long duration, 2) subject to a habit or disease for a lengthy period." Synonyms are continuing, lingering, persistent, prolonged, protracted. *Subacute* is another term often used nearly synonomously in the medical literature.[2]

The term *chronic* is sometimes used as in definition 2 above — "A history of CO inhalation and an awareness of the typical distributions of lesions are important for recognition of the effects of CO poisoning, especially when patients are in the chronic stage,"[3] or in a title, "Magnetic resonance imaging of carbon monoxide poisoning in chronic stage."[4] For this discussion of chronic CO poisoning, the word *chronic* is used to describe how long the insult (exposure) lasts, not how long the resulting effects last. A damaging effect of CO poisoning or, in fact, any change that persists should be referred to as a residual effect.

The possible scenarios of chronic CO exposure are almost limitless (Figure 18.1). The CO concentration may start out high or low, and then fall or rise with time. The CO concentration may remain low or high throughout the period of exposure. CO concentration may change, or not change, in far less regular ways throughout the period of exposure.

18.1.2 COMPARISONS WITH ACUTE CARBON MONOXIDE POISONING

Chronic CO poisoning often does not elicit the typical symptoms of (acute) CO poisoning, such as headache, nausea, weakness, and dizziness. Symptoms are often said to be "flu-like."[5] Mucous membranes are almost never cherry pink and COHb is usually not excessively elevated. Computed tomography (CT) and magnetic resonance imaging (MRI) approaches are generally not useful. Chronic CO poisoning is often misdiagnosed as chronic fatigue syndrome, a viral or bacterial pulmonary or gastrointestinal infection, a "run-down" condition, immune deficiency, or a psychiatric condition. Patients may present with polycythemia, increased hematocrit, erythrocyte count, etc. In one epidemic of occult CO poisoning, patients were misdiagnosed with atypical pneumonia, myocardial infarction, cholecystitis, epilepsy, or viral illness.[6] It is reported that in chronic CO exposure there is no difference in tissue CO content of the heart and skeletal muscles, whereas in acute exposure there is.[7]

According to the World Health Organization (WHO),[8] long-term exposure to CO may cause headache, memory defects, falling work output, sleep disturbances, vertigo, emotional distress, central and peripheral nervous system damage, and increased concentrations of serum cholesterol, lipoproteins, and glucose.

Chronic CO poisoning is frequently referred to as "occult" CO poisoning, because it is "hidden from view, secret, concealed, not divulged." Most chronic CO poisoning is of this type, at least at first, in both adults[9] and children.[10] Symptoms associated with occult CO poisoning are headache, fatigue, dizziness, paresthesias, chest pains, palpitations, and visual disturbances.[11] Paresthesias are "abnormal or morbid sensations, such as burning, prickling, etc. feelings, but without objective symptoms."

Chronic CO poisoning is, in fact, extremely difficult to diagnose and is very frequently mistaken for other conditions noted above. Similar symptoms seen simultaneously in more than one person,[12] and which disappear upon removal from a specific site/environment are tip-offs that CO is involved. COHb is usually not excessively elevated. More often than not, by the time air CO or blood CO levels are measured, the condition has been corrected, making air and/or blood detection impossible. CT and MRI generally show no lesion, even when psychological/psychiatric and neurologic evaluations detect functional deficits. The characteristics of chronic CO poisoning are presented in Table 18.1. Clues to its discovery are shown in Table 18.2.

18.1.3 MISDIAGNOSIS

It is well known that the misdiagnosis rate for CO poisoning is very high. It is even higher for chronic CO poisoning. Some common misdiagnoses are listed in Table 18.3.

There are probably a number of reasons chronic CO poisoning is not better recognized by the medical community:

1. It almost invariably presents with too many disparate, seemingly unrelated, and often nonspecific symptoms. This tends to confuse physicians who act mainly on pattern recognition of one or a few symptoms to come up with a probable diagnosis, or at least a "short list." The result of being presented with 5, 10, 15, or more symptoms is likely to yield a diagnosis of hypocondriasis (faking), psychiatric condition, or both.

TABLE 18.1
Characteristics of Chronic Carbon Monoxide Poisoning

- Often goes long undetected
- Masquerades as flu, fatigue, etc.
- Often many people "sick" simultaneously
- Goes away upon leaving poisoning site (to work, on vacation, etc.)
- Nearly always misdiagnosed by physicians
- May involve pets "sick," dead at same time
- Rarely involves sinus congestion, cough (when present, it may be due to other compounds, e.g., NO_x, SO_2 in exhaust gases)

TABLE 18.2
Clues to Discovery of Chronic Carbon Monoxide Poisoning

- Lethargy, headache, etc. of long duration
- Long-standing "illness" intractable to medical solutions
- "Illness" that suddenly improves when leaving site
- Multiple cases at one location
- Morbidity/mortality of pets

TABLE 18.3
Common Misdiagnoses for Carbon Monoxide Poisoning

- Chronic fatigue syndrome
- Viral or bacterial pulmonary or gastrointestinal infection
- "Run-down" condition
- Endocrine problem
- Immune deficiency
- Psychiatric/psychosomatic problem
- Allergies
- Bad/tainted food

2. Presentation in urgent care settings that usually appears not to require emergency measures — absence of unconsciousness, no obvious provoking agent, low or normal COHb values, skin/mucous membranes not pink, etc.
3. It has been difficult to study in animal models because rats, mice, etc. are far more resistant to CO than are humans, and also are unable to report the many psychological, cognitive, and emotional changes that result. Thus, there is little understanding of the underlying cellular mechanisms at play.
4. Lack of training in the area, thus a low index of suspicion for the condition and the resultant shockingly high rate of misdiagnosis.

There are a number of problems in dealing with chronic CO poisoning. Table 18.4 lists some of these. It is this author's belief that, "For every single case of chronic CO

TABLE 18.4
Problems in Dealing with Chronic Carbon Monoxide Poisoning

- Fact of exposure usually only recognized later
- Good COHb level measurements rarely obtained
- Air CO level measurements not usually obtained
- Long-term effects recognized by few physicians, etc., thus difficult to prove
- Seldom produces damage recognizable by high-tech scanning techniques (MRI, CT, SPECT)
- Changes seen by neuropsychological testing usually most useful
- Considerable variability of effects from one individual to the next

---

poisoning reported/successfully diagnosed, there are 10 cases that go unreported/undiscovered/undiagnosed."

This is a subject about which very little human experimental data are available. Until recently, those data that were available were mainly anecdotal. A small body of animal data are available which are of extremely limited value in understanding and predicting human physical, cognitive, and emotional responses. For additional information on chronic CO poisoning, readers should see Chapters 10, 12, 19, and 23.

18.1.4 R397397ESIDUAL EFFECTS

One of the most controversial questions concerning chronic CO poisoning is whether or not it can result in residual (i.e., long-lasting) effects (i.e., damage). As studies described below in this and in other chapters (e.g., Chapter 19) show, this often does occur.

On a theoretical basis, it is suggested that damage begins to develop immediately with both acute and chronic CO exposure (Figure 18.2). Damage is manifest when CO exposure ends with both forms. With chronic CO, however, exposure lasts so much longer that the symptoms expressed by victims are a combination of acute symptoms (those seen in the main during very brief acute exposures) and increasingly with duration of exposure, residual symptoms caused by long-term damage (e.g., neurocognitive, behavioral). Thus, in chronic CO exposure, common physical symptoms (e.g., headache, nausea, dizziness) many times fail to disappear during intervals of nonexposure after some period of CO exposure, blurring the effect of the site-specific and temporal relationships to CO exposure, which are often said to be important clues to the fact of CO poisoning itself.

It is a paradox of biology that, while CO is deleterious to the body in limiting oxygen delivery, binding to the intracellular energy generating system, killing cells, causing damage to tissues and organs, and killing people, it is now known that CO is generated by the human body as a by-product of hemoglobin metabolism and, as such, is "natural/helpful" along with nitric oxide (NO), as an integrator of vascular control, and part and parcel of a mechanism by which blood vessels dilate as COHb increases, providing increased blood perfusion.[13]

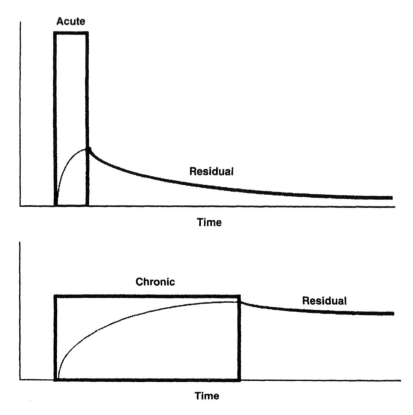

FIGURE 18.2   Theoretical approach to the long-term effects of chronic carbon monoxide exposure.

18.1.5   Case Reports

Hypothetical

A typical case report of chronic CO poisoning is as follows:

Mrs. Betty Jones is a 35-year-old homemaker. She and her husband George, 37 years old, live in a city in the Midwest. She has an associate degree in accounting, while her husband has a master's degree in business administration. Neither of them smokes.

In early 1995, they purchased a home in a suburban community through a real estate brokerage company. The home was built in 1958. It was inspected and the major appliances in the home were guaranteed for 5 years. The home has three bedrooms, a living room, family room, and a glassed-in back porch. It is heated by a forced-air, natural gas furnace in the basement. Hot water is provided by a gas-fired water heater, also in the basement.

Beginning in the autumn of 1995, Betty Jones began having headaches and feeling very tired. Her two children, John (12 years of age) and Cathy (9 years of age), and her husband George occasionally awoke in the morning with headaches, dizziness, and nausea. They believed that they all had a touch of "flu" or had eaten tainted food.

Mrs. Jones continued to feel "out of it" for the remainder of 1995 and into the spring of 1996. Her physician, Dr. Blackstone, gave her a "physical," obtaining chest radiographs, blood for complete CBC, and samples for a Pap smear test. He found nothing wrong, saying that "flu" has been going around. A furnace company who regularly serviced the heating system found "everything in good working order."

During the summer of 1996, Betty Jones and the whole family felt much better, although she and the children continued to have frequent headaches and to feel slightly fatigued. They felt better when they went away for vacation for 2 weeks.

In late October 1996, Betty Jones again began to have frequent severe headaches and to become extremely fatigued. She was becoming so lethargic that she could not accomplish her normal housework. She was forgetting tasks that needed doing, and finding it increasingly difficult to maintain the family checkbook. She was also feeling depressed and defeated in her daily life.

On several visits to Dr. Blackstone she was told that there was nothing wrong with her. He said her perceived state was psychosomatic, and that she should seek counseling or schedule regular visits with a psychiatrist.

Actual

The following is an actual case study of a family who generously allowed the author to reproduce it here:

"My family and I suffered through carbon monoxide poisoning from a gas heater. We had numerous health problems, always worse in the winter, before discovering the faulty heater in February 1997, with a near catastrophy. I had severe body flu symptoms in the mornings. I constantly was checking my temp, but never had a fever. It was usually below normal during the worst times.

"Doctors either didn't know why or tried in the dark. One gave me antidepressants; one said it might be my high cholesterol! I've had some sore, stiff neck pain in the past, but it became severe during the period. My small son suffered *bad* headaches, and woke up crying with them many mornings. The doctor didn't know why, and guessed allergies. My daughter had body aches and lethargy. My wife would many times sleep until noon. She developed anxiety and panic attacks. She had her thyroid checked and it was normal. We even sought help from chiropractors.

"Turns out, the furnace was recalled years earlier for the problem. We obviously had lived with it for 3 to 4 years at least, never knowing anything except these health problems were worse when at home and in the winter. I thought it must be allergies in my case. When CO was discovered, the headaches, flu symtoms, etc. were explained.

"Now the problem is my neck pain is constant. My short-term memory and concentration is a problem. The memory problem is about to drive me crazy. I write myself notes and forget the notes. I'm not just talking age either (43). I have bad moods like never before, and realize I have been much more pleasant around my family in the past.

"My wife's panic attacks, mostly at night, are serious. Her most frequent episodes result in her leaping from the bed in the middle of the night and running to the hall air return to listen to the heater, absolutely horrified, thinking something is wrong.

She has now been rechecked and has thyroid problems. She is on strong antidepressants, and has been to a psychologist. She, many days, has trouble functioning.

"My son, now 8, has been diagnosed with ADD-HA (Attention Deficit Disorder–Hyperactivity) and is on Ritalin. My daughter has bad concentration problems, does poorly on tests, and has attention problems. Searching for answers, I decided to check out carbon monoxide on the Internet. As you can imagine, I was shocked after reading your site. I ran to my wife with prints of the info!

"No doctors around know anything about chronic CO poisoning. If I bring it up, they think only of acute poisoning, and apparently don't know about chronic poisoning. I talked to my insurance health-care advisor. She was lost, didn't know any doctor she could send me to, and encouraged me to contact you. I hope this has not been too time-consuming. I need help and would love the opportunity to talk to you. May I call? If so, when are good times for you.

"Thank you for anything you can do."

18.2 PUBLISHED CHRONIC CARBON MONOXIDE STUDIES

18.2.1 INTRODUCTION

Harvey Beck,[14] writing in 1927, states that "chronic carbon monoxide poisoning is widespread and far more prevalent than is generally supposed...." The most frequent symptoms according to him are headache, dizziness, muscular weakness, disturbance of gait, parasthesia, breathlessness on exertion, and nervous and emotional instability. The muscular symptoms include asthenia and twitching of muscle fiber bundles. A "hypertonicity of the involuntary muscles occurs which may give rise to spasmodic contraction producing symptoms of oesophageal obstruction with inability to swallow food, ureteral spasm simulating renal colic and enterospasm with symptoms of acute abdominal obstruction or peritonitis." Beck found that in the majority of his patients, the symptoms were slow to subside. He published the results of a large case series study on chronic CO poisoning in 1936 (see below).

Readers are asked to consult Chapter 10 in this book, where Donnay examines the hypothesis that CO poisoning from the use of illuminating gas in the 19th and early 20th centuries may have been a major unrecognized cause of neurasthenia, a condition that involved symptoms very similar to those described here.

18.2.2 OCCUPATIONAL STUDIES

Lindgren[15] in 1961 studied 970 workers occupationally exposed to CO and 432 controls, paired to them by age. He found an increased frequency of headache among the CO-exposed group. More recently, secondhand smoke has been suggested as the cause of irritability, headache, blurred vision, slowed reaction time, and decreased concentration in nonsmokers.[16]

Minor, transient changes in the electrocardiogram (ECG), mainly in the repolarization phase, were observed in 1966 in 7 of 12 workmen occupationally exposed to CO.[17] The changes were not considered of pathological importance. In a controlled study in 1980, Davies and Smith[18] reported ECG changes in young, healthy, non-smoking Navy personnel exposed continuously to CO for 8 days. P-wave changes were seen in 6 of 15 subjects at 50 ppm, and in 3 of 15 subjects at 15 ppm. One subject showed marked S-T changes at 15 ppm. A related pilot study by this group revealed significant ECG changes in 7 of 10 subjects.

Wilson and Schaeffer[19] in 1979 reported significant increases in hemoglobin concentration, hematocrit, and erythrocyte counts in men exposed to elevated CO (15 to 20 ppm) and carbon dioxide (0.9%) during a submarine patrol. This is not unlike the increase in these blood parameters reported in cigarette smokers.[20]

Ocular effects following chronic, intermittent CO exposure over a 16-month period from a defective furnace were reported by Trese et al.[21] The 57-year-old woman experienced headaches and somnolence, and on examination had peripheral neuropathy, constricted visual fields, tortuous retinal vessels, and an abnormal visual-evoked response. She was a nonsmoker and her COHb was 2.1% 36 h after the last CO exposure. CT showed a "wedge-shaped defect in the right occipital lobe."

Kowalska[22] in 1981 studied 78 men occupationally exposed to CO using audiometric and electronystagmographic approaches. Hearing impairment was found to exist in 66.6% of cases, and in 79.5% of that group vestibular changes were present. These effects were regarded by the investigators as typical of CO exposure.

Barrowcliff[23] reported in 1978 on the case of a middle-aged man who had worked for 3 years in a poorly ventilated room using large quantities of methylene chloride (i.e., dichloromethane). He presented with unsteady gait, a peculiar dysarthria, and a loss of memory. It was presumed that damage resulted from the CO generated during catabolism of the solvent, although a direct effect of the methylene chloride was possible. The writers were unable to find a "natural" disease to explain his condition.

A Danish study in 1989 examined garage employees who worked around diesel-powered vehicles.[24] Of the workers, 11 presented with acute symptoms in the form of headache, vertigo, fatigue, irritation of mucous membranes, nausea, abdominal discomfort, or diarrhea. Six of the seven workers who had been employed for more than 5 years complained of deteriorating memory, difficulty in concentration, irritability, increased sleep requirements, psychological changes, or reduced libido. Neuropsychological evaluation found that five of these had slight organic brain damage.

Tvedt and Kjuus[25] recently presented two cases of chronic CO exposure. In one case a crane operator at a smelter developed permanent symptoms after 20 years. In a second case, four members of a family incurred long-lasting (i.e., residual) symptoms as the result of a faulty oil-fired central heating system.

18.2.3 EPIDEMIOLOGICAL (LARGE-SCALE) STUDIES

Excess deaths due to periodic elevated ambient CO level in Los Angeles were noted by Hexter and Goldsmith[26] in 1971, who found a highly significant correlation between air CO level and mortality. They estimated that the contribution of CO to mortality in L.A. County for an average CO concentration of 20.2 ppm was 11 deaths, during the 24-h period that the CO level was present.

Increased morbidity caused by "ultra low"-level CO exposure, rather than mortality, has been revealed in a number of recent epidemiologic studies. This has included studies of birth weight,[27] congestive heart failure (CHF)[28] (see Chapter 17), and coronary artery disease (CAD).[29] Interestingly, hospital admissions for psychiatric conditions were shown to be correlated with ambient CO level in St. Louis during 1972.[30] The maximum CO concentration recorded at that time was 23 ppm.

It could be argued that the health effects of CO on patients with CHF and CAD result from spikes, and thus constitute "acute," not chronic CO effects. This probably cannot be said about the effects of ultralow levels of CO on birth weight. Ritz and Yu[27] evaluated the effect of CO exposures during the last trimester of pregnancy on the frequency of low birth weight among neonates born from 1989 to 1993 to women living in the L.A. area, a study cohort of 125,573 singleton children. Within the cohort, 2813 (2.2%) were low in birth weight (1000 to 2499 g). Exposure to higher levels of ambient CO (>5.5 ppm 3-month average) during the last trimester was associated with a significantly increased risk for low birth weight, after adjustment for potential confounders, including commuting habits in the monitoring area, sex of the child, level of prenatal care, and age, ethnicity, and education of the mother. The inescapable conclusion is that levels of environmental CO, previously thought to be extremely low, lower the birth weight of children born to women exposed to CO during the last trimester of pregnancy.

18.2.4 PEDIATRIC STUDIES

Please consult Chapter 21, where White discusses two studies on chronic CO poisoning in children. In one instance, 4- and 5-year-old siblings developed hyperactivity, mood swings, headache, and lethargy, although psychometric testing in these children was normal.[31] In another case, progressive neurologic disability in a 2½-month-old infant was linked to chronic CO exposure. The presence of bilateral basal ganglia hypodensities led to the diagnosis.[32]

18.2.5 BECK CASE SERIES

In the 1920s and 1930s Beck[33] was convinced that chronic CO poisoning could be hazardous to health and was a significant public menace (see above). He noted that deaths from CO poisoning at that time, both accidental and through suicide, were second only to those from automobile accidents. In New York City during the 5-year period from 1928 to 1932, 5289 deaths involved CO.

He went on to say, "No noxious gas so potent when inhaled in atmospheric dilutions of 1% or even less as to cause almost instantaneous death can be incapable of producing symptoms if inhaled in lesser concentration over a longer period of time." Thus, he maintained a belief in the toxic effects of CO even at lower concentrations.

He conducted a retrospective study of patients coming to him for CO poisoning over a period of time. The study included 97 patients — 67 men and 30 women; 49 were from West Virginia, 37 from Maryland, and the rest from five other states.

The sources of CO encountered by his patients were natural gas (43), illuminating gas (28), automobiles/engines (24), blast furnace (1), and coke oven (1). Exposure modes were continuous or intermittent, while duration was from several months to 18 years. The signs and symptoms observed by Beck are presented in Table 18.5, listed in order of frequency of occurrence.

TABLE 18.5
Signs and Symptoms Observed by Beck
Listed in Order of Frequency of Presentation

Symptom	No. Reporting
Headaches	58
Weakness	52
Vertigo/disturbance of gait	46
Spastic constipation	45
Eythrocyte count elevated	44
Dyspnea	36
Paresthesia	36
Weakness/tremulousness/ataxia	30
Palpitation	27
Muscle spasm	17
Basal metabolic rate low	16
Coughing	15
Precordial distress	14
Glycosuria	13
Albuminuria	12
Dysuria	10
Gastrointestinal spasm	7
Speech defects	7
Cardiospasm	6
Visual disturbances	5
Diarrhea	5
Yawning	5
Paraplegia	3
Angina pectoris	2
Heart block	1

Source: Modified from Beck.[33]

Other signs and symptoms reported by Beck in his study but without notation of frequency were as follows:

1. *Nervous/mental* — Depression, restlessness, anxiety, fears, introspection, emotional upheavals, mental retardation/memory defects, confusion, drowsiness, insomnia, paresthesia, speech defects, paraplegia, vasomotor instability/morbid flushing, local sweating, cold extremities, purplish congestion of hands and feet, tinnitus aurium, visual disturbances, and change in sense of smell.
2. *Neuromuscular* — Pain in back, shoulders, epigastrium, lower abdomen, and chest, cardiospasm, gastrointestinal spasm, dysuria, cramping of toes, and muscle twitching.
3. *Gastrointestinal manifestations* — Glossitis, dysphagia, anorexia, nausea, and vomiting.

4. *Cardiorespiratory manifestations* — Decreased blood pressure and bradycardia were seen in 30% of the patients, and hypotension in 50%.
5. *Genitourinary manifestations* — Vesical irritability, nocturia, dysuria, pol-
lakuria, incontinence (several), dysmenorrhea, menorrhagia, amenorrhea,
and decreased libido in women.

It might be noted how similar the complaints and conditions observed by Beck[33]
are to those that appear in studies conducted recently, i.e., some 60 and more years
later (see below and Chapter 19). This includes the physical effects (headache,
nausea, confusion, weakness, etc.), the sensory effects (visual disturbances, tinnitus,
etc.), the cognitive and memory effects (mental retardation, memory defects), and
the emotional and mood effects (depression, anxiety, etc.).

It is unfortunate that Beck did not separate the acute effects/symptoms that were
occurring during the CO exposure from the long-term/residual effects that endured
for weeks, months, and years after CO exposure ended. The study hints at residual
effects, but nowhere are the symptoms clearly stated as continuing after CO exposure.

18.2.6 KIRKPATRICK CASE SERIES STUDY

The data for 26 patients who sought medical care for CO poisoning between 1982
and 1984 were analyzed by Kirkpatrick.[11] Inclusion criteria were (1) presence of
symptoms for at least 1 month and (2) improvement in their condition after a source
of CO was eliminated. According to questionnaire entries, 15 patients were exposed
to nine defective gas furnaces, 3 to faulty oil furnaces, 7 to malfunctioning automobile
exhaust systems, and 1 to a defective bus exhaust system. Included were 18 females
and 8 males, with ages ranging from 5 to 62 years. Only 3 of the patients were
admitted smokers.

Mean COHb level was 15.1%, with a range of 8.8 to 36.8% for 14 patients. The
mean COHb levels for patients exposed to automobile exhaust vs. those exposed to
furnace fumes were 14.0 and 15.8%, respectively. Three patients with gas furnaces
presented in coma.

The frequency of acute symptoms are seen in Table 18.6. Fatigue was the most
frequently reported symptom, closely followed by headache, trouble thinking, and
dizziness, in that order. These symptoms had been experienced by patients for periods
from 1 month to 4 years.

According to the report, residual symptoms "resolved" 1 to 12 weeks after
discovery and medical intervention. There is no indication that possible cognitive
and/or emotional changes in the patients were further explored through referral to
a neurologist, psychiatrist, or neuropsychologist.

18.2.7 RYAN CASE REPORT

In this case study,[34] the patient was a 48-year-old married woman, with high school
education, including 2 years of business school. She has a 3-year history of constant
headaches, lethargy, and memory problems. She had trouble recalling new information, but no problem with recall of the distant past. Also, she had difficulties with

TABLE 18.6
Frequency of Symptoms in 26 Patients
· Exposed to Carbon Monoxide

Symptoms	No. of Patients	% of Patients
Fatigue	24	92
Headaches	22	85
Trouble thinking	20	77
Dizziness	19	73
Nausea	15	65
Trouble sleeping	15	65
Heart pounding	14	54
Shortness of breath	14	54
Numbness or tingling	12	46
Chest pain	9	35
Decreased vision	9	35
Diarrhea	9	35
Unusual spells	9	35
Abdominal pain	7	27

Source: Modified from Kirkpatrick.[11]

depression and anxiety, and was having problems in her marriage. She had no history of alcohol or drug abuse, no history of head trauma, no family history of psychiatric problems, and no known exposure to other toxic chemicals.

The patient had been exposed to CO while running a typing service out of her basement in a tightly insulated home. Her furnace was found to be releasing 180 ppm CO, possibly for up to 3 years. No COHb measurement was available. She had never been unconscious. The woman's headaches stopped when the furnace was replaced, but her memory difficulties continued.

Neuropsychological evaluation revealed that the woman was of average intelligence. She performed at the 4th percentile on the Recurring Word test. She had trouble with the Verbal Learning test, scoring in the lowest 25th percentile, and had special difficulty with delayed recall. Both short- and long-term visual memory was also found to be impaired. Administration of the MMPI gave scores consistent with the woman's complaints of chronic subjective distress, her feelings of depression and anxiety, her complaints of memory problems and mental confusion, and her complaints of nonspecific physical symptoms including dizziness and headaches.

18.2.8 MYERS ET AL. SHORT CASE SERIES STUDY

Myers et al.[35] published a case series of seven patients who had been exposed to CO 1 to 18 months, constantly or intermittently. This included six women and one man. Ages ranged from 23 to 67 years. Cases were reviewed for length of exposure, the patient's age, whether smoker or nonsmoker, CO levels, and symptoms and physical signs of illness. Before treatment with hyperbaric oxygen (HBO) therapy, patients were given a CO neuropsychological screening battery and specific neuropsychological tests. It is not clear from the report how long or short a time after termination of CO exposure the evaluations were done. It is from this work-up that symptoms associated with chronic CO exposure were tabulated (Table 18.7).

They concluded that comprehensive neuropsychological assessment is an essential component of the formal diagnostic workup, and is more valuable than measurement of COHb. They say, furthermore, that HBO treatment, even when it is delayed weeks after presentation, is of value in improving functional status.

While the study did not clearly differentiate acute/immediate symptoms during CO exposure from residual/long-term symptoms after exposure, it is possible from the narrative to tabulate those symptoms that persisted (i.e., residual) for some time following termination of CO exposure (Table 18.8).

18.2.9 Bayer et al. Retrospective Hospital Study

In 1998, Bayer and his collaborators[36] presented a paper on the residual effects of chronic CO poisoning at a meeting held in Dijon, France. The study was a retrospective chart review. The objective was to demonstrate substantial persistent neurological sequelae (PNS) in a case series following chronic CO exposure. The data collected included CO air samples, initial COHb levels, initial treatment rendered, subjective symptoms, and objective findings on neuropsychological testing (NPT). Chronic exposure was defined as lasting more than 1 week. PNS were defined as CNS symptoms and/or abnormalities on NPT persisting for more than 3 months after the CO exposure.

In all, 94 patients with CO poisoning were evaluated from January 1995 to December 1997. Of these, 38 patients had acute exposures and 56 had chronic exposures to CO. Of the 56 patients with chronic exposures, 23 developed PNS. This included 8 men and 15 women, ages 8 to 58 years. Initial COHb levels of the 23 patients with PNS were 0.4 to 5.8%. Only 1 of the 23 patients was treated with hyperbaric oxygen. Of the 23 patients, 17 had NPT done, and 15 of 17 were found to be abnormal. PNS included headache, fatigue, irritability, difficulty concentrating, and impaired memory. Of the 23 patients, 15 were followed for at least 12 months, and all 15 were still symptomatic 1 year postexposure.

This study strongly suggests that PNS does occur following chronic exposure to CO, even with low apparent blood CO levels.

18.3 NEW RETROSPECTIVE CARBON MONOXIDE INVESTIGATIONS

18.3.1 Carbon Monoxide Support Study

A well-designed questionnaire study of victims of CO poisoning in the United Kingdom was recently carried out by CO Support, a registered charity headed by Debbie Davis. See Chapter 19 for a thorough discussion of this study.

TABLE 18.7
Patients' Self-Report of Chronic CO Poisoning Symptoms From a Neuropsychological Questionnaire

Symptom	No. Reporting
Headaches	7
Dizzy spells	7
Trembling hands	7
Body weight change	6
Memory change	6
Changes in sleep	6
Forget what is being said	6
Worried/anxious	6
Cannot understand what is read	6
Changes in vision	5
Changes in walking	5
Difficulty remembering	5
Muscles jump/twitch	5
Things drop from hand	5
Others mumbling	5
Slur words	5
Lose balance easily	5
Parts of body numb	5
Sense of direction changed	4
Sense of smell changed	4
Part of body hurts	4
Ringing in ears	4
Speech changed	4
Handwriting changed	3
Eyelids drop	3
Smell strong odors	3
Dejá vu	2
Ruminative thoughts	2
Bladder/bowel control loss	2
Temporary blindness	2
Drinks more water	2
Reached but hand misses item	2
Change in libido	1
Passed out	1

Note: It is unclear from the published report how long after the end of CO exposure these symptoms were reported or whether, indeed, exposure had actually ended at the time the symptoms were reported.

Source: Modified from Myers et al.[35]

TABLE 18.8
**Residual Symptoms Extracted From Cases 1 to 7 of
the Myers et al.[35] Study, Listed in Alphabetical Order**

Anxiety[a]
Breathing difficulties
Cognitive ability decreased
Cramps (arms)
Depression
Driving at night difficult
Emotional lability/emotional distress
Headaches[a]
Hypocondriasis
Hysteria
Irritability
Joint pain
Language problems
Math skills rudimentary
Motor (movement) problems
Muscle twitches[a]
Muscle weakness
Panic attacks
Paranoia
Personality disorder
Photophobia
Reading difficult[a]
Remembering things difficult (2 cases)[a]
Tremors[a]
Urinary incontinence[a]
Vision blurred[a]

[a] Items that also appear on the published list of symptoms.

18.3.2 STUDY A — PENNEY

The following are data obtained from a survey of electronic mail received by this investigator over a period from late 1997 through early 1999. Included are reports from 66 correspondents who indicated they had sustained chronic CO exposure (defined as CO exposure lasting more than 24 h). The source(s) of the poisoning and the symptoms were recorded and plotted by the investigators. The duration of CO exposure as stated by 29 correspondents ranged from 0.18 to 120 months. Mean duration was 30.76 ± 6.00 months (S.E.M.). Air CO concentration reported in 15 instances was 427.8 ± 115.2 ppm. Blood COHb saturation reported in 11 instances was $9.65\% \pm 2.46\%$. The sources of CO are presented in Figure 18.3.

The symptoms reported by correspondents during chronic CO exposure are seen in Table 18.9. They were many and varied. The most frequently reported symptoms are displayed in order of frequency in Figure 18.4.

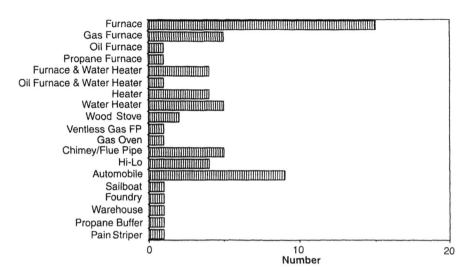

FIGURE 18.3   Sources of chronic carbon monoxide exposure reported, Study A.

The symptoms reported by correspondents following CO exposure are seen in Table 18.10. Like the initial symptoms, they are many and varied. The most frequently reported symptoms are displayed in order of frequency in Figure 18.5.

18.3.3   STUDY B — PENNEY

A questionnaire study was carried out in which questionnaires were prepared and distributed to CO exposed persons between February and October 1999. A four-page questionnaire, henceforth to be referred to as the "Long" CO Questionnaire, was distributed by mail and by fax. See the sample in Figure 18.6. A second questionnaire requesting information about approximately 80% of the items contained in the Long CO Questionnaire, and henceforth to be referred to as the "Short" CO Questionnaire, was available on the World Wide Web at Carbon Monoxide Headquarters (COHQ) (*http://www.phymac.med.wayne.edu/facultyprofile/penney/COHQ/CO1.htm*). Viewers were asked to download and print it, fill it out, and fax it to the author.

By the end of October 1999, 103 questionnaires of the two types had been received. Of these, 82 questionnaires were from people who had suffered chronic CO exposure, and 21 were from people who had suffered acute CO exposure. Of the chronic exposure respondents, 58 were women and 24 were men. The mean age of the respondents was 37.4 years (±1.5 years, S.E.M.), with a range of 7 to 73 years. The mean school grade completed was 13.4 years (±0.4 years), with a range of 1 to 21 years (high school = 12 years; associate degree, master's degree, technical program = 2 years; baccalaureate = 4 years; Ph.D. = 4 years). The mean duration of CO exposure was 28.4 months (±4.4 momths), or nearly 2.5 years, with a range of 0.75 to 120 months. No respondent reported being unconscious as a result of CO at any time during exposure. The mean period of time after termination of CO exposure that responses were given was 21.4 months (±2.2 months), or nearly 2 years, with a range of 0.25 to 60 months.

TABLE 18.9
**Symptoms Exhibited during Chronic Carbon Monoxide Exposure, Study A,
Listed Alphabetically**

Agitation	Dysarthria	Lethargy	Skin, dryness
Anxiety	Ear problems	Libido loss	Sleep problems
Apathy	Emotional problems	Lightheadedness	Sleepiness
Appetite loss	Energy level	Lips red	Smile, convulsive
Ataxia	Extremities cold	Liver pain	Speaking problems
Attention loss	Eye pain/ache	Memory loss	Spelling problems
Back pains	Fatigue	Mood changes	Suicidal
Balance problems	Fibromyalgia	Moodiness	Sweats
Body ache	Flu-like symptom	Muscle ache/pain	Syncope, part/all
Bronchitis	Flushed	Nausea	Tachycardia
Chest tightening/pain	Forgetful	Neck pain	Throat, burning, sore
Choking	Gastrointestinal problems	Nerve deafness	Tingling legs/arms
Chronic fatigue	Hair loss	Numbness	Tingling lips
Concentration problems	Hallucinations	Palpitations	Tinnitus
Confusion	Handwriting problems	Panic attack	Tiredness
Constipation	Headache	Paralysis	Tongue, thickened
Coolness	Hearing problems	Parathesias	Tremor
Coordination problems	Hypertension	Personality change	Twitching fingers
Cough, spells	Hypoglycemia	Pressure in head	Vertigo
Cramps	Ill, violently	Shortness of breath	Vision problem
Depression	In "a fog"	Seasick	Vomiting
Diaphragm pain	Incontinence	Seizure	Walk, inability to
Diarrhea	Insomnia	Shoulder pain	Weakness
Disorientation	Iron level low	Sick feeling	Weight loss
Dizziness	Irritability	Sinusitis	Word-finding problems
Drop things	Learning problems	Skin, cherry red	

Responses came from 23 of the U.S. states, and from four Canadian provinces (Table 18.11). The majority of the exposures occurred in the home (houses, apartments, etc.), while a substantial number took place in various work settings (offices, shops, hospital). The source of CO in most cases was a furnace (boiler, heater, etc.), while the source in a substantial number of cases was a water heater or a range (cooking stove). Faulty chimneys and flues and internal combustion engines were the CO source in a few cases. In several cases more than a single source was identified, e.g., furnace and water heater.

Air CO concentration had a mean value of 220.4 ppm (±28.1) for 25 cases. Thus, only a fraction of the respondents knew the measured CO value in their case. The mean COHb saturation reported was 9.0% (±1.6) for 29 cases. Again, only a fraction the respondents knew their measured blood CO value. The mean estimated time from the termination of CO exposure until a blood sample was drawn for COHb assay was 20.9 h (±8.0) in 9 cases. Estimation of the "time-zero" COHb value was done, where possible, using the reported delay. Other measured COHb values were

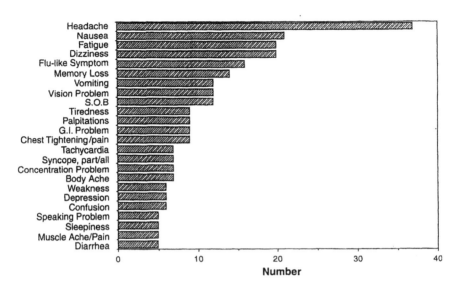

FIGURE 18.4 Symptoms reported during chronic carbon monoxide exposure in order of frequency, Study A (first 24 only).

used "as is," if the delay was unknown, while other values were not used if the delay was greater than 12 h. This approach gave a mean corrected COHb value of 14.0% (±2.4) in 22 cases. Because more than half of the measured COHb values could not be corrected, the actual mean "time-zero" value was certainly higher than 14.0%.

Respondents were asked to check "yes" or "no" for a number of symptoms they might have experienced *after termination of CO exposure*. The physical symptoms tabulated were: headache, muscle/joint pain, chest pain, tiredness/fatigue, dizziness/balance problems, numbness/tingling, visual/eye problems, sleep problems, change of taste or smell, and physical limitations. The cognitive and intellectual symptoms tabulated were difficulty making decisions, difficulty/problems at work, difficulty following directions, difficulty finding familiar places, problems balancing a checkbook, memory problems, attention/concentration difficulties, and difficulty sorting/organizing. The emotional symptoms tabulated were mood changes, increased use of illicit drugs or alcohol, temper problems, emotional swings, social problems, family problems, and school problems, and whether others believe they are different since being exposed to CO and and whether they believe they have undergone a change in personality and memory. Finally, respondents were asked to comment on their medical treatment following discovery of the CO poisoning — whether they had received 100% oxygen and whether they had been admitted to a hospital (not simply visited an emergency center).

Over 80% of respondents indicated they continued to have the residual physical symptoms of headache, muscle/joint pain, tiredness/fatigue, dizziness/balance problems, and sleep problems nearly 2 years after termination of CO exposure (Figure 18.7). Of the respondents, 100% continued to suffer tiredness/fatigue. The effects of chronic CO poisoning share many symptoms in common with chronic fatigue syndrome.

TABLE 18.10
Residual Symptoms Following Chronic Carbon Monoxide Exposure, Study A,
Listed Alphabetically

Academic problems	Emotional problems	Learning problems	Skin, hypersitivity/touch
ADD	Energy level	Libido loss	Sleep problems
Aggression	Executive function	Math, difficulty	Spasm
Altered consciousness	Eye, feels puffy	Memory loss	Speaking problems
Amnesia	Fatigue	Mood changes	Spelling problems
Anxiety	Fatigue, chronic	Motivation, lack of	Staring spells
Arthitis	Fear	Muscle ache/pain	Stiffness
Ataxia	Flu-like symptom	Nausea	Stroke
Attention, loss	Forgetful	Neck pain	Tachycardia
Balance problems	Gastrointestinal problems	Nervous	Talkative
Body ache	Hand control	Numbness	Temper, short
Body temperature control	Headache	Palpitations	Thinking problems
Chest tightening/pain	Hearing problems	Panic attack	Tingling legs/arms
Choking	Heart murmur	Paraphasias, literal	Tingling, hands
Concentration problems	Hyperactivity	Paraphasias, verbal	Tinnitus
Confusion	Hypertension	Parkinsonism	Tiredness
Coordination problems	Hypersensitivity/MCS[a]	Peripheral neuropathy	Tremor
Cramps	I.Q. loss	Personality change	Vision problems
Depression	Impulsiveness	Phonophobia	Vocabulary down
Disorientation	Information processing/slow	Photophobia	Vomiting
Dizziness	Irrational behavior	PMS, heightened	Weakness
Dysarthria	Itching	Reading problems	Word-finding problems
Dystonia	Joint pain	Shortness of breath	Writing problems
Ear problems	Kidney problems	Sinusitis	

[a] MCS = Multiple chemical sensitivity.

Over 50% of respondents continued to suffer chest pain, numbness and tingling, visual/eye problems, and claimed to have physical limitations. Over 45% of respondents indicated that they experienced a change in their perception of or sensitivity to smell or taste. This is consistent with the known role of CO in inducing multiple chemical sensitivity.[37]

Over 70% of respondents indicated they continued to have the residual cognitive/intellectual symptoms of difficulty making decisions, difficulty/problems at work, difficulty following directions, memory problems, attention/concentration problems, and believed they were different since the CO exposure nearly 2 years after termination of CO exposure (Figure 18.8). Over 40% were still having difficulty finding familiar places, balancing a checkbook, and doing organizing and sorting operations.

Over 80% of respondents indicated they had residual emotional symptoms of mood change, temper problems, emotional swings, and believed their personality and memory had changed since the CO exposure nearly 2 years before (Figure 18.9). Over 40% were still having social and family problems related to the CO exposure.

Only 41.2% of respondents had been given 100% oxygen upon discovery of the CO poisoning, and only 11.9% had been admitted to a hospital for a period of time longer than 2 h.

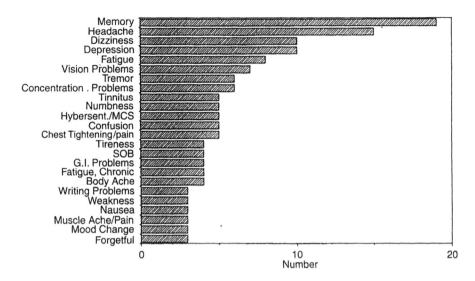

FIGURE 18.5 Symptoms reported following chronic carbon monoxide exposure in order of frequency, Study A (first 24 only).

The smaller cohort of 21 acute CO-exposure respondents was not significantly different from the chronic CO respondents in terms of age (44.8 ± 3.3 years) or education (13.1 ± 0.7 years); however, men (13) predominated over women (8) in this group. Average duration of CO exposure was 8.2 h (±2.2 h), while the time since termination of CO exposure was 41.9 months (±18.3 months). Mean air CO concentration at exposure was 312 ppm (±111 ppm). Mean COHb saturation was 18.7% (±4.9%) for 5. With a reported "delay time" of 2.6 h (±0.7 h) for 4, the mean "time-zero" COHb value was 26.3% (±5.7%) for 5. This value is more than 12% COHb higher than the estimated corrected value in the chronic CO exposure cohort.

Residual headache and dizziness/balance problems, physical symptoms reported by nearly every chronic CO respondent, were experienced by only 76 and 57% of acute CO respondents, respectively. In general, nearly all of the physical, cognitive, and some of the emotional symptoms were experienced by a smaller percentage of the acute CO respondents than by the chronic CO respondents.

Of the acute CO respondents, 50% received 100% oxygen when their CO poisoning was discovered, and 33% were admitted for hospital care.

This study suggests that a multitude of physical, cognitive, and emotional symptoms persist for very long periods of time following chronic exposure to CO. The CO exposure need not produce altered consciousness at any time for this to occur. In fact, the CO exposure concentrations and COHb saturations are usually quite low, in the range previously thought incapable of producing lasting health harm in humans.

While a study such as described here has the inherent weaknesses of self-reporting errors and biases, it provides valuable clues to a condition until recently all but ignored by the medical community.

Carbon Monoxide Questionnaire

Name: _____   Date: _____

Home address: _____

City: _____   State: _____   Zip code: _____

Home phone: _____   Work phone: _____

Age: _____   Date of birth: _____   E-mail: _____

Highest grade in school completed: _____

Name of nearest relative: _____

Address: _____

Phone #: _____

Date of exposure: _____   How long was your exposure? _____

Place and circumstances where you were exposed: _____

_____

What was the source (eg. automobile, furnace, water heater)? _____

Was CO measured in the air?   Yes   No   in your blood?   Yes   No

    If in the air, what was the value, if you know _____

    If in your blood, what was the value, if you know _____

Was anyone else exposed to the CO you received?   Yes   No

    If yes, who _____

Medications at time of exposure: _____

_____

Current medications: _____

_____

FIGURE 18.6   Sample carbon monoxide questionnaire, page 1 of 4, Study B.

18.4   CONCLUSIONS

The amount of data and the resulting conclusions regarding chronic CO poisoning in this chapter (and book) is probably more than has been published in any other work in the past 40 years. Clearly, prolonged exposure to this insidious poison, even at what were previously thought to be ultralow levels, is capable of producing many and varied residual health effects. Furthermore, the incidence of such unpleasant and often debilitating effects is far higher than was previously believed by the medical

TABLE 18.11

Locales, Sites, and Sources in Questionnaire Study (Study B) of Chronic Carbon Monoxide Poisoning

State/Province				Site			Source		
AK	1	NJ	1	Home	59	72%	Furnace	48	58.5%
Alberta	1	NM	1	Work: Office	18	22%	Water heater	18	22%
AL	3	Nova Sc.	3	Shop	1	1%	Propane W.H.	1	1%
Brit. Col.	1	NV	2	Hospital	1		Range/stove	12	14.6%
CA	5	NY	3	School	1		Engine(s)	4	4.9%
CO	2	OH	1	Vehicle: Auto	1		Chimney/flue	3	3.7%
FL	2	Ontario	8	Truck	1		Wood stove	1	
IL	2	PA	6				Roof heater	1	
IN	1	SC	12						
KY	2	TN	1						
MA	1	TX	5						
ME	1	VA	1						
MI	12	WA	2						
MN	1								

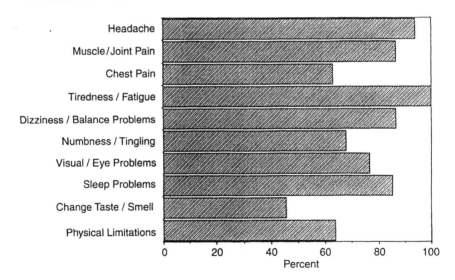

FIGURE 18.7   Residual physical symptoms following chronic carbon monoxide exposure, Study B.

and public health community, and can continue for a very long period of time. It is not believed that this body of work answers all of the questions relevant to chronic CO poisoning, but rather will act to spur others to continue investigations of the problem, which will provide a better understanding of what occurs and improve treatment for victims in the future.

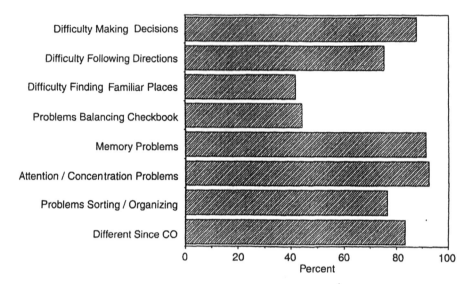

FIGURE 18.8   Residual cognitive symptoms following chronic carbon monoxide exposure, Study B.

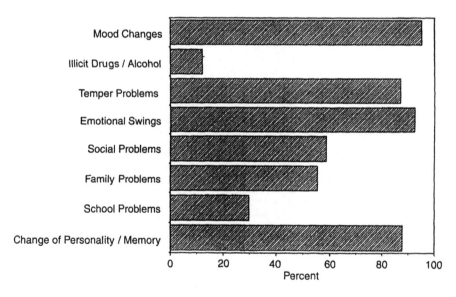

FIGURE 18.9   Residual emotional symptoms following chronic carbon monoxide exposure, Study B.

ACKNOWLEDGMENT

I wish to thank my wife, Linda Mae Penney, for her invaluable assistance in developing most aspects of this chapter.

REFERENCES

1. *Webster's New College Dictionary*, Houghton Mifflin, Boston, 1986.
2. Grace, T.W. and Platt, F.W., Subacute carbon monoxide poisoning, *J. Am. Med. Assoc.*, 246, 1698–1700, 1981.
3. Uchino, A., Hasuo, K., Shida, K., Matsumoto, S., Yasumori, K., and Masuda, K., MRI of the brain in chronic carbon monoxide poisoning, *Neuroradiology*, 36, 399–401, 1994.
4. Kojima, S., Kawamura, M., Shibata, N., Takahashi, N., and Hirayama, K., Magnetic resonance imaging of carbon monoxide poisoning in chronic stage, *Rinsho Shinkeigaku* (Japan), 26, 291–299, 1986.
5. Dolan, M.C., Haltom, T.L., Barrows, G.H., Short, C.S., and Ferriell, K.M., Carboxyhemoglobin levels in patients with flu-like symptoms, *Ann. Emerg. Med.*, 16, 782–786, 1987.
6. Fisher, J., Occult carbon monoxide poisoning, *Arch. Intern. Med.*, 142, 1270–1271, 1982.
7. Iffland, R., Klose, H., Eiling, G., and Balling, P., Measuring carbon monoxide content in muscles, *Arch. Kriminol.*, 186, 75–84, 1990.
8. WHO, Diseases Caused by Asphyxiants: Carbon Monoxide, Hydrogen Cyanide and Its Toxic Derivatives, and Hydrogen Sulfide. Early Detection of Occupational Diseases, World Health Organization, Geneva, 1986, 154–164.
9. Candura, S.M., Fonte, R., Finozzi, E., Guarnone, F., Taglione, L., Manzo, L., and Biscaldi, G., [Indoor pollution: a report of 2 clinical cases of occult carbon monoxide poisoning] (in Italian), *G. Ital. Med. Lav.*, 15, 67–70, 1993.
10. Foster, M., Goodwin, S.R., Williams, C., and Loeffler, J., Recurrent acute life-threatening events and lactic acidosis caused by chronic carbon monoxide poisoning in an infant, *Pediatrics*, 104, e34, 1999.
11. Kirkpatrick, J.N., Occult carbon monoxide poisoning, *West. J. Med.*, 146, 52–56, 1987.
12. Heckerling, P.S., Leikin, J.B., and Maturen, A., Occult carbon monoxide poisoning: validation of a prediction model, *Am. J. Med.*, 84, 251–256, 1988.
13. Snyder, S.H., Jaffrey, S.R., and Zakhary, R., Nitric oxide and carbon monoxide: parallel roles as neural messengers, *Brain Res. Rev.*, 26, 167–175, 1998.
14. Beck, H.G., The clinical manifestations of chronic carbon monoxide poisoning, *Ann. Clin. Med.*, 5, 1088–1096, 1927.
15. Lindgren, S.A., A study of the effect of protracted occupational exposure to carbon monoxide: with special reference to the occurrence of so-called chronic carbon monoxide poisoning, *Acta Med. Scand.*, 167 (Suppl. 356), 1–135, 1961.
16. Kachulis, C.J., Second-hand cigarette smoke as a cause of chronic carbon monoxide poisoning, *Postgrad. Med.*, 70, 77–79, 1981.
17. Zanardi, S., Villa, A., and Monti, G., [Electrocardiographic changes in workers usually exposed to inhalation of carbon monoxide] (in Italian), *Med. Lav.*, 57, 761–770, 1966.
18. Davies, D.M. and Smith, D.J., Electrocardiographic changes in healthy men during continuous low-level carbon monoxide exposure, *Environ. Res.*, 21, 197–206, 1980.
19. Wilson, A.J. and Schaefer, K.E., Effect of prolonged exposure to elevated carbon monoxide and carbon dioxide levels on red blood cell parameters during submarine patrols, *Undersea Biomed. Res.*, Submarine Suppl., 6, S49–S56, 1979.
20. Smith, J.R. and Landaw, S.A., Smokers' polycythemia, *N. Engl. J. Med.*, 298, 6–10, 1978.

21. Trese, M.T., Krohel, G.B., and Hepler, R.S., Ocular effects of chronic carbon monoxide exposure. *Ann. Ophthalmol.*, 12, 536–538, 1980.
22. Kowalska, S., [State of the hearing and equilibrium organs in workers exposed to carbon monoxide] (in Polish), *Med. Pr.*, 32, 145–151, 1981.
23. Barrowcliff, D.F., Chronic carbon monoxide poisoning caused by methylene chloride paintstripper, *Med. Sci. Law*, 18, 238, 1978.
24. Jensen, L.K., Klausen, H., and Elsnab, C., Organic brain damage in garage workers after long-term exposure to diesel exhaust fumes, *Ugeskr. Laeg.*, 151, 2255–2257, 1989.
25. Tvedt, B. and Kjuus, H., [Chronic CO poisoning. Use of generator gas during the second world war and recent research] (in Norwegian), *Tidsskr. Nor. Laegeforen.*, 117, 2454–2457, 1997.
26. Hexter, A.C. and Goldsmith, J.R., Carbon monoxide: association of community air pollution with mortality, *Science*, 172, 265–276, 1971.
27. Ritz, B. and Yu, F., The effect of ambient carbon monoxide on low birth weight among children born in southern California between 1989 and 1993, *Environ. Health Perspect.*, 107, 17–25, 1999.
28. Morris, R., Naumova, E.N., and Munasinghe, R.L., Ambient air pollution and hospitalization for congestive heart failure among elderly people in seven large U.S. cities, *Am. J. Public Health*, 85, 1361–1365, 1995.
29. Allred, E.N., Bleecker, E.R., Chaitman, B.R., Dahms, T.E., Gottlieb, S.O., Hackney, J.D., Pagano, M., Selvester, R.H., Walden, S.M., and Warren, J., Short-term effects of carbon monoxide exposure on the exercise performance of subjects with coronary artery disease, *N. Engl. J. Med.*, 321, 1426–1432, 1989.
30. Strahilevitz, M., Strahilevitz, A., and Miller, J.E., Air pollutants and the admission rate of psychiatric patients, *Am. J. Psychiatr.*, 136, 205–207, 1979.
31. Khan, K. and Sharief, N., Chronic carbon monoxide poisoning in children, *Acta Paediatr.*, 84, 742, 1995.
32. Piatt, J.P., Kaplan, A.M, Bond, G.R., and Berg, R.A., Occult carbon monoxide poisoning in an infant, *Pediatr. Emerg. Care*, 6, 21, 1990.
33. Beck, H.G., Carbon monoxide asphyxiation: a neglected clinical problem, *J. Am. Med. Assoc.*, 107, 1025–1029, 1936.
34. Ryan, C.M., Memory disturbances following chronic, low-level carbon monoxide exposure, *Arch. Clin. Neuropsychol.*, 5, 59–67, 1990.
35. Myers, A.M., DeFazio, A., and Kelly, M.P., Chronic carbon monoxide exposure: a clinical syndrome detected by neuropsychological tests, *J. Clin. Psychol.*, 54, 555–567, 1998.
36. Bayer, M.J., Orlando, J., McCormick, M.A., Weiner, A., and Deckel, A.W., Persistent neurological sequelae following chronic exposure to carbon monoxide, in *Carbon Monoxide: The Unnoticed Poison of the 21st Century*, Satellite Meeting, IUTOX VIIIth International Congress of Toxicology, Dijon, France, July 3–4, 1998, 179.
37. Hartman, D.E., Missed diagnoses and misdiagnosis of environmental toxicant exposure, *Psychiatr. Clin. North Am.*, 21, 659–670, 1998.

19 Chronic Carbon Monoxide Exposure: The CO Support Study

Alistair W.M. Hay, Susan Jaffer, and Debbie Davis

CONTENTS

19.1 INTRODUCTION

Exposure to carbon monoxide (CO) and the consequences for health of those exposed acutely are well documented.[1-5] The effects of chronic exposure to the gas are not nearly as well known and are a subject of far less research. The effects of repeated exposure to sublethal concentrations of CO on 97 individuals over a prolonged period was reported over 60 years ago.[6] In this report, Beck[6] recorded the fact that individuals had been exposed to CO over periods which ranged from several months to 18 years. Beck recorded seven principal symptoms in his cohort, all of whom were over 40 years of age. Symptoms included headaches, vertigo, nervousness, palpitations, and neuromuscular pain. Most of these symptoms are also to be found in individuals who have been acutely poisoned by CO. When poisoning is discussed, it is usually assumed that the exposure has been acute.

For those exposed to CO, coping with what is assumed to be an acute exposure is one thing, dealing with perceptions about the effects of such exposure in the medical profession is another. A view all too prevalent among clinicians is that, unless exposure to CO renders an individual unconscious, the consequences for that person's health are minimal. Sadly, the evidence indicates that exposure to concentrations of CO insufficient to render an individual unconscious will still cause damage to the brain, and may result in delayed neurological sequelae.[7,8]

If acute CO exposure will cause delayed neurological sequelae in individuals, then the consequences of repeated exposure to the gas may be of even more consequence. A principal factor that will determine the outcome for those poisoned is the peak blood CO concentration. In the vast majority of studies published in this field, peak CO concentrations (as measured by carboxyhemoglobin, COHb) are not known. For individuals chronically exposed to CO, there is even less likelihood of some measurement of COHb concentrations. The absence of these measurements has frequently frustrated individuals who have attempted to discover the cause of their continuing ill-health. More problematic still for those chronically exposed to CO is the lack of awareness in the medical profession that such exposures can have consequences for health. To address some of these gaps in what is known about the effects of chronic CO exposure, the authors conducted a questionnaire study on 77 individuals who were known to have been exposed to CO over a period ranging from several days to more than 20 years. Results are reported separately for two groups who were chronically exposed, a cohort of 65 who were never unconscious and a cohort of 12 who were unconscious.

19.2 METHODS

Participants in the study were those individuals who contacted a charity — Carbon Monoxide Support — in 1996 and the first 3 months of 1997. Individuals were asked if they would be willing to complete a questionnaire and given a guarantee that the anonymity of all respondents would be respected.

To test the suitability of the questionnaire, a pilot group of 17 individuals were asked to complete the questionnaire. The responses to this informed the design of the final questionnaire, the results of which are reported in this chapter.

For the main study, 120 questionnaires were sent out, including 17 sent to the participants in the pilot study. A total of 87 replies were received, of which 10 were rejected because of inadequate information. Of the 43 subjects who either did not return questionnaires or did not supply adequate information, 24 were female, 15 male, and 4 were unknown. The questionnaires for very young children were completed by their parents.

19.2.1 Control Subjects

For each respondent, a control subject was obtained of the same sex and similar age and income bracket. Nearly half of the control subjects (36) were identified by participants in the study who had been exposed to CO. Matching controls for the other 41 respondents were identified by Carbon Monoxide Support.

19.2.2 Questionnaires

Questionnaires sent to subjects and controls were too lengthy to reproduce in this chapter. The full questionnaire can be found in the detailed, separately published report of this study.[9] Where possible, questions to be completed were designed in the form of "tick box" answers with space for explanatory comment where appropriate. Control subjects were asked questions identical to those in the exposed group, in 20 areas. Slight rephrasing of questions was required for 10 categories of question used for the control group, as these related to the specific circumstances on the cause, method of discovery, and treatment of CO exposure, circumstances which were clearly not, or unlikely to be, appropriate for the control group.

The questionnaire asked for some biographical details on subjects and controls, as well as income, type of residence, nature of heating appliance in the home, and details about servicing of appliances. Individuals were also asked to indicate whether they had experienced any of the symptoms listed in the questionnaire, and how long they had experienced these prior to and after the diagnosis of CO poisoning had been made. Details about the medical treatment received were also sought, as well as information about prescribed medication for treatment. Respondents were asked about previous medical history, as well as the medical history of family members. Information on alcohol consumption, smoking history, ethnicity, educational qualifications and training, and current working status were also sought.

19.2.3 Statistics

The chi square test was used to assess the significance at the 5 and 2.5% level for a number of the variables.

19.3 RESULTS

Biographical details and other answers to questions of subjects and their matching controls are shown in the accompanying tables. Details for the 65 subjects chronically exposed to CO and information from their matching controls are recorded in Tables 19.1 through 19.5. Information from the 12 individuals who were unconscious as a result of their exposure to CO and their controls are documented in Tables 19.6 through 19.10.

19.3.1 Chronically Exposed

It is evident from the results in Table 19.1 that the controls and the subjects exposed to CO were closely matched in many categories. Biographical details for the exposed and control subjects are similar for sex, age, marital status, ethnicity, and smoking history. Some differences were noted in the types of accommodation of both groups. More of the exposed subjects than controls lived in rented accommodation. Estimated alcohol consumption was approximately 7.3 units/week in the control group, but

TABLE 19.1
Biographical Details of 65 Subjects Chronically Exposed to CO and Controls

	Number (Percentage)	
	CO Group	Control Group
Number	65 (100)	65 (100)
Sex: Female	41 (63.1)	40 (61.5)
Male	24 (36.9)	25 (38.5)
Age: Mean	35.9	36.1
Median	36	36
Max	76	78
Min	2	3
Household income (prior)	£11,300	£14,400
Residential property[a]		
House	26 (40.0)	39 (60.0)
Terrace	16 (27.7)	21 (32.8)
Apartment	10 (15.4)	3 (4.6)
Office/store	3 (4.6)	0
Other	8 (12.3)	0
N/A	2 (3.1)	1 (1.5)
Owner occupied	15 (23.1)	39 (60.0)
Rented	43 (66.2)	26 (40.0)
N/A	7 (10.8)	0
Marital status		
Never married	12 (18.5)	16 (24.6)
Divorce/widowed	9 (13.8)	10 (15.4)
Married/with partner	36 (55.4)	36 (55.4)
Not applicable or N/A	8 (12.3)	3 (4.6)
Ethnicity		
White	62 (95.4)	64 (98.5)
Other	2 (3.1)	1 (1.5)
N/A	1 (1.5)	0
Smoking history		
Never	27 (41.5)	33 (50.8)
Past smoker	17 (26.2)	16 (24.6)
Current: 5–10 per day	9 (13.8)	5 (7.7)
11–20 per day	8 (12.3)	8 (12.3)
21–30 per day	0 (0.0)	1 (1.5)
>30 per day	1 (1.5)	2 (3.1)
Not known	3 (4.6)	0
Alcohol consumption; average units per week	4.0	7.3
Coffee/tea average cups		
Coffee	2.3	1.9
Tea	3.1	3.1

N/A = Not available.

[a] For a chronic group, type of property where exposure took place. For a control group, type of residence.

TABLE 19.2

Fuel Use, Source of CO Exposure, and Dwellings for 65 Subjects Chronically Exposed to CO and Controls

	Number (Percentage)	
	CO Group	Control
Number of households	38	65
Heating appliances[a]		
Fireplace	23 (60.5)	38 (58.5)
Water heater	4 (10.5)	32 (49.2)
Central heating	12 (31.6)	44 (67.7)
Other	4 (10.5)	11 (16.9)
Flue involved	42 (60.5)	
Type of fuel		
Gas	31 (81.6)	43 (66.2)
Oil	1 (2.6)	5 (7.7)
Solid	4 (10.5)	14 (21.5)
Other	0 (0.0)	12 (18.5)
Not known	2 (5.3)	0
Dwelling involved		
Owner occupied	12 (31.6)	39 (60.0)
Rented	21 (55.3)	26 (40.0)
Not known	5 (13.2)	0
Location		
City	8 (21.1)	9 (13.9)
Town	11 (29.0)	22 (33.9)
Village	8 (21.1)	16 (24.6)
Rural	4 (10.5)	8 (12.3)
Not known	7 (18.4)	10 (15.4)

[a] Number of households with each appliance as a source of exposure in CO group. For control, number of households with each type of appliance. Percentages may add to more than 100 due to multiple appliances.

only 4 units/week in the 65 exposed subjects. Household income per annum was also greater in the control than in the exposed group.

Table 19.2 documents the heating appliances and type of fuel used by those who were chronically exposed to CO, in addition to the location of dwellings and the source of exposure to CO. The distribution of dwellings between urban and rural locations was similar for the two groups. Use of gas was greater among the exposed subjects than among controls. Table 19.2 also documents the source of exposure to CO and this indicates that a flue or chimney was the principal problem as far as exposure to the gas was concerned. The waste gases from whatever fuel was being burned were not properly vented from the flue or chimney. Fireplaces that fed into the flue were the principal source of CO, followed by central heating boilers.

The prevalence and duration of five symptoms commonly associated with exposure to CO are shown in Table 19.3. Some 80% or more of the subjects experienced

TABLE 19.3
Prevalence and Duration of Five Common Symptoms Experienced During and
After Exposure to CO in 65 Chronically Exposed Subjects and Controls

| | CO Exposed | | | | Control | |
Symptoms	During (%)	After (%)	Duration during (months)[a]	Duration after (months)[b]	Current (%)	Duration (months)[c]
Nausea	81.5	29.2	20.8	18.1	6.2	7.3
Vomiting	44.6	16.9	24.2	25.4	3.1	2.1
Severe headaches	87.7	61.5	21.0	23.1	6.2	12.5
Dizziness	78.5	52.3	20.7	23.8	3.1	5
Drowsiness	84.6	36.9	18.6	15.7	3.1	6

[a] Duration of symptoms according to respondents.
[b] Duration of symptoms after CO exposure was identified or diagnosed.
[c] Prevalence of symptoms in control subjects.

nausea, drowsiness and severe headaches; 44% of subjects experienced vomiting during their exposure. The average duration for the five symptoms ranged from 18.6 to 24.2 months, indicating a significant period of exposure before the source of the problem was identified. More significantly, symptoms persisted after exposure to the gas had been stopped. The number of subjects who continued to experience symptoms after exposure was less than during it, but the persistence of these symptoms was not much altered, ranging from 16 to 25 months. Table 19.3 indicates that there is a much lower prevalence of these symptoms in the general population, with the symptoms persisting for much shorter periods of time.

Table 19.4 documents a wider range of symptoms experienced during exposure to CO. The replies to 81 questions were grouped into the 23 categories listed in the table. The figures indicate a much greater prevalence of symptoms in the exposed group than in the controls, both during exposure to CO and afterward. In every category, the prevalence of symptoms is much greater among the exposed than among the controls. It is not just physical symptoms that differ between groups, personality and emotional problems and depression are also much more prevalent in the exposed than in the control subjects, and these symptoms persist well after exposure ceases.

Individuals exposed to CO frequently complain of muscle pain. In the report of Beck,[6] severe muscular pain, particularly in the neck and back, was a frequent complaint of those chronically exposed to CO. The type of pain experienced both during and after exposure is documented in Table 19.5 for the 65 subjects. The number of control subjects reporting this type of pain at any time during the 2 years prior to completion of the questionnaire is also recorded in Table 19.5. It is notable that the prevalence of symptoms experienced during and after exposure in the exposed group is much greater than in the controls. For most categories of pain there is a reduction in the prevalence of symptoms after exposure to CO had ceased. It is important to note, however, that two categories, pain in the neck and the back and deep muscle pain, were as prevalent after exposure ceased as they were during exposure.

TABLE 19.4

Symptoms Experienced during and after Exposure by 65 Subjects Chronically Exposed to CO and Current Symptoms in a Control Population

Symptoms	CO Group, Number (%) During	Continuing	Control, Number (%) Reported
Flu symptoms	46 (70.8)	26 (40.0)	4 (6.15)
Pains/cramps	57 (87.7)	47 (72.3)	20 (30.8)
Pins and needles/stiffness	42 (64.6)	38 (58.5)	11 (16.9)
Headaches	56 (86.2)	40 (61.5)	5 (7.7)
Tiredness/weakness	58 (89.2)	44 (67.7)	11 (16.9)
Nausea/sickness	55 (84.6)	20 (30.8)	4 (6.2)
Lack of concentration/confusion	50 (76.9)	43 (66.2)	5 (7.7)
Memory loss	40 (61.5)	41 (63.1)	3 (4.6)
Dizziness	51 (78.46)	34 (52.3)	2 (3.1)
Appetite, digestion problems, weight loss	50 (76.9)	30 (46.2)	11 (16.9)
Cardiac symptoms	48 (73.9)	38 (58.5)	6 (9.2)
Unable to walk far, do chores or work	36 (55.4)	30 (46.15)	5 (7.7)
Hearing problems	23 (35.4)	15 (23.1)	7 (10.8)
Vision problems	42 (64.6)	33 (50.8)	5 (7.7)
Mouth/throat problems	37 (56.9)	22 (33.9)	6 (9.2)
Depression	29 (44.6)	25 (38.5)	5 (7.7)
Personality/emotional problems	40 (61.5)	32 (49.3)	0
Hallucinations/zombie state	30 (46.2)	13 (20.0)	2 (3.1)
Panic attacks	26 (40.0)	19 (29.2)	2 (3.1)
Trembling	26 (40.0)	25 (38.5)	1 (1.5)
Clumsiness	34 (52.3)	28 (43.1)	0
Difficulty breathing	46 (70.8)	34 (52.3)	10 (15.4)
Sleep disturbance	38 (58.5)	31 (47.7)	8 (12.3)

19.3.2 UNCONSCIOUS SUBJECTS

Table 19.6 documents the biographical details of 12 individuals who became unconscious as a result of CO exposure, and details on their matching controls. The figures indicate good matching between subjects and controls with similar age distribution, marital status, and smoking history.

Table 19.7 details the use of fuel for heating appliances in their homes, and the location of their property for the unconscious group. There are no differences of note for any categories between the unconscious group and their controls. Table 19.7 also documents the source of exposure to CO for those who were unconscious. In the majority, exposure seems to have been through failure to remove CO at the fireplace. Failures with the central heating system were also responsible for a number of individuals becoming unconscious. Two of the questionnaire returns did not document the source of exposure to the gas.

Table 19.8 indicates the prevalence and duration of five significant symptoms reported by those who were unconscious as a result of CO. Symptoms experienced

TABLE 19.5
Pain Experienced During and After Exposure by 65 Subjects
Chronically Exposed to CO and Current Symptoms in a
Control Population

Type of Pain	CO Group, Number (%) During	CO Group, Number (%) Continuing	Control, Number (%) Reported
Deep muscle	28 (43.1)	29 (44.6)	3 (4.6)
Chest pains	41 (63.1)	29 (44.6)	3 (4.6)
General flu-like aches	46 (70.8)	26 (40.0)	4 (6.2)
Neck and back pain	34 (52.3)	35 (53.9)	12 (18.5)
Sharp shooting pains	27 (41.5)	20 (30.8)	4 (6.2)
Joint pain	27 (41.5)	26 (40.0)	7 (10.8)
Neuralgic pain	27 (41.5)	23 (35.4)	1 (1.5)
Pain in jaw/tongue	15 (23.1)	12 (18.5)	3 (4.6)
Arms and/or legs	36 (55.4)	31 (47.7)	4 (6.2)
Cramps in arms/legs	29 (44.6)	25 (38.5)	2 (3.1)
Tendon pain	14 (21.5)	11 (16.9)	1 (1.5)

both during exposure to the gas and afterward are documented. Over two thirds of subjects experienced all five symptoms, which included vomiting, nausea, severe headaches, dizziness, and drowsiness. Apart from vomiting, which is not recorded after exposure to CO had been identified, all the other symptoms persisted for many months after exposure ended. At first sight, there appears to be a contradiction in the figures, with subjects not reporting vomiting after exposure stopped, yet vomiting also persisting for 2.8 months when they were no longer exposed. This is explained by the subjects reporting that they had no vomiting at the time they completed the questionnaire, but that they had suffered from this complaint for, on average, some 2.8 months after CO had been identified.

A significant feature of the reported duration of the symptoms in the unconscious group is the persistence of their symptoms that preceded the unconscious state. The duration of the symptoms clearly indicates that there was a significant period of exposure to CO (in some cases over many months) prior to the date on which their exposure was sufficient to render them unconscious. The figures in the two columns for duration, during and after exposure, are the average durations reported by the 12 subjects. The percentage of control subjects experiencing the same symptoms is also documented in Table 19.8.

Table 19.9 documents the group's replies to 81 questions in 23 categories of symptoms. The figures again indicate marked differences in the prevalence of symptoms between the unconscious group and the controls.

Table 19.10 records the pain experienced during and after exposure to CO by the unconscious group. Figures again indicate marked differences in the prevalence of these types of pain between the exposed and the control subjects. In most categories of pain, there is a reduction in the prevalence after exposure to CO stops. However, for a number of categories, individuals report continuing pain at the time

TABLE 19.6
Biographical Details of 12 Subjects Unconscious
Due to CO Exposure and Controls

	Number (%)	
	CO Group	Control
Number	12	12
Sex: Female	8 (66.7)	8 (66.7)
Male	4 (33.3)	4 (33.3)
Age: Mean	47.8	48.2
Median	45.5	46
Max	81	80
Min	27	28
Household income (prior)	£11,200	£16,900
Residential property[a]		
House	7 (58.3)	7 (58.3)
Terrace	1 (8.3)	3 (25.0)
Apartment	1 (8.3)	2 (16.7)
Office/store	0	0
Other	2 (16.7)	0
N/A	1 (8.3)	0
Owner occupied	1 (8.3)	9 (75.0)
Rented	9 (75.0)	3 (25.0)
N/A	2 (16.7)	0
Marital status		
Never married	0	1 (8.3)
Divorced/widowed	2 (16.7)	1 (8.3)
Married/with partner	8 (66.7)	10 (83.3)
Not applicable	2 (16.7)	0
Ethnicity		
White	10 (83.3)	12 (100.0)
Not known	2 (16.7)	0
Smoking history		
Never	3 (25.0)	4 (33.3)
Past smoker	2 (16.7)	4 (33.3)
Current: 5–10 per day	4 (33.3)	2 (16.7)
11–20 per day	1 (8.3)	1 (8.3)
21–30 per day	0	0
>30 per day	0	1 (8.3)
Not known	2 (16.7)	0
Alcohol consumption,	0.9	6.6
average units per week		
Coffee/Tea, cups per day		
Coffee	4.6	2.9
Tea	2.3	3.4

N/A = Not available.

[a] For chronic group, type of property where exposure took place.
For control group, type of residence.

TABLE 19.7
Fuel Use, Source of CO Exposure, and Dwellings for
12 Subjects Unconscious Due to CO Exposure and Controls

	Number (%)	
	CO Group	Control
Number of households	10	12
Heating appliances[a]		
Fireplace	6 (54.6)	7 (58.3)
Water heater	2 (18.2)	5 (41.7)
Central heating	4 (36.4)	9 (75.0)
Other	2 (18.2)	2 (16.7)
Flue involved	0	
Type of fuel		
Gas	7 (63.6)	11 (91.7)
Oil	0	0
Solid	1 (9.1)	3 (25.0)
Other	1 (9.1)	0
Not known	2 (18.2)	0
Dwelling involved		
Owner occupied	1 (9.1)	9 (75.0)
Rented	8 (72.7)	3 (25.0)
Not known	1 (9.1)	0
Location		
City	1 (9.1)	1 (8.3)
Town	6 (54.6)	5 (41.7)
Village	1 (9.1)	2 (16.7)
Rural	1 (9.1)	0
Not known	1 (9.1)	4 (33.3)

[a] Number of households with each appliance as a source of exposure in CO group. For control, number of households with each type of appliance.

they completed the questionnaire. For some categories the prevalence of pain was the same both during and after exposure. More-marked reductions in prevalence was noted for "flu-like" aches and chest pains after exposure ceased.

19.3.3 FEATURES ABOUT THE EXPOSURE TO CARBON MONOXIDE

Subjects exposed to CO were asked to indicate how long they had experienced symptoms. Figure 19.1 records the length of time subjects reported that they had experienced symptoms prior to confirmation that they had been exposed. Figures indicate that 58% of subjects who reported chronic exposure to CO had experienced symptoms for between 3 months and 3 years prior to the cause of their symptoms being identified. For those individuals who were eventually rendered unconscious, one third were exposed to the gas for between 3 months and 2 years. Some other subjects claimed exposure for longer periods.

TABLE 19.8
Prevalence and Duration of Five Common Symptoms Experienced during and after Exposure to CO in 12 Subjects Rendered Unconscious and Controls

	CO Exposed				Control	
Symptoms	During (%)	After (%)	Duration during (months)	Duration after (months)	Current (%)	Duration (months)
Nausea	75	33.3	9.5	19.9	16.7	—
Vomiting	66.7	0	10.7	2.8	16.7	—
Severe headaches	83.3	58.3	11.9	26.0	8.3	—
Dizziness	83.3	50	13.3	14.9	8.3	—
Drowsiness	91.7	58.3	12.1	12.7	16.7	3

TABLE 19.9
Symptoms Experienced during and after Exposure by 12 Subjects Unconscious Due to CO and Current Symptoms in a Control Population

	CO Group, Number (%)		Control, Number (%)
Symptoms	During	Continuing	Reported
Flu symptoms	9 (75.0)	5 (41.7)	1 (8.3)
Pains/cramps	11 (91.7)	9 (75.0)	4 (33.3)
Pins and needles/stiffness	11 (91.7)	8 (66.7)	3 (25.0)
Headaches	10 (83.3)	7 (58.3)	0
Tiredness/weakness	11 (91.7)	9 (75.0)	1 (8.3)
Nausea/sickness	9 (75.0)	4 (33.3)	0
Lack of concentration/confusion	10 (83.3)	9 (75.0)	1 (8.3)
Memory loss	9 (75.0)	8 (66.7)	0
Dizziness	10 (83.3)	6 (50.0)	0
Appetite, digestion, weight loss	9 (75.0)	7 (58.3)	1 (8.3)
Cardiac symptoms	9 (75.0)	8 (66.7)	3 (25.0)
Unable to walk, do chores, work	10 (83.3)	9 (75.0)	3 (25.0)
Hearing problems	5 (41.7)	4 (33.3)	0
Vision problems	10 (83.3)	7 (58.3)	0
Mouth/throat problems	8 (66.7)	6 (50.0)	2 (16.7)
Depression	9 (75.0)	6 (50.0)	2 (16.7)
Personality/emotional problems	7 (58.3)	6 (50.0)	0
Hallucinations/zombie state	7 (58.3)	8 (66.7)	0
Panic attacks	6 (50.0)	3 (25.0)	1 (8.3)
Trembling	6 (50.0)	4 (33.3)	0
Clumsiness	5 (41.7)	6 (50.0)	0
Difficulty breathing	8 (66.7)	6 (50.0)	1 (8.3)
Sleep disturbance	7 (58.3)	6 (50.0)	2 (16.7)

TABLE 19.10

Pain Experienced during and after Exposure by 12 Subjects Unconscious Due to CO and Current Symptoms in a Control Population

Symptoms	CO Group, Number (%) During	CO Group, Number (%) Continuing	Control, Number (%) Reported
Deep muscle pain	5 (41.7)	4 (33.3)	0
Chest pains	6 (50.0)	4 (33.3)	3 (25.0)
General flu-like aches	9 (75.0)	5 (41.7)	1 (8.3)
Neck and back pain	9 (75.0)	8 (66.7)	0
Sharp shooting pains	6 (50.0)	6 (50.0)	0
Joint pain	9 (75.0)	7 (58.3)	1 (8.3)
Neuralgic pain	3 (25.0)	3 (25.0)	0
Pain in jaw/tongue	2 (16.7)	2 (16.7)	0
Arms and/or legs	8 (66.7)	6 (50.0)	2 (16.7)
Cramps in arms/legs	6 (50.0)	5 (41.7)	2 (16.7)
Tendon pain	5 (41.7)	5 (41.7)	0

Figure 19.2 shows how subjects were identified as having been exposed to CO. The figure documents identification of CO for both the group who were chronically exposed to the gas and for those who were unconscious. In only 5 out of the 65 cases who were chronically exposed to the gas was a diagnosis of exposure to CO made by a physician. In the majority of cases, identification of CO as the cause of the patient's symptoms was made through servicing and subsequent condemnation of appliances as dangerous. In a few cases, a CO alarm or detector warned of the presence of the gas. For subjects who were unconscious, diagnosis of CO poisoning was generally made by a physician on examination of the patients. For a few of the cases, the cause of the problem was identified by engineers servicing equipment.

In over 40% of the cases of chronic exposure to CO, it was clear that the appliance had been serviced regularly; however, a problem of fugitive emissions had not been identified through the service or problems, may have occurred after the service. The details of this are shown in Figure 19.3. In over 20% of the cases of chronic CO exposure, it would appear that the appliance had not been serviced regularly. For over 25% of the cases of chronic exposure, the respondent had no idea whether the equipment had been serviced regularly or not. Similar responses were provided by the group of individuals who were unconscious as a result of exposure to the gas.

19.3.4 INITIAL DIAGNOSIS

With few cases of CO poisoning being recognized by physicians, it is interesting to note the diagnosis of the patient's condition that had been made by physicians prior to discovery of exposure to CO. Figure 19.4 records the reported diagnosis for both the 65 individuals chronically exposed and the 12 who were rendered unconscious by the exposure. In many cases, no diagnosis of the problem was made, and in some

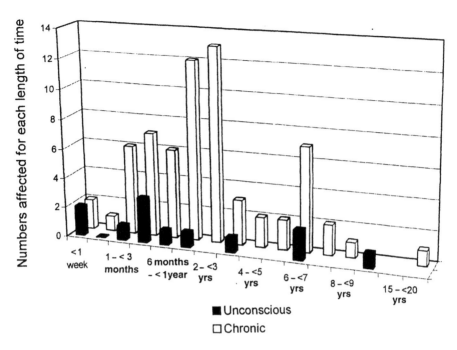

FIGURE 19.1 Duration of exposure to CO.

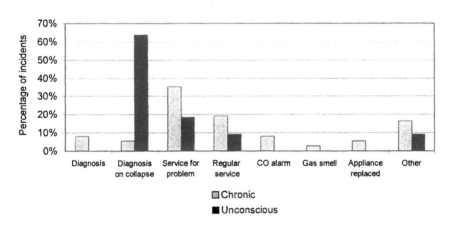

FIGURE 19.2 How the CO poisoning was discovered.

15% of cases flu or a viral infection was considered to be the cause. Depression on its own is identified as the problem in some 8% of the group who were chronically exposed. Tables 19.4 and 19.8 indicate that depression is a significant problem in individuals who are exposed to CO.

In a proportion of cases, patients found that their physicians were supportive. However, a significant number of patients indicated that their physicians did not know how to help them, or were of the opinion that exposure to CO would have no

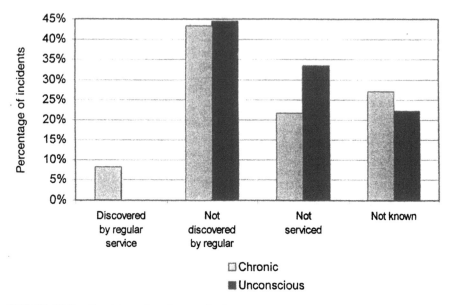

FIGURE 19.3   Servicing of equipment for known CO poisoning incidents.

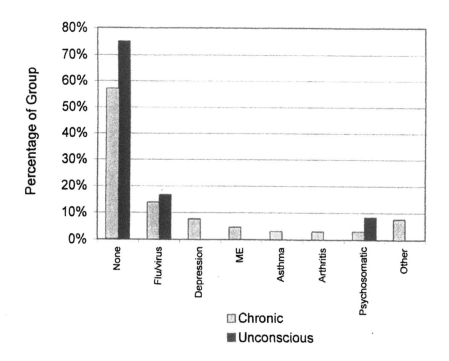

FIGURE 19.4   Diagnosis prior to discovery of CO.

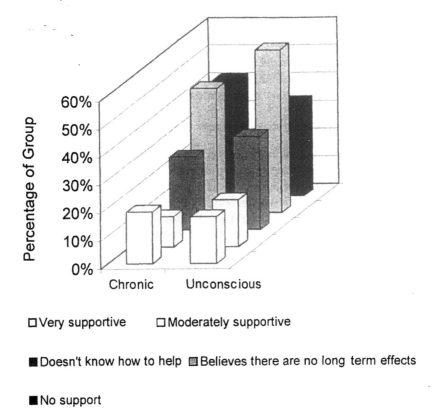

□ Very supportive   □ Moderately supportive

■ Doesn't know how to help ▨ Believes there are no long term effects

■ No support

FIGURE 19.5   Patient's perceptions of physician's support.

long-term effects. The patients' perception of their doctor's attitude is shown in Figure 19.5.

19.3.5   ABILITY TO WORK, HOUSEHOLD INCOME, AND PAIN RELIEF

During their exposures to CO, 55% of those chronically exposed to the gas reported that they were unable to work, or to walk any distance. When they completed the questionnaire after their exposure to CO had stopped, 45% of the group reported that they still had these problems. Only 8% of the control group reported similar problems.

Of the patients who were unconscious as a result of exposure, 82% reported an inability to work or walk any distance during their exposure, with some 75% reporting this to be a continuing problem at the time they completed the questionnaire, and after exposure to the gas had stopped. Of the control group for the unconscious patients, 25% (three subjects) reported a similar difficulty.

Both the chronic group of exposed subjects and those who were unconscious reported a significant reduction in household income following their exposure to CO. Reduction in income is supported by the patients who reported an inability to work.

Patients found that medication was of little help. Out of the 65 individuals chronically exposed to the gas, 15 reported taking painkillers to relieve symptoms, but only 3 found relief; 9 patients reported only partial relief of symptoms. Seven patients took antidepressants; only two found these to be effective. Three reported some partial relief of symptoms. Steroids and inhalers given to five and six individuals, respectively, were only reported to be effective for one person taking each class of drug.

19.3.6 Current Symptoms and Previous Medical History

Those subjects who reported continuing symptoms, particularly chest pain and breathing problems, were assessed to investigate the relationship between their current problems and their previous medical history. In the group who were chronically exposed to CO, 29 subjects reported continuing chest pain, with 36 reporting that the pain did not persist. Of the cohort, 3 reported a previous history of heart problems, with 17 having some family history of heart disease. A chi square analysis at both the 0.05 and 0.025 level of significance indicated that continuing chest pain was independent of both previous personal history and previous family history of heart problems.

Levels of significance also applied for subjects complaining of continued breathing difficulties. The statistical tests indicated that the continued breathing problems were independent of either the patient's family history or the subject's previous history of lung problems.

19.4 DISCUSSION

More than 60 years after Beck[6] reported the consequences of chronic exposure to CO, the situation appears little changed in the United Kingdom. The results of this survey indicate that there is a continuing and unrecognized problem associated with chronic exposure to CO. Most doctors do not recognize the symptoms of CO poisoning and, as a consequence, do not diagnose it. Many individuals suffer for years as a result of their exposure to the gas and, as this survey indicates, many people continue to suffer symptoms years after the exposure has stopped. Respondents of the questionnaire indicate that they have experienced a wide range of symptoms for up to 2 years after exposure ended.

Finding a control group for the study was difficult. It was decided, therefore, to identify control individuals living in the same street, where possible, to the person who had experienced CO poisoning. Controls were matched as closely as possible to subjects on the basis of sex, age, residential accommodation, and income. The data indicate good matching.

One weakness of this study is in relation to the evidence about the duration of symptoms during exposure to CO. Most respondents complained of a range of symptoms, and some stated that they had these symptoms for many years. Without any objective evidence of actual exposure to CO, it is difficult to be certain that individuals who complained that they had experienced these symptoms for, say, 20 years prior to CO being discovered, were, in fact, suffering the effects solely of CO exposure. For the majority of respondents, however, the symptoms were identifiably ascribed to exposure to CO as the problem was identified by gas engineers inspecting gas fires, central heating boilers, and the flues and chimneys that vented waste gases into the atmosphere. In the majority of the cases reported in this study, it was a gas engineer who condemned a fireplace or a flue as dangerous, and as the source of CO exposure. Thus, there is confirmation for the vast majority of the respondents, that they were indeed exposed to CO. However, it is not possible to say how much of the gas they inhaled, and what their likely peak COHb concentrations would have been. For many, COHb concentrations are likely to have been over 20%, given the reports of headaches; however, the range of concentrations, as well as the duration, is uncertain. For the subjects who were rendered unconscious as a result of their exposure, the evidence suggests that their COHb concentrations would have been 60% or greater.[1,2]

In his report on the investigation of clinical symptoms in 97 individuals repeatedly subjected to sublethal concentrations of CO, over periods ranging from several months to 18 years, Beck[6] noted that the principal symptoms suffered by this group of individuals included headaches, vertigo, nervousness, neuromuscular pains, digestive disturbances, dypsnea, and palpitations. In addition to these symptoms, individuals also complained of nervous and mental symptoms, including feelings of depression, restlessness, anxiety, and fears. Also reported were experiences of introspection, emotional upheaval, mental retardation with memory defects and confusion.

Feelings of weakness and an inability to walk properly were mentioned by some. Drowsiness and insomnia were recorded frequently, and approximately one third of patients complained of parathesia, chiefly in the extremity. In the 97 individuals concerned, 7 complained of speech defects, but the exact nature of the defects is not specified. Disturbances of the vasomotor system were also reported, and these resulted in morbid flushing, local sweating, cold extremities, and a purplish congestion of the hands and feet. The principal neuromuscular complaint was a pain that was either felt as a dull pain or as acutely spasmodic in nature. A dull aching pain often occurred in the back, shoulders, epigastrium, lower abdomen, and chest. Of 97 patients, 10 complained of dysuria. Beck would have no difficulty recognizing the spectrum of symptoms in the cohort studied here. Complaints of the respondents to the questionnaire catalog a spectrum of symptoms all too similar to those reported by Beck's own patients. It is clear that the problem has not disappeared.

Few new publications in the literature discuss the issue of chronic CO poisoning. Recently, however, some articles have begun to appear. A recent report[10] documents the case of a crane driver at a smelting works, who developed permanent symptoms after 20 years of exposure to CO. The report also documents the effects of exposure from a faulty oil-fired central heating system, and the long-lasting symptoms in four members of one family. The authors also note that symptoms of chronic CO poisoning, which include headache, dizziness, and tiredness, usually disappear after some weeks or months in individuals who have been exposed, but that in some patients they probably become permanent.

The failure to recognize CO as the cause of symptoms in the cohort described here means that few of the subjects received treatments that were of much benefit. Although some severe cases of CO poisoning can be identified radiologically, there are many instances where this is not the case. Neuropsychological tests can help to identify individuals who have been poisoned. These tests are now under experimental evaluation for subjects chronically exposed to CO.[11] It is reported that there is a protocol for treatment with the possible use of hyperbaric oxygen (HBO) that can benefit a patient and may help functional cognitive and psychiatric capacity; this protocol is currently under evaluation.

Only one third of the patients who were unconscious as a result of their exposure to CO in the cohort described here received HBO treatment. Two of the subjects who were never unconscious, but were chronically exposed to CO, also received some HBO treatment a period of time after their exposure to CO was recognized.

Treatment of the symptoms in patients chronically exposed to CO and those who were unconscious, following what would appear to have been a prolonged exposure prior to their collapse, was far from satisfactory. Subjects complained of continuing pain. There is an urgent need to identify treatment protocols, in addition to the psychological, for subjects suffering from the effects of CO poisoning. As well as the need for more effective treatment, there is a more pressing requirement: the need to help doctors recognize patients who are exposed to CO. It is important to explain to the medical community that although many individuals who are acutely exposed to CO may recover relatively quickly from their ordeal, there are some who will experience continuing ill-health for a long period of time. For those who are chronically exposed, recovery will be slow and prolonged.

The manner in which the source of CO was identified in the cohort described here bears further examination. Apart from the small role that the medical profession played in the diagnosis of the problem, it is evident that regular servicing of appliances will not always guarantee that individuals escape poisoning by CO. Figure 19.3 indicates that for over 40% of the respondents in both the chronically exposed group and in the group of unconscious subjects regular servicing of an appliance did not identify that there was a problem with it or the problem occurred after the equipment was serviced. How long after the servicing the problem appeared is not known. It is also apparent from the responses to the questionnaire, that many appliances were not serviced routinely. The best advice that can be given to individuals to prevent exposure to CO is to ensure that all appliances which use fuel are serviced regularly, and that they ensure that the flues or chimneys that vent waste gases to the atmosphere are cleaned regularly — at least once a year. Use of CO detectors and alarms will provide a third tier of protection. Ultimately, however, equipment needs to be designed so that it will cease operation when there is any risk of significant concentrations of CO building up in the living space. It is not sufficient to have equipment cease functioning simply if there is a failure of oxygen to ensure complete combustion. Detectors that record excessive CO production could be wired to shut down heating appliances. This would prevent further production of CO and should significantly reduce the problem of CO exposure.

19.5 CONCLUSIONS

The survey described here has been reported in full in a separate publication[9] which has already stimulated some debate about the effects of chronic exposure to CO in the United Kingdom. The authors are now aware of significant research programs that have been instituted by the gas industry in the United Kingdom on measures to reduce CO exposures in the home. These research programs, which total some £2,000,000 (U.K. sterling) were implemented some months after the publication of the authors' report. Whether the report helped to prompt these programs is not clear; however, that this research is in progress is gratifying. Only continuing discussion with victims of poisoning, the medical community, and the fuel industry that provides the much-needed heating for homes, will solve the problem.

ACKNOWLEDGMENTS

The authors are grateful to the individuals who answered the extensive questionnaires and, thus, made this report possible.

REFERENCES

1. World Health Organization, *Carbon Monoxide. Environmental Health Criteria, No. 13*, WHO, Geneva, 1979.
2. Winter, E.M. and Miller, J.N., Carbon monoxide poisoning, *J. Am. Med. Assoc.*, 236, 1502, 1976.
3. Lowe-Ponsford, F.L. and Henry, J.A., Clinical aspects of carbon monoxide poisoning in *Adverse Drug Reaction, Acute Poisoning Review*, Vol. 8, Oxford University Press, Oxford, 1989, 217–240.
4. Smith, J.S. and Brandon, S., Morbidity from acute carbon monoxide poisoning at three year follow-up, *Br. Med. J.*, 1, 318, 1973.
5. Garland, H. and Pierce, J., Neurological complications of carbon monoxide poisoning, *Q. J. Med.*, New ser. 36, 144, 445, 1967.
6. Beck, H.G., Carbon monoxide asphyxiation: a neglected clinical problem, *J. Am. Med. Assoc.*, 107, 1025, 1936.
7. Choi, H.S., Delayed neurological sequelae in carbon monoxide intoxication, *Arch. Neurol.*, 40, 434, 1983.
8. Min, S.K., A brain syndrome associated with delayed neuropsychiatric sequelae following acute carbon monoxide intoxication, *Acta Psychiatr. Scand.*, 73, 80, 1986.
9. Carbon Monoxide Support, Effects of Chronic Exposure to CO: A Research Study Conducted by CO Support, Technical Paper, October 1977, 47 pp., appendices.
10. Tvedt, B. and Kjuus, H., Kronisk CO-Forgiftning. Eruken av. Generator gass under den annen verdenskrig og nyere forskning [Chronic CO poisoning. Use of generator gas during the Second World War and recent research], *Tidsskr. Nor. Lageforen.*, 117, 2454, 1997.
11. Myers, R.A., Defazio, A., and Kelly, M.P., Chronic carbon monoxide exposure: a clinical syndrome detected by neuropsychological tests, *J. Clin. Psychol.*, 54, 555, 1998.

20 Neuropsychological Evaluation of the Carbon Monoxide-Poisoned Patient

Dennis A. Helffenstein

CONTENTS

20.1 INTRODUCTION

Carbon monoxide (CO) is well established to be one of the most common causes of poisoning in the United States. It has a high degree of mortality and, for those individuals who survive, a high percentage experience permanent physical, cognitive, emotional, and psychological sequelae. White[1] notes that numerous studies have documented delayed neuropsychiatric sequelae in a significant percentage of individuals who survive exposure to CO. Neuropsychological testing can be helpful in both the acute and post-acute diagnosis and treatment of individuals exposed to CO. A multidisciplinary team approach to the diagnosis and treatment of individuals exposed to CO is clearly indicated.[2] Acutely, the treatment team will often include the emergency room physician, neurologist, radiologist, hyperbaric oxygen (HBO) physician, and neuropsychologist. Early on, brief neuropsychological testing can be used to help guide treatment as well as to document recovery of cognitive function. In the post-acute phase or in cases of long-term, low-dose exposure to CO, more comprehensive neuropsychological testing can help to identify and quantify the significance of cognitive dysfunction. It is also critical in developing a comprehensive rehabilitation treatment plan in the post-acute phase. The multidisciplinary team often includes the primary care physician, neurologist, toxicologist, psychiatrist, ophthalmologist/optometrist, neuropsychologist, and a variety of allied health care professionals.

Whether during the acute or post-acute phase of diagnosis and treatment, it is generally accepted that the Gross Mental Status Exam is an insufficient measure of cognitive functioning and is potentially the weakest portion of neurological examinations.[3] The Gross Mental Status Exam is not standardized and it lacks the sensitivity to identify subtle cognitive dysfunction. Because so much of the damage caused by CO is microscopic, imagery techniques such as computer tomography (CT) and magnetic resonance imaging (MRI) are often negative and cannot rule out CO poisoning.[2] It is now generally accepted that standardized psychometric testing which compares the individual's performance to demographically corrected norms is the most sensitive tool in identifying and quantifying cognitive dysfunction associated with toxic exposure.[4]

Comprehensive neuropsychological testing will typically assess intellectual functioning, academic skills, language and communication abilities, visual and auditory attention and concentration, visual and auditory short-term memory, speed of information processing, motor and sensory skills, visual perceptual and constructional abilities, planning and organization, problem solving, abstract reasoning, and cognitive flexibility. A comprehensive neuropsychological evaluation should also include a clinical interview, an assessment of emotional and psychological functioning, and an assessment of the patient's level of effort.

20.2   ACUTE ASSESSMENT/HYPERBARIC SETTING

20.2.1   INITIAL PRESENTATION

CO poisoning may be difficult to diagnose upon initial presentation to the emergency room. This is because multiple systems may be involved and the presenting problems may mimic other medical or psychological disorders. When CO poisoning is suspected, carboxyhemoglobin (COHb) levels are often the first diagnostic test performed as part of the initial work-up. However, it is now well established that COHb levels frequently do not correspond to other symptoms of CO poisoning, nor do they correlate with subsequent residual deficits.[1,2] Also, as noted above, imaging techniques such as CT and MRI are often normal, especially early postexposure.

20.2.2   Brief Testing and Objectives

Patients presenting with vague, multisystem complaints and without a history of loss of consciousness can be easily misdiagnosed with other medical or psychiatric problems. Over the last several years, neuropsychologists have become more involved in the initial phases of diagnosing and treating CO poisoning, especially in centers that provide HBO treatment. In cases of acute high-dose exposure to CO when HBO treatment may be indicated, only brief testing is typically performed. Results from this brief testing can be used to help determine which patients are experiencing significant neurological disturbance and thus require aggressive HBO treatment vs. those who can be effectively treated with surface oxygen.

Screening batteries have been developed and specifically designed for use in emergency settings. These batteries have been developed to meet the time and administration restrictions imposed by an emergency room situation. One such battery is the Carbon Monoxide Neuropsychological Screening Battery (CONSB) developed by Messier and Myers.[5] This screening battery includes a general orientation section as well as the Digit Span Test, Trail Making Test, Digit Symbol Test, Aphasia Screening Test, and Block Design Test. The results of this screening battery have been found to be helpful in emergency settings in making differential treatment decisions. While quite brief (estimated time of administration is 45 to 60 min), the battery has been found to identify accurately patients with cerebral dysfunction associated with CO poisoning.

If time permits, more extensive testing can also be performed. Areas of functioning that are often assessed include motor skills (e.g., Fingertip Tapping, Grooved Pegboard), executive functioning (e.g., Stroop Test, Trails B Test), attention and concentration (e.g., Digit Span and/or Digit Symbol), visual tracking (e.g., Digit Vigilance Test), speed of information processing (e.g., Paced Auditory Serial Addition Test), and tests of short-term memory. The results of the tests can assist in differential diagnosis and are an important way to assess rate and degree of cognitive recovery across HBO treatments. The initial testing also provides a baseline with which later testing is compared.

Many HBO units will do follow-up neuropsychological testing at 1 month, 6 months, and 1 year as a way to monitor the patient's progress. This testing is also critical in developing ongoing rehabilitation treatment plans. In cases of chronic (long-term, low-dose) exposure to CO, HBO treatment has also now been established to be effective and medically indicated.[6] However, in cases of long-term, low-dose exposure to CO, more comprehensive neuropsychological testing is indicated.

20.3   POST-ACUTE NEUROPSYCHOLOGICAL ASSESSMENT

20.3.1   Referral

Should the CO-exposed patient continue to demonstrate deficits on the neuropsychological screening batteries, then referral should be made for a comprehensive neuropsychological evaluation. Even if performance on the screening test is within normal limits but the patient is reporting cognitive difficulties in his or her day-to-day work and/or academic functioning, referral for comprehensive testing is also indicated. If possible, the comprehensive testing should be done between 2 and 4 months postexposure. Certainly, if the patient is reporting ongoing cognitive dysfunction by 6 months post-exposure, a comprehensive battery of neuropsychological tests is indicated. As a rule of thumb, however, the sooner the referral to the treating neuropsychologist, the better. While comprehensive testing may not be undertaken immediately, the treating neuropsychologist may have treatment recommendations that could be of extreme benefit to the patient and family which could begin early postexposure.

20.3.2 CLINICAL NEUROPSYCHOLOGICAL INTERVIEW

The initial phase of a neuropsychological evaluation typically includes a clinical interview with the neuropsychologist. Obtaining a history of the event is an important aspect of this interview. In instances of long-term, low-dose exposure, the patient is often able to provide a relatively clear and accurate history. However, in cases of higher-dose exposure, the patient will often have only vague memories for the event or no memories at all. In either case, it is important to gather information about the event from family members, colateral sources, and/or records that are available. Aspects of the event that would be important to the neuropsychologist include the intensity, duration, and frequency of exposure.[7]

When available, important data points include parts per million of CO in the environment and COHb level if one was taken. However, it is important for the neuropsychologist to understand that if the COHb level is taken sometime after the exposure and/or after the patient has received oxygen, the maximum COHb level at the time of the exposure would have been much greater. The half-life of COHb is approximately 5 h, and 20 min when breathing normal air at sea-level pressure. Breathing oxygen, forced oxygen, or HBO will substantially shorten the half-life of COHb. In cases of higher-dose exposure, it is important to gather information about the patient's actual memories for the events leading up to, during, and following the exposure. This provides important data about retrograde and post-traumatic amnesia. It is generally accepted that the greater the length of either type of amnesia, the greater the likelihood of permanent cognitive impairment.

In the course of the clinical interview, it is also important to gather data regarding the acute symptoms associated with the exposure. Acute symptoms are many and varied and often include "flu-like" symptoms such as nausea, vomiting, headache, muscle and joint aches, malaise, and fatigue/lethargy. Chest pain, heart palpitations, and cough may also be reported. Additional acute symptoms include lightheadedness (dizziness) and somnolence, and blackouts are often present. Acutely exposed patients can experience a variety of motor and sensory problems including motor weakness, incoordination, muscle spasms or tremors, paresthesias, and decreased muscle tone. The patient may experience a variety of visual problems including light sensitivity, phonophobia (sensitivity to noise), and sleep disturbance. From an emotional and psychological standpoint, feelings of depression are most common, although anxiety, social isolation, and irritability have also been reported. From a cognitive standpoint, generalized mental confusion and disorientation, as well as problems with attention and concentration and short-term memory are frequently reported.[8,9] When patients are exposed in such a manner that they enter and leave the toxic environment (e.g., faulty furnace in a home and the person leaves for work), the acute symptoms may subside and then worsen when they reenter the environment.

A detailed review of chronic symptoms is also an important aspect of the neuropsychological interview. Information regarding persistent physical, cognitive, psychological, visual, and behavioral problems should be obtained. Many of these symptoms are contained in checklists such as the Toxic Symptom Checklist. This checklist includes common chronic physical complaints including ongoing problems with headaches, dizziness/vertigo, motor incoordination, balance problems, paresthesias, alterations in sense of taste and/or metallic taste (dysguesia), tinnitus, fatigue, phonophobia, and sensitivity to a variety of chemicals or substances.[10]

Developing a sensitivity to a variety of substances appears to be quite common following CO exposure. Most often, patients will report sensitivity to car exhaust, smoke (cigarette or cigar), paint or gas fumes from spray cans, heavy colognes, products containing formaldehyde, and herbicides/pesticides. Whether this sensitivity is a classically conditioned response (which is unlikely since CO is odorless, colorless, and tasteless), an actual physical sensitivity due to biochemical abnormalities or to an autoimmune abnormality, post-traumatic stress disorder (PTSD), or other psychological disturbances (e.g., somatoform disorder) remains unclear.[8] However, a diagnosis of multiple chemical sensitivity (MCS) will often follow a documented exposure and symptoms will occur and subside in response to even low-level exposure to a variety of substances.[11] A survey of peer-reviewed scientific articles finds that more than half support an organic explanation for MCS and many more support a mixed organic and psychogenic explanation.[12] While this is a topic that clearly requires more research, it is important for neuropsychologists to screen for sensitivity to chemicals/substances as this will, at least in some instances, have functional and/or vocational implications. This finding also has treatment implications in that the individual may be sensitive to or react adversely to medications.

A review of visual functions should also be conducted as part of the clinical interview. Post-traumatic vision syndrome is common following cerebral injury,[13,14] including toxic encephalopathy. Symptoms include double vision, blurred vision, decreased acuity (especially for night vision), difficulties with visual scanning (e.g., losing place when reading, missing things one is looking for), problems with depth perception, bumping into things more frequently, problems with accommodation (e.g., eyes are slow to focus), and photophobia (sensitivity to light). Photophobia is extremely common following CO exposure. Decreased peripheral vision, hemispatial inattention, seeing stationary objects move, and perception of movement in the peripheral fields when there is nothing there (unilateral saccades) are also common. Eye fatigue and the development of headaches while performing visually loaded tasks can also occur. Hemispatial inattention, visual scanning deficits, slowed speed of visual processing, and suppression of visual information on bilateral simultaneous stimulation are often seen on testing.

During the clinical interview, it is important to gather information about the patient's subjective experience of his or her cognitive deficits. Particular attention should be paid to attention and concentration, cognitive set maintenance, and ability to multitask. Short-term memory deficits are extremely common following CO exposure and complete review of verbal and visual memory functions should be conducted. It is common that cueing will not assist in retrieval of information from memory. Because of the frequency of executive dysfunction following CO exposure, these abilities should also be thoroughly reviewed. Functions such as planning, organizing and sequencing, problem solving and decision making, abstract reasoning, self-monitoring, verbal fluency, and initiation should all be discussed. Slowed speed of information processing is also a very common sequela. Patients will often report "slower thinking" and taking longer to perform tasks. Paraphasias, difficulties with language comprehension, reading, spelling, and math, and transpositions when writing or speaking can also occur and should be discussed with the patient.

From an emotional standpoint, depression and irritability (reduced frustration tolerance) are the most common psychological sequelae following CO exposure. Emotional lability (crying more frequently and easily), generalized anxiety, apathy, and impulsivity are also fairly common. Based on the exposure event, a thorough review of PTSD symptoms should also be performed.

Family members and friends will often describe a "change in personality" postexposure. Common changes that they observe include mood swings, flat affect, decreased motivation, anhedonia, and reduced frustration tolerance with verbal and physical outbursts. Becoming upset and angry over small things that would not have bothered the patient previously is common. Disinhibition and impulsivity can occur. For example, people may say or do things that they would not have done previously or they may respond impulsively (e.g., acting without thinking, impulsive spending).

With regard to the individual's depression, to some degree this may be reactive and is associated with the physical and cognitive deficits that he or she experiences and the limitations that these impose. However, it is also important to be aware that, to some degree, the depression (as well as other emotional and behavioral changes) is organically based and associated with an organic mood and/or behavior disorder. As will be discussed later, psychotherapeutic intervention and psychotropic medications may both be indicated to address these symptoms. It is important to be aware that, in addition to mood and behavior changes, CO-induced psychotic symptoms are also possible. The development of hallucinations or delusions following CO exposure constitutes a substance-induced psychotic disorder.

20.3.3 Neuropsychological Testing

Numerous studies have been conducted to identify those structures of the brain most sensitive to and affected by CO poisoning. Ginsburg[15] suggested that the globus pallidus, cerebral cortex, hippocampus, cerebellum, and substantia nigra are some of the brain structures most vulnerable. Other studies have suggested that virtually all lobes of the cerebral cortex can be affected by CO poisoning, which would be consistent with any hypoxic event in that diffuse cortical involvement is possible. As a result, there is no consistent pattern of permanent damage following CO exposure.[2]

While one might expect a rather diffuse pattern of impairment, it is important to note that lateralized cerebral dysfunction can occur as a result of CO poisoning.

Reitan and Wolfson[16] present a case study (Case 26) that involves an attempted suicide by running the engine of a car in a closed garage. On testing, the woman demonstrated multiple deficits which clearly indicated more severe lateralized dysfunction in the left cerebral hemisphere. These included such notable deficits as a right homonymous hemianopia, right hemiparesis, mild dysstereognosis of the right hand, fine motor tremor in the right upper extremity, a mild dysphasia, and pronounced dyscalculia. An electroencephalographic study demonstrated abnormalities in the entire left cerebral hemisphere. Clinically, this author has also identified lateralized deficits in CO-poisoned patients such as a significant verbal learning and retention deficits with intact nonverbal learning and retention abilities. Indeed, animal studies have demonstrated that destruction of the pyramidal cells of the hippocampus can be asymmetric.[17] Therefore, if such a pattern is identified, this should not dissuade the neuropsychologist from diagnosing cognitive impairment associated with CO poisoning.

Because virtually all cortical structures and many of the subcortical structures can be affected, a wide variety of permanent cognitive deficits may exist. The World Health Organization[18] identified common cognitive deficits following CO poisoning. These included deficits in sustained and alternating attention and concentration, short-term memory, decision making, motor coordination and speed, speed of information processing, sensory-motor deficits, deficits in the higher cortical functioning, and visual problems. While no classic CO pattern has been identified[19] in this author's clinical experience, deficits in executive functioning, attention and concentration, memory functioning, visual-perceptual abilities, and information-processing speed are most common, although any deficit or combination of deficits is possible.

There are four general approaches to neuropsychological assessment: the standard or fixed batteries, flexible batteries, the "process" approach, and microcomputer-based neuropsychological batteries. The two most widely used standard batteries are the Halstead–Reitan Battery (HRB) and the Luria Nebraska Neuropsychological Battery (LNNB). The HRB is the most widely used standardized battery. The tasks were originally developed by Dr. Ward Halstead and were later modified and standardized by his graduate student, Dr. Ralph Reitan.[20] As it exists today, the expanded HRB consists of five tests producing seven scores. Several supplemental tests and one of the Wechsler Intelligence Scales as well as academic testing are also commonly used in conjunction with the HRB. The LNNB is based on methods developed by the Russian psychologist Dr. Alexander Luria. These items were first introduced by Christenson[21] in 1975 and further developed by Golden et al.[22] in 1980. The LNNB produces 11 clinical scales entitled "Motor, Rhythm, Tactile, Visual, Receptive Language, Expressive Language, Writing, Reading, Arithmetic, Memory, and Intelligence".

Hartman[8] discusses limitations related to each of these batteries. He also discusses the importance of conducting a comprehensive evaluation that assesses a wide variety of cognitive, emotional, and neurobehavioral functions when evaluating any patient exposed to a neurotoxic substance. He also points out that it is important to assess specifically those cognitive abilities and brain functions known to be affected by specific neurotoxins. For example, because of the significant impact that CO has on the hippocampus and memory functioning, the HRB by itself is inadequate because it lacks specific measures of verbal and nonverbal learning and retention of memories over time. It is also weak in the assessment of frontal functioning. Hartman[8] also notes that "most research to date finds Lurian methods insensitive to neurotoxic exposure" (p. 54). He goes on to point out that the LNNB does not contain complex or multicomponent tasks, which are abilities often affected by CO. It also contains a limited assessment of memory functioning. However, one major advantage to the use of standard batteries is that they typically have good validity and normative data.

When utilizing a flexible battery approach, a neuropsychologist will choose existing standardized neuropsychological tests that are proved to be valid and reliable in the diagnosis of brain dysfunction. The neuropsychologist will typically choose tests that have been found to be sensitive to cognitive abilities most affected by neurotoxic exposure. This type of battery can range from an extremely comprehensive assessment of brain function to more minimal screening-type batteries. Obviously, screening batteries are less sensitive and accurate than comprehensive testing. Hartman[8] has conducted an extremely thorough review of flexible batteries that have been developed to assess neurotoxic exposure from both a clinical and epidemiological standpoint. A complete review of these batteries is beyond the scope of this chapter, but several of the more comprehensive batteries include the World Health Organization Neurobehavioral Core Test Battery (NCTB), the Pittsburgh Occupational Exposures Test (POET) battery,[23] and the California Neuropsychological Screening-Revised (CNS-R).[24]

For CO-poisoned patients it would be most appropriate to perform a standard battery such as the HRB and then supplement that testing with additional standardized tests to evaluate more fully frontal function (which includes many of the executive functions), memory function, visual-perceptual abilities, and information-processing abilities which are commonly affected by CO poisoning. The obvious disadvantage to such a battery is the length of time involved to administer the tests, which would typically range from 6 to 8 h. However, in this author's clinical experience, this is by far the most desirable approach to evaluating a patient exposed to CO. It is the most comprehensive, sensitive, and accurate method of neuropsychological assessment. Such an approach provides a high degree of validity and standardized norms are available. Many of the tests are normed demographically for age, gender, and education.[25,26] In addition, the wealth of data obtained is extremely helpful in developing a cognitive rehabilitation program and in making vocational and other treatment recommendations.

The "process approach" to neuropsychological assessment places greater emphasis on the patient's approach to the neuropsychological tasks than on standardized quantitative performance. This approach incorporates both standardized neuropsychological tests as well as tasks developed by the clinician during the process of the evaluation to evaluate specific problem areas qualitatively. While this type of approach does utilize some standardized testing, it also includes substantial subjective interpretation of performance by the examiner. In addition, the content of the testing is often based on the patient's self-report of deficits. Unfortunately, there are often times when the patient has impaired self-monitoring and, therefore, lacks insight into his or her deficits. Overall, the process approach has multiple disadvantages and should be avoided in the assessment of patients exposed to CO.

Several computerized cognitive test batteries are also available, including the NES (Neurobehavioral Evaluation System) battery,[27] the MANS (Milan Automated Neurobehavioral System) battery (a computerized version of the World Health Organization NCTB), the Psychometric Assessment System/Dementia Screening Battery,[28] and the MTS (Micro-Computer Based Testing System) battery.[29] While the automated batteries have several advantages, such as ease of transport, standardized administration, consistency of administration, and cost efficiency related to administering, scoring, and analyzing the test data, they also present several disadvantages. Hartman[8] reviews these disadvantages in detail. In summary, these batteries often lack clinical norms; they do not thoroughly assess all areas of cognitive functioning related to specific exposures; they may encourage inappropriate interpretation; important behavioral and observation data are missed; and a subset of the population remains computer phobic.

20.3.4 CLINICAL PSYCHOLOGICAL TESTING

In addition to conducting a clinical interview related to the patient's current emotional and psychological status, most neuropsychologists will include inventories or questions related to personality and emotional status as part of their test battery. Several instruments are available that assess a range of emotional and personality factors, such as the Minnesota Multiphasic Personality Inventory-2 (MMPI-2), the Personality Assessment Inventory (PAI), the Millon Clinical Multiaxial Inventory-2 (MCMI-2), and the Symptom Checklist-90-R (SCL-90-R). In addition, symptom-specific self-report inventories can be used, such as the Beck Depression Inventory, Beck Anxiety Inventory (BDI), and the Burns Anxiety Inventory.

The MMPI and MMPI-2 are the most commonly used of all paper-and-pencil personality tests and are widely utilized by neuropsychologists as part of a comprehensive neuropsychological test battery. The patient reports current or long-standing symptoms or traits by responding true or false to 567 statements. The inventory produces 14 scales and a variety of subscales are also available. Four validity scales provide information about the patient's approach to the inventory and its validity. They also provide information about malingering, symptom denial, or symptom magnification. The profile produced by the 10 clinical scales is compared with those of normal control subjects and a wide variety of diagnostic groups of psychiatric patients.

It is now well established in the professional literature that individuals who have sustained some type of neurological injury such as a toxic exposure will often demonstrate elevations on several of the clinical scales. Many of the items contained in the MMPI-2 represent a wide variety of problems associated with neurological injury. It has been found that these items load heavily on several of the clinical scales, specifically Scales 1, 2, 3, 7, and 8 (1 = Hypocondriasis; 2 = Depression; 3 = Hysteria; 7 = Psychesthenia; and 8 = Schizophrenia).[7,30–32] Scale 4 (Psychopathic Deviant) has also been found to be elevated in patients with head trauma.[7] Because so many of the clinical scales are elevated by neurological symptoms, clinicians are encouraged to conduct individualized interpretation of the MMPI-2 profile taking the neurological injury and symptoms into account. They are cautioned not to interpret the MMPI-2 in the traditional fashion, as that would most likely lead to the misdiagnosis of a psychiatric disorder where none exists. Lezak[7] states:

> Since so many MMPI items describe symptoms common to a variety of neurological disorders, self-aware and honest patients with these symptoms may produce MMPI profiles that can be misinterpreted as evidence of psychiatric disturbance when they do not have a psychiatric or behavioral disorder. (page 782)

Also related to this issue, Cripe[31] states:

> The MMPI profiles of well validated neurologic patients consistently indicate that, as a group, they elevate on scales 1, 2, 3, 7 and 8 to varying degrees, dependent on the particular study and the neuropathology investigated. Similar elevations are seen on MMPI profiles in a large group of medical patients.

Reitan and Wolfson[32] also review this topic and conclude:

> The findings of these studies justify the admonition for neuropsychologists to consider carefully and cautiously any deviant MMPI results produced in brain damaged patients because their responses may reflect at least in part legitimate neurological symptoms and complaints rather than psychiatric problems.

Several studies have been conducted specifically with individuals exposed to neurotoxic substances. Morrow et al.[33] identified workers coming to an occupational health clinic following toxic exposures and demonstrated elevations on scales 1, 2, 3, and 8. Bowler et al.[24] studied a group of workers exposed to organic solvents and the predominant MMPI-2 profile that emerged was 1-2-3-7-8. Therefore, extreme caution needs to be utilized when interpreting the MMPI-2 profiles of individuals exposed to CO. Indeed, any self-report questionnaire that contains items that pertain to neurological symptoms needs to be evaluated on an individual basis within the context of the toxic exposure. Scores on even brief self-report questionnaires such as the BDI can be elevated related to neurological symptoms. For example, the BDI contains items related to emotional lability, irritability, decision making, sleep disturbance, and fatigue, all of which are common sequelae of CO exposure.

20.3.5 Assessing Level of Effort

Assessment of level of effort has become a standard part of neuropsychological testing, especially when the evaluation is performed within a medical-legal context. Multiple factors should be taken into account when assessing a patient's level of effort during the neuropsychological testing. These include review of records, style of presentation during the clinical interview, and behavioral observations during the testing process. Experienced test administrators will monitor the patient's level of effort during the testing, observing such factors as persistence to the task at hand, level of encouragement required, and whether the patient attempted to utilize strategies to maximize his or her performance. As noted above, the MMPI-2 contains validity scales which can provide information about potential response bias (e.g., symptom magnification vs. symptom minimization). The pattern of neuropsychological strengths and weaknesses can also provide important data about the validity of the testing.[34] Lateralized deficits in motor, sensory, or memory tests would be unlikely in patients attempting to feign deficits. A pattern of dysfunction suggesting localized impairment would also tend to rule out malingering, as no layperson could intentionally create such a pattern.

In addition to the above, specific techniques for detecting dissimilation and feigning of cognitive dysfunction have been developed.[35,36] Specific tests designed to detect symptom magnification have been developed. Typically, these are memory tasks although several incorporate other cognitive abilities. Many utilize a forced-choice paradigm in which the patient chooses from only two multiple choice alternatives. With this format, the individual would be expected to get 50% correct by chance alone. Scores that are notably worse than chance performance suggest that the individual accurately perceived the correct response and chose to respond incorrectly. Several of the most commonly used tests are the Rey 15-Item Memory Malingering Test, the Hiscock Forced Choice Digit Recognition format,[37] the Portland Digit Recognition Test (PDRT),[38,39] and the Test of Memory Malingering (TOMM).[40] The Computerized Assessment of Response Bias (CARB) and the Multi-Digit Memory Test are computerized versions of the PDRT and Hiscock Forced Choice Digit Recognition format.

It should be pointed out that some of the tests of symptom validity are quite tedious. Lezak[7] notes that in some cases patients who have given their best effort on all other tests may become annoyed, bored, or insulted by symptom validity tests which are lengthy, repetitive, and monotonous. In some instances, abbreviated versions of the tests have been developed. For example, the original Hiscock Forced Choice Digit Recognition Procedure which contains 72 items was reduced to 36. Guilmette[41] compared the two versions of the test and found that the abbreviated version was equally sensitive in distinguishing brain-damaged subjects, individuals simulating memory dysfunction, and depressed subjects. In summary, the assessment of a patient's level of effort is a complex one and a diagnosis of malingering should be made cautiously and carefully using all available clinical data.[42] Guilmette[41] as well as other researchers in this field recommend that sources other than performance on dissimulation tests be taken into account when making this diagnosis.

20.4 DIAGNOSTIC AND REHABILITATION CONSIDERATIONS

20.4.1 DSM-IV DIAGNOSES

The Diagnostic and Statistic Manual of Mental Disorders, fourth edition (DSM-IV)[43] provides guidelines regarding several diagnoses appropriate for patients who have been exposed to CO. These diagnoses generally fall in the area of substance-related disorders, which generally includes disorders associated with the taking of a drug of abuse, with the side effects of a medication, or with toxic exposure. CO is listed as a specific toxic substance that can result in a variety of diagnosable disorders. When a patient demonstrates cognitive deficits manifested by both memory impairment and impairment in one other cognitive area (e.g., aphasia, apraxia,.agnosia, or disturbance in executive functioning), then the diagnosis of "Substance Induced Persisting Dementia" (p. 152) may be appropriate.[43] This diagnosis would be charted as "292.82, Carbon Monoxide Induced Persistent Dementia." It is important to note that, by definition, dementia may be progressive, static, or remitting (p. 137) and, therefore, this diagnosis would be appropriate at any point following CO poisoning. Giving this diagnosis early postexposure would not rule out the possibility for future recovery.

If memory impairment in the form of difficulty learning new information or in the ability to retain that information over time is the sole cognitive deficit identified, then the diagnosis of "292.83, Carbon Monoxide Induced Persisting Amnesic Disorder" (p. 162) may be appropriate. As previously discussed, exposure to a variety of toxins, including CO, can result in psychotic symptoms including delusions and hallucinations. If these symptoms can be attributed to the CO exposure, then the appropriate diagnosis would be "Carbon Monoxide Induced Psychotic Disorder" (p. 310). If the predominant presentation is of delusions, then this would be coded as 292.11, and, if the predominant symptom presentation is that of hallucinations, then the code would be 292.12.

As indicated previously, exposure to CO can also produce a variety of mood disorders. If the predominant mood disturbance relates to depression and/or elevated expansive or irritable mood, then the appropriate diagnosis would be "292.84, Carbon Monoxide Induced Mood Disorder" (p. 370). If the predominant mood disturbance is that of anxiety in the form of either generalized anxiety or panic attacks, then the appropriate diagnosis would be "292.89, Carbon Monoxide Induced Anxiety Disorder" (p. 439).

20.4.2 PSYCHOTHERAPY

As noted above, individuals who have been exposed to CO will most often experience a variety of emotional and psychological sequelae including depression, anxiety, irritability, sleep disturbance, anger, and in some cases PTSD. To some degree, the psychological sequelae can be reactive to the losses and trauma associated with the exposure, but in many cases these problems are organically based. In either event, individual psychotherapy can be highly beneficial in addressing these issues. Education about the effects of CO exposure can help patients and their family understand and better cope with the changes that they have experienced. Training in relaxation and stress management techniques can be helpful in managing their overall level of anxiety, tension, and irritability. Biofeedback can also help in achieving these goals. Organically based fatigue is common following CO exposure, and training in energy conservation techniques can be beneficial.

PTSD and/or specific phobia symptoms are often seen following CO exposure and a variety of desensitization techniques are available to address this disorder. Techniques such as Eye Movement Desensitization and Reprocessing (EMDR)[44] and Thought Field Therapy (TFT)[45] have been found to be particularly effective techniques. Note that, when using EMDR with any neurologically impaired individual, it is usually best not to use the eye movement method of stimulation. Because so many of these individuals experience post-traumatic vision syndrome symptoms they will often have difficulty with visual tracking, or it may lead to headaches, dizziness, or nausea. Therefore, it is best to utilize the auditory and/or tactile methods of stimulation.

20.4.3 PSYCHIATRIC INTERVENTION

Given that many of the emotional, psychological, and behavioral sequelae of CO exposure relate to organic mood and personality disorders, referral to a psychiatrist familiar with organic injury is often indicated as a variety of psychotropic medications may be helpful in addressing these symptoms. Antidepressant medications, particularly in the selective serotonin reuptake inhibitor class (e.g., Prozac, Zoloft, or Celexa) have been found to be particularly helpful for individuals who have sustained toxic exposures in stabilizing their moods and in maximizing their energy level and frustration tolerance. Antianxiety medications (e.g., BuSpar or Xanax) may be indicated for patients experiencing high levels of anxiety and tension. Sleep disturbance is common following CO exposure and a variety of sleep aids are available (e.g., Ambien or Trazadone). For those patients experiencing significant chronic fatigue, stimulant medications such as Ritalin and Dycipromine can be utilized in small doses to maximize the patient's energy level. Development of psychotic symptoms postexposure is not uncommon and, therefore, antipsychotic medications may also be indicated.

20.4.4 NEUROCOGNITIVE REHABILITATION

There are two theoretical approaches to cognitive rehabilitation, restorative and compensatory.[46] Early postexposure, restorative exercises can be helpful in maximizing the patient's basic cognitive abilities such as sustained and alternating attention, short-term memory, information-processing speed, and visual-processing abilities. Cognitive therapists may also focus on more specific cognitive deficits in speech, language, and academics. This approach focuses on rebuilding impaired cognitive abilities. Another important aspect of cognitive rehabilitation focuses on teaching the patient strategies to compensate for his or her residual cognitive deficits in day-to-day work and/or academic activities. The goal here is to improve functioning or level of performance through the use of external aids or environmental supports. Given the prevalence of memory problems following CO exposure, particular emphasis should be placed on helping the patient develop specific memory strategies such as memory notebooks, "to-do" lists, and use of dry-erase boards. A more complete review of approaches to memory rehabilitation can be found in Raymond et al.,[47] Parenté and Herrmann,[48] and Hutchinson and Marquardt.[49] If individuals experience difficulties in their activities of daily living, then occupational therapy may also be indicated.

20.4.5 NEURO-OPHTHALMOLOGY/BEHAVIORAL OPTOMETRY

Post-Traumatic Vision Syndrome is common following any neurotrauma, including toxic exposure.[13,14] Exotropia, exophoria, convergence and accommodation insufficiency, and ocular dysfunction can lead to a wide variety of vision problems which have been previously reviewed in this chapter. When these types of vision problems have been identified, referral to a neuro-ophthalmologist or behavioral optometrist for further evaluation and possible treatment is indicated. Neuro-optometric rehabilitation has been found to be extremely helpful in addressing many of these vision problems and at times corrective lenses such as the use of prisms or yoked prisms can be beneficial in correcting visual field cuts and/or hemispatial inattention.[50] Of course, dark "wrap-around"-type sunglasses would be important for those patients with photophobia.

20.4.6 VOCATIONAL REHABILITATION

When a patient is unable to return to his or her prior occupation, yet it appears that the patient has residual vocational potential, then referral for vocational rehabilitation is indicated. The goal of vocational rehabilitation is to identify alternative vocations that the person is able to perform successfully either on a full-time or part-time basis. Vocational rehabilitation experts will take into account the patient's residual physical, cognitive, fatigue, visual, and emotional coping problems when identifying vocational alternatives. Unfortunately, there will be times when a combination of the above deficits will result in the person being totally and permanently disabled from a vocational standpoint. In those cases the neuropsychological evaluation will provide important data in helping the individual obtain disability benefits.

20.4.7 PHONOPHOBIA

Following any neurological injury, including CO poisoning, patients will often report sensitivity to loud noises or noisy environments. They will often report that it is difficult for them to filter out background noise and that such situations cause them to feel irritable, anxious, and fatigued. This problem will often lead to the person avoiding such environments, which can lead to or exacerbate social isolation. Should this problem persist, prescription of noise-attenuation earplugs is often helpful. The ER15/25 earplugs (also called musician's earplugs) have been found to be particularly beneficial. These or similar earplugs can be provided by most audiologists.

20.5 CASE STUDY

Ms. D.R. was a 47-year-old woman referred for neuropsychological testing following a 5-day exposure to CO. The first 4 days of the exposure were relatively low dose followed by a higher-dose exposure on the fifth day. The exposure occurred in her home as a result of unvented trench blasting being conducted near her home. The CO migrated to an area underneath her home and entered through the floor drains. CO levels tested in the home following the exposure ranged from 500 parts per million (ppm) in her bedroom to as high as 2800 ppm in the garage. Her COHb level taken approximately 8 h after leaving the home environment was 7.5%.

Education/Employment: Ms. D.R. was a high school graduate and had a bachelor's degree in business administration. She also had a master's of business administration degree and worked professionally on a full-time basis.

Acute Symptoms: During the 5 days of exposure, acute symptoms included fatigue, headaches, nausea, vomiting, muscle aches, trouble waking up in the morning when her alarm went off, dizziness, and sore throat. She noted generalized mental confusion as well as difficulty with attention and concentration, multitasking, and was making more frequent errors as a result of memory inefficiencies. She found that she was highly "ineffective" in her day-to-day work activities. She observed that on the days that she left the home environment, she felt better and that by the end of the day her symptoms were somewhat improved. When she would return home, the symptoms would worsen. On the morning of the higher-dose exposure, she did not hear her alarm go off and she continued to sleep. She was awakened by a friend calling her, at which point she proceeded to her son's room to wake him up. She found her son comatose as CO levels were found to be higher in his bedroom. His eyes were open and fixed and his body was stiff. He was not moving and black emesis was coming from his mouth. He began to experience muscle spasms and abdominal contractions which may have been seizurelike activity. Ms. D.R. was notably confused and disoriented and her daughter had to instruct her to call 911. After the emergency personnel arrived, she was removed from the environment.

Chronic Subjective Symptoms: Ms. D.R. was evaluated at 22 months postexposure. At that time she was reporting a variety of ongoing physical problems and concerns. All of her problems developed at the time of exposure or shortly thereafter. These included problems with headaches, occasional problems with nausea and shortness of breath, dizziness, motor incoordination, tingling and numbness in her hands, decreased sense of taste, occasional metallic taste in her mouth, fatigue, phonophobia, and sensitivity to a variety of substances including car exhaust, cigar and cigarette smoke, and heavy colognes.

She was continuing to experience a variety of vision problems including difficulties with blurred vision, maintaining her visual focus, visual scanning, and difficulties with depth perception (e.g., judging distances). She noted difficulties bumping into things more frequently as well as perception of movement in her peripheral vision when there was nothing there (a classic symptom of unilateral saccades). She noted that her eyes were slow to focus and she also experienced photophobia.

From a cognitive standpoint, she experienced subjective problems with attention and concentration, cognitive set maintenance, multitasking, short-term memory, verbal fluency, and paraphasias. She also noted difficulties with language comprehension, reading comprehension, math, and slowed speed of information processing. Within her day-to-day activities and work, she noted difficulties with problem solving and decision making, planning, organizing, and logical sequencing, and also noted frequent transpositions of letters and numbers when writing.

From an emotional and psychological standpoint, she reported ongoing feelings of depression which varied in intensity and emotional lability. She was also aware of ongoing problems with irritability and decreased frustration tolerance. She noted difficulty getting to sleep at night for no specified reason. She also was experiencing ongoing anxiety about a possible second exposure. She would often wake up frightened and scared and also noted intrusive thoughts about the incident. She purchased multiple CO detectors and placed them around her home.

Ms. D.R. had returned to work by the time of evaluation but only with great difficulty. She was having to utilize a wide variety of strategies to compensate for her residual cognitive deficits and, despite the use of strategies, she was making more errors in her work. Her overall productivity was notably reduced as a result of needing to use strategies, double-check her work, work more slowly, and redo work when errors were made. She made the observation that, "I am having to work twice as hard to accomplish half as much." She was having to put in extra hours at night and on weekends to meet even minimal productivity standards.

Behavioral Observations: Testing was conducted over a 2-day period and all tests were administered in the standardized fashion. Ms. D.R. was highly cooperative with the evaluation process and demonstrated good effort throughout the evaluation. Indeed, at times she attempted to utilize strategies to maximize her performance. A verbal fluency deficit as well as multiple paraphasias were noted in her style of speech. Comprehension of test instructions was good. Response style was at times impulsive. Confabulation was noted on a verbal short-term memory test. No fatigue or pain were reported or noted during the days of testing. She was not taking any prescription medications the days of testing. Neither her performance on the symptom validity tests nor on her MMPI-2 profile suggested any symptom magnification or exaggeration. The testing was considered to be an accurate and valid reflection of her neuropsychological status.

Test Results

Age = 47; Education = 18 years; Dominant hand = Right

Indexes

	Raw	T	Percentile	Heaton Level
Halstead Impairment Index (HII)	0.4	43	25th	Below average/borderline
Average Impairment Rating (AIR)	1.08	42	21st	Below average/borderline
General Neuropsychological Deficit Scale (GNDS)	33			Mild impairment

Intelligence
Wechsler Adult Intelligence Scale–Revised

WAIS-R	IQ	WAIS-R Percentile	Range	Heaton Percentile	Heaton Range
FSIQ	104	61st	Average	18th	Borderline/below average
VSIQ	102	56th	Average	16th	Borderline/below average
PSIQ	107	68th	Average	28th	Borderline/below average

WAIS-R Subtests

	SS	Age	Percentile	Heaton Level		SS	Age	Percentile	Heaton Level
INFO	10	10	16th	Borderline	PC	6	8	5th	Mild/moderate
DIGIT	6	7	1st	Moderate	PA	10	11	50th	Average
VOCAB	12	13	38th	Average	BD	10	12	50th	Average
ARITH	13	13	69th	Above Average	OA	11	13	66th	High Average
COM	11	11	28th	Borderline	DS	10	12	38th	Average
SIM	9	10	13th	Mild	(DS	Memory 3/9)			

Percentile range = 1st to 69th; SS Range = 7.

Academic
Peabody Individual Achievement Test – Revised (PIAT-R)

	Raw	Grade Equivalent	Percentile
Math	83	11.5	42nd
Reading recognition	85	8.9	27th
Reading comprehension	86	11.7	47th
Spelling	81	7.5	18th

Halstead Reitan Tests

	Raw		T	Percentile	Heaton Level
Category	Errors: 60		32	3rd	Mild/moderate
TPT Dom	Time 4.2 min.	Blocks: 10	59	82nd	Above average
TPT Ndom	Time 3.3 min.	Blocks: 10	53	62nd	High average
TPT Both	Time 1.2 min.	Blocks: 10	67	96th	Above average
TPT Total	Time 8.7 min.		58	79th	Above average
TPT Memory	Correct: 7		42	21st	Borderline
TPT Location	Correct: 5		49	46th	Average
Rhythm	Errors: 7		35	7th	Mild
Speech	Errors: 6		40	16th	Borderline
Tapping	Dom: 39.2		40	16th	Borderline
Tapping	Ndom: 38.2		44	28th	Borderline
Trails A: 32 s	Errors: 0		39	13th	Mild
Trails B: 89 s	Errors: 1		36	8th	Mild
Pegs D: 63 s	Drops: 0		45	31st	Low average
Pegs ND: 88 s	Drops: 0		32	3rd	Mild/moderate

Speech and Language

	Raw	T	Percentile	Heaton Level
Reitan-Indiana Aphasia Screening	0	57	76th	Above average
Boston Aphasia Screening Test				
Complex	12/12	53	62nd	High Average
Nonverbal ability	7/12			
Verbal ability	13/14			
Thurstone Verbal Fluency				
Part A: 29 Part B: 8 Total: 37		35	7th	Mild

Visual Perceptual

	Raw	T	Percentile	Level
Spatial relations	2	54	66th	High average
Rey-Osterreith copy	32	40	16th	Borderline
Hooper Visual Organization Test	25.5	49	46th	Average
Key Copy (RKSPE)	WNL			

WNL = Within normal limits.

Sensory/Perceptual
Reitan/Klove Sensory Perceptual Examination

	Right	Left		Right	Left
FA	0	3	Tactile suppressions	2	0
FNW	4	3	Auditory suppressions	2	0
FORM	1	0	Visual suppressions	0	0

	Raw	T	Percentile	Heaton Level
SP-R	10	28	1st	Moderate
SP-L	6	27	1st	Moderate
Total	16	27	1st	Moderate

	T	Percentile	Heaton Level
Tactile Form Recognition: Right: 6.9 s	68	96th	Above average
Tactile Form Recognition: Left: 6.1 s	67	96th	Above average
Visual Fields	WNL		

Digit Vigilance Test

	T	Percentile	Heaton Level
Time: 270 s	71	98th+	Above average
Errors: 19	28	1st	Moderate

Line Bisection Test

No. greater than 1 cm off center to right	1
No. greater than 1 cm off center to left	1

Motor
Behavioral Dyscontrol Scale

	Motor Psv.	Disinhibition	Right/Left Confusion	Echopraxic	Sequencing	Learning
Errors	1	1	3	2	3	0

Score: 14; Range: Mild impairment.

Dynamometer

	Raw	T	Percentile	Heaton Level
Dominant:	25.5 kg	41	18th	Borderline/below average
Nondominant:	22.5 kg	44	28th	Borderline/below average

Memory

	Raw	T	Percentile	Heaton Level
Story Memory				
Learning	7.75	34	5th	Mild/moderate
Recall	% Loss: 45.2	17	<1st	Severe
Figure Memory				
Learning	8.0	37	10th	Mild
Recall	% Loss: 0	57	76th	Above average

Buschke Verbal Selective Reminding Test (Larabee)

	Raw	Percentile	Level
Long-term recall	89	5th	Mild/Moderate
Long-term storage	92	3rd	Mild/Moderate
Cumulative LTR	68	7th	Mild
1/2-h delayed recall	8	2nd	Mild/Moderate

Cued recall = 11; MC recall = 12; Intrusive errors = 1

Rey–Osterreith Complex Figure (Meyers and Meyers)

	Raw	Percentile	Level
Copy	32	16th	Borderline
Immediate recall	13	7th	Mild
1/2-h delayed recall	13.5	10th	Mild
Percent loss	0		WNL

Other
PASAT (Levin Version)

	Raw	Percentile	Level
Trial 1	35	3rd	Mild/moderate
Trial 2	22	2nd	Moderate
Trial 3	Discontinued — too frustrated		

STROOP

	Raw	Percentile	Level
Interference	10	8th	Above average

Wisconsin Card Sorting Test

	Raw	T	Percentile	Heaton Level
PSV responses	18	41	16th	Borderline/below average

Categories 6/6; Lost sets 0.

Symptom Validity Test

TOMM	50/50	Nonsignificant
Abbreviated Hiscock Forced Choice Digit Recognition Test	36/36	Nonsignificant

Summary of Results: Ms. D.R.'s performance on the Halstead Impairment Index and the Average Impairment Rating were both in the below average/borderline range[51] when age, gender, and education factors were considered. Her performance on the General Neuropsychological Deficit Scale was in the mildly impaired range. A review of the data reveals that her most significant neurocognitive deficits were in the areas of executive or frontal functioning, verbal and nonverbal short-term memory, and tactile sensitivity bilaterally. Note the discrepancy between her ability to recall spontaneously nonverbal information (WNL) compared with her ability to recall verbal information (<1st to the 2nd percentile). This is a good example of the type of lateralized dysfunction discussed previously which can occur with CO poisoning. She experienced a clear temporal onset of cognitive dysfunction just following the exposure. The cognitive dysfunction that she was experiencing could not be accounted for by other factors such as pain, fatigue, medications, or depression. Malingering/symptom magnification were taken into account and ruled out as a possible explanation for her deficits. The type of deficits that she demonstrated on testing were consistent with CO poisoning, which was ultimately determined to be the cause of her residual cognitive deficits. She was evaluated at 22 months postexposure and these deficits were considered to be permanent. A variety of treatment recommendations were made following the guidelines presented above.

DSM-IV Diagnoses:

Axis I	292.82	Carbon Monoxide Induced Persisting Dementia
	292.84	Carbon Monoxide Induced Mood Disorder with Depression and Irritability
	300.29	Specific Phobia
Axis II	V71.09	No Diagnosis or Condition on Axis II
Axis III		Status Post Carbon Monoxide Exposure
Axis IV		Psychosocial stressors: Permanent residual physical and cognitive deficits. Permanent residual Organic Mood and Personality changes.
Axis V		Current GAF = 60
		Highest GAF past year = 60

GAF = Global Assessment of Functioning.

REFERENCES

1. White, S.R., Treatment of carbon monoxide poisoning, in *Carbon Monoxide*, Penney, D.G., Ed., CRC Press, Boca Raton, FL, 1996.
2. Millington, J.T. and Schiltz, K.L., A multidisciplinary approach to the evaluation and treatment of carbon monoxide poisoning, paper presented at the 17th Annual Natl. Acad. of Neuropsych. Conference, Nov., 1997.
3. Prigatano, G.P. and Redner, J.E., Uses and abuses of neuropsychological testing in behavioral neurology, *Behav. Neurol.*, 2, 219–231, 1993.
4. Lezak, M.D., Neuropsychological assessment in behavioral toxicology — developing techniques and interpretive issues, *Scand. J. Work Environ. Health*, Suppl. 1, 10, 25–29, 1984.
5. Messier, L.D. and Meyers, R.A.M., A neuropsychological screening battery for emergency assessment of carbon monoxide poisoned patients, *J. Clin. Psychol.*, 47, 675–684, 1991.
6. Schiltz, K.L., Millington, J.T., Wilmeth, J.B., and Satz, P., The role of neuropsychological testing in the hyperbaric management of the carbon monoxide patient, *J. Clin. Exp. Neuropsychol.*, 15(1), 98–99, 1993.
7. Lezak, M.D., *Neuropsychological Assessment*, 3rd ed., Oxford University Press, New York, 1995.
8. Hartman, D., *Neuropsychological Toxicology, Identification and Assessment of Human Neurotoxic Syndromes*, 2nd ed., Plenum Press, New York, 1995.
9. Myers, R.M, Snyder, S.K., and Emhoff, T.A., Subacute sequelae of carbon monoxide poisoning, *Ann. Emerg. Med.*, 14, 1163–1167, 1985.
10. Bowler, R.M. and Hartney, C.J., *Toxic Symptom Check List: Manual*, Department of Psychology, San Francisco State University, San Francisco, CA, August 1998.
11. Cullen, M.R., The worker with multiple chemical sensitivity: an overview, in *Workers with Multiple Chemical Sensitivities*, M.R. Cullen, Ed., Hanley and Belfus, Philadelphia, 1987, 655–662.
12. Donnay, A., On the recognition of multiple chemical sensitivity in medical literature and government policy, *Int. J. Toxicol.*, 18, 383–392, 1999.
13. Padula, W.V. and Argyris, S., Post-traumatic vision syndrome and visual midline shift syndrome, *NeuroRehabilitation*, 6, 165–171, 1996.

14. Zost, M.G., Diagnosis and management of visual dysfunction in cerebral injury, in *Diagnosis and Management of Special Populations*, Maino, D.M. Editor, Mosby Yearbook, St. Louis, 1995, chap. 4.

15. Ginsberg, M.D., Carbon monoxide intervention: clinical factors, neuropathology, and mechanism of injury, *Clin. Toxicol.*, 23, 281–288, 1985.

16. Reitan, R.M. and Wolfson, D., *The Halstead-Reitan Neuropsychological Test Battery: Theory and Clinical Interpretation*, 2nd ed., Neuropsychology Press, Tucson, AZ, 1993.

17. Hiramatsu, M., Kamayama, T., and Nabishima, T., Carbon monoxide-induced impairment of learning, memory and neurological function, in *Carbon Monoxide*, Penney, D.G., Ed., CRC Press, Boca Raton, FL, 1996.

18. World Health Organization, *Carbon Monoxide Environmental Health Criteria 13*, Published under the joint sponsorship of the United Nations Environment Program and the World Health Organization, Geneva, 1979.

19. Ryan, C., Memory disturbances following chronic, low level carbon monoxide exposure, *Arch. Clin. Neuropsychol.*, 5, 59–67, 1990.

20. Reitan, R.M. and Wolfson, D., *The Halstead-Reitan Neuropsychological Test Battery: Theory and Clinical Interpretation*, Neuropsychology Press, Tucson, AZ, 1985.

21. Christensen, A., *Lurias Neuropsychological Investigation*, Spectrum, New York, 1975.

22. Golden, C.J., Hammeke, T.A., Purisch, A.D., *The Luria-Nebraska Neuropsychological Battery: Manuals*, Western Psychological Press, Los Angeles, 1980.

23. Ryan, C.M., Morrow, L.A., Bromet, E.J., and Parkinson, D.K., The assessment of neuropsychological dysfunction in the workplace: normative data from the Pittsburgh Occupational Exposures Test Battery, *J. Clin. Exp. Psychol.*, 9, 665–679, 1987.

24. Bowler, R.M., Mergler, D., and Rauch, S.S., Affective and personality disturbances among female former microelectronics workers, *J. Clin. Psychol.*, 47, 41–52, 1991.

25. Heaton, R.K., Grant, I., and Mathews, C.G., *Comprehensive Norms for an Expanded Halstead–Reitan Battery*, Psychological Assessment Resources, Odessa, FL, 1991.

26. Heaton, R.K., *Comprehensive Norms for an Expanded Halstead–Reitan Battery: A Supplement for the Wechsler Adult Intelligence Scale — Revised*, Psychological Assessment Resources, Odessa, FL, 1992.

27. Baker, E.L., Letz, R.E., Fidler, A.T., Shalat, S., Plantamura, D., and Lyndon, M., A computer-based neurobehavioral evaluation system for occupational and environmental epidemiology: methodology and validation studies, *Neurobehav. Toxicol. Teratol.*, 7, 369–378, 1985.

28. Branconnier, R.J., Dementia in human populations exposed to neurotoxic agents: a portable microcomputerized dementia screening battery, *Neurobehav. Toxicol. Teratol.*, 7, 379–386, 1985.

29. Eckerman, D.A., Carroll, J.B., Force, C.M., Gullian, M., Lansman, E.R., Long, E.R., Waller, M.B., and Wallsten, T.S., An approach of brief testing for neurotoxicity, *Neurobehav. Toxicol. Teratol.*, 7, 387–394, 1985.

30. Gass, C.S., MMPI-2 interpretation and closed head injury: a correction factor, *J. Consulting Clin. Psychol.*, 3, 27–31, 1991.

31. Cripe, L.l., The MMPI in a neuropsychological assessment: a murky measure, *Appl. Neuropsychol.*, 3/4, 97–103, 1996.

32. Reitan, R.M. and Wolfson, D., *Detection of Malingering and Invalid Test Scores*, Neuropsychology Press, Tucson, AZ, 1997.

33. Morrow, L.A., Kamis, H., and Hodgsson, M.J., Psychiatric symptomatology in persons with organic solvent exposure, *J. Consulting Clin. Psychol.*, 61, 171–174, 1993.

34. Reitan, R.M. and Wolfson, D., Emotional disturbances and their interaction with neuropsychological deficits, *Neuropsychol. Rev.*, 7, 3–19, 1997.

35. Rogers. R., *Clinical Assessment of Malingering and Deception*, The Guilford Press, New York, 1997.

36. Reynolds, C.R., *Detection of Malingering during Head Injury Litigation*, Plenum Press, New York, 1998.

37. Hiscock, M. and Hiscock, C.K., Refining the forced-choice method for the detection of malingering, *J. Clin. Exp. Neuropsychol.*, 11, 967–974, 1989.

38. Binder, L.M., Malingering following minor head trauma, *Clin. Neuropsychol.*, 4, 25–36, 1990.

39. Binder, L.M. and Willis, S.C., Assessment of motivation after financially compensable minor head trauma. Psychological Assessment: A, *J. Consulting Clin. Psychol.*, 3, 175–181, 1991.

40. Tombough, T.N., *Test of Memory Malingering (TOMM) Manual*, Multi-Health Systems, New York, 1996.

41. Guilmette, T.J., Hart, K.J., Giuliano, A.J., and Leininger, B.E., Detecting simulated memory impairment: comparison of the Rey fifteen-item test and the Hiscock forced choice procedure, *Clin. Neuropsychol.*, 8, 283–294, 1994.

42. Frederick, R.I., Sarfaty, S.O., Johnston, J.D., and Powel, J., Validation of a detector of response bias on a forced-choice test of non-verbal ability, *Neuropsychology*, 8, 118–125, 1994.

43. *Diagnostic and Statistical Manual of Mental Disorders*, 4th. ed. (DSM-IV), American Psychiatric Association, Washington, D.C., 1994.

44. Shapiro, F., *Eye Movement Desensitization and Reprocessing: Basic Principles, Protocols, and Procedures*, The Guilford Press, New York, 1995.

45. Gallo, F.P., *Energy Psychology: Explorations of the Interface of Energy, Cognition, Behavior, and Health*, CRC Press, Boca Raton, FL, 1999.

46. Corrigan, P.W. and Yudofsky, S.C., *Cognitive Rehabilitation for Neuropsychiatric Disorders*, American Psychiatric Press, Washington, D.C., 1996.

47. Raymond, R.J., Bewick, K.C., Malia, K.B., and Bennett, T.L., A comprehensive approach to memory rehabilitation following brain injury, *J. Cogn. Rehabil.*, Nov/Dec., 18–23, 1996.

48. Parenté, R. and Herrmann, D., Retraining memory strategies, in *Retraining Cognition: Techniques and Application*, Aspen Publishers, Gaithersburg, MD, 1996, 105–114.

49. Hutchinson, J. and Marquardt, T.P., Functional treatment approaches to memory impairment following brain injury, in *Memory and Language Impairment in Children and Adults*, Gillman, R.B., Ed., Aspen, Gaithersburg, MD, 1998.

50. Rossetti, Y., Rode, G., Pisella, L., and Farne, A., Prism adaptation to a rightward optical deviation rehabilitation left hemispatial neglect, *Nature*, 395, 166–169, 1998.

51. Jarvis, P.E. and Barth, J.T., *The Halstead-Reitan Neuropsychological Test Battery: A Guide to Interpretation and Clinical Application*, Psychological Assessment Resources, Odessa, FL, 1994.

21 Pediatric Carbon Monoxide Poisoning

Suzanne R. White

CONTENTS

21.1 EPIDEMIOLOGY OF CARBON MONOXIDE POISONING IN CHILDREN

Carbon monoxide (CO) poisoning is the leading cause of poisoning death in the United States. On average, 5600 death records annually in this country list CO either as the underlying or contributing cause of death.[1] Once burns, fires, suicides, and homicides are excluded, approximately 1150 deaths per year unintentionally related to CO inhalation remain. To put this into perspective, 4000 unintentional deaths per year occur on average from all other poisonings combined. Therefore, no other single, unintentional toxic exposure is so frequently fatal.

In the United States, adolescents are at greatest risk for death from CO poisoning, with the highest death rate found in 15 to 24 year olds.[1] In fact, a recent 15-year analysis of adolescent poisoning deaths in the United States attributed over

0-8493-2065-8/00/$0.00+$.50
© 2000 by CRC Press LLC

200 deaths per year in the 10 to 19 year age group to CO.[2] Two thirds of these deaths were suicides. The death rate from CO poisoning in children aged 0 to 14 years is somewhat lower (0.1 to 0.17/100,000 compared with 0.6/100,000 for all other age groups).[1] This would suggest that 54 children per year die from unintentional CO inhalation in this youngest age category. Outside the United States, a higher death rate for CO poisoning both in children aged 0 to 4 years (11 to 15/100,000) and in adolescents (14.7/100,000) has been noted, with reported fatal to nonfatal exposure ratios of 1:3.6 in 0 to 4 year olds and 1:19.5 in 10 to 14 year olds.[3]

Across all age groups, roughly 50% of fatalities from CO are unintentional, and of those 50% are related to motor vehicle exhaust inhalation.[1] In adolescents, motor vehicle exhaust accounted for 65.6 and 84.8% of CO-related accidental deaths and suicides, respectively.[2]

Nonfatal CO exposures have historically been implicated in 10,000 poisonings per year in the United States that are significant enough for the victims to seek medical attention or lose 1 day of normal activity. More recent estimates extrapolate this number to an excess of 40,000 emergency department visits annually.[4] How many of these visits are by children is not known. The incidence of acute CO poisoning was 76.5 to 109.4/100,000 in Spain during 1993, with 50% of those affected less than age 17 years.[5] In Korea, the incidence of CO intoxication in children younger than 15 years old is 7.8%, with the highest incidence in the 12 to 14 year old group (28%), as compared with the general population (36%). This extremely high prevalence is related to the use of anthracite coal briquettes for domestic fuel.[6]

21.2   SOURCES OF CARBON MONOXIDE EXPOSURE IN CHILDREN

CO is colorless, odorless, tasteless, nonirritating, and slightly less dense than air. Because of these properties, exposure may go undetected until severe toxicity occurs. Natural sources of CO include volcanic gases, marsh gases, and coal mine emissions. It is also ubiquitous wherever there is incomplete combustion of carbonaceous material, and, after carbon dioxide, CO is the most abundant air pollutant.[7] It is estimated that 90% of CO in the environment is generated by transportation-related combustion of fossil fuels. For urban children, therefore, air pollution is a potentially significant source of exposure, although this has not been well documented. One cohort of children living in an air-polluted area of Austria were assessed yearly for COHb levels, pulmonary function, and immune function testing over a 5-year period. The concentration of COHb in the 10 year olds studied was 3% on average (no control group) and declined to a significantly lower level within 1 week of relocation to a cleaner air environment.[8] The effects of passive cigarette smoke exposure were not accounted for in this study.

Tobacco smoke is the most common indoor air pollutant to which children are exposed.[9] While it is suggested that exposed children have diminished lung capacity and higher incidence of respiratory tract reactivity and infection,[9] further research is needed to determine to what extent CO in tobacco smoke contributes to these or other acute and chronic illnesses.

One particularly disturbing trend in pediatrics is the reporting of severe injury or death resulting from exposure to exhaust while riding in the back of motor vehicles. Hampson and Norkool[10] reported on 20 children accidentally poisoned by this method in pickup trucks. In this case series, severe poisoning, as evidenced by loss of consciousness in 75%, death, and permanent neurological sequelae were common. In this series, 17 children were riding under a rigid closed canopy on the rear of the truck while three were beneath a tarpaulin. Others report similar tragedies inside air-conditioned station wagons,[11] fully enclosed transit vans,[12] school buses,[13] and camper-trucks.[14] Delays in diagnosis in these situations were common since initial symptoms were consistent with motion sickness and many parents believed their children were sleeping, while in fact they are unconscious. Negative pressure is created in moving vehicles with vertical tailgates or doors, and opening the rear window can result in the drawing of exhaust into the rearmost compartments. Specific configurations of vehicle exhaust systems (exits beneath the rear bumper) or those that are worn or have improperly replaced mufflers have been most frequently implicated.[10] Holes in the car body, floorboards, or leaks around windows or doors may also allow fumes to enter the passenger compartment. Finally, a disproportionate number of children as compared with adults are poisoned by CO during winter storms while inside vehicles with snow-obstructed exhaust pipes.[15] Many of these injuries occur in parked cars in which the motor was running to provide heat.

As with adults, exposure to CO in children commonly occurs from defective or alternative forms of heating or cooking systems.[6,16-19] Documented exposures of this type in children date back to the 18th century in Europe, when infants were "pacified" by holding them over a fire.[7] Beyond faulty furnaces, sources of exposure reported in children include kerosene heaters,[20] failure of mobile gas containers,[3] methane/butane/propane-burning heating units,[21,22] the indoor use of charcoal briquettes,[6,23-25] the use of stoves as heating sources,[18] and faulty fireplaces.

CO poisoning may occur outside the home or vehicle. Recreationally, children may be exposed at ice rinks,[26] while boating[27] or water skiing,[28] in camping tents, and as spectators at sporting events.[29] A high index of suspicion must be maintained when children are brought from these settings with complaints of flu-like illness, headache, or syncope.

Children may be exposed to CO, hypoxia, and other toxic insults following smoke inhalation. This has been extensively reviewed by Parish.[30] Zikria et al.[31] found COHb levels of over 50% in 24% of children dying within the first 12 h postburn. In addition, COHb levels were elevated in 39% of those dying within 12 h who did not have burns, making CO the primary cause of death in these fire victims.

Dihalogenated methanes such as methylene bromide, chloride, or iodide are widely used as industrial solvents, but are also found in household products such as paint strippers and varnish strippers. These agents are converted by hepatic P-450 enzymes to CO. Children coming into direct contact with these solvents or inhaling the fumes, either intentionally or unintentionally, may develop CO toxicity. While pediatric methylene chloride exposures and resultant CO toxicity are not specifically tracked by the American Association for Poison Control Centers, it is known that household products account for most childhood accidental poisonings.[32] Therefore, it is likely that this exposure will be encountered by health-care professionals who risk underdiagnosis of resultant CO toxicity if unaware of the metabolic fate of dihalomethanes.

As implied by the discussion above, CO has primarily been regarded as a dangerous toxin to which children are exposed through exogenous sources. It is now known that CO is produced endogenously, primarily through the degradation of hemeproteins by heme oxygenase to biliverdin IXα and CO. It has recently been discovered that these products of heme degradation serve as active biochemical and physiological regulators. This process is increased with hemolytic, sideroblastic, or sickle cell anemias, hematomas, exertional hemoglobinuria, and thalessemias. Other minor sources of endogenous CO include the auto-oxidation of phenols, flavinoids, photo-oxidation of organic compounds, and lipid peroxidation of membranes.[33] CO is involved in neuronal signaling and produces presynaptic enhancement of long-term potentiation, a mechanism thought to be important in learning. CO most likely acts in this capacity as a neural messenger in the hippocampus, cerebellum, and olfactory areas.[34] Like nitric oxide (NO), CO binds to soluble, heme-containing guanylyl cyclase and elevates cyclic guanosine monophosphate (cGMP) levels. This interaction is essential to normal cerebral blood flow autoregulation. Additionally, CO and bilirubin production is increased in response to hypoxia, hemodynamic stress, and endotoxic shock and may play a protective role in these disease states.[33] The activation of soluble guanylyl cyclase *in vitro* has also been associated with the inhibition of platelet aggregation[35] and mediates the production of cytokines by pulmonary macrophages. For example, tumor necrosis factor-alpha (TNFα) or interleukin-1 (IL-1) exposure substantially increases the production of CO by human type II pneumocytes, which corresponds to decreased surfactant synthesis.[36] Further investigation is needed to clarify the exact role of CO in normal pulmonary physiology.

21.3   MECHANISMS OF CARBON MONOXIDE TOXICITY

Hypoxic ischemia plays a significant role in the neurotoxicity of CO, which binds slowly to hemoglobin, but with extremely high affinity (240 times that of oxygen). Oxygen-binding sites are occupied by CO at very low partial pressures of the gas, decreasing the oxygen-carrying capacity of the blood and subsequently decreasing the usual facilitation phenomenon for further unloading of oxygen at the tissue level. The net result is an abnormally hyperbolic oxygen dissociation curve that is shifted to the left. Those tissues most susceptible to the hypoxic effects of CO are those that are the most metabolically active. Oxygen delivery may be further impaired through the alteration of erythrocyte diphosphoglycerate concentration.[37] In adults, COHb half-life is dependent upon the concentration of inspired oxygen, and is most commonly reported to be approximately 4.5 h on room air, 90 min on normobaric 100% oxygen, and 20 min with oxygen applied at hyperbaric concentrations. It should be noted that reported half-lives are extremely varied in the literature. In children, the half-life of COHb has not been well studied, but is reported by one author to be 44 min on 100% oxygen at normobaric pressure, based on measurements performed on 26 school-aged children.[38] The half-life of COHb in the fetus is approximately 7 h.[39] In addition to hypoxia, CO induces ischemia secondary to hypotension. The degree of central nervous system (CNS) damage correlates well with degree of hypotension.[40] Hypotension may be mediated through increased cGMP, resulting in vascular smooth muscle dilatation or myocardial suppression.

Additional mechanisms of toxicity have been sought following observations that: (1) COHb levels did not correlate with toxicity, (2) COHb formed by noninhalational routes did not produce the same lethal consequences as inhalational exposure to CO, and (3) that delayed neuropsychiatric sequelae were common after apparent complete recovery from the initial CO insult. Early researchers such as Haldane,[41] Drabkin et al.,[42] and Goldbaum[43-45] suggested intracellular uptake of the gas as a possible mechanism for toxicity. In competition with oxygen, CO will bind iron- or copper-containing proteins such as myoglobin, mixed-function oxidases, and cytochrome c oxidase *in vitro*. The binding to the cytochrome c oxidase (a,a3) has been proposed as the mechanism for intracellular CO toxicity and has been demonstrated in animals.[46] It has also been demonstrated that during recovery, the ultimate restoration of mitochondrial function lags behind clearance of COHb.[47] However, the Warburg constant for cytochrome oxidase is unfavorable for CO binding relative to the other hemeproteins.[48] Furthermore, only reduced cytochrome a,a3 binds CO. It is likely, then, that other hemeproteins act as "CO buffers," thus preventing significant binding to cytochrome c oxidase at COHb levels of less than 50%. At high levels of COHb, depletion of high-energy stores and intracellular neuronal acidosis occurs, which may favor CO–cytochrome binding. On the other hand, *in vivo* data from Smithline et al.[49] supports hemoglobin binding with impaired oxygen delivery, rather than mitochondrial poisoning as the etiology of the metabolic acidosis in CO poisoning. In this model, even at extremely high COHb levels, dogs were able to extract and utilize oxygen fully, indicating a lack of mitochondrial effect. The role of iron as a promoter or attenuator of CO toxicity is not clear. Iron deficiency, a condition to which children are extremely susceptible during various stages of development, clearly results in lowered hemoglobin levels. This state also causes lowered concentrations of cytochrome and myoglobin in the animal model.[50] These combined effects could potentially predispose children to CO toxicity. Ironically, neuronal tissues high in iron content, such as the basal ganglia, seem particularly vulnerable to the effects of CO.

Myoglobin binding may play a role in CO-mediated toxic effects. Like hemoglobin, myoglobin is a hemeprotein with similar three-dimensional configuration that can bind CO reversibly. Myoglobin binds CO more slowly and with greater affinity than does hemoglobin *in vivo*. Normally, myoglobin is an O_2 carrier protein that facilitates oxygen diffusion into skeletal or cardiac muscle cells and serves to place oxygen stores in close proximity to mitochondria. The clinical significance of high carboxymyoglobin levels is not yet clear but may in part explain the cardiac and skeletal muscle toxicity seen after CO poisoning. These effects likely come into play at COHb levels of 20 to 40%.[33,51]

The mechanism for delayed effects of CO poisoning have been a medical conundrum. An increasing body of research suggests that brain ischemia/reperfusion injury, lipid peroxidation, vascular oxidative stress, neuronal excitotoxicity, and apoptosis play significant roles. Reactive oxygen species (ROS) are generated during CO poisoning and appear to be involved in the development of delayed toxicity in animals. Thom[52-54] has demonstrated CO-induced perivascular oxidative injury, elaboration of endothelial intercellular adhesion molecules (ICAMS), activation of polymorphonuclear (PMN)

CD18 receptors and beta-2 integrins, PMN degranulation, PMN protease conversion of xanthine dehydrogenase to xanthine oxidase, superoxide formation, prolonged lipid peroxidation reactions, vascular injury, and neuronal death. Even low-level CO can produce vascular oxidative stress as evidenced by platelet-mediated deposition of peroxynitrate, a highly oxidative substance.[55] ROS generation can be attributed to several other sources including mitochondria and cycloxygenase. ROS production increases notably during CO hypoxia, with the highest oxidative stress occurring in the most vulnerable brain regions.[56] This stress may result from lower antioxidant capacity or higher tissue concentrations of iron. In mitochondria, decreased ratios of reduced to oxidized glutathione are seen following CO poisoning, and may reflect decreased ability to detoxify ROS.[57]

Neuronal excitation may also play a role in the development of delayed toxicity following CO poisoning. These effects have been extensively reviewed by Piantadosi[58] and are paraphrased here. Excitatory amino acids (EAA), such as glutamate, accumulate in synaptic clefts during neuronal depolarization due to both excessive presynaptic release and failure in ATP-dependent reuptake mechanisms.[59] Interstitial glutamate concentration increases in the hippocampus during and after CO exposure.[57] Postsynaptic binding to at least three glutamate receptors including N-methyl-o-aspartate (NMDA), kainic acid (KA), and (S)-alpha-amino-3-hydroxy-5-methylisoxazole-4-propionate (AMPA) occurs with a secondarily increased influx of calcium into postsynaptic neurons. This hypercalcemia is associated with neuronal death. ROS production follows increased EAA release as well. NMDA receptor antagonism attenuates delayed neuronal degeneration in the hippocampus after CO poisoning.[60] Another modulating factor may be local NO production; however, the exact role of NO as an attenuator of CO toxicity remains unclear.[61]

Catecholamine excess may also be detrimental following CO exposure. EAA release causes excessive surges of norepinephrine and dopamine release with synaptic accumulation. This effect appears to be related to NO production, as both events can be prevented by NOS inhibition. Therefore, EAA release, catecholamine release, and NO production are closely regulated *in vivo* and can be influenced by CO hypoxia. Furthermore, auto-oxidation or oxidative deamination of catecholamines occurs during ischemia and reperfusion by type B monoamine oxidase. ROS production after CO exposure can be inhibited by partially blocking type B monoamine oxidase, located predominantly in glia.[62] Gliosis, a known neuronal response to injury, and a condition found in Alzheimer's disease, may play a role in the delayed toxicity seen after CO hypoxia.

ROS are capable of triggering programmed cell death or apoptosis. Apoptotic cell death requires activation and/or expression of specific cellular processes, some of which may act through oxidant pathways. In animals, CO-induced neuronal loss was slight at Day 3, increased at Day 7, and persistent at Day 21 following exposure. Neuronal apoptosis was observed at histopathological examination in this model.[63]

As eloquently summarized by Piantadosi,[58] "impaired mitochondrial energy provision in CO hypoxia/ischemia leads to neuronal depolarization, EAA and catecholamine release, and failure of re-uptake until energy metabolism is restored during reoxygenation. These processes, normally modulated by NO production, could contribute to degeneration of neurons in vulnerable regions, possibly by enhancing mitochondrial ROS generation which can initiate apoptosis."

21.4 CLINICAL PRESENTATION

Symptoms of CO poisoning are notoriously vague, and may result in misdiagnosis.[16,64] In adults, the poisoning goes undetected in 30% of cases presenting to the emergency department.[65-67] In children, Baker[17] observed that of those with flu-like illness screened for CO poisoning in the emergency department, 28% had COHb levels greater than 5%.[17] Other investigators[68] found 22 out of 1200 children screened had COHb levels above 5% (mean 11%, range 5 to 33%). In none of these cases had the diagnosis of CO poisoning been entertained by the pediatric resident. Clearly, a high index of suspicion must be maintained by all health-care providers who see children.

Signs and symptoms of CO poisoning in adults classically result from energy crises in those tissues most metabolically active, such as the brain and myocardium. Neurological signs and symptoms predominate and include lethargy, headache, nausea, vomiting, coma, seizures, ataxia, nystagmus, increased deep tendor reflexes (DTRs), pyramidal tract signs, decerebrate posturing, peripheral neuropathy (related to compression), visual complaints, vestibular and auditory dysfunction, tinnitus, and subtle personality or cognitive changes. Cardiovascular manifestations may include tachycardia, electrocardiographic (ECG) changes, alterations in blood pressure, dysrhythmias, and, in severe, end-stage toxicity, bradycardia or agonal cardiac rhythms. Other less common complications include pulmonary edema, rhabdomyolysis, renal failure secondary to myoglobinuria, excessive sweating or fever, and bullae, related to sweat gland necrosis. A pink to cherry color to the blood and skin is described, but rarely seen in survivors.

Generalizations have been made regarding the range of toxicity seen in adults. For example, some authors propose that, with COHb levels less than 10%, patients will be asymptomatic; at 10 to 20% headache develops; at 20 to 40% weakness, dizziness, nausea vomiting, syncope, and visual complaints occur; at 40 to 60% symptoms progress to cardiovascular collapse, coma, seizures, and Cheyne–Stokes respirations; and above 60% apnea, bradycardia, hypotension, and death ensue. Similar correlations between CO levels and symptoms have not been made in the pediatric population. Illustrative of the reasons such correlations are not useful is one interesting observation of a family poisoned by CO from automobile exhaust. Two of the three children in the vehicle (aged 27 months and 7 years) were asymptomatic on presentation, despite high COHb levels of 34.7 and 33.6%. A third child, who had been lying on the floor in the back of the vehicle was initially unresponsive, with a COHb level of 35%. This was in contrast to the adults in the vehicle who exhibited symptoms of headache and dizziness, despite lower levels of 18.4 and 16.1%. This description highlights the dissimilarity of symptoms manifested among children despite similar COHb levels (as well as dissimilarity in levels between children and adults riding in the same vehicle).[69] It is the opinion of this author, along with others,[70] that COHb levels do not correlate well with toxicity. Other clinical parameters may be better predictors of outcome, such as neurological status upon presentation,[71,72] duration of exposure, presence of cerebral edema, syncope, elevated lactate or glucose levels, based on clinical experience with adult patients.

Peculiarities in pediatric presentation may exist. Some authors suggest that children are more "sensitive" to CO, in that symptoms appear at lower levels relative to adults.[18] This is proposed to result from the more rapid uptake of CO, higher metabolic needs, and smaller blood volume. Observations regarding pediatric presentation are limited, but are reviewed here. Crocker and Walker[73] reported on 16 pediatric patients with COHb levels of 17 to 44%. All patients were symptomatic with nausea at levels of >15%; 75% experienced vomiting, 93% headache, and 21% had visual complaints. One patient presented with a major motor seizure. An unusually high incidence of syncope (57%) and lethargy (69%) were noted. In fact, all patients with levels >24.3% had syncope.

Kim and Coe[6] described 107 pediatric patients with CO poisoning. The most common presenting symptoms were vomiting, seizures, headache, irritability, speech disturbance, blindness, and dizziness. Carrasco Gomez et al.[74] noted the following presenting symptoms in 17 children: dizziness (53%), headache (53%), nausea and vomiting (35%), and syncope or coma (17%). The average COHb level in this series was 28.6% ± 9.0. Hampson noted syncope in 15 of 20 acutely CO-poisoned children.[10] Binder and Roberts[75] studied 17 children with CO poisoning and found the most common presenting symptoms, in decreasing order of frequency, were nausea and vomiting, headache, unconsciousness, dizziness, confusion/decreased level of consciousness, seizures, fever, blurred vision, and bilateral sensory hearing loss. One patient was dead on arrival. A limitation of this report is that COHb levels were only obtained on eight patients.

Lattere et al.[76] reported the largest review of pediatric CO exposures to date; 233 Italian children with acute CO poisoning were studied. The most common symptoms in the 0 to 4 year old group were vomiting, drowsiness, headache, pallor, syncope, hypotonia, and coma. Children in the 5 to 13 age group had a higher incidence of headache and syncope. Other symptoms noted were dysequilibrium, cyanosis, agitation, tachycardia, abdominal pain, nausea, vertigo, incontinence, seizures, and tremor (Table 21.1). Similar constellations of presenting symptoms including altered mental status, nausea, vomiting, headache, and syncope were noted in other case series as well.[64,77] In summary, presentation in children seems commonly to involve nausea, vomiting, and headache, and is associated with a relatively higher incidence of syncope, altered mental status, and seizures than in adults.

Little is known about CO poisoning in infants. Interestingly, case reports of asymptomatic infants with relatively high levels of COHb exist.[69,78] This paradoxical lack of symptoms in infants is illustrated by one series of 14 patients less than 2 years of age, in which five were asymptomatic at COHb levels up to 34%. In another series of pediatric CO exposures, the youngest patient, 6 months of age, presented with vomiting alone, and the next youngest, 2 years old, presented with headache and dizziness (COHb level 18.9 and 32.8%, respectively). Apparent age-dependent symptom differences on presentation were also recently noted by Lattere et al.,[76] who observed that the younger children presented atypically with relatively nonspecific symptoms, most commonly, nausea, vomiting, and lethargy. A possible explanation for this apparent age-related disparity between COHb levels and symptoms is that subtle findings such as irritability or feeding difficulty may easily be missed in infants or very young children. In contrast to the above observations, however, is the report of a 13-week-old infant brought from a house fire exhibiting intermittent opisthotonus, hypotonia, severe metabolic acidosis, and a surprisingly low COHb level of 2%,

TABLE 21.1
Carbon Monoxide Poisoning:
Presenting Symptoms According to Age

Age Group	0–4	5–9	10–13
Number of cases	82	72	70
Vomiting	37.6	33.3	14.3
Somnolence/confusion	34.1	26.4	24.3
Headache	23.2	45.9	57.1
Pallor	22.0	22.2	14.3
Loss of consciousness	15.9	25.0	25.7
Hypotonia	14.6	6.9	10.0
Coma	12.2	12.5	15.7
Dysequilibrium	9.8	9.7	20.0
Cyanosis	9.8	4.2	4.3
Agitation	6.1	—	5.7
Tachycardia	6.1	1.4	—
Abdominal pain	2.4	5.6	1.4
Nausea	—	6.9	8.6
Vertigo	1.2	4.2	10.0
Incontinence	—	—	7.1
Convulsions	1.2	—	5.7
Tremor	1.2	4.2	10.0

Source: Kim, J.K. and Coe, C.J., *Yonsei Med. J.*, 28, 266, 1987.

drawn 85 min following arrival. The infant recovered completely and was doing well at 6-month follow-up, while the mother (COHb level 26%) died. In this case, the infant's presentation was quite dramatic, yet the outcome was good (the low COHb level at presentation introduces the possibility of laboratory error, however).[79]

21.5 FINDINGS AT PHYSICAL EXAMINATION

Acute neurological signs in children presenting with CO poisoning are common, and are seen in up to 69%.[78] The most common signs of exposure in one large series were altered mental status, increased DTRs, positive Babinski sign, neck stiffness, ankle clonus, nystagmus, ataxia, amnesia, hemiplegia, and tremor.[6] Signs on presentation noted by Binder and Roberts[75] included ataxia, extrapyramidal signs, posturing, neck stiffness, irregular respiratory pattern, conjunctival irritation, and excessive sweating. Similar presenting signs were noted in a large series of smoke inhalation victims with CO poisoning who presented with acute lethargy, coma, syncope, ataxia or tremors, irritability or confusion, bradypnea or apnea, vital sign derangements, and metabolic acidosis.[77]

As indicated above, most signs presented following CO exposure are neurological in nature. Cardiac signs, although not uncommon in adults, are rarely described in children. Tachycardia may be noted[6,11,75] and, rarely, bradycardia or shock has been described.[77] Hypertension has been reported as has hypotension, which was uniformly associated with death or persistent neurological deficit.[77] A case of severe, reversible cardiac dysfunction, diagnosed by radionuclide imaging and ECG was described in an adolescent following CO exposure. The patient responded to inotropes and intra-aortic balloon counterpulsation, and fully recovered without neurological deficit.[80] Myocardial enzymatic changes without ECG changes has recently been noted in adults with CO poisoning,[81] and has also been noted in children.[79] It is not clear whether this "occult" evidence of myocardial injury is clinically significant. Finally, as seen in adults, myglobinuric renal failure, compartment syndrome, and pulmonary edema have been described in pediatric cases of CO poisoning.[82]

Of special interest in the pediatric population are the ophthalmologic findings surrounding CO exposure. Blindness, papilledema, and retinal hemorrhage have been described.[83-86] The finding of retinal hemorrhage is particularly worrisome in children, since it most commonly stems from child abuse, and may pose a serious diagnostic dilemma if discovered in a patient with an elevated COHb level or history of exposure. In general, retinal hemorrhage related to CO exposure is reported with levels greater than 10% or exposures of more than 12 h duration.[86] Overall, this is a very rare condition and it is prudent to consider other more likely inciting events such as shaking, CPR-related chest compressions, or sudden increases in intrathoracic pressure related to trauma. Of note, retinal hemorrhages from CO have been noted to have white centers, and are associated with arterial and venous engorgement like those seen with altitude hypoxemia.[85,86]

21.6 OUTCOME

Mortality following CO exposure has been reported to be higher in children than in adults (3.7%[6] vs. 0.3 to 1.8%[87-89]). Lattere et al.[76] reported that 3 out of 233 children died, with a 1.2% mortality. Regarding long-term outcome in survivors, experience with adults has revealed that persistent neurological symptoms (PNS) occur in 5.3 to 20%.[89,90] Weaver et al.[90] noted that a high percentage (up to 52%) of patients with acute CO poisoning exhibit cognitive and subjective complaints for at least 1 year. Nearly 26% of those patients had abnormal neuropsychological evaluations.[90] PNS in adults most commonly involve the higher cortical functions such as memory impairment, personality alterations, parietal lobe dysfunction (visual agnosia, dyspraxia, dysnomia, and dysgraphia). Less commonly, blindness, deafness, seizures, increased reflexes, hemiplegia, muscular hypertonia, extrapyramidal rigidity, akinesia, muscle atrophies, skin hyperesthesia, disorientation, confusion, decreased libido, incontinence, and psychotic changes are seen.[72,75,91-93] In one series, over half of survivors of CO poisoning reported residual problems such as numbness, ageusia, memory, attention span, task frustration, blurred vision, anosmia, attention deficit, anxiety, sleep disruption, headache, concentration difficulty, depression, and family problems.[94]

Unlike other forms of hypoxic/ischemic brain injury, CO-hypoxia may be associated with delayed onset of neurological symptoms (DNS) in 2.8 to 43% of the exposed adult population.[72,89,95-97] An apparent lucid interval is peculiar to CO poisoning and averages 22 days in adults. Symptoms of DNS in adults include the full spectrum of those seen with other forms of hypoxic ischemic damage and include memory deficits, mental confusion, personality alterations, speech impairment, forgetfulness, parietal lobe dysfunction (apraxia, agnosia, dysgraphia), and motor dysfunction (parkinsonism, hemiplegia).[98,99] Risk factors for the development of DNS in adults are reported by some experts to include loss of consciousness, a "soaking injury," and cerebral edema.[72] Other investigators note that loss of consciousness is not predictive.[100]

Analogous to the adult experience, persistent neurological sequelae in children have been described. These sequelae were formerly thought to be exceedingly rare.[84] More recent observations by Crocker and Walker[73] and Kim and Coe[6] indicate that the incidence of neurological sequelae in 24.3% of children is similar to that of adults, with delayed onset of neurologic deficits in 10.3%. Outcome has also been assessed by Lattere et al.,[76] who found that 8.6% of children were left with severe symptoms, 23.2% with moderate symptoms, 57.1% with mild symptoms, and 6.4% were asymptomatic at follow-up. The characterization of these symptoms, criteria for grading of their severity, time period elapsed prior to onset, and duration of follow-up are not specified.[76]

PNS in children include temporary or permanent cortical blindness, involuntary movements, chorea, persistent memory deficits, convulsions, peripheral neuropathy, speech disturbance, hearing disturbance, hemiplegia, poor reading and writing skills, and apraxia.[6,82,84,92,101] Klees et al.[19] studied the neuropsychological and cognitive faculties of 20 children from 25 months to 15 years old in short-term follow-up and 14 children in long-term follow-up. Significant neuropsychiatric abnormalities noted at short-term follow-up 3 months postexposure are summarized in Table 21.2. Although neuropsychiatric testing was performed in these children at the time of intoxication, the data are not reported. Therefore, it is not clear whether the neurological abnormalities reported are best characterized as "persistent" or "delayed." Serious neurological abnormalities discovered 2 to 11 years postexposure in 14 other children were also reported by Klees. These findings are found in Table 21.3. Again, the time to onset of these abnormalities is not known. Meert et al.[77] noted "gross" neurological abnormalities in 8.5% of survivors of CO poisoning from smoke inhalation, with seven of those persistent at various stages of follow-up, which ranged from 2 months to 3.3 years. Four of these children presented in cardiac or respiratory arrest, however. In a large series of 111 hyperbaric oxygen (HBO)-treated children with CO poisoning,[102] 1.4% of patients had neurological sequelae. The time to onset of symptoms is not reported.

DNS reported in children include chronic headaches, memory impairment, and decreased school performance in 19% of patients (3/16). Two of these patients reported complete resolution of their symptoms within 9 months of exposure, and the remaining child was felt to have a preexisting reading disability.[73] Risk factors for DNS in this group included alteration in level of consciousness on arrival, syncope, and COHb levels > 24.5%. Binder and Roberts[75] described a 5½-year-old child who recovered from CO poisoning, but later manifested signs of severe visual-perceptual abnormalities, expressive and receptive language difficulties, and emotional problems. Delayed onset of mental retardation, epilepsy, mutism, urinary and fecal incontinence, hemiplegia, paraplegia, monoplegia, facial palsy, and psychosis are described in 10% of those patients studied by Kim and Coe[6]. The lucid interval

TABLE 21.2
Findings at Short-term Follow-up of Children 3 Months after Carbon Monoxide Intoxication

Age	25 months to 3 years	4–9 years	10–15 years
Number	6	8	6
Severity of Symptoms			
Light/medium	Standstill in development by 2 months; tantrums; irritability; anxiety; psychomotor instability and fear; amplification of negative behavior	Intellectual capacities not altered; ¾ had altered amnestic (auditory/visual-spatial memory) and instrumental aspects of cognitive development; visual-spatial perturbations; dysorthography; conceptualization intact	Difficulty with perceiving and organizing material for memorization (3/6); balance impairment (1/6); slowness and instability (2/6); anxious or depressive state
Severe	Major disturbance in developmental and adaptive capacities; spectacular loss of acquired skills		

Source: Klees, M. et al., Sci. Total Environ., 44, 165, 1985.

TABLE 21.3
Findings at Long-term Follow-up of Children Following CO Poisoning

N	Severity of Intoxication	Age at Intoxication	Symptoms
1	Severe	9 years	None
7	4 light/3 medium	6 patients > 6 years old; 1 patient 3.5 years old	Slight impairment of visual memory and concentration; all have IQ over 98
6	3 severe, 2 light, 1 medium	3 patients > 6 years old 3 patients < 6 years old	Serious disorders of spatial organization; constructive apraxia; deterioration of lexical activity, spelling and arithmetic

Source: Klees, M. et al., Sci. Total Environ., 44, 165, 1985.

prior to the development of DNS in this group was somewhat shorter than in adults and averaged 7 days. Analogous to adults, the risk of developing DNS correlated with level of consciousness on admission and a duration exposure of greater than 8 h. Of the nine patients noted by Meert et al.[77] to have PNS, three patients developed delayed neurological syndromes that included tremors, hallucinations, seizures, occipital lobe infarction, and defects in cognitive and interpersonal skills. Poor prognostic factors in these children included hypotension and altered mental status. Limitations of this study are discussed below.

Acute presentation with seizures or the development of epilepsy as a sequela of CO exposure is of particular concern, since an association between toxin-induced seizures and subsequent learning disabilities in children has been made.[103] Whether this relationship also holds true for pediatric victims of CO exposure remains to be seen.

21.7 OUTCOME IN THE UNBORN CHILD

The fetus is especially vulnerable to the toxic effects of CO. Fetal hemoglobin has a higher affinity for CO, which results in a longer half-life of fetal COHb of 7 h vs. 4.5 h. In addition, fetal COHb levels peak at 10 to 15% higher than do maternal levels.[39] The fetus normally has a lower oxygen tension, a higher rate of oxygen consumption, and an oxyhemoglobin dissociation curve which is shifted leftward and is additively affected by CO. Uterine artery constriction in response to hypoxia even further compromises the availability of oxygen for transfer to the fetus. Reduced placental oxygen transport results in a drop in fetal umbilical vein oxygen tension, with a similar drop in the uterine veins, with which umbilical vein partial pressure of oxygen is equilibrated.[104–106]

CO is a known potent fetotoxin and teratogen.[104,107–110] The effects of CO on the fetus is extensively reviewed by Penney.[111] Animal studies link low-birth-weight infant (LBW), cardiomegaly, behavioral and cognitive impairment to CO exposure. Most recently, an association with the CHARGE syndrome, a constellation of defects including coloboma of the eye, heart defects, atresia or stenosis of the choanae, retarded growth and development, CNS anomalies, genital hypoplasia, and ear anomalies and/or deafness has been proposed.[112,113] In humans, 51 case reports of accidental CO poisoning during pregnancy have been associated with stillbirth or fetal brain damage. Overall, the estimated fetal mortality is 36 to 67% following serious exposures.[114–116]

Only one prospective, long-term follow-up study with regard to acute CO poisoning during pregnancy exists. Koren et al.[117] noted that in severe maternal CO poisoning, three out of five pregnancies had adverse outcome. In contrast, 31 babies born to mothers with mild or moderate CO poisoning had normal physical and neurobehavioral development. Based on these encouraging data, mild accidental exposure is likely to result in good fetal outcome.

The fact that cigarette smoking during pregnancy has an effect on birth weight has been recognized for three decades, and CO is thought to be an important cause of this growth restriction. Reduction of cigarette consumption or exhaled CO levels less than 5 ppm were associated with a decreased risk of having a LBW.[118]

21.8 ROLE OF CARBON MONOXIDE IN SUDDEN INFANT DEATH SYNDROME

Some features of sudden infant death (SIDS), such as a peak time of death or conception occurring during the winter months and an excess of deaths among young women of low socioeconomic status who smoke, have led investigators to propose

a causal link to CO.[119,120] Theoretically, CO could exert its effect through hypoxia
or have a direct effect on the developing respiratory center, which would render the
infant susceptible to further insults such as hyperthermia or infection. COHb measurements in 50 unexplained infant deaths showed no difference between those and
hospitalized controls.[121] Among infant deaths at home in Michigan, CO toxicity was
not a significant factor.[122] Investigation carried out on 54 SIDS cases with age-
matched controls showed no greater concentration of COHb in SIDS cases compared
with controls.[123]

21.9 CHRONIC CARBON MONOXIDE EXPOSURE

As with adults, this type of exposure to CO is the least well studied (see Chapter 18).
It is reported in children only twice in the literature. In one report, 4- and 5-year-old
siblings were noted to develop hyperactivity, mood swings, headache, and lethargy
as a result of chronic CO exposure. Psychometric testing in these children was
normal.[124] Progressive neurological disability in a 2½-month-old infant was linked
to chronic CO exposure. The diagnosis in this situation was suspected when the
neurological work-up revealed bilateral basal ganglia hypodensities.[125]

21.10 DIAGNOSIS

21.10.1 Carboxyhemoglobin Levels

While the COHb level is the most accessible tool for diagnosing CO exposure, it
does not reflect the severity of poisoning or the potential for development of delayed
sequelae. Important factors known to influence toxicity such as the duration of
exposure, the peak level, any possible delay between exposure and treatment, and
oxygen therapy rendered are simply not reflected in a single blood measurement.
One important potential pitfall for the clinician to remember is that blood gas analysis
by methods other than co-oximetry will not detect COHb. It is therefore critical to
request that COHb levels be performed by co-oximetry, or oxygen saturation could
be falsely reported as normal, if calculated from the pO_2 measurement. Another
caveat is that pulse oximetry overestimates oxyhemoglobin levels by the amount of
COHb present, and therefore cannot be used.[126,127] Work by Touger et al.[128] suggests
that venous COHb levels correlate well with arterial COHb levels, differing by only
1 to 2% when COHb levels are below the 20 to 25% range. This is a clinically
valuable observation in the pediatric setting, where the technique of arterial puncture
for blood gases is invasive and often technically difficult. Interference of fetal
hemoglobin in fetal blood containing oxyhemoglobin with COHb measurement
using the Il282 and CCD2500 co-oximeters by CIBA has been noted.[129,130] This
same interference has not been seen with newer co-oximeters such as the CCD270.
The normal newborn has approximately 70% fetal and 30% adult hemoglobin. This
ratio reverses by age 3 months, and nearly all fetal hemoglobin dissipates by 1 year
of age.[131] Other interfering substances in measuring COHb are bilirubin, methylene
blue, and sulfhemoglobin. Regarding other methods of analysis, spectophotometry
is less accurate than gas chromatography in detecting low levels of COHb, those

less than 5%.[132] Breath sampling correlates well with blood COHb levels both in children and adults.[133] One potential limitation in the very young child is the requirement to breath-hold for approximately 20 s.

21.10.2 NEUROPSYCHIATRIC TESTING

The realization that COHb levels are not predictive of toxicity has prompted a search for other diagnostic and prognostic tools in adults. The CO neuropsychological (NP) screening battery is one such diagnostic modality.[134] While it is generally accepted as the most sensitive indicator of CO toxicity in adults, its predictive value in children is uncertain. Problems with NP testing have historically included a lack of baseline scoring, lack of test validity and/or reliability, tester/testee bias, and placebo effect. These issues are compounded with the difficulty in administration of such detailed testing to children. A battery of tests in children may be more appropriate than one single test.[19]

Critics of NP evaluation of children poisoned with CO cite lack of testing amenable to an emergency department setting. It may in fact be impractical to test children in the emergency department setting. Admission to the hospital for complete NP evaluation prior to discharge is more feasible. CO-poisoned patients should be longitudinally followed and testing should be repeated at several intervals over the year following exposure, ideally at time of discharge, 6 weeks, 6 months, and 12 months.[90]

21.10.3 ANCILLARY STUDIES

Other laboratory studies will not be diagnostic of CO exposure, although some may provide clues or be useful prognostically. Glucose abnormalities are well described in adults and also commonly reported in pediatric CO poisonings.[75,80,82,135,136] Hyperglycemia, in fact, is an independent risk factor for poor outcome following adult CO exposure. Hypoglycemia is much less commonly seen.[75,77] Lactate levels are commonly elevated,[136] and have been used as clinical markers of both the chronicity and the severity of poisoning. Lactate accumulation results from cellular anaerobic metabolism, and generally accounts entirely for the often severe metabolic acidosis typically seen in patients prior to resuscitation and oxygen therapy. A persistent acidosis despite adequate resuscitation and declining COHb levels should prompt an investigation for other toxins or medical conditions (i.e., concurrent cyanide poisoning in victims of smoke inhalation, sepsis, rhabdomyolysis).

Abnormalities in blood urea nitrogen (BUN) and creatinine should be sought and may reflect renal compromise related to CO exposure. Other laboratory abnormalities reported in children include leukocytosis, hemoconcentration, elevated LDH, CPK, transaminases, proteinuria, myoglobinuria, and thrombocytopenia.[75,82,135–138]

Chest radiography may reveal infiltrates or pulmonary edema in children.[82,136,139] Other findings reported include a "ground glass" appearance, perihilar haze, or peribronchial/vascular cuffing. Most chest radiographs reported in one series of victims of CO poisoning were normal (44/67).[139] The mechanism for CO-induced pulmonary edema when present is not clear. Pathogenesis may include the inhalation of other toxic gases, a direct CO effect on alveolar membranes, hypoxia, aspiration of gastric contents, neurogenic factors, decreased cardiac output, or a CO-mediated decrease in surfactant production. Intracellular cGMP increase due to endogenous generation of CO mediates cytokine-induced inhibition of surfactant synthesis of type II pneumocytes.[140]

ECG changes are well documented in adults and include ischemic ST segment depression, T-wave inversion, ventricular ectopic beats, or atrial fibrillation.[141-143] Similar findings have been noted in only two pediatric cases.[82,84]

In adults, EEG shows a slow-wave low-voltage, frontal preponderance pattern which somewhat mirrors the clinical neurological picture, but is not a sensitive marker of toxicity.[69,81,95] Other investigators report abnormal EEGs in 90% of victims of CO poisoning, showing patterns characteristic of anoxic encephalopathy.[95] Evoked potentials would theoretically be useful in young patients with CO poisoning unable to perform neuropsychological testing. There is, however, no reported pediatric experience to date, and this method has not proved to be a sensitive screening tool in adults.[144]

21.10.4 NEUROIMAGING STUDIES

In adults, pathological lesions seen on computed tomography (CT) due to CO intoxication are variable, including cerebral edema, symmetrical low-density areas in the basal ganglia, symmetrical and diffuse white matter low-density areas, ventricular dilation, and sulcal widening as late changes. The characteristic finding of bilateral symmetrical hypodensity in the basal ganglia, especially the globus pallidus, when seen, most typically occurs within 24 to 48 h of exposure.[6,84,145-148] In adults, CT findings were prognostic with regard to outcome in several studies.[149-151] With regard to pediatric CT findings, Kim and Coe[6] reports that in 16 patients, 37.5% had diffuse cerebral edema, 12.4% had bilateral low density in the globus pallidus, 6.3% had cortical atrophy with moderate ventricular dilatation, and 43.8% were normal. Also reported on a CT done 48 h after severe CO exposure in a 2½-year-old child, was cerebral edema, acute hydrocephalus, and bilateral basal ganglia infarcts.[152] The differential diagnosis for radiographic basal ganglia lesions in children include methanol, cyanide, or hydrogen sulfide toxicity; hypoxia; hypoglycemia; the hemolytic uremic syndrome; osmotic myelinolysis; encephalitis; inborn errors of metabolism; and Huntington's disease.[153]

Magnetic resonance imaging (MRI) may be superior to CT in detecting cerebral, cerebellar, and basal ganglia lesions following CO poisoning.[154-157] MRI demonstrated cerebellar white matter damage in one child 6 years following acute CO exposure.[158] A year following severe exposure in a 17 year old with residual parkinsonism, MRI showed bilateral pallidal and substantia nigra lesions.[159] MRI performed 3 days following CO poisoning revealed bilateral diffuse high signal in the centrum semiovale and in the anterior thalami, not seen on CT.[160] Quantitative MRI may be more sensitive in evaluating the hippocampal regions in patients with DNS following CO poisoning, and corresponds to memory deficits on neuropsychological evaluation, but this method has only been studied in adults.[161-163]

SPECT scanning provides an indicator of the severity of cerebral damage and correlates with outcome.[164] The combination of EEG with SPECT scanning provided greater sensitivity in detecting anomalies than EEG alone.[165] In a cohort of adult patients with acute, severe CO poisoning, treated with HBO, PET scan findings of globally increased oxygen extraction ratio (OER) and decreased blood flow in the frontal and temporal cortex were most severe in those patients with DNS or PNS. These changes are temporary in patients who appear normal following CO exposure and in those with temporary neurological and psychiatric deficits. This suggests that ischemia is ongoing after CO intoxication, even after normalization of the clinical status.[166] Others have noted prolonged abnormalities in patients with neurological deficits following exposure,[167] but experience with this modality in children with CO poisoning is lacking. More recent work suggests that newer techniques in SPECT scanning allow for earlier detection of regional CO-induced anomalies.[168] There is extremely limited data on SPECT scanning in children, with only three cases reported, all older adolescents. Two of the three had normal initial scans, but abnormalities were detected with added surface three-dimensional display.[168]

21.11 TREATMENT

Treatment of CO poisoning in children is as controversial as in adults, and has not been subject to investigation through prospective, randomized controlled trials. A few issues of particular relevance to pediatrics are discussed here.

As with adults, early aggressive resuscitation of the child is critical. Ensuring a patent, protected airway, adequate ventilation, and institution of 100% normobaric oxygen (NBO) is paramount in all cases of suspected CO poisoning, even before COHb levels become available. Circulatory status and body temperature should be stabilized, and cardiorespiratory monitoring should be instituted. In those patients with altered mental status, treatment of rapidly reversible causes of coma should be carried out through the bedside assessment of a finger-stick glucose and the administration of naloxone, a narcotic antagonist. Supplemental glucose may be needed to correct hypoglycemia, but every attempt should be made to maintain euglycemia, and avoid iatrogenic hyperglycemia. A thorough physical assessment for burns, odors, toxidromes, skin findings, and signs of smoke inhalation, trauma, or abuse is indicated. A careful history regarding the circumstances surrounding the exposure must be obtained once the child is stabilized. Considerations should be made to gastric or skin decontamination and activated charcoal administration in the setting of suspected intentional drug abuse, suicide attempt, or dermal chemical exposure (e.g., methylene chloride).

NBO therapy is best administered by a tight-fitting, continuous positive airway pressure face mask in the older child, but this route may be impractical in the infant or newborn, and a respiratory tent may be better tolerated. Generally, 100% oxygen therapy is continued for 4 to 6 h if the COHb level is <20%. Longer treatment periods may need to be considered in infants, given the presence of fetal COHb, which has a longer half-life than adult COHb. With regard to the fetus, pregnant women are generally treated five times longer than is necessary to reduce maternal COHb to <5%.[169]

Mild metabolic acidosis should not be corrected with sodium bicarbonate, as this will shift the hemoglobin–oxygen dissociation curve to the left and impair unloading of oxygen at the tissue level. Concurrent smoke inhalation places the patient at greater risk for pulmonary complications and systemic toxicity from toxins such as cyanide. Patients with significant smoke inhalation typically deteriorate over the first 12 to 24 h and may develop upper airway obstruction, bronchospasm, or pulmonary edema. They must be admitted and monitored closely. Empiric treatment for cyanide toxicity may be considered in those patients with smoke inhalation who remain acidotic, despite adequate resuscitative efforts, since confirmatory cyanide levels are generally not readily available. The safest method of treatment in this setting involves the use of only the thiosulfate component of the cyanide antidote kit. Administration of nitrites is generally not advisable in patients with elevated COHb levels, as they may not tolerate the additive effects of resultant methemoglobinemia.

Treatment of cerebral edema is supportive. Currently, there is no role for steroids, hypothermia, or hyperventilation-induced hypocarbia in this setting. The role of barbiturates and mannitol in treating CO-induced cerebral edema in children has not been studied.

The use of HBO therapy in the treatment of CO poisoning is controversial, and recommendations vary, even among pediatric toxicology experts.[170,171] In a retrospective analysis of 11 comatose children, the development of neurological sequelae occurred in 28.6% of those treated with HBO vs. 75% of those given NBO. Subgroup analysis showed that those patients semicomatose on arrival had neurological sequelae in 24.2 and 43.8%, respectively (HBO vs. NBO). In those who were stuporous, 7.7 and 33.3% respectively, had sequelae. Another investigator showed that neurological sequelae were more frequent in patients who did not receive HBO therapy (29.4%) compared with those who did (19.6%).[6] Meert et al.[77] looked retrospectively at the outcome of children treated with NBO, and concluded that acute neurological manifestations resolve rapidly without HBO therapy. However, neurological outcome was assessed from nursing and physician records, physical and occupational therapy evaluations, and unspecified neuropsychological reports in an unspecified number of patients. Therefore, the major limitation of this study is the lack of detailed neurological assessment both at presentation and at follow-up that would allow detection of those specific neuropsychiatric changes known to result from CO poisoning. Nonetheless, the authors noted "gross" neurological abnormalities in nine survivors (8.5%), with seven of those persistent at various stages of follow-up (2 months to 3.3 years). Three patients developed delayed neurological syndromes including tremors, hallucinations, seizures, occipital lobe infarctions, and defects in cognitive and interpersonal skills. A final limitation of the study was the presence of serious comorbidities. Since 90% of children had smoke inhalation, 40.5% had burns, 50% were mechanically ventilated, and 20% underwent surgical procedures, outcome was confounded by the complications resulting from smoke inhalation, and did not reflect that of pure CO exposure.

HBO therapy is proposed to be safe in children and during pregnancy.[78,102,172-179] Potential complications include tympanic rupture, which occurs in less than 1%. Because of this concern, 35% of patients require myringotomy in the infant age group, primarily if they are unconscious or unable to suck a bottle to clear the ears. Other complications include tension pneumothorax, which occurs in less than 1%, and oxygen-induced seizures, which occur in 1 in 10,000 patients. Exposure to

TABLE 21.4
Proposed Indications for HBO Therapy in Children

- Abnormal level of consciousness or mental status examination
- COHb level >25 to 40% with no other symptoms
- Any neurological symptom other than isolated headache, including brief loss of consciousness at the scene
- Coma
- Recurrent symptomatology
- Refractory symptoms despite 100% oxygen

Other situations in which the use of HBO could be considered:
"Soaking" exposure, based on history, metabolic acidosis, or elevated lactate level condition of other
exposed family members when assessing an infant (COHb levels may be unreliable secondary to fetal
hemoglobin interference)

pharmaceuticals such as disulfiram, cisplatinum, adriamycin, and bleomycin may predispose the patient to oxygen toxicity.[180] Patients with cystic lung disease are at theoretically increased risk for pneumothorax.[180] Other considerations in infants treated with HBO therapy are that burping must be carried out before ascent to avoid regurgitation with expansion of the gastric air bubble, and that hypothermia may occur during ascent.[78,181]

Actual safety experience with HBO therapy in children has been variable. Use in children in one series was associated with no morbidity. However, all children underwent prophylactic tympanotomy.[73] In another series of critically ill, mechanically ventilated children, 9/32 with CO poisoning (age range 3 days to 11.3 years) had complications.[180] Those reported were hypotension (63%), bronchospasm (34%), hemotympanum (13%), progressive hypoxemia (6%), seizures (3%), and accidental extubation during transport (3%). Most complications were "easily" manageable by the treatment team, but of concern were two patients with refractory bronchospasm and two others with apparent pulmonary edema. These complications were dealt with in a multiplace chamber with a staff experienced with pediatric HBO therapy, whereas monoplace chambers may not allow the team to attend directly to the victim. Concerns regarding complications during transport of critically ill children for HBO therapy have also been raised. While there is no consensus at this time regarding the use of HBO in children, indications have been proposed and mirror those applied to adults (Table 21.4).

The proposed mechanisms for HBO therapy include the inhibition of PMN CD18 and beta-2 integrin activation which prevents the tight adherence to endothelial ICAMs and degranulation activity. This prevents xanthine oxidase formation and further lipid peroxidation in the animal model.[182] Decreased production of ROS is also observed. HBO therapy also shortens the half-life of COHb. The hyperoxygenated bloodstream bypasses the COHb-blocked delivery of oxygen to tissues and restores mitochondrial respiration. Other theoretical benefits include protection against cerebral edema, induction of protective stress proteins (SP-72), and antagonism of NMDA excitotoxic neuronal injury. HBO therapy is most beneficial in animal models if given within 6 h at 2.8 to 3.0 atm for 90 min. Many investigators routinely perform additional treatments if lack of improvement is noted after the first treatment.[183] Others report no benefit with multiple HBO treatments.[184] Given that HBO therapy has not yet proved to be effective in treating PNS or preventing DNS in adults[185,186] and since there are no prospective, randomized trials in children, the treatment of CO poisoning will likely continue to vary tremendously. In the future, perhaps greater research emphasis will be placed upon CO-poisoning prevention and non-HBO therapies, such as insulin, allopurinol, N-acetylcysteine or other antioxidants, and NMDA receptor antagonists, currently investigational.[60]

21.12   CRITERIA FOR ADMISSION

Criteria for admission of patients with CO poisoning have been proposed as follows:[136]

1. COHb level of 25% or greater,
2. COHb levels over 15% in a patient with known cardiac disease,
3. COHb levels over 10% in the pregnant patient,
4. Abnormal ECG,
5. Metabolic acidosis,
6. Thermoregulatory difficulties,
7. Abnormal CONSB,
8. Hypoxemia,
9. Myoglobinuria,
10. Abnormal CXR, and
11. History of syncope.

While these criteria may be useful, in reality, a more conservative approach is warranted. Given the high incidence of abnormal mental status on presentation and potential for the development of neurological sequelae, all children should have careful baseline neuropsychiatric assessments performed. Since this type of age-appropriate testing is not feasible in the emergency department, nearly all children will require admission. Close neurological follow-up for repeat neuropsychological testing at 6 weeks, 6 months, and 1 year is then indicated once discharged from the hospital. This will, it is hoped, allow prompt intervention if deficits are uncovered.

21.13   PREVENTION

It is evident that treatment options for CO poisoning are both limited and controversial. Consequently, the most important role for health-care providers may be one of primary prevention. Health-care providers should educate their patients about measures to prevent CO poisoning (Table 21.5). Further, to assure prompt diagnosis and intervention, health-care providers should remain ever vigilant for potential historical clues that suggest CO poisoning, such as nonspecific symptoms, headaches, viral-like illness with absence of significant fever, occurrence of illness during winter months, illness in other family members, recent travel in a vehicle

TABLE 21.5
Prevention of Carbon Monoxide Poisoning

- Place CO detectors in the home and workplace
- Clean and maintain chimneys and fireplaces yearly
- Have furnaces inspected, maintained, and checked for CO leaks yearly
- Have automobiles maintained regularly
    — Muffler and tailpipes should be checked regularly
    — Avoid high risk vehicles such as those with rear exit of tailpipes or vertical tailgates
    (pickups, station wagons)
    — Do not allow children to ride in the back of such vehicles or in the back of pickup trucks
- Check gas appliances yearly for CO leaks
- Do not use gas ranges as heating sources
- Vent all indoor heaters to the outside
- Never use charcoal grills or hibachis indoors
- Only use paint strippers such as methylene chloride outdoors, in well-ventilated areas with gloves
- Do not leave automobiles running inside garages
- Do not sit in parked cars while running; if this is necessary, open all windows
- Contact your local poison control center if you think that you have been poisoned by CO

Source: Cobb, N. and Etzel, R.A., *J. Am. Med. Assoc.*, 266, 659, 1995.

(especially those with vertical tailgates), and the presence of dead or ill small animals at home. Widespread screening through breath sampling of emergency department and clinic patients will further allow detection of low-level exposures and allow for prompt interventions to reduce hazards.

REFERENCES

1. Cobb, N. and Etzel, R.A., Unintentional carbon monoxide-related deaths in the United States, 1979 through 1988, *J. Am. Med. Assoc.*, 266, 659, 1995.
2. Shepherd, G. and Klein-Schwartz, W., Accidental and suicidal adolescent poisoning deaths in the United States, 1979–1994, *Arch. Pediatr. Adolesc. Med.*, 152, 1181, 1998.
3. Wilson, R.C. and Saunders, P.J., An epidemiological study of acute carbon monoxide poisoning in the West Midlands, *Occup. Environ. Med.*, 55, 72, 1998.
4. Hampson, N.B., Emergency department visits for carbon monoxide poisoning in the Pacific Northwest, *J. Emerg. Med.*, 16, 695, 1998.
5. Revert, M., Brotons, C., Navarro, J., Gutierrez, C., Doz, J.F., Cervantes, M., and Bonfill, X., Winter epidemic of carbon monoxide poisoning in Badia, *Atencion Primaria*, 16, 261, 1995.
6. Kim, J.K. and Coe, C.J., Clinical study on carbon monoxide intoxication in children, *Yonsei Med. J.*, 28, 266, 1987.
7. Jaffe, F.A., Pathogenicity of carbon monoxide, *Am. J. Forensic Med. Pathol.*, 18, 406, 1997.
8. Marth, E., Haselbacher, S., and Schaffler, K., A cohort study with children living in an air-polluted region — a model for public health, *Toxicol. Lett.*, 88, 155, 1996.
9. Little, D.N., Children and environmental toxins, *Primary Care*, 22, 69, 1995.

10. Hampson, N.B. and Norkool, D.M., Carbon monoxide poisoning in children riding in the back of pickup trucks, *J. Am. Med. Assoc.*, 267, 538, 1992.

11. Mofenson, H.C. and Caraccio, T.R., Poisoning by pick-up truck, *Pediatrics*, 89, 1268, 1992.

12. Smith, R.A. and Ball, R.J., Carbon monoxide poisoning in two children riding in the back of a van, *Arch. Dis. Childhood*, 71, 482, 1994.

13. Johnson, C.J., Moran, J., and Pekich, R., Carbon monoxide in school buses, *Am. J. Public Health*, 65, 1327, 1975.

14. Fatal carbon monoxide poisoning in a camper-truck-Georgia, *MMWR*, 40, 154, 1991.

15. Carbon monoxide poisonings associated with snow-obstructed vehicle exhaust systems — Philadelphia and New York City, January, 1996, *MMWR*, 45, 1, 1996.

16. Paul, R.I. and Tanz, R.R., The case of the slandered hamburgers, *Pediatr. Emerg. Care*, 4, 189, 1988.

17. Baker, M.D., Henretig, F.M., and Ludwig, S., Carboxyhemoglobin levels in children with nonspecific flu-like symptoms, *J. Pediatr.*, 113, 501, 1988.

18. Gemelli, F. and Cattani, R., Carbon monoxide poisoning in childhood, *Br. Med. J.*, 291, 1197, 1985.

19. Klees, M., Heremans, M., and Dougan, S., Psychological sequelae to carbon monoxide intoxication in the child, *Sci. Total Environ.*, 44, 165, 1985.

20. Selbst, S.M., Kulick, R., Henretig, F., and Baker, M.D., Kerosene heater-related injuries in children, *Pediatr. Emerg. Care*, 12, 81, 1996.

21. Baron, R.C., Backer, R.C., and Sopher, I.M., Fatal unintended carbon monoxide poisoning in West Virginia from non-vehicular sources, *Am. J. Public Health*, 79, 1656, 1989.

22. Wagner, B.A. and Orrison, W.W., Clinical outcome and brain MRI four years after carbon monoxide intoxication, *Acta Neurol. Scand.*, 87, 205, 1993.

23. Katafuchi, Y., Nishimi, T., Yamaguchi, Y., Matsuishi, T., Kimura, Y., Otaki, E., and Yamashita, Y., Cortical blindness in acute carbon monoxide poisoning, *Brain Dev.*, 7, 516, 1985.

24. Hampson, N.B., Kramer, C.C., Dunford, R.C., and Norkool, D.M., Carbon monoxide poisoning from indoor burning of charcoal briquets, *J. Am. Med. Assoc.*, 271, 52, 1994.

25. Houck, P.M. and Hampson, N.B., Epidemic carbon monoxide poisoning following a winter storm, *J. Emerg. Med.*, 15, 469, 1997.

26. Carbon monoxide poisoning at an indoor ice arena and bingo hall-Seattle, 1996, *J. Am. Med. Assoc.*, 275, 1468, 1996.

27. Silvers, S.M. and Hampson N.B., Carbon monoxide poisoning amoung recreational boaters, *J. Am. Med. Assoc.*, 274, 1614, 1995.

28. Jumbelic, M.I., Open air carbon monoxide poisoning, *J. Forensic Sci.*, 43, 228, 1998.

29. Carbon monoxide levels during indoor sporting events — Cincinnati, 1992–1993, *J. Am. Med. Assoc.*, 271, 419, 1994.

30. Parish, R.A., Smoke inhalation and carbon monoxide poisoning in children, *Pediatr. Emerg. Care*, 2, 36–39, 1986.

31. Zikria, B.A., Wistin, G.C., and Chodoff, M., Smoke and carbon monoxide poisoning in fire victims, *J. Trauma*, 12, 641, 1972.

32. Litovitz, T.L., Klein-Schwartz, W., Dyer, S.K. et al., 1997 annual report of the American Association of Poison Control Centers, *J. Emerg. Med.*, 16, 443, 1998.

33. Marilena, G., New physiological importance of two classic residual products, carbon monoxide and bilirubin, *Biochem. Mol. Med.*, 61, 136, 1997.

34. Verma, A., Carbon monoxide: a putative neural messenger, *Science*, 259, 381, 1993.

35. Brune, B. and Ullrich, V., Inhibition of platelet aggregation by carbon monoxide is mediated by activation of guanylate cyclase, *Mol. Pharmacol.*, 32, 497, 1987.

36. Arias-Diaz, J., Villa, N., Hernandez, J., Vara, E., and Balibrea, J.L., Carbon monoxide contributes to the cytokine-induced inhibition of surfactant synthesis by human type II pneumocytes, *Arch. Surg.*, 132, 1352, 1997.

37. Dinman, B., Eaton, J., and Brewer, G., Effects of carbon monoxide on DPG concentrations in the erythrocyte, *Ann. N.Y. Acad. Sci.*, 1974, 246, 1970.

38. Klasner, A.E., Smith, S.R., Thompson, M.W., and Scalzo, A.J., Carbon monoxide mass exposure in a pediatric population, *Acad. Emerg. Med.*, 5, 992, 1998.

39. Longo, L.D., The biological effects of carbon monoxide on the pregnant woman, fetus, and newborn infant, *Am. J. Obstet. Gynecol.*, 129, 69, 1977.

40. Okeda, R., Funata, N., Song, S.J. et al., Comparative study of selective cerebral lesions in carbon monoxide poisoning and nitrogen hypoxia in cats, *Acta Neuropathol.*, 56, 265, 1982.

41. Haldane, J.B.S., Carbon monoxide as a tissue poison, *Biochem. J.*, 21, 1086, 1927.

42. Drabkin, D.L., Lewey, F.H., Bellet, S. et al., The effect of replacement of normal blood by erythrocytes saturated with carbon monoxide, *Am. J. Med. Sci.*, 205, 755, 1943.

43. Goldbaum, L.R., Studies on the relationship between carboxyhemoglobin concentration and toxicity, *Aviat. Space Environ. Med.*, 48, 969, 1977.

44. Goldbaum, L.R., Mechanism of the toxic action of carbon monoxide, *Ann. Clin. Lab. Sci.*, 6, 372, 1976.

45. Goldbaum, L.R., Orellano, T., and Dergal, E., What is the mechanism of carbon monoxide toxicity? *Aviat. Space Environ. Med.*, 46, 1289, 1975.

46. Brown, S. and Piantadosi, C., *In vivo* binding of CO to cytochrome oxidase in rat brain, *J. Appl. Physiol.*, 68, 604, 1990.

47. Brown, S. and Piantadosi, C.A., Recovery of energy metabolism in rat brain after CO hypoxia, *J. Clin. Invest.*, 89, 666, 1992.

48. Coburn, R.F. and Forman, H.J., Carbon monoxide toxicity, in *Handbook of Physiology*, Sect. 3: *The Respiratory System*, American Physiological Society, Washington, D.C., Vol. IV, 1987, chap. 21, 439.

49. Smithline, H.A., Rivers, E.P., Chiulli, D.A., Rady, M.Y., Baltarowich, L.L., Blake, H.C. et al., Systemic hemodynamic and oxygen transport response to graded carbon monoxide poisoning, *Ann. Emerg. Med.*, 5, 203, 1993.

50. Dallman, P. and Schwartz, H., Distribution of cytochrome C and myoglobin in rats with dietary iron deficiency, *Pediatrics*, 35, 677, 1965.

51. Wittenberg, B.A. and Wittenberg, J.B., Effects of carbon monoxide on isolated heart muscle cells, *Res. Rep. Health Effects Inst.*, 62, 1, 1993.

52. Thom, S.R., Carbon monoxide-mediated brain lipid peroxidation in the rat, *J. Appl. Physiol.*, 68, 997, 1990.

53. Thom, S.R., Leukocytes in carbon monoxide-mediated brain oxidative injury, *Toxicol. Appl. Pharmacol.*, 123, 234, 1993.

54. Thom, S.R., Dehydrogenase conversion to oxidase and lipid peroxidation in brain after carbon monoxide poisoning, *J. Appl. Physiol.*, 73, 1584, 1992.

55. Thom, S.R., Garner S., and Fisher D., Vascular oxidative stress from carbon monoxide exposure, *Undersea Hyperb. Med.*, 25 (Suppl.), 47, 1998.

56. Hall, E.D., Andrus, P.K., Althaus, J.S. et al., Hydroxyl radical production and lipid peroxidation parallels selective post-ischemic vulnerability in gerbil brain, *J. Neurosci. Res.*, 34, 107, 1993.

57. Zhang, J. and Piantadosi, C.A., Mitochondrial oxidative stress after carbon monoxide hypoxia in the rat, *J. Clin. Invest.*, 90, 1193, 1992.

58. Piantadosi, C.A., Toxicity of carbon monoxide, hemoglobin vs. histotoxic mechanisms, in *Carbon Monoxide*, Penney, D.G., Ed., CRC Press, Boca Raton, FL, 1996, 163.

59. Amara, S.G., Neurotransmitter transporters, *Nature*, 360, 420, 1992.

60. Ishimaru, H., Katoh, A., Suzuki, H. et al., Effects of NMDA receptor antagonists on carbon monoxide-induced brain damage in mice, *J. Pharmacol. Exp. Ther.*, 261, 349, 1992.

61. Shinomura, T., Nakao, S., and Mori, K., Reduction of depolarization-induced glutamate release by hemoxygenase inhibitor, possible role of carbon monoxide in synaptic transmission, *Neurosci. Lett.*, 166, 131, 1994.

62. Piantadosi, C.A., Tatro, L., and Zhang, J., Hydroxyl radical production in the brain after CO hypoxia in rats, *Free Radical Biol. Med.*, 18, 603, 1995.

63. Piantadosi, C.A., Zhang, J., Levin, E.D. et al., Apoptosis and delayed neuronal damage after carbon monoxide poisoning in the rat, *Exp. Neurol.*, 147, 103, 1997.

64. Roy, B. and Crawford, R., Pitfalls in diagnosis and management of carbon monoxide poisoning, *J. Accid. Emerg. Med.*, 13, 62, 1996.

65. Barrett, L. et al., Carbon monoxide poisoning, a diagnosis frequently overlooked, *Clin. Toxicol.*, 23, 309, 1985.

66. Heckerling, P.S., Occult carbon monoxide poisoning, a cause of winter headache, *Am. J. Emerg. Med.*, 5, 201, 1987.

67. Dolan, M.C., Haltom, T.L., Barrows, G.H. et al., Carboxyhemogloin levels in patients with flu-like symptoms, *Ann. Emerg. Med.*, 16, 782, 1987.

68. Bouton, J.M. and Steppe, M., Particularities in children poisoning, *Acta Clin. Belg.*, (Suppl.), 1, 51, 1990.

69. Sanchez, R., Fosarell, P., Felt, B., Greene, M., Lacovara, J., and Hackett, F., Carbon monoxide poisoning due to automobile exposure, disparity between carboxyhemoglobin levels and symptoms of victims, *Pediatrics*, 82, 663, 1988.

70. Hopkins, R.O., Weaver, L.K., Larson-Lohr, V., and Howe, S., Neuropsychological performance of patients with acute carbon monoxide (CO) poisoning does not correlate with carboxyhemoglobin (COHb) or carboxyhemoglobin half-life, *Undersea Hyperb. Med.*, 22 (Suppl.), 71 (abstr.), 1995.

71. Smith, J. and Brandon, J., Acute carbon monoxide poisoning, three years experience, *Post. Grad. Med.*, 46, 65, 1970.

72. Smith, J. and Brandon, J., Morbidity from acute carbon monoxide poisoning at three year follow-up, *Br. Med. J.*, 1, 318, 1973.

73. Crocker, P.J. and Walker, J.S., Pediatric carbon monoxide toxicity, *J. Emerg. Med.*, 3, 443, 1985.

74. Carrasco Gomez, J.A., Lopez-Herce, C.J., Bernabe de Frutos, M.C. et al., Carbon monoxide poisoning, a home accident to remember, *An. Esp. Pediatr.*, 39, 411, 1993.

75. Binder, J.W. and Roberts, R.J., Carbon monoxide intoxication in children, *Clin. Toxicol.*, 16, 287, 1980.

76. Lattere, M., Raspino, M., Vietti, R.M. et al., Acute carbon monoxide poisoning in children, *Pediatr. Med. E Chir.*, 16, 565, 1994.

77. Meert, K.L., Heidemann, S.M., and Sarnaik, A.P., Outcome of children with carbon monoxide poisoning treated with normobaric oxygen, *J. Trauma*, 44, 149, 1998.

78. Rudge, F.W., Carbon monoxide poisoning in infants, treatment with hyperbaric oxygen, *South. Med. J.*, 86, 334, 1993.

79. Gasche, Y., Unger, P.F., Berner, M., Roduit, C., Jolliet, P.H., and Chevrolet, J.C., Le nourrisson est-il resistant a la toxicite du monoxyde de carbone? *Schweiz Med. Wochenschr.*, 123, 2413, 1993.

80. Diltoer, M.W., Colle, I.O., Hubloue, I., Ramet, J., Spapen, H.D.M., Nguyet, N., and Huyghens, L.P., Reversible cardiac failure in an adolescent after prolonged exposure to carbon monoxide, *Eur. J. Emerg. Med.*, 2, 234, 1995.

81. Aurora, T., Chung, W., Dunne, R., Martin, G., Ward, K., Rivers, E., Knoblich, B., Nguyen, H.B., and Tomlanovich, M.C., Occult myocardial injury in severe carbon monoxide poisoning, *Acad. Emerg. Med.*, 6, 394 (abstr.), 1999.

82. Zimmerman, S.S. and Truxal, B., Carbon monoxide poisoning, *Pediatrics*, 68, 215, 1981.

83. Rosenberg, N.M., Singer, J., Bolte, R. et al., Retinal hemorrhage, *Pediatr. Emerg. Care*, 10, 303, 1994.

84. Lacey, D.J., Neurologic sequelae of acute carbon monoxide intoxication, *Am. J. Dis. Child*, 135, 145, 1981.

85. Dempsey, L.C, O'Donnell, J.J., and Hoff, J.T., Carbon monoxide retinopathy, *Am. J. Ophthalmol.*, 82, 691, 1976.

86. Kelly, J.S. and Sophocleus, G.J., Retinal hemorrhages in subacute carbon monoxide poisoning, *J. Am. Med. Assoc.*, 239, 1515, 1978.

87. Yun, D.R. and Cho, S.H., A study of the incidence and therapeutic measures on carbon monoxide poisoning in Seoul, *J. Kor. Med. Assoc.*, 20, 705, 1977.

88. Song, D.B., Epidemiology of carbon monoxide poisoning in Korea, *J. Kor. Med. Assoc.*, 28, 1059, 1985.

89. Choi, I.S., Delayed neurologic sequelae in carbon monoxide intoxication, *Arch. Neurol.*, 40, 433, 1983.

90. Weaver, L.K., Hopkins, R.O., Howe, S., Larson-Lohr, V., and Churchill, S., Outcome at 6 and 12 months following acute CO poisoning, *Undersea Hyperb. Med.*, 23 (Suppl.), 9 (abstr.), 1996.

91. Garland, H. and Pearce, J., Neurological complaints of carbon monoxide poisoning, *Q. J. Med.*, 144, 445, 1967.

92. Raskin, N. and Mullaney, O.C., The mental and neurological sequelae of carbon monoxide asphyxia in a case observed for fifteen years, *J. Nerv. Ment. Dis.*, 92, 640, 1940.

93. Veil, C.L., Bartoli, D., and Baume, S., Bilan psychologique, socioprofessionel, psychopathologique et physiopathologique a un an de distance de l'intoxication exycarbone aigue, *Ann. Med.*, 2, 343, 1970.

94. Hopkins, R.O. and Weaver, L.K., Long-term outcome in subjects with carbon monoxide poisoning, *Undersea Hyperb. Med.*, 21 (Suppl.), 17 (abstr.), 1994.

95. Ginsburg, R. and Romano, J., Carbon monoxide encephalopathy, need for appropriate treatment, *Am. J. Psychiatr.*, 133, 317, 1976.

96. Lee, S.D., A clinical observation and the therapeutic effect of hyperbaric oxygenation in acute carbon monoxide poisoning, *Hum. Sci.*, 7, 465, 1983.

97. Plum, F., Posner, J.B., and Hain, R.F., Delayed neurological deterioration after anoxia, *Arch. Intern. Med.*, 110, 56, 1962.

98. Ajuriaguerra, J., de Zazzo, R., and Granjon, N., Le phenomene d'accolement au modele dans un syndrome d'apraxie oxycarbone, *Encephale*, 1, 1, 1949.

99. Fau, R., Andrewy, B., and LeMen, J., Syndrome apraxo-agnostique oxycarbone, *Rev. Neurol.*, 94, 375, 1956.

100. Hopkins, R.O., Weaver, L.K., Larson-Lohr, V., and Howe, S., Loss of consciousness is not required for neurological sequelae due to CO poisoning, *Undersea Hyperb. Med.*, 22 (Suppl.), 14, 1995.

101. Katafuchi, Y., Nishimi, T., Yamaguchi, Y., Matsuishi, T., Kimura, Y., Otaki, E., and Yamashita, Y., Cortical blindness in acute carbon monoxide poisoning, *Brain Dev.*, 7, 516, 1985.

102. Waisman, D., Shupak, A., Weisz, G., and Melamed, Y., Hyperbaric oxygen therapy in the pediatric patient: the experience of the Israel Naval Medical Institute, *Pediatrics*, 102, E53, 1998.

103. Angle, C.R., McIntire, M.S., and Meile, R.L., Neurological sequelae of poisoning in children, *J. Pediatr.*, 73, 531, 1968.

104. Kopelman, A.E. and Plaut, T.A., Fetal compromise caused by maternal carbon monoxide poisoning, *J. Perinatol.*, 18, 74, 1998.

105. Christensen, P., Gronlund, J., and Carter, A.M., Placental gas exchange in the guinea-pig, fetal blood gas tensions following the reduction of maternal oxygen capacity with carbon monoxide, *J. Dev. Physiol.*, 8, 1, 1986.

106. Brown, D.B., Mueller, G.L., and Golich, F.C., Hyperbaric oxygen treatment for carbon monoxide poisoning in pregnancy, a case report, *Aviat. Space Environ. Med.*, 63, 1011, 1992.

107. Farrow, J.R., Davis, G.J., Roy, T.M., McCloud, I.C., and Nichols, G.R., Fetal death due to nonlethal maternal carbon monoxide poisoning, *J. Forensic Sci.*, 35, 1448, 1990.

108. Okeda, R., Matsuo, T., Kuroiwa, T., Tajima, T., and Takahashi, H., Experimental study on pathogenesis of the fetal brain damage by acute carbon monoxide intoxication of the pregnant mother, *Acta Neuropathol.*, 69, 244, 1986.

109. Woody, R.C., Telencephalic dysgenesis associated with presumptive maternal carbon monoxide intoxication in the first trimester of pregnancy, *Clin. Toxicol.*, 28, 467, 1990.

110. Cramer, C.R., Fetal death due to accidental maternal carbon monoxide exposure, *J. Toxicol. Clin. Toxicol.*, 19, 297, 1982.

111. Penney, D.G., Effects of carbon monoxide on developing animals and humans, in *Carbon Monoxide*, Penney, D.G., Ed., CRC Press, Boca Raton, FL, 1996, 109–144.

112. Courtens, W., Hennequin, Y., Blum, D., and Vamos, E., Charge association in a neonate exposed in utero to carbon monoxide, *Birth Defects*, 30, 407, 1996.

113. Hennequin, Y., Blum, D., Vamos, E. et al., *In utero* carbon monoxide poisoning and multiple fetal abnormalities, *Lancet*, 341, 240, 1993.

114. Seger, D. and Welch, L., Carbon monoxide controversies, neuropsychologic testing, mechanism of toxicity, and hyperbaric oxygen, *Ann. Emerg. Med.*, 24, 242, 1994.

115. Caravati, E.M., Adams, C.J., Joyce, S.M. et al., Fetal toxicity associated with maternal carbon monoxide poisoning, *Ann. Emerg. Med.*, 17, 714–717, 1988.

116. Hill, E.P., Hill, J.R., Power, G.G. et al., Carbon monoxide exchanges between the human fetus and mother, a mathematical model, *Am. J. Physiol.*, 232, 311, 1977.

117. Koren, G., Sharav, T., and Pastuszak, A., A multicenter, prospective study of fetal outcome following accidental carbon monoxide poisoning in pregnancy, *Reprod. Toxicol.*, 5, 397, 1991.

118. Secker-Walker, R.H., Vacek, P.M., and Flynn, B.S., Smoking in pregnancy, exhaled carbon monoxide, and birth weight, *Obstet. Gynecol.*, 89, 648, 1997.

119. Hutter, C.D. and Blair, M.E., Carbon monoxide — does fetal exposure cause sudden infant death syndrome? *Med. Hypotheses*, 46, 1, 1996.

120. Watkins, C.G. and Strope, G.L., Chronic carbon monoxide poisoning as a major contributing factor in the sudden infant death syndrome, *Am. J. Dis. Child*, 140, 619, 1986.

121. Variend, S. and Forrest, R.W., Carbon monoxide concentrations in infant deaths, *Arch. Dis. Childhood*, 62, 417, 1987.

122. Smialek, J.E. and Monforte, J.R., Toxicology and sudden infant death, *J. Forensic Sci.*, 22, 757, 1977.

123. Altoff, H., Wehr, K., Michels, S. et al., Toxic environmental factors in sudden infant death (SIDS), *J. Legal Med.*, 98, 103, 1987.

124. Khan, K. and Sharief, N., Chronic carbon monoxide poisoning in children, *Acta Paediatr.*, 84, 742, 1995.
125. Piatt, J.P., Kaplan, A.M, Bond, G.R., and Berg, R.A., Occult carbon monoxide poisoning in an infant, *Pediatr. Emerg. Care*, 6, 21, 1990.
126. Buckley, R.G, Aks, S.E., and Eshom, J.L., The pulse oximetry gap in carbon monoxide intoxication, *Ann. Emerg. Med.*, 24, 252, 1994.
127. Vegfors, M. and Hennmarket, C., Carboxyhaemoglobin and pulse oximetry, *Br. J. Anaesth.*, 66, 625, 1991.
128. Touger, M., Gallagher, E.J., and Tyrell, J., Relationship between venous and arterial carboxyhemoglobin levels in patients with suspected carbon monoxide poisoning, *Ann. Emerg. Med.*, 25, 481, 1995.
129. Vreman, H.J., Ronquillo, R.B., Ariagno, R.L. et al., Interference of fetal hemoglobin with the spectrophotometric measurement of carboxyhemoglobin, *Clin. Chem.*, 34, 975, 1988.
130. Perrone, J. and Hoffman, R.S., Fetal hemoglobin interference with carboxyhemoglobin determination, *Clin. Toxicol.*, 33, 548, 1995.
131. Pearson, H.A., Diseases of the blood, in *Nelson Textbook of Pediatrics*, Behrman, R.E. and Vaughan, V.C., Eds., 13th ed., W.B. Saunders, Philadelphia, 1987.
132. Vreman, H.J., Mahoney, J.J., and Stevenson, D.K., Carbon monoxide and carboxyhemoglobin, *Adv. Pediatr.*, 42, 303, 1995.
133. Kurt, T.L., Anderson, R.J., and Reed, W.G., Rapid estimation of carboxyhemoglobin by breath sampling in an emergency setting, *Vet. Hum. Toxcol.*, 32, 227, 1990.
134. Messier, L.D. and Myers, R.A., A neuropsychological screening battery for emergency assessment of carbon monoxide-poisoned patients, *J. Clin. Psychol.*, 47, 675, 1991.
135. Brown, D.B., Golich, F.C., Tappel, J.J., Dykstra, T.A., and Ott, D.A., Severe carbon monoxide poisoning in the pediatric patient: a case report, *Aviat. Space Environ. Med.*, 67, 262, 1996.
136. Mofenson, H.C., Caraccio, T.R., and Brody, G.M., Carbon monoxide poisoning, *Am. J. Emerg. Med.*, 2, 254, 1984.
137. Finck, P.A., Exposure to carbon monoxide, review of the literature and 567 autopsics, *Mil. Med.*, 131, 1513, 1966.
138. Stonesifer, L.D., Bone, R.C., and Hiller, F.C., Thrombotic thrombocytopenia purpura in carbon monoxide poisoning, *Arch. Intern. Med.*, 140, 104, 1980.
139. Sone, S., Higashihara, T., and Kotake, T., Pulmonary manifestations in acute carbon monoxide poisoning, *Amer. J. Roentgenol.*, 120, 865, 1974.
140. Arias-Diaz, J., Villa, N., Hernandez, J., Vara, E., and Balibrea, J.L., Carbon monoxide contributes to the cytosine-induced inhibition of surfactant synthesis by human type II pneumocytes, *Arch. Surg.*, 132, 1352, 1997.
141. Aronow, W.S. and Isbell, M.W, Carbon monoxide effects on exercise induced angina pectoris, *Ann. Intern. Med.*, 79, 392, 1973.
142. Anderson, R.F., Allensworth, D.C., and deGroot, W.J., Myocardial toxicity from carbon monoxide poisoning, *Ann. Intern. Med.*, 67, 1172, 1967.
143. Shafer, N., Smiley, M., and MacMillan, F.P., Primary myocardial disease in man resulting from acute carbon monoxide poisoning, *Am. J. Med.*, 38, 316, 1965.
144. Emerson, T.S. and Keiler, J., Pattern shift visual evoked potential screening for HBO2 in mild-to-moderate carbon monoxide poisoning, *Undersea Hyperb. Med.*, 25, 27, 1998.
145. Ikeda, T., Kondo, T., Mogami, H., Mima, T., Mimoyo, M., Shimazaki, S., and Sugimoto, T., Computerized tomography in cases of acute carbon monoxide poisoning, *Med. J. Osaka Univ.*, 29, 253, 1978.

146. Kim, K.S., Weinburg, P.E., Suh, J.H., and Ho, S.U., Acute carbon monoxide poisoning computed tomography of the brain, *Amer. J. Neuroradiol.*, 1, 399, 1980.
147. Nardizzi, L., Computerized tomographic correlate of carbon monoxide poisoning, *Arch. Neurol.*, 36, 38, 1979.
148. LaPresle, J. and Fardeau, M., The central nervous system and carbon monoxide poisoning, II. Anatomical study of brain lesions following intoxication with carbon monoxide (22 cases), *Prog. Brain Res.*, 24, 31, 1967.
149. Sawada, Y., Ohashi, N., Maemura, K. et al., Computerized tomography as an indication of long-term outcome after carbon monoxide poisoning, *Lancet*, 1, 783, 1980.
150. Pracyk, J.B., Stolp, B.W., Fife, C.E. et al., Brain computerized tomography after hyperbaric oxygen therapy for carbon monoxide poisoning, *Undersea Hyperb. Med.*, 22, 1, 1995.
151. Lee, M.S. and Marsden, C.D., Neurological sequelae following carbon monoxide poisoning clinical course and outcome according to the clinical types and brain computed tomography, *Movement Disorders*, 9, 550, 1994.
152. So, G.M., Kosofsky, B.E., and Southern, J.F., Acute hydrocephalus following carbon monoxide poisoning, *Pediatr. Neurol.*, 17, 270, 1997.
153. Ho, V.B., Fitz, C.R., Chuang, S.H. et al., Bilateral basal ganglia lesions: Pediatric differential considerations, *Radiographics*, 13, 269, 1993.
154. Horowitz, A.L., Kaplan, R., and Sarpel, G., Carbon monoxide toxicity: MR imaging in the brain, *Radiology*, 162, 787, 1987.
155. Taverni, N., Dal Pozzo, G., Bartolozzi, C., Caramelli, L., and Boddi, P., Magnetic resonance imaging in the study of brain changes due to carbon monoxide poisoning, *Radiol. Med.*, 76, 289--292, 1988.
156. Silverman, C.S., Brenner, J., and Murtagh, F.R., Hemorrhagic necrosis and vascular injury in carbon monoxide poisoning: MR demonstration, *Amer. J. Neuroradiol.*, 14, 168, 1993.
157. Uchino, A., Hasuo, K., Shida, K., Matsumoto, S., Yasumori, K., and Masuda, K., MRI of the brain in the chronic phase of carbon monoxide poisoning, *Neuroradiology*, 36, 399, 1994.
158. Mascalchi, M., Petruzzi, P., and Zampa, V., MRI of cerebellar white matter damage due to carbon monoxide poisoning: case report, *Neuroradiology*, 38, S73, 1996.
159. Kawanami, T., Kato, T., Kurita, K., and Sasaki, H., The pallidoreticular pattern of brain damage on MRI in a patient with carbon monoxide poisoning, *J. Neurol. Neurosurg. Psychiatr.*, 64, 282, 1998.
160. Tuchman, R.F., Moser, F.G., and Moshe, S.L., Carbon monoxide poisoning: bilateral lesions in the thalamus on MR imaging of the brain, *Pediatr. Radiol.*, 20, 478, 1990.
161. Gale, S.D., Hopkins, R.O., Weaver, L.K., Bigler, E.D., Booth, E.J., and Blatter, D.D., MRI, quantitative MRI, SPECT, and neuropsychological findings following carbon monoxide poisoinng, *Brain Injury*, 13, 229, 1999.
162. Hopkins, R.O., Weaver, L.K., and Kesner, R.P., Long term memory impairments and hippocampal magnetic resonance imaging in carbon monoxide poisoned subjects, *Undersea Hyperb. Med.*, 20 (Suppl.), 15 (abstract), 1993.
163. Hopkins, R.O., Weaver, L.K., and Kesner, R.P., Quantitative MRI analysis of the hippocampus corresponds with persistent memory impairments in carbon monoxide poisoned subjects, *Brain Cognition*, 28, 215 (abstr.), 1995.
164. Turner, M. and Kemp, P.M., Isotope brain scanning with Tc-HMPAO: a predictor of outcome in carbon monoxide poisoning? *J. Accid. Emerg. Med.*, 14, 139, 1997.
165. Denays, R., Makhoul, E., Dachy, B., Tondeur, M., Noel, P., Ham, H.R., and Mols, P., Electroencephalopgrahic mapping and 99mTc HMPAO single-photon emission computed tomography in carbon monoxide poisoning, *Ann. Emerg. Med.*, 24, 947, 1994.

166. DeReurck, J., Decoo, D., Lemahieu, I. et al., A positron emission tomography study of patients with acute carbon monoxide poisoning treated by hyperbaric oxygen, *J. Neurol.*, 240, 430-434, 1993.

167. Shimosegawa, E., Hatazawa, J., Nagata, I., and Okudera, T., Cerebral blood flow and glucose metabolism measurements in a patient surviving one year after carbon monoxide intoxication, *J. Nucl. Med.*, 33, 1696, 1992.

168. Kao, C.H., Hung, D.Z., ChangLai, S.P., Liao, K.K., and Chieng, P.U., HMPAO brain SPECT in acute carbon monoxide poisoning, *J. Nucl. Med.*, 39, 769, 1998.

169. Margulies, J.L., Acute carbon monoxide poisoning during pregnancy, *Am. J. Emerg. Med.*, 4, 516, 1986.

170. Rosenberg, N.M., Tenenbein, M. et al., Controversial issues, *Pediatr. Emerg. Care*, 11, 45, 1995.

171. Shannon, M., Index of suspicion, case presentation, *Pediatr. Rev.*, 13, 395, 1992.

172. VanHoesen, K.B., Camporesi, E.M., and Moon, R.E., Should hyperbaric oxygen be used to treat the pregnant patient for acute carbon monoxide poisoning? *J. Am. Med. Assoc.*, 261, 1039, 1989.

173. Brown, D.B., Mueller, G.L., and Golich, F.C., Hyperbaric oxygen treatment for carbon monoxide in pregnancy: a case report, *Aviat. Space Environ. Med.*, 63, 1011, 1992.

174. Silverman, R.K. and Montano, J., Hyperbaric oxygen treatment during pregnancy in acute carbon monoxide poisoning: a case report, *J. Reprod. Med.*, 42, 309, 1997.

175. Ekert, P., Tibballs, J., and Gorman, D., Three patients with carbon monoxide poisoning treated with hyperbaric oxygen therapy, *Aust. Paediatr. J.*, 24, 197, 1988.

176. Gabrielli, A., Layon, A.J., and Gallagher, T.J., Carbon monoxide intoxication during pregnancy: a case presentation and pathophysiologic discussion, with emphasis on molecular mechanisms, *J. Clin. Anesth.*, 7, 82, 1995.

177. Elkharrat, D., Raphael, J.C., Korach, J.M. et al., Acute carbon monoxide intoxication and hyperbaric oxygen in pregnancy, *Intensive Care Med.*, 17, 289, 1991.

178. Santamaria, J.P., Williams, E.T., and Desautels, D.A., Hyperbaric oxygen therapy in pediatrics, *Adv. Pediatr.*, 42, 335, 1995.

179. Rhodes, R.H., Skolnick, J.L., and Roy, T.M., Hyperbaric oxygen treatment for carbon monoxide poisoning: observations based on 8 years experience, *Kentucky Med. Assoc. J.*, 89, 61, 1991.

180. Keenan, H.T., Bratton, S.L., Norkool, D.M., Brogan, T.V., and Hampson, N.B., Delivery of hyperbaric oxygen therapy to critically ill, mechanically ventilated children, *J. Crit. Care*, 13, 7, 1998.

181. Martorano, F.J. and Hoover, D., The child hyperbaric patient, *J. Hyperb. Med.*, 1, 15, 1986.

182. Thom, S.R., Functional inhibition of leukocyte B-2 integrins by hyperbaric oxygen in carbon monoxide-mediated brain injury in rats, *Toxicol. Appl. Pharmacol.*, 123, 248, 1993.

183. Gorman, D.F., Clayton., D., and Gilligan, J.E., A longitudinal study of 100 consecutive admissions for carbon monoxide poisoning to the Royal Adelaide Hospital, *Anesth. Intensive Care*, 20, 311, 1992.

184. Raphael, J.C., Elkharrat, D., Jars-Guincestre, M.C. et al., Trial of normobaric and hyperbaric oxygen for acute carbon monoxide intoxication, *Lancet*, 1, 414, 1989.

185. Tibbles, P.M. and Perrotta, P.L., Treatment of carbon monoxide poisoning: a critical review of human outcome studies comparing normobaric oxygen with hyperbaric oxygen, *Ann. Emerg. Med.*, 24, 269, 1994.

186. Weaver, L.K., Carbon monoxide poisoning, *Crit. Care Clin. North Am.*, 15, 297, 1999.

22 Carbon Monoxide Production, Transport, and Hazard in Building Fires

Frederick W. Mowrer and Vincent Brannigan

CONTENTS

22.1 INTRODUCTION

Most fire-related deaths in buildings are attributed to "smoke inhalation" rather than burns or other thermal effects. Berl and Halpin[1] conducted a seminal study identifying the role of smoke inhalation vs. burns in 1978. After analyzing fire-related deaths in Maryland during the period from 1972 through 1977, they found that approximately one half of the victims studied had carboxyhemoglobin (COHb) levels of at least 50%. Another quarter of the victims had COHb levels between 30 and 50%. These levels of blood carbon monoxide (CO) were assumed to contribute to death when combined with other conditions, such as elevated cyanide levels or preexisting health conditions. Based on these figures, Berl and Halpin estimated that smoke inhalation, specifically CO poisoning, accounts for three fourths of all fire deaths. In 1989, Harwood and Hall[2] reviewed fire death data and concluded that approximately two thirds of fire-related deaths can be attributed to the presence of

CO. In 1995, Hall and Harwood[3] revisited the issue and concluded that smoke inhalation was the leading cause of fire deaths by a factor of 3 to 1 over burn deaths. They also noted that the share of deaths attributable to smoke inhalation was steadily increasing by approximately 1% per year since at least 1979. Even if these numbers are only approximate, the data clearly demonstrate that CO is the dominant toxicant in fire-related deaths.[4]

Fire safety issues associated with smoke inhalation involve complex relationships among CO production and transport, the effects of CO on humans both in the short and the long term, and potential additive and synergistic effects with other toxicants and irritants present in building fire environments. Unlike relatively low-dose chronic exposure to CO, fatal fires tend to involve high doses delivered to people in stressful situations that demand complex choices and actions. CO concentrations in excess of 1% (10,000 ppm) are not unusual in serious building fires; CO concentrations in excess of 10% (100,000 ppm) have been measured in a number of enclosure fire simulations.[5,6] Such concentrations can render incapacitating and lethal doses within seconds to a few minutes of exposure.

Because of the importance of CO production and transport to life safety from fire, considerable research has been conducted to develop an understanding of CO generation mechanisms during enclosure fires and to predict the transport of CO and other combustion products throughout a building under fire conditions. Despite the level of research, however, there is still significant uncertainty regarding CO production and transport due to the complexity of uncontrolled building fires.

Current developments in fire safety regulation depend on advanced understanding of CO production, transport, and effects on humans. Unlike older prescriptive regulatory systems, the emerging practice in fire safety engineering is to control the environment around persons escaping from a fire. Determining the production and movement of toxic materials and the acceptable limits for exposure are critical to this emerging fire safety regulatory process. This chapter provides a summary of current knowledge regarding the production and transport of CO during building fires, and discusses some current strategies for using that knowledge for performance-based fire safety regulation.

Relative Quantities and Hazards of Carbon Monoxide: In general, it is useful to consider the relative quantities and hazards related to CO production and transport in building fires in terms of the following matrix:

	Type of Fire	
Location of Exposed Persons	**Smoldering**	**Flaming**
Room of fire origin	Moderate–high	Low
Beyond room of origin	Low	High

This matrix is limited to CO; a person in a room subject to flaming combustion may be exposed to other hazards as well. As indicated in this matrix, levels of CO in the fire environment are generally most significant for smoldering fires within the room of fire origin and for flaming fires beyond the room of origin. Reasons for this general observation are discussed below.[3]

22.2   CARBON MONOXIDE PRODUCTION

CO is a product of incomplete combustion of carbonaceous fuels. It is produced in large quantities in some building fires primarily due to an imbalance between the production of combustible vapors and the availability of oxygen in the combustion zone to completely burn the fuel. The production and subsequent transport of CO during building fires depend on a number of variables. Key variables include the combustion mode, whether smoldering or flaming, and ventilation, particularly for flaming fires. Other factors include temperature and suppression effects as well as fuel type and geometry.

The concentration of CO in the near-fire environment depends on its rate of production. The rate of CO generated in building fires is typically addressed in terms of two factors: a *yield* factor and a *fuel mass loss* rate. This is represented as

$$\dot{m}_{CO} = f_{CO} \cdot \dot{m}_f \qquad (22.1)$$

The yield factor, f_{CO}, represents the conversion of fuel to CO on a mass basis (g CO/g fuel). In a sense, this term describes the "efficiency" with which a fire produces CO instead of CO_2. The fuel mass loss rate refers roughly to the size of the fire. For fires that are referred to as "fuel-limited," i.e., fires that are supplied with sufficient air to burn all the available fuel, the intensity of the fire, represented in terms of heat release rate, $\dot{Q}_f$, is directly proportional to the mass loss rate as

$$\dot{Q}_f = \dot{m}_f \cdot \Delta H_{c,\text{eff}} \qquad (22.2)$$

The proportionality constant is referred to as the effective heat of combustion. Thus, for fuel-limited fires, the rate of CO production is proportional to the yield of CO and the fire intensity:

$$\dot{m}_{CO} = f_{CO} \cdot \frac{\dot{Q}_f}{\Delta H_{c,\text{eff}}} \qquad (22.3)$$

This represents the rate at which CO is being produced by a fire, but it does not describe the concentration of CO in the atmosphere. The concentration of CO in the fire environment is frequently represented in terms of its mass fraction, Y_{CO}. For a fixed volume of gases, the mass fraction of CO is represented as

$$Y_{CO} = \frac{m_{CO}}{m_{tot}} = \frac{\int_0^t \left(\dot{m}_{CO}\right)_{net} dt}{\int_0^t \left(\dot{m}_{tot}\right)_{net} dt} \qquad (22.4a)$$

Under quasi-steady-flow conditions, the mass fraction of CO is represented more simply as

$$Y_{CO}(t) = \frac{\dot{m}_{CO}(t)}{\dot{m}_{tot}(t)}$$
(22.4b)

Equations 22.4a and b demonstrate that to evaluate the concentration of CO in a particular atmosphere it is necessary to know both the CO production rate and the total ventilation rate. The CO production rate is frequently related directly to the ventilation rate of the fire, particularly for flaming fires, so these two terms are not necessarily independent of each other.

22.2.1 SMOLDERING FIRES

Not all fires involve flaming combustion. Carbon can also burn directly on the surface of a material, as it does on a cigarette ember or on a charcoal briquette. This relatively slow surface combustion is called "smoldering." Smoldering fires tend to have relatively high CO yield factors but relatively low intensities when compared with well-ventilated flaming fires. In other words, they produce higher percentages of CO but smaller absolute quantities. Consequently, smoldering fires tend to generate CO at relatively low rates. But if this CO accumulates in a poorly ventilated space, such as a bedroom or house with closed doors and windows, hazardous concentrations of CO can develop despite the slow generation rate. Such hazardous concentrations can occur before typical residential smoke alarms will activate to alert sleeping occupants of the developing danger. In large and well-ventilated buildings the dilution of the CO with fresh air will tend to mitigate the hazard from small smoldering fires.

Representative CO yield factors for smoldering fires have been reported in the literature. Ohlemiller[7] summarizes work that has been done on smoldering fires. He reports molar percentages in evolved gases for two fuels. For a flexible polyurethane foam, the CO molar yield was 6 to 7% at an air velocity through the porous fuel of 1.5 mm/s, while for a cellulosic insulation material the CO molar yield ranged from approximately 10 to 22% for air velocities through the fuel that ranged from less than 1 to about 8 mm/s.

Quintiere et al.[8] performed a review of smoldering fire experiments conducted in closed rooms and buildings. The objectives of their investigation were twofold: (1) to consolidate for analysis the data of available full-scale smoldering experiments and (2) to develop a model for the prediction of CO concentration in an enclosure subject to a smoldering fire source. They analyzed data from 40 full-scale experiments conducted under a range of conditions. As part of their analysis, Quintiere et al. determined CO yield factors of approximately 0.1 (g CO/g fuel) for both polyurethane and cotton, with a range of 0.06 to 1.1 for the polyurethane and 0.085 to 0.11 for the cotton. These materials represent the most widely used padding materials in upholstered furniture and mattresses.

Quintiere et al.[8] also developed an idealized mathematical model for the prediction of CO concentrations in a closed room due to a smoldering fire source. They selected a hazard criterion of 4.5% CO·min (45,000 ppm·min) as the critical dose of CO for incapacitation, based on 20% COHb as representative of the threshold for incapacitation. They note that this dose criterion is somewhat arbitrary. It is based on some approximations to the equation developed by Coburn, Foster, and Kane[9] for predicting the conversion of CO concentration in the air to COHb in the blood. It also assumes that the exposed humans are in a "resting" state; an increase in activity level would increase the respiration rate and consequently decrease the critical value for the dose. Finally, they note that the analysis holds for times less than about 150 min. As discussed below, knowledge of the relationship between COHb and incapacitation is developing rapidly.

For the 40 large-scale experiments they analyzed, Quintiere et al.[8] compared the time to the onset to flaming with the time to achieve a critical dose of CO from smoldering. For a number of these experiments, one or the other time was never achieved, but in many cases, the time to achieve a critical CO dose was very close to the time when the smoldering transitioned to flaming. In 17 of the experiments, the critical CO dose occurred during smoldering. Both the transition to flaming and the achievement of the critical CO dose seem to have a similar chance to occur in a period of 1 to 2.5 h. Which will occur first cannot be predicted at present. It should be noted, however, that transition to flaming occurs in only a relatively small fraction of cases, while CO concentrations from smoldering will continue to increase as long as the smoldering continues.

From this analysis, it would appear that the times to achieve critical doses of CO from smoldering fires in closed rooms is on the order of an hour or more. Consequently, this hazard is of most concern for people who are asleep, when they would be less likely to be alerted to the smoldering condition by the odor or obscuration of the smoke. Unfortunately, smoke detectors may not be present in bedrooms because until recently they were usually installed only in hallways or spaces outside of rooms used for sleeping, not within bedrooms. Furthermore, even if smoke detectors are installed within a room with a smoldering fire, the current technology of residential smoke detectors may not respond promptly to the quantities and sizes of smoke particles being released by smoldering fires.[10]

Beyond the room of fire origin, the hazards associated with smoldering fires diminish due to the larger volumes and ventilation rates of air available to dilute the CO concentration. These hazards have not been studied as extensively as the "room of origin" hazards, but modeling techniques are available to calculate the influences of natural and mechanical building ventilation on the transport of CO and other contaminants throughout a building. These techniques also apply to flaming fire sources. They are discussed below in Section 22.3.

Finally with respect to smoldering fires, it is noted that most work has only considered smoldering as the initial mode of combustion, with the possible transition from smoldering to flaming. The potential for transition from flaming back to smoldering has not been addressed in detail. Such a transition might occur due to oxygen depletion, fuel depletion, or partial suppression of a flaming fire. Such a "reverse transition" fire may represent a significant hazard since the mass loss rates associated with smoldering in a previously flaming fire would be expected to be higher than when smoldering is the initial mode of combustion. As a result, CO generation rates would be higher if the yield factors stay the same.

22.2.2 Flaming Fires

Flaming fires represent a different process for CO production. Pitts[11] and Gottuk and Roby[12] provide excellent reviews of the available literature and data on the production of CO and other combustion products from flaming fires and on the effects of combustion conditions on the production of major species, particularly CO. As noted in these reviews, the production of CO in flaming fires is influenced strongly by the relative ventilation of the fire. This strong dependence on ventilation is to be expected, since CO should generally burn to CO_2 when sufficient ventilation is available. To a lesser extent, the production of CO is influenced by the nature of the fuel, with partially oxidized fuels generally more prone to CO production than pure hydrocarbons. Perhaps more importantly, the "volatility" of a fuel will govern its release rate in a fire and consequently the relative ventilation of the fire. Because of the importance of relative ventilation to the production of CO in enclosure fires, the emphasis here is on ventilation rather than fuel types.

22.2.3 Equivalence Ratios

The relative ventilation of a fire can be represented in terms of an *equivalence ratio*, which is the ratio between the amount of air required for complete (stoichiometric) combustion of the fuel and the amount of air available for combustion. On a flow basis, the equivalence ratio Φ, is represented as

$$\Phi = \frac{\left(\dot{m}_f/\dot{m}_a\right)_{avail}}{\left(\dot{m}_f/\dot{m}_a\right)_{req}} = \frac{\left(\dot{m}_a/\dot{m}_f\right)_{req}}{\left(\dot{m}_a/\dot{m}_f\right)_{avail}} = r\left(\dot{m}_f/\dot{m}_a\right)_{avail} \qquad (22.5)$$

The ratio between the mass of air required for complete combustion of a unit mass of fuel and the mass of the fuel itself is referred to as the *stoichiometric ratio, r*. The stoichiometric ratio can be calculated from the chemical composition of a fuel. It generally ranges from about 5 to 10 (g air/g fuel) for partially oxidized fuels, such as wood, alcohol, or polyurethane, to about 15 for pure hydrocarbons, such as propane, octane, or polyethylene. Through rearrangement of Equation 22.2 and substitution into Equation 22.5, the equivalence ratio can also be expressed in terms of the fire heat release rate, Q_f, based on the fuel flow rate, and the available airflow rate, $\dot{m}_a$:

$$\Phi = r\frac{\dot{m}_f}{\dot{m}_a} = \frac{\dot{Q}_f}{\dot{m}_a\left(\Delta H_c/r\right)} \qquad (22.6)$$

The ratio shown in parentheses in the denominator of Equation 22.6 can be considered as the "air heat of combustion." This term represents the amount of heat released per unit mass of air consumed rather than per unit mass of fuel consumed. For most common fuels, it has been observed that the air heat of combustion is virtually constant, with a value of 3.0 (±5%) MJ/kg air.[13] This suggests that for a given fire intensity, based on the fuel release rate, the equivalence ratio can be considered as a strong function of the ventilation rate and as only a weak function of the actual fuel type. Of course, the fire intensity does depend on the nature of the fuel, so the type of fuel is not irrelevant, but the purpose of this analysis is to demonstrate the critical role of the airflow rate on the relative ventilation of a fire with a given intensity.

This analysis is important for several reasons. Unlike many other toxic materials that might be released into the environment, CO is a product of the fire. It is not produced in constant quantities, but varies depending on the ventilation status of the fire. Even with perfect knowledge of the fuel load prior to a fire, the ventilation status might not be known adequately to predict the CO production.

22.2.4 OVERVENTILATION AND UNDERVENTILATION

Equivalence ratios are defined so that a ratio of less than unity implies an "over-ventilated," "fuel-lean," or "fuel-limited" fire, while an equivalence ratio greater than 1 implies an "underventilated," "fuel-rich," or "ventilation-limited" fire. In general, underventilated flaming fires tend to produce considerable quantities of CO, while significantly overventilated flaming fires tend to produce relatively little CO. While in theory CO production should be small if there is adequate ventilation, in practice CO production typically begins to increase at an equivalence ratio of about 0.5, suggesting that imperfect mixing plays a role in CO production from the diffusion flames associated with most building fires.

Unlike in furnaces, automobile engines, and other engineered combustion devices, where the fuel/air mixture and hence the equivalence ratio is regulated, in building fires the equivalence ratio is uncontrolled. Phillip Thomas, formerly of the Fire Research Station at the British Research Establishment and one of the world's leading fire scientists, has referred to fire as "combustion without taps"[12] to highlight this unregulated and inefficient mixing of fuel and air that occurs. In such a fire even if oxygen is present in the room, it may not be in sufficiently close proximity to the fuel and heat to complete combustion. This is the reason a fire can produce substantial quantities of CO even though there is nominally enough air present for complete combustion.

In enclosure fires, the equivalence ratio becomes somewhat self-regulating around its equivalence ratio; the flow of fuel is regulated by heat feedback from the fire to fuel surfaces while the flow of air is regulated in part by fire-induced pressures caused by the expansion and buoyancy of the heated gases. Serious building fires frequently go through a transitional phase known as "flashover." Flashover occurs when the fire becomes intense enough for heat feedback from the flame and accumulated hot gases to cause ignition of virtually all other combustible surfaces within the room. As a result of flashover, fuel production rates typically increase dramatically, driving the equivalence ratio well into the fuel-rich regime. Hence, Thomas's combustion without taps analogy is particularly appropriate for postflashover fires. For the purposes of this discussion, the critical point is that the production of CO will also increase dramatically in this fuel-rich environment. It is therefore important to understand CO production in the fuel-rich environments associated with postflashover fires.

Equivalence ratios for enclosure fires can be evaluated in a number of ways. The concept of a global equivalence ratio has been used as a convenient way to express the overall ventilation conditions within the fire enclosure. Even with this simplification, a unique definition for the global equivalence ratio does not exist and two common definitions are used. The upper-layer global equivalence ratio is the ratio of the mass of the upper layer originating from fuel sources to the mass of the upper layer originating from airstreams relative to the stoichiometric fuel/air ratio at any given time. The term "upper layer" refers to the relatively homogenous cloud of hot, buoyant smoke that accumulates under the ceiling during enclosure fires. The plume global equivalence ratio is the ratio of the mass of fuel burning to the mass of air entrained into the fire plume, normalized by the stoichiometric fuel/air ratio. Experimentally, the plume equivalence ratio is normally evaluated on a quasi-steady-flow basis from the fuel and airstream flow rates. Under quasi-steady conditions, the upper-layer and plume global equivalence ratios become equal. For the present review, differences between these definitions are not of consequence.

22.2.5 How Much Carbon Monoxide Will a Flaming Fire Produce?

Pitts[11] and Gottuk and Roby[12] note that two primary types of experiments have been conducted to evaluate specie yields from flaming fires. These are hood experiments and compartment experiments. The hood experiments have been conducted in open laboratory spaces, with a variety of fuels burned beneath instrumented exhaust hoods. Compartment experiments have been conducted in reduced-scale and large-scale enclosures under natural ventilation conditions. Some compartment experiments have considered only single ventilated compartments, while others have included two connected compartments to represent a room–corridor arrangement.

Hood experiments have been conducted by Beyler,[15,16] by Toner et al.,[17] and by Morehart et al.[18] While details of the experiments differ, the concept was the same for all. A burner was situated beneath an exhaust hood. The fuel supply rate to the burner was varied along with the hood exhaust rate and the distance between the burner and the smoke layer interface in the exhaust hood. This in turn caused the air entrainment rate into the fire plume and hence the equivalence ratio to vary. Concentrations of combustion products were measured by continuous sampling from the exhaust duct into a battery of gas analyzers.

Results of the hood experiments conducted by Beyler[15,16] and Toner et al.[17] are similar with respect to CO production. CO yields on the order of 0.01 (g CO/g fuel) or less were observed for plume equivalence ratios less than about 0.5, at which point the CO yield began to increase markedly to values typically between 0.2 and 0.3 g CO/g fuel at equivalence ratios above 1. Data are reported for equivalence ratios up to approximately 2. One reason for this upper limit on the equivalence ratio was the ignition of the smoke layer interface observed by Beyler at equivalence ratios above about 1.4. It is interesting to note that the yield factors tend to stabilize at values between about 0.2 and 0.3 at the higher equivalence ratios. It should be recognized, however, that this does not mean the total yield of CO remains constant, because the total yield is also directly proportional to the fuel flow rate and hence to the equivalence ratio.

Building fires typically occur within an enclosure or compartment, not beneath an exhaust hood. Some potentially significant differences exist between the compartment and hood configurations, particularly with respect to temperature, ventilation, and flame structure. Consequently, a number of experimental studies have been conducted to evaluate CO production in compartment fires.

Gottuk et al.[19] and Gottuk[20] conducted experiments in a 1.5 by 1.2 by 1.2 m compartment to investigate the burning of hexane, polymethylmethacrylate (PMMA), spruce, and flexible polyurethane foam and the production of CO. Bryner et al.[21,22] used a compartment 1.5 by 1.0 by 1.0 m with a single doorway 0.5 m wide by 0.8 m high to evaluate CO production with and without wood lining the compartment, respectively. They used a natural gas burner as the primary heat source, varying the flow of natural gas to the burner to achieve a wide range of heat release rates and equivalence ratios. These same investigators also conducted experiments in a full-scale enclosure in an effort to validate their reduced-scale results. Lattimer et al.[6] used the same facility as Gottuk[20] to determine the effects of wood in the upper layer of a postflashover compartment fire on CO levels.

The results of the compartment fire experiments are qualitatively similar to the hood fire experiments with respect to CO production. At equivalence ratios less than approximately 0.5, the yield of CO is very low, on the order of 0.01 g CO/g fuel. At this equivalence ratio, the CO yield begins to increase markedly, reaching values in the range of 0.20 to 0.25 g CO/g fuel at equivalence ratios above about 1.5. At equivalence ratios between about 1.5 and 6, the CO yield remains in the range of about 0.17 to 0.25 g CO/g fuel. Equivalence ratios for these experiments were always below 3, with the exception of two experiments reported by Lattimer et al.[6] with wood in the upper layer. For these two experiments, the equivalence ratio was reported to be between 5 and 6.

Pitts[11] and Gottuk et al.[23] have also considered the role of temperature and chemical kinetics on the production of CO in compartment fires. Gottuk et al.[23] note that the effect of changing temperature on species composition is twofold: (1) the generation of species in the fire plume changes and (2) the oxidation of postflame gases in the upper layer is affected. At temperatures less than 800 K (approximately 527°C), the gas mixture is relatively unreactive. Consequently, combustion within the fire plume controls the final CO levels expected. The upper layer becomes chemically reactive at temperatures above about 900 K; this permits nearly complete oxidation of CO to CO_2 for overventilated and slightly underventilated conditions. For more severely underventilated conditions, both chemical kinetics modeling and experimental results indicate that higher-temperature environments will result in slightly higher CO yields. At these higher temperatures, two mechanisms affecting the net formation of CO compete. On the one hand, the higher temperature still promotes the complete oxidation of CO to CO_2. But this is more than offset by the increased CO production caused by incomplete oxidation of unburned hydrocarbons in the fuel-rich mixture. Hydrocarbon oxidation is much faster than CO oxidation, so net CO levels increase with equivalence ratios in high-temperature environments.

Converting these laboratory results to real-world scenarios requires some assumptions, but these results clearly demonstrate that very large quantities of CO can be produced in an underventilated room fire. These and other toxic products of combustion then can move far from where they are produced in the fire to other parts of a building.

22.3   CARBON MONOXIDE TRANSPORT IN FIRES

Most people who die from "smoke inhalation" are located beyond the room of fire origin.[3] Consequently, the mechanisms of CO transport are of considerable importance and interest, particularly in large buildings where hundreds or even thousands of people may be exposed. The 1980 MGM Grand Hotel fire in Las Vegas is a case in point. In this fire, the actual thermal damage was largely confined to the main casino level on the first floor, but 66 of the 85 deaths resulting from this fire occurred on the 17th through 24th floors.[24]

CO is transported by the typical forces influencing air movement in buildings as well as by forces unique to the fire environment. As discussed by Klote and Milke,[24] these forces include

* Expansion and buoyancy of fire gases;
* Stack effect;
* Wind effect;
* Forced (mechanical) ventilation;
* Elevator piston effect.

Because the forces that transport CO through a building are the same forces that influence the ventilation of the fire itself, the transport of CO in flaming fires is coupled directly to its production.

The expansion and buoyancy of fire gases influences flows both within the room of origin and beyond the room of origin. Within the room of fire origin, the buoyancy induced by expansion of the fire gases manifests in a number of ways. First, a coherent plume rises from the fire source, entraining surrounding ambient air as it rises toward the ceiling. This is shown in Figure 22.1a. Second, the buoyant combustion products and entrained air rising in the plume accumulate beneath the ceiling, as shown in Figure 22.1b, forming a layer of smoke that descends as additional combustion products flow into it via the plume. This relatively uniform layer of accumulated combustion products is known variously as the "smoke layer," the "upper layer," or the "hot gas layer." Finally, these accumulating gases will begin to flow out of the smoke layer once the smoke layer has descended to the level of available vent openings, such as open doors or windows, as shown in Figure 22.1c. Under quasi-steady-flow conditions, a balance develops between the rate of airflow into the room of fire origin, the rate of entrainment into the fire plume, and the rate of smoke flow out of the room. When this balance occurs, the smoke layer stops descending within the room.

Unfortunately, smoke flowing from one room may be flowing into an adjacent room or corridor and into stairways and ventilation systems. Evacuation routes can

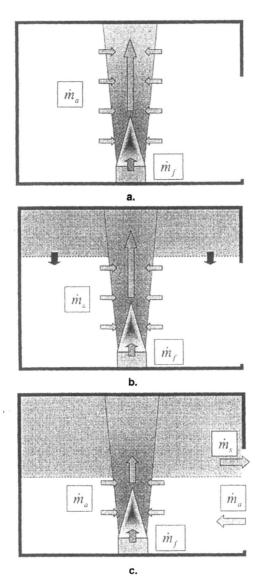

FIGURE 22.1   (a) The fire plume stage of enclosure fires; (b) the smoke filling stage of enclosure fires; (c) the quasi-steady vented stage of enclosure fires.

become untenable under these conditions. Forces other than the fire-induced buoyancy and expansion can dominate flow patterns and can act to spread smoke to remote areas of a large building.

Stack effect, also known as chimney effect, is caused by differences between the inside and outside temperature of ambient air. Stack effect is a form of buoyancy-induced flow. Normal stack effect occurs when the ambient inside temperature is warmer than the outside temperature; it is most prevalent during the winter months

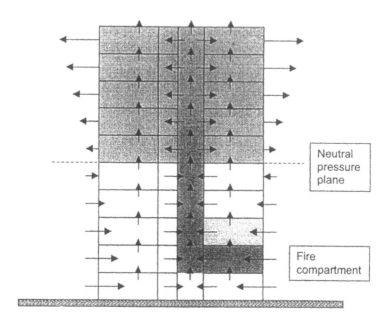

FIGURE 22.2   Flow patterns for normal stack effect in a tall building.

when buildings are heated. Reverse stack effect occurs when the outside temperature is hotter than the inside temperature; it is most significant in air-conditioned buildings in hot climates. Stack effect occurs due to the differences that result from density differences between the hot and cold air columns.

In normal stack effect, outside pressures exceed inside pressures at lower floors of a building, while inside pressures exceed outside pressures at higher floors. This causes a general circulation pattern where outside air enters a building through openings and leakage paths at lower floors, rises through vertical paths such as stairs and air-handling shafts, and flows out of the building at upper floors. This pattern is reversed for reverse stack effect. Flow patterns for normal and reverse stack effect are shown schematically in Figures 22.2a and b, respectively.

Stack effect is most pronounced in tall buildings in cold climates, but it is a factor in any size building where there is a difference between the inside and outside ambient temperatures. Together, buoyancy and stack effect account for why people are more likely to die due to "smoke inhalation" on the upper floors of buildings even when the direct effects of the fire are restricted to lower levels, as in the MGM Grand Hotel fire.

Pressures caused by wind can have a significant impact on air and smoke flows in buildings, particularly in buildings with operable windows. In general, wind causes a positive relative pressure on windward facades and a negative relative pressure on leeward facades, as shown in Figure 22.3. This tends to cause flows to move from the windward side(s) of a building toward the leeward sides. In congested areas, such as downtown high-rise districts, wind patterns and pressures become more complicated; scale models of a building and surrounding structures are sometimes used to determine experimentally the effects of wind on facade pressure distributions.

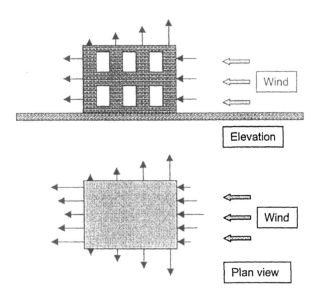

FIGURE 22.3   Typical influence of wind on facade pressure distributions.

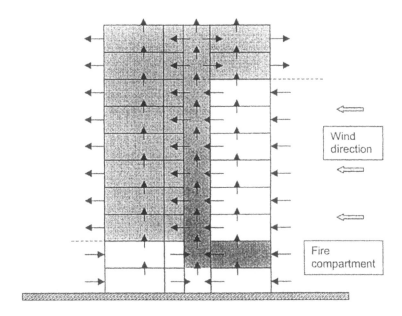

FIGURE 22.4   Combined influence of stack effect and wind on flow patterns in a tall building.

Forces produced by mechanical ventilation systems also influence pressure distributions and flow patterns in buildings. Traditionally, large mechanical ventilation systems were required to be shut down upon detection of smoke to prevent such systems from actively circulating smoke throughout a building. More recently, mechanical ventilation systems are sometimes designed to manage smoke actively, either by extracting smoke directly from a building or by developing pressure differentials between spaces to prevent smoke spread from the lower-pressure space to the higher-pressure space. For example, pressurized stairways are now typically required in high-rise buildings to reduce the potential for these critical exit paths to become "smoke-logged." Similar pressurization systems are being contemplated for more widespread use in elevator shafts and lobbies, particularly since elevators may be needed to evacuate people with disabilities (Figure 22.4).

The combined influences of expansion, buoyancy, stack effect, wind effects, and mechanical ventilation give rise to very complex pressure and flow distribution patterns within buildings. In all but the simplest cases, computers are needed to model these influences. For evaluation of smoke movement and management, a public-domain computer model developed by Walton[25] is becoming more widely used. This program, known as CONTAM (current version is CONTAM96), was developed for indoor air quality evaluation purposes, but is equally applicable to smoke movement. The model permits evaluation of the transport and dispersion of airborne contaminants within a building as a result of the different user-specified forces and flow paths.

22.4   THE REGULATORY RESPONSE TO CARBON MONOXIDE FROM FIRE

With few exceptions, traditional fire safety regulations have not dealt directly with the problem of CO exposure, despite its central role in fire-related deaths. These traditional regulations have dealt with the problem indirectly by mandating requirements for barriers to the spread of fire and smoke. The traditional regulatory framework relies on a largely empirical combination of some or all of the following fire protection features:

1. Compartmentation or the limitation of fires by barriers (fire walls, fire doors);
2. Limitations on the surface flame spread of building materials;
3. Controls on some ultrahazardous contents and operations;
4. Detection, alarm, and evacuation;
5. Suppression by automatic or manual means.

Unfortunately, in many buildings, construction standards are not up to the legal requirements, and smoke barriers are often defeated by occupants who block open the smoke barriers with wedges, trash cans, or even fire extinguishers themselves. Any such breaches in the compartmentation will result in more widespread and often unpredictable smoke movement. The normal response to these types of problems has been to design buildings with multiple safety systems. However, there is little evidence that these systems are coordinated in any organized way.

In general, modern fire safety systems work reasonably well in highly engineered code-complying buildings. Most of the lives lost in the United States are in small residential structures and are due to burning contents, rather than the structure. However, since the MGM Grand Hotel fire in 1980, regulators have become much more sensitive to the need to preserve a safe atmosphere for people in the building while they are exiting or awaiting rescue. Requirements have been developed for "smoke proof tower exits" and pressurized stairways. However, until relatively recently there was no real means to analyze the problem of toxic atmospheres separate from the general problem of fire development.

22.4.1 PERFORMANCE-BASED CODES

During the past decade, there has been a worldwide change in the focus of fire protection engineering. New technological systems, the demands of international trade, and the development of new scientific understanding of fires have combined to create a new approach to building fire safety, the "performance-based" approach to analysis, design, and regulation. While there are many different variations on performance-based methods,[26] they all include estimating the fires that could occur in the building, determining the hazard to the occupants and contents, and creating various engineered responses to reduce the hazard below some acceptable level.

One of the major debates about performance-based design is the suggestion that advanced engineering design can build safe structures without some of the traditional fire protection systems. In other cases structures are desired that are much larger and often more open than would have been possible under traditional codes. Almost all the advances in building technology and design increase the "openness" of structures. On the one hand, shopping malls, megastores, open-plan offices, escalator foyers, atriums in hotels and offices, and huge enclosed spaces such as stadiums and casinos all reduce the number of barriers between inhabitants of a building and the CO and other products of combustion. On the other hand, these larger spaces may provide sufficient ventilation to reduce the likelihood that hazardous quantities of CO will be developed.

To implement performance-based design, it is critical to determine the effect of products of combustion on the occupants of a building. Since at the time of the MGM Grand Hotel fire there was little comprehensive understanding in fire engineering of the role of CO in fire safety, the toxicology literature was reviewed and a number of research projects were performed by a series of distinguished scientists. Several slightly different systems were developed for estimating the life safety risk of CO and the other products of combustion. This work was summarized for the fire engineering community by Purser.[27]

22.4.2 CARBOXYHEMOGLOBIN HYPOTHESIS

CO presents fire safety engineers with a very special problem. Traditionally, safety engineering has dealt overwhelmingly with deterministic processes with well-defined thresholds or values. To apply deterministic fire safety engineering to life safety, it was necessary to come up with "tenability limits" for CO and other products of combustion. There are many types of tenability limits that might have been chosen. While the precise path is not easy to reconstruct, the fire protection engineering community came to accept a model of CO toxicity based on predicting COHb levels.

In these models, specific COHb levels were described as lethal or incapacitating. Typical levels were 50% for lethality and 30% for incapacitating. The COHb approach did not ignore other gases, since Purser[27] pointed out that the effects of several different fire gases are additive, so such gases can be easily combined into a fractional effective dose model.

   As Purser notes, COHb had the advantage of being relatively easily measured in laboratory tests, and had been traditionally used in research projects to describe the cause of death in fire. In addition, at an upper bound a high COHb is clearly inconsistent with life, so that it was not irrational to focus on COHb as a measure of fire safety risk. At the high levels of CO typical of the near-fire environment, (1000 to 8000 ppm CO, or more) there is a roughly linear relationship between exposure and COHb levels. Once certain assumptions are made about the size of the person and the level of activity, COHb levels in the blood could be reliably predicted from the product of the concentration of CO in the atmosphere multiplied by the duration of exposure. This product, referred to as Ct, is the simplest estimation of dose for inhalation toxicology and it is commonly described in units of ppm times minutes of exposure. Physiological researchers expressed a number of cautions regarding the applicability of this method of calculating doses, since the variables of size and activity both affect the rate of uptake of CO. Purser specifically cautions users on many of these points. However, there was little investigation in the early days of the COHb hypothesis itself. Instead, the researchers focused on problem of prediction of Ct.

   Ct is in many respects ideal from an engineering perspective because it is expressed in terms of engineering, rather than physiological, variables. Time to exit and atmospheric levels of CO were the type of engineering variables that were easily incorporated into the sophisticated computer models of fire being developed in the early 1990s. These models tended to treat CO as a "bright line" hazard. In accordance with the COHb hypothesis, the models set a level below which CO was assumed not to cause harm. The leading National Institute of Standards and Technology (NIST) model, HAZARD, treats CO as a deterministic hazard:

> TENAB estimates the hazard, as determined by a set of tenability measures, to which each occupant is exposed as he or she performs designated actions. TENAB uses the occupant time and location data from EXITT along with the environmental data from CFAST to determine the tenability conditions for each occupant or compartment. When a measure exceeds a certain level, the occupant is considered incapacitated or dead.[28]

   Ct-based levels were quickly used in performance-based analyses, and found their way into various regulatory documents. For example, the British standard is based on Ct:

| Carbon Monoxide | 5 min | 6000 ppm incapacitation | 12,000 ppm death |
| Carbon Monoxide | 30 min | 1400 ppm incapacitation | 2,500 ppm death |

   It was commonly accepted in the fire community that tenability limits expressed in ppm·min represented "safe" levels of exposure to CO. Basically, fire safety researchers believed that COHb was an accurate marker of the risk of injury from CO, and COHb levels below a certain level (often 50%) did not represent lethal conditions. This impression was reinforced by publications describing levels of COHb below 50% as representing dizziness and other nonfatal conditions. But in the process of setting tenability limits several cautions from the researchers were often lost. The first was that the Ct levels only predicted COHb for a person of selected size and activity. Even if the COHb hypothesis is correct, the same exposure does not generate the same COHb in a person of different size and activity.

This is roughly similar to the error made in designing air bags. Automobile air bags were designed to protect a person of a specific height and weight. Unfortunately, they turned out to be lethal for persons of shorter stature than the "normal" person the designers had in mind. Even if the COHb hypothesis is correct, the designs would produce widely different levels of COHb in the same exposure.

More importantly, there is growing evidence that the COHb hypothesis is of dubious validity. A review article in the *New England Journal of Medicine* expresses the norm in the current medical literature:

> It is important to recognize that COHb levels do not correlate well with the severity of symptoms in a substantial number of cases.[29]

It had long been known that people routinely died in fires at much less than 50% COHb. A great deal of effort had been expended to find other toxicants that could account for these deaths. But the research kept coming back to CO, albeit at much lower levels than had previously been thought hazardous. It was not until the late 1990s that a few fire researchers began to state explicitly that there was simply no justification for a deterministic "safe" level of COHb at 50%, or at any other level. Instead, it began to be accepted that the hazard of CO was probabilistic in nature, and depended heavily on the characteristics of the person exposed to the fire. Children, the elderly, persons with asthmatics, and others with impaired cardiopulmonary systems were clearly at risk of injury from much lower levels of COHb than had been accepted in the earlier models.

As a result, levels of exposure that were assumed to be safe clearly pose substantial dangers to major portions of the population. How had it gone wrong? Dedicated engineers had relied on sophisticated work by distinguished scientists. However, in the process of incorporating the science into the models, there was no real provision for continuous updating of the model with further advances in science. This probably illustrates a structural flaw in the method of developing safety regulations.

In the same vein it was always assumed that CO represented only an acute hazard. But in addition to the direct asphyxiating effects of CO, the medical literature has numerous references to the neurological and other debilitating effects of CO exposure. At the present time this literature has not been directly incorporated into the safety level set out in computer models or codes and standards.

22.4.3 MULTIPLICATION OF UNCERTAINTIES

Determining the hazard from CO in fire requires resolving extremely complex uncertainties. The production of CO in any given fire environment is not easily predicted. It can be represented as a probability distribution over a wide variety of fires that might occur in the building. The transport of CO is likewise a probability distribution that is a product of both the condition of the building and the size of the fire. Finally, the hazard to individuals can also be described as a probabilistic distribution, both of the locations and numbers of individuals and their individual susceptibility to CO.

These multiple interlocking probabilities pose special problems for regulators in the fire safety field, since margins of safety in fire designs are typically 50 to 100%. However, the range of reasonably expected events can be far wider than this margin of safety, and society may be incurring far larger risks than it expects based on the current engineering analysis.

22.5 SUMMARY AND CONCLUSIONS

CO has been identified as the leading toxicant in fire-related deaths. CO is produced in significant quantities both in smoldering combustion and in underventilated flaming combustion. CO poisoning from smoldering combustion is of primary concern within the room of fire origin, while hazards related to CO production from flaming combustion are of concern throughout a building. Flaming fires that reach hazardous proportions frequently undergo a transition known as flashover. Postflashover fires tend to be poorly ventilated; they can produce CO at extremely high rates as a consequence of this underventilation. As noted by Pitts,[11] the most important scenario for which CO formation has been implicated in fire deaths is a fully developed, flashed-over enclosure fire where victims are located in compartments remote from the fire. For this scenario, toxic gases, including CO, are transported from the room of fire origin to the locations of the victims.

The transport of CO and other combustion products through a building with a fire follows the same physical laws governing the flow of air through the building. The buoyancy of hot products of combustion causes them to tend to rise, but as temperatures cool away from the fire, the flow of combustion products becomes controlled more by stack, wind, and mechanical ventilation effects than by fire-induced buoyancy and expansion.

Traditional building regulations have dealt indirectly with CO and other combustion products, despite the direct and overwhelming relationship between "smoke inhalation" and fire-related deaths and injuries. Emerging performance-based approaches to building fire safety generally address toxicity explicitly in terms of specified tenability limits for different combustion products and different methods to consider potential additive and synergistic effects. It is not clear that the current models of injury based on the COHb hypothesis accurately predict the injuries that could occur to especially vulnerable populations. In addition, the multiple probabilistic systems and their attendant uncertainties make calculation of an appropriate safety margin much more difficult.

More research is needed to incorporate uncertainties and variability in exposed populations properly into these analyses. Preventing or reducing fire deaths and injuries from CO will require substantial increases in the understanding of CO production, transport, and injuries, and a firm commitment to integrate that understanding properly into the design and regulation of buildings.

REFERENCES

1. Berl, W.G. and Halpin, B.M., *Human Fatalities from Unwanted Fires*, APL/JHU FPP TR 37, Johns Hopkins University, December, 1978.
2. Harwood, B. and Hall, J.R., What kills in fires: smoke inhalation or burns? *Fire J.*, 83, 29, 1989.
3. Hall, J.R. and Harwood, B., Smoke or burns — which is deadlier?, *NFPA J.*, 89, 38, 1995.
4. Babrauskas, V., Levin, B.C., Gann, R.G., Paabo, M., Harris, R.H., Jr., Peacock, R.D., and Yusa, S., Special Publication 827, National Institute of Standards and Technology, December, 1991.
5. Pitts, W.M., Johnsson, E.L., and Bryner, N.P., Carbon monoxide formation in fires by high-temperature anaerobic wood pyrolysis, presented at 25th Symposium (International) on Combustion, The Combustion Institute, 1994.
6. Lattimer, B.Y., Vandsburger, U., and Roby, R.J., Carbon monoxide levels in structure fires: effects of wood in the upper layer of a post-flashover compartment fire, *Fire Technol.*, 34, 325–355, 1998.
7. Ohlemiller, T.J., Smoldering combustion, in *SFPE Handbook of Fire Protection Engineering*, Sect. 2, 2nd ed., DiNenno, P.J., Ed., National Fire Protection Association, Quincy, MA, 1995, 2/171–179.
8. Quintiere, J.G., Birky, M., Macdonald, F., and Smith, G., An analysis of smoldering fires in closed compartments and their hazard due to carbon monoxide, *Fire Mater.*, 6, 99–110, 1982.
9. Coburn, R.F., Foster, R.E., and Kane, P.B., Considerations of the physiological variables that determine blood carboxyhemoglobin concentration in man, *J. Clin. Invest.*, 44, 1899–1910, 1965.
10. Mulholland, G.W., Smoke production and properties, in *SFPE Handbook of Fire Protection Engineering*, Sect. 2, 2nd ed., DiNenno, P.J., Ed., National Fire Protection Association, Quincy, MA, 1995, 2/217–227.
11. Pitts, W.M., Global equivalence ratio concept and the formation mechanisms of carbon monoxide in enclosure fires, *Prog. Energ. Combustion Sci.*, 21, 197–237, 1995.
12. Gottuk, D.T. and Roby, R.J., Effects of combustion conditions on species production, in *SFPE Handbook of Fire Protection Engineering*, Sect. 2, 2nd ed., DiNenno, P.J., Ed., National Fire Protection Association, Quincy, MA, 1995, 2/64–84.
13. Huggett, C., Estimation of rate of heat release by means of oxygen consumption measurements, *Fire Mater.*, 4, 61–65, 1980.
14. Thomas, P.H., Perceptions and reflections on fire science, in *Interflam 99 — Proceedings of the Eighth International Conference*, Interscience Communications, London, 1999, 835.
15. Beyler, C.L., Major species production by diffusion flames in a two-layer compartment fire environment, *Fire Safety J.*, 10, 47–56, 1986.
16. Beyler, C.L., Ignition and burning of a layer of incomplete combustion products, *Combustion Sci. Technol.*, 39, 287–303, 1984.
17. Toner, S.J., Zukoski, E.E., and Kubota, T., Entrainment, Chemistry and Structure of Fire Plumes, NBS-GCR-87-528, National Institute of Standards and Technology, 1987.
18. Morehart, J.H., Zukoski, E.E., and Kubota, T., Characteristics of large diffusion flames burning in a vitiated atmosphere, in *Fire Safety Science — Proceedings of the Third International Symposium*, Cox, G. and Langford, B., Eds., Elsevier, New York, 1991, 575–583.
19. Gottuk, D.T., Roby, R.J., Peatross, M.J., and Beyler, C.L., Carbon monoxide production in compartment fires, *J. Fire Prot. Eng.*, 4, 133–150, 1992.

20. Gottuk, D.T., Generation of Carbon Monoxide in Compartment Fires, NIST GCR 92-619, National Institute of Standards and Technology, December, 1992, 265 pp.

21. Bryner, N.P., Johnsson, E.L., and Pitts, W.M., Carbon Monoxide Production in Compartment Fires: Reduced-Scale Enclosure Test Facility, NISTIR 5568, National Institute of Standards and Technology, December, 1994, 214 pp.

22. Bryner, N.P., Johnsson, E.L., and Pitts, W.M., Carbon Monoxide Production in Compartment Fires: Full-Scale Enclosure Burns, NISTIR 5499, National Institute of Standards and Technology, September, 1994.

23. Gottuk, D.T., Roby, R.J., and Beyler, C.L., Role of temperature on carbon monoxide production in compartment fires, *Fire Saf. J.*, 24, 315–331, 1995.

24. Klote, J.H. and Milke, J.A., Design of Smoke Management Systems, American Society of Heating, Refrigerating, and Air-Conditioning Engineers, Inc., Atlanta, GA, and Society of Fire Protection Engineers, Boston, MA, 1992, 236 pp.

25. Walton, G.N., CONTAM96: User Manual, NISTIR 6056, National Institute of Standards and Technology, Gaithersburg, MD, September, 1997, 77 pp.

26. Hadjisophocleous, G.V., Benichou, N., and Tamin, A.S., Literature review of performance-based fire codes and design environment, *J. Fire Prot. Eng.*, 9, 12–40, 1998.

27. Purser, D.A., Toxicity assessment of combustion products, in *SFPE Handbook of Fire Protection Engineering*, Sect. 2, 2nd ed., DiNenno, P.J., Ed., National Fire Protection Association, Quincy, MA, 1995, 2/85–146.

28. Jones, W.W., Evolution of HAZARD, the fire hazard assessment methodology, *Fire Technol.*, 33, 167–182, 1997.

29. Ernst, A. and Zibrak, J., Current concepts: carbon monoxide poisoning, *N. Engl. J. Med.*, 339, 1603–1608, 1998.

23 Approaches to Dealing with Carbon Monoxide in the Living Environment

Thomas H. Greiner and Charles V. Schwab

CONTENTS:

23.1　INTRODUCTION

Carbon monoxide (CO) poisoning has likely occurred since humans began using fire. Historians believe that Byzantine Emperor Jovian, who ruled from C.E. 334 to 364, died from CO produced by burning coal in a brazier, a usual method of indoor heating during that epoch. At an early age, and in perfect health, he died suddenly one cold night. The historian Ammianus Marcellinus concluded that the unhealthy fumes produced by burning a great quantity of coal in the brazier were the main cause of the emperor's death.[1]

There are still nonintentional deaths from burning fires and from appliances that burn carbon fuels. In 1992, the U.S. Consumer Product Safety Commission estimated that 38 U.S. deaths were associated with charcoal grills.[2] Yearly, there are thousands of other CO deaths from suicides, homicides, house fires, and unintentional exposures.[3]

CO production occurs when carbon-based fuels (e.g., coal, wood, charcoal, fuel oil, kerosene, gasoline, and diesel fuel) burn. When these fuels burn completely, the carbon combines with oxygen to form carbon dioxide (CO_2). During incomplete combustion, some of the carbon combines with only a single atom of oxygen to form CO, a poisonous gas. The concentration of CO in combustion products varies widely, from less than 1 part per million (ppm) to over 100,000 ppm. Anything that disrupts the burning process or results in a shortage of oxygen increases CO production.

Potential sources are everywhere, since so many homes have appliances that burn carbon fuels: furnaces, boilers, clothes dryers, ranges, ovens, water heaters, space heaters, fireplaces, charcoal grills, and wood-burning stoves. From outside the home, CO produced by vehicles and engines in attached garages can enter the house. In a study of 86 CO investigations conducted over a 5-year period, 94 sources producing elevated concentrations of CO were identified.[4] Gas-fired furnaces were the largest single source, with 31 furnaces producing excessive concentrations. Furnaces also caused the largest number of deaths (4) during the 5-year period. Other sources of CO exposures identified were gas-fired water heaters (14), kitchen ranges (10), and gas boilers (9). Internal combustion engines were identified as a source in 12 cases in the home (Table 23.1).

The majority (63) of the exposures occurred in dwellings with sleeping quarters (Table 23.2). More than half (45; 51.5%) were in single-family homes, while 20% occurred in rental homes, apartments, mobile homes, college dormitories, hotels, motels, and campers. A California study found that in 5 to 10% of homes, indoor CO concentrations exceeded the federal outdoor air standard of 9 ppm.[5-7] A comparison with the outdoor air standard was made because no indoor residential air quality standard exists. In 30 to 40% of the California homes, indoor CO levels were measurably higher than outdoor levels, indicating that many homes have indoor sources.

The combustion of wood, coal, and charcoal in open, residential fireplaces always produces high concentrations of CO, typically several thousand ppm. Combustion of natural gas, liquefied petroleum gas (LPG, or propane), and fuel oil in properly designed and operating residential heating appliances will produce lesser concentrations of CO. Maximum allowable CO concentrations in flue gases are 200 ppm for water heaters, 400 ppm for furnaces, and 800 ppm for household cooking gas appliances.[8-10]

The large number of residential CO detectors sounding alarms confirms that many persons are exposed to elevated levels of CO in their homes. CO is ubiquitous and exposures can occur anywhere.[11] In a study of 86 CO investigations, the largest number exposed was 6000 at an indoor motorcycle race.[4] The second largest number exposed was 300 attending a church wedding, followed by 80 in a hotel (Table 23.3). In 1995, MidAmerica Energy (an Iowa utility) responded to 2023 requests for CO checks.[12] They found elevated concentrations of CO (above 20 ppm) in 490 buildings. In 1996, the number of calls increased to 5017, with 1012 CO positive, and in 1997 they had 5794 calls, with 1327 CO positive. Minnegasco Utility, in Minneapolis/St.Paul, responded to nearly 14,000 CO calls during the 1995/96 heating season.[13] It found CO in approximately 2800 dwellings, representing a serious problem. They were concerned about homes where they did not find a source after a CO detector had alarmed. They conducted a follow-up study of 50 homes where they originally had found no CO.

TABLE 23.1
Sources of Unintentional CO in the Living Environment[a]

Sources	No. of Instances	% of Instances	No. of Deaths
Home			
Gas furnace	31	33	4
Gas water heater	14	15	1
Kitchen range	10	11	0
Gas boiler	9	10	2
Vehicle running in closed garage	7	8	3
Vehicle started in open garage	3	3	0
Unvented gas fireplace	2	2	0
Unvented gas water heater	1	1	0
Wood-burning fireplace	1	1	0
Wood-burning air-tight stove	1	1	0
Gasoline-powered carpet cleaner used in open garage	1	1	0
Fumes from vehicle outside open garage	1	1	0
Other Living Environments			
Propane-powered forklift	3	3	0
Faulty vehicle exhaust	2	2	0
Charcoal cooking grill	1	1	2
Vehicle in closed car wash	1	1	2
Unvented gas pressure washer	1	1	0
Gas-fired air conditioner	1	1	0
Waste oil burner	1	1	0
Gasoline forklift	1	1	0
Motorcycles operated in open auditorium	1	1	0
Propane floor buffer	1	1	0
Total	94	100	14

[a] Suicides involving intentional poisoning by CO are not included. This table lists multiple sources that spilled CO into the living environment for a single event. A combination could include spillage from a furnace, range, and a car in the garage.

Data collected by Greiner at Iowa State University.

In 49 of the homes the in-depth follow-up research study found sources of CO missed during the initial utility technician response. Utility companies respond to thousands of CO incidents every year, but are not required to report these incidents.

The presence of CO can be located in many living environments. Two conditions must exist for a CO hazard to be present in the living environment:

1. CO must be produced.
2. Combustion gases must be released into the structure.

TABLE 23.2
Location of CO Exposures from an Investigative Sample[a]

Type of Structure	No. of Cases	% of Cases
Single family home	45	51.5
Mobile home	5	5.7
Rental home	5	5.7
Office	4	4.5
Apartment	3	3.4
Church	3	3.4
Manufacturing	3	3.4
Commercial auto garage	2	2.3
Day care	2	2.3
Hotel	2	2.3
Residential garage	2	2.3
Retail sales	2	2.3
University classroom	2	2.3
Vehicle	2	2.3
Auditorium	1	1.1
Camper	1	1.1
Car wash	1	1.1
College dormitory	1	1.1
Motel	1	1.1
Warehouse	1	1.1
Total	89	100

[a] Suicides involving intentional poisoning by CO are not included.

Data collected by Greiner at Iowa State University.

TABLE 23.3
CO Exposure Levels and Number of Victims by Locations[a]

Location	No. of Exposed	Conc. of CO (ppm)
Auditorium	6000	92
Church	300	600
Hotel	80	600
University classrooms	75	70
Day-care center	69	180
Manufacturing plant	34	Unknown[b]
University laboratory	20	20–30
Warehouse	20	267
College dormitory	16	Unknown[c]
Rural day care	12	56

[a] Suicides involving intentional poisoning by CO are not included.
[b] Levels unknown, but workers required medical treatment.
[c] Levels unknown, but students required hyperbaric oxygen treatment.

Data collected by Greiner at Iowa State University.

If either condition is present, 50% of the hazard exists and one factor of safety is removed. There are a variety of reasons CO problems occur: failure of heating systems, poor maintenance of heating systems and vents, tighter houses, depressurization of homes, failure of vent systems, and operating vehicles in garages.

As houses become tighter to conserve energy, natural draft heating appliances that rely on buoyant forces are failing to vent. Tight houses have fewer air leaks, with less air infiltration/exfiltration. Vented appliances and appliances that exhaust air must have adequate combustion and makeup air from outdoors. In most houses, air needed for proper combustion, for proper venting of the appliance, to replace exhausted air, and to dilute contaminants is assumed to be supplied by naturally occurring air leaks. These natural leaks are often not sufficient.

Building codes, including the Uniform Mechanical Code, "assume" houses have a minimum of 1.0 air change per hour (ACH) from infiltration/exfiltration. They further "assume" that, unless the house is of "unusually tight construction," adequate combustion air will enter through building leaks.[14] Previously, the assumption that houses are "loose" and have 1.0 ACH may have been valid, but it is not valid today. The California study found that single-family detached homes with electric ranges had an average of only 0.36 ACH over the testing period.[7] An average air change rate does not ensure adequate air change during all conditions. In the California houses, 90% had less than 1.0 ACH, 52% had less than 0.5 ACH, 30% had less than 0.35 ACH, 7% had less than 0.2 ACH, and two houses had 0.1 ACH. Other studies confirm that houses are tighter than 1.0 ACH.[15]

Natural draft appliances depend on extremely small pressure differences to draft correctly. If the house is depressurized (operates under a negative pressure or suction), the vent system can fail. Adding an exhaust fan or recessed ceiling lights, sealing basement windows, closing a supply register, or adding a return air can upset the balance, and cause combustion products to spill into the home. When depressurization is less than 5.0 Pa, properly designed chimney exhaust stacks will usually draw adequately. Combustion gases will vent to the outdoors through the exhaust stack (Figure 23.1). An exhaust fan can cause house depressurization greater than 5.0 Pa. Operation of a roof-mounted attic exhaust fan, causing a house depressurization of 15.0 Pa, will cause a complete reversal in the exhaust stack and spillage of all combustion products produced by the water heater into the basement. This example is from an actual investigation.[4]

In many instances no one takes responsibility for ensuring a house functions correctly during design, construction, and modification. Anything that changes airflow (such as weather; house orientation; exterior obstructions; vent location, height, size, and type; attic ventilation; and operation of wood-burning fireplaces) can adversely affect the safe operation of natural draft heating appliances. These factors interact in complex ways to cause vent failure. When the vent fails, combustion products enter living quarters. In most (65%) of the 86 cases of CO summarized by Greiner,[4] vent failure or lack of venting (vented directly into the building) was the primary reason for CO in a building. In 23 cases, the pressure differences between outdoors and indoors caused vent failure, with spillage or downdrafting of combustion products. In 13 cases the appliances were unvented, and in 5 cases the vents were blocked with debris, including rust, leaves and branches, and dead birds (Table 23.4).

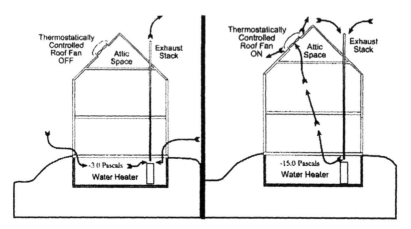

FIGURE 23.1   Downdrafting of water heater caused by activation of thermostatically controlled roof fan.

TABLE 23.4
Conditions for Release of CO into the Living Environment[a]

Description	No. of Instances	% of Instances
Vent Problems		
Negative house pressure	23	27
Combination of improper venting and negative pressure	8	9.2
Disconnected or improper design	6	7
Blocked (debris or birds)	5	6
Exhaust reentrained into house	1	1.1
Other Problems		
Internal combustion engines	19	22
Unvented (no vent installed)	13	15
No CO exposure or CO	3	3.5
Unknown cause of CO[b]	3	3.5
Cracked heat exchanger	3	3.5
False alarm by CO detector[c]	1	1.1
Low battery or bad sensor on CO detector[c]	1	1.1
Total	86	100

[a] Suicides involving intentional poisoning by CO are not included.

[b] A thorough investigation of living environment did not determine the source of CO.

[c] These problems did not yield an exposure to CO.

Data collected by Greiner at Iowa State University.

Nationally, almost 5000 people are injured each year from unintentional CO poisoning–related incidents at a residential societal cost of over $1 billion annually.[16] From 1979 through 1988, 203 Iowans died due to inhalation of CO.[3] The public has expressed concern about CO. Of 2198 responders to the 1992 Iowa Farm and Rural Life Poll, 74% rated CO poisoning in homes a major to moderate risk.[17] Only 7% thought CO poisoning was of little importance. In a survey of 106 Iowa fire departments, 94 agreed or strongly agreed with the statement that CO poisoning is a serious problem.[18] Fire departments respond to thousands of CO incidences every year, but their reporting system does not provide a separate category for CO.

Nonintentional deaths from CO are preventable through education, installation of new fuel-burning equipment, and use of CO alarms. Education is needed to warn the public of the dangers of CO. Education is needed for all professionals involved with equipment that burns fossil fuel and for medical staff responding to CO exposures. New equipment designed for use in tight houses is available. And last, placing CO alarms in all locations where there is a potential danger of CO gives early warning of elevated concentrations.

23.2 SOURCES AND CHARACTERISTICS OF UNINTENTIONAL CARBON MONOXIDE

23.2.1 GAS FURNACES

Vent failure is a common reason for combustion products from natural draft gas-fired furnaces to enter buildings. In 1986, Canada Mortgage and Housing Corporation, the Canadian government housing agency, conducted a field survey of gas- and oil-heated houses.[19] Of the 606 gas-heated houses surveyed, 2.2% experienced a major venting failure. An additional 7.8% had prolonged venting failure on at least one occasion. In total, approximately 10% of the gas-heated houses had experienced prolonged and unusual amounts of combustion gas spillage over the monitoring period. Another 65% experienced either start-up spillage or prolonged spillage of small quantities of combustion gas. Only 24% experienced no combustion gas spillage. Fortunately, most of the furnaces were not producing significant amounts of CO.

Vent spillage or failure allows the CO produced to enter the building. In a study of 86 CO exposures, vent failure was the most common cause of CO release into a building, occurring in 65% of the cases.[4]

Failure of the heat exchanger in a furnace, typically thought to be a major cause of CO production and release, was found in only 3 of the 86 cases. Two of the furnaces were old and had rust holes in the exchangers. The third furnace, a high-efficiency forced-draft unit, was less than 2 years old. The safety system had been modified by a heating contractor, the vent system was undersized, and an improperly adjusted gas pressure regulator allowed too much gas flow to the burner. The resulting overheating, without an operating safety system, cracked the heat exchanger.

Heat exchanger failure is extremely dangerous. Large holes or cracks in a heat exchanger allow air from the circulating fan to disrupt the flame pattern, increasing the production of CO. The air blowing into the burner area also disrupts the flow of combustion gases through the furnaces, increasing combustion gas spillage from the burner chamber and/or the draft diverter. There are several methods used to detect holes in heat exchangers, including direct observation, pressure testing, smoke bombs, odor tracing, gas tracing, salt sprays, CO measurement in combustion gases, and observation of the flame. Various specialized equipment is used for the inspections, including mirrors, microvideo cameras, pressure gauges, and gas measurement instruments.

Burners can produce excess CO for a number of reasons, including:
Insufficient combustion air,
Insufficient primary air,
Insufficient secondary air,
Cooling of the flame,
Impingement of the flame on cold metal,
Rust or dirt on the burner,
Excess gas flow,
Disruption of the flame.

Properly adjusted gas burners in residential heating appliances produce little CO, typically less than 20 ppm. Incorrectly operating burners can produce CO in extremely high concentrations, with units in excess of 8000 ppm found.

Visual inspection of the burner will reveal obvious problems including rust, scale, or soot. Obvious flame pattern disruptions or improper color indicates a problem with combustion. Unfortunately, visual inspection is not sufficient to verify proper combustion. Burners producing high concentrations of CO can burn blue. Conversely, burners producing little CO can burn yellow.

Without instruments, it is difficult to determine reliably if a burner is producing excessive CO. The American Society of Heating, Refrigerating, and Air Conditioning Engineers, Inc. states, "It is desirable through the use of suitable indicators to determine whether or not CO is present in flue gases. For safe operation, CO should not exceed 0.04 percent (400 ppm) air-free basis in the flue."[20] Use of an instrument to measure CO concentrations outdoors, inside the structure, and in the flue products is critical to ensure the heating appliance is operating safely.

Representative case studies of CO exposures from gas furnaces demonstrate the complexity and difficulty of accurately identifying the source(s) of CO.[4] Many times there exist multiple sources of CO. Identifying and correcting one source may not remove the exposure to CO, as illustrated by case 981029. This case also depicts the interconnectedness of operating several appliances and how backdrafting can occur.

Case 981029

Two residents were exposed to CO from a residential furnace. The family installed a CO detector in the spring of 1997, which alarmed. The utility company determined the gas kitchen range produced excessive amounts of CO. It was replaced with an electric range. In the fall, the CO detector again alarmed. The family purchased a second detector and had two events of CO. The fire department and utility company determined that operating the gas fireplace and forced air furnace concurrently resulted in downdrafting. The family was told not operate both simultaneously. The family, while following the advice previously given them, was still being exposed to CO. Additional investigation found the furnace and water heater downdrafting resulted when either the gas fireplace or clothes dryer operated. The downdrafting furnace produced 4424 ppm CO, far in excess of industry standards. Due to the age and low efficiency of the furnace, the family was advised to replace the furnace with a direct-vent sealed-combustion unit. Their gas log insert vented to the outdoors produced 299 ppm CO in the flue products. The family was cautioned about the need to ensure adequate drafting and to monitor CO levels in the room carefully when using the log.

Case 9402161 illustrates that when a CO prevention solution is implemented and CO detectors continue to alarm, a common response is to discount the accuracy of the CO detectors.

Case 9402161

Five people, a woman and her four daughters, were repeatedly exposed to CO from furnaces. The family purchased a 5500-ft^2 home. Heating appliances were inspected before purchase and placed on a maintenance contract. In October all five occupants were poisoned by CO and treated with hyperbaric oxygen (HBO). The heating contractor found an improperly installed flue damper on the water heater and replaced it. The family installed four battery-operated CO detectors. They intermittently alarmed. The heating contractor told the family the cause was the extremely difficult weather in the area. He said the weather had caused the outside combustion air intakes to freeze over and also caused downdrafting of the gas fireplace. He extended the exhaust roof vent higher, changed the combustion air opening to reduce the likelihood of freezing, added an additional combustion air intake, and advised the homeowner to add glass doors on the fireplace. All the changes were made, and still the CO detectors would intermittently sound. The contractor told the homeowner the CO problem was fixed, that she should get rid of the CO detectors, and that he was not willing to respond to any more alarm calls. He loaned her his chemical draw detection equipment and advised her to do her own test. She asked for assistance. Further investigations found the furnaces and water heater produced high concentrations of CO and would downdraft, spilling CO into the house in a number of situations, including operation of bathroom exhaust fans, clothes dryers, kitchen exhaust fan, and/or operation of the gas fireplace. The house was "loose," and when a strong west wind blew over the roof, air leaking out of recessed lights could also downdraft the furnaces and water heater located in the basement. The recommended solution was to replace the natural draft furnaces with direct-vent sealed-combustion units. Replacement of the furnaces solved the CO problems caused by the furnaces, although the gas fireplace will still downdraft and raise CO concentrations under certain conditions.

Often the source of CO is not identified in the first investigation as reported in case 9512211. The "false alarm" poses a hasty solution to transient CO exposures that result from downdrafting.

Case 9512211

A family of five became ill from CO from a residential furnace in a rented house. The family moved in on December 1 and all family members became ill. The utility company and fire department responded, and told the family it was a "false alarm," caused by a faulty sensor. The fire department replaced the sensor. A newspaper article about the "false alarm" prompted another check of the home for possible sources of CO. The furnace was emitting over 4600 ppm CO in the flue gases with evidence that downdrafting often occurred. The furnace burners were rusty and sooty, with paint burned off the furnace cabinet. The vent system was severely undersized. The owner was contacted, but was reluctant to repair the furnace since the fire department said there was "no problem." His suggested solution was to get rid of the CO detector and do nothing more. After the city inspector and fire department were contacted, the owner made the repairs. No more detector alarms were noted.

23.2.2 GAS WATER HEATERS

Gas water heaters can be the cause of CO problems. Extensive indoor air quality tests conducted for the Minneapolis/St. Paul Metropolitan Airports Commission found that 10% of the water heaters in the homes evaluated produced more than 100 ppm CO in the flue gases during downdrafting conditions. Almost half of the water heaters failed a worst-case condition spill test.[21] In a summary by Greiner[4] of 86 cases causing 14 deaths, water heaters were the source of CO in 14 instances, and caused one death. Rescue workers and police officers failed to recognize the presence of CO and were also poisoned. In another instance, an 8-month-old water heater filled a six-story hotel with CO, and required the hotel to be evacuated.

Three case studies of CO exposures from water heaters demonstrate the high concentration of CO that can be produced by water heaters and the significance of effective venting of combustion gases from water heaters. Water heater investigations must examine both the amount of CO produced and the venting system.[4]

Case 9808051

CO poisoning from a water heater in a mobile home resulted in one death and seven sickened. An adult male experienced breathing problems and died. Two other family members were ill. Three rescue workers and two police officers responded to a 911 call but failed to recognize the presence of CO and were also poisoned. The fire department arrived and discovered over 660 ppm of CO. All six survivors, including one pregnant female, were treated with HBO. The fire department determined the source of the CO was from a disconnected water heater vent pipe resulting when a water heater leg dropped through a floor softened by leaking water.

Case 98062501

One woman was poisoned by CO from a residential water heater in an 8-year-old home. She experienced headaches for 2 to 3 weeks, tiredness, sore muscles, dizziness, and was nauseous. She installed a CO detector, which alarmed, indicating 50 ppm CO, while she was doing laundry and washing dishes. Later investigation by the utility company and a heating contractor correctly determined the source of the CO was the water heater. In an attempt to correct the problem, the heating contractor planned to extend the vent higher above the roof, which would not have corrected the problem. Further investigation determined that the water heater produced excessive CO (3277 ppm) and failed to vent due to furnace return air leaks in the basement. Cleaning and adjusting the water heater did not correct the CO emissions. The recommendation was to replace the water heater and vent with a new direct-vent sealed-combustion unit.

Case 9708161

Over 60 guests of a six-story downtown hotel were evacuated after exposure to CO from a gas water heater. Four guests required oxygen. In December of 1996 a new water heater had been installed in the basement of the hotel. On Friday night, August 15, 1997, a family of four staying in a sixth-story room became ill. Saturday afternoon they sought medical treatment. The cause of their illness was determined to be CO. Police, fire, and utility investigators found up to 600 ppm CO in the hotel and it was evacuated. On Sunday, a heating contractor worked on the water heater and declared it safe. The hotel was reoccupied, but water heater use was delayed until it could be rechecked. On Monday, the water heater, although burning with a blue flame, still produced over 4000 ppm CO in the flue. Additionally, the vent system, which on Sunday had been declared safe by the heating contractor, was rusted and had fallen apart at the connection to the main stack. All combustion products from the water heater were spilling into the hotel. In approximately 45 min of water heater operation, the CO produced was detectable on the fourth, fifth, and sixth floors, with levels of 100 ppm in the sixth floor room where the family had been poisoned the previous Friday and Saturday. The heating contractor, by not testing combustion products for CO and by not thoroughly inspecting the vent system, left the water heater and vent in a dangerous condition.

23.2.3 GAS-FIRED OVENS/RANGES

CO from kitchen ranges can elevate CO concentrations in homes. Kitchen ranges are allowed to produce up to 800 ppm CO in an air-free sample of the flue gases.[10] Most residential gas-fired oven/ranges vent directly into the space surrounding the oven. Continued operation of a kitchen range producing 800 ppm, without operation of an exhaust fan to remove combustion gases or adequate ventilation to remove and dilute combustion gases, will cause CO levels to rise. In a study of 60 residential gas ovens in apartments, 6 ovens produced peak CO concentrations in flue gases of over 2000 ppm.[22] The CO emitted quickly raised CO concentrations in the apartments.

The highest steady-state concentration reached was 350 ppm, after only 40 min of operation. Two other apartments reached concentrations of 200 ppm or more in 65 min or less. In 51% of the apartments, the ovens raised CO concentrations in the room above the EPA standard of 9 ppm.

Not all agencies agree that 800 ppm CO in flue gases is an allowable emission level. The Metropolitan Airports Commission, Minneapolis/St. Paul, instituted a Sound Insulation Program in homes surrounding the airport and contracted with the Center for Energy and Environment to conduct extensive indoor air quality evaluations. The program selected 100 ppm as the maximum allowable CO concentration in flue gases of gas appliances. Of the natural gas ovens tested, 52% exceeded 100 ppm CO in the flue gases.[21]

A summary of 86 CO investigations found kitchen ranges were the source of CO in 10 instances.[4] The highest steady-state CO concentration measured in the flue of a gas oven was 6566 ppm. Previous to the investigation, after a day of baking, a pet lovebird in the home had died. Suspecting CO, the family installed a CO alarm, which verified that CO levels in the home rose after the oven was used. The lowest emission rate from a gas oven in the 10 instances was 232 ppm. This compares with lowest emissions rates for furnaces of 0 ppm, water heaters of 0 ppm, and tailpipe emissions of an automobile of 15 ppm.

The basis, rationale, and assumptions for the 800-ppm CO allowance in ANSI Z21.1 were first formulated in 1925.[23] In 1995, the Gas Research Institute contracted with Battelle (Columbus, OH) to critique the ANSI Z21.1 Standard. Battelle concluded,

> The basis for and limit set by the original CO standard remain valid today, compared with estimates derived from Consumer Product Safety Commission (CPSC) and Underwriters Laboratories (UL) models and codes, respectively. Although the Z21.1 protocol to determine compliance generates CO levels that are much higher than those expected during normal oven/range use, they are still much lower than the Z21.1 limit of 800 ppm CO, O_2-free. These facts, and recent field data, compellingly indicate that if the 800 ppm limit is met, ambient CO levels should remain below possibly threatening limits (≤25 ppm for 1-hour). As such, gas ovens/ranges should not pose a public safety or health threat with regard to CO emissions, a characteristic that can be reliably validated in the lab, factory, or field using the ANSI Z21.1 protocol, and its specified limit of 800 ppm CO, O_2-free.

Still, there remain concerns that an allowable CO emission rate of 800 ppm CO, O_2-free does not adequately protect the safety and health of the general public. In a paper presented at the 1999 Affordable Comfort Conference, the validity of the 1925 assumptions was questioned. The assumptions regarding the acceptable concentrations of CO in a room (100 ppm) and the number of air changes per hour (4) were thought too high. The duration of use (1 h) was thought too low.[24] Note that buying a new range might not correct a CO problem, since new ranges are allowed to emit up to 800 ppm CO.

Electric ovens can also be a source of elevated CO levels in a home. Electric heating elements do not produce CO, but burning food, or burning food residue, does. A utility company educator reported elevated CO during the self-cleaning cycle of electric kitchen ranges.[25]

Two representative case studies of gas oven/ranges illustrate that extended use or failure to follow recommended procedures could result in CO exposures.[4] Gas oven/ranges can produce levels of CO that over time can cause health effects as identified in case 9510301. Exhaust hoods and other supplemental ventilation should be considered when operating a gas oven/range (case 9712101).

Case 9510301

Two women were exposed to CO from a new gas-fired kitchen range. They had just moved into a new mobile home and wanted to add moisture to the air. They placed a large pan of water on a top burner of the gas kitchen range, and left the burner on all night. The next morning they were ill, vomited, and had severe headaches. They opened the doors and windows, and called the utility company, who found elevated concentrations of CO. Neither the furnace nor water heater, both new direct-vent sealed-combustion units made for mobile homes, produced any CO or spilled CO into the mobile home. The gas kitchen range was checked and found to produce up to 232 ppm CO. It was identified as the likely source of the CO exposure.

Case 9712101

A family was alerted to CO from the gas-fired kitchen range oven by a CO detector located in the kitchen. During baking, CO concentrations rose to 90 ppm and the detector alarmed. The woman was pregnant and concerned about exposure to CO. The family was advised to always use the kitchen exhaust range hood, to have the oven cleaned, adjusted, and tested, to open windows when using the oven, and to minimize use. The woman reported she was able to prevent any CO readings on the detector by following the above procedures.

An exhaust hood vented to the outdoors should be used whenever a gas-fired oven/range is used. Even when a gas-fired oven/range is properly tuned, there will be some CO produced along with CO_2, nitrogen dioxide, and water vapor. Kitchen range manufacturers recommend installation of a range hood to exhaust the combustion products along with cooking odors, grease, and moisture produced during cooking. Failure to use the range hood exhaust fans results in indoor air pollution.

Adequate makeup air into the house must be provided for the kitchen exhaust hood. Exhaust fans can depressurize a house and cause downdrafting of vented furnaces, water heaters, boilers, fireplaces, and vented room heaters. A qualified heating contractor should install the exhaust hood and run a "worst-case" downdrafting test to ensure that all the systems work correctly.

Ovens should never be used to heat the home. The broiler and oven burners are designed to burn with the door closed. Opening the oven door disrupts the airflow pattern, and high concentrations of CO may be produced. The oven burner is not designed to operate continuously. CO emissions standards are based on operation for only 1 h. Continuous operation could produce higher levels of CO in the home.

23.2.4 VEHICLES

Many vehicle engines produce high concentrations of CO. This is not a new problem. In 1923, Henderson and Haggard reported commonly finding 100 ppm CO on New York City streets, with concentrations of 460 ppm occasionally present for brief periods. Van Deventer remarked that dangerous mixtures of CO could be found in a second car closely following another in traffic.[26] In 1973, Ayres[27] reported, "A 90-minute Los Angeles freeway exposure produced electrocardiographic (ECG) abnormalities in 40% of patients with preexisting cardiovascular disease and decreased exercise tolerance. Expressway CO levels approached 25–100 ppm."

Typical spark-ignited gasoline engines, even when well tuned, will produce CO. Generally, power output is maximum when CO is from 15,000 to 25,000 ppm. A rapid power decline occurs for lean limits (CO < 10,000 ppm), which is the primary reason gasoline engines usually are tuned on the rich side.[28] An effective way to reduce CO emissions is by means of a catalytic converter. Field measurements have found CO emissions as low as 15 ppm from the tailpipe of a gasoline engine equipped with a catalytic converter.[4]

In 1968 the U.S. Environmental Protection Agency began regulating CO emissions from some on-road motor vehicles. The levels of CO emitted have been reduced by over 95%. Still, CO remains a major pollutant. As of 1989, more than 33 million people resided in 41 areas that failed to meet national air quality standards for CO.[29] Some states do not require emission checks. It is common to encounter individual vehicles emitting excessive amounts of CO and leaving a plume (cloud) of CO. Today, just as Van Deventer reported in 1935, following a vehicle emitting high concentrations of CO can cause CO exposure in the second vehicle.

The lethal consequences of CO in engine exhaust is tragically illustrated by the hundreds of persons who die each year from CO poisoning caused by running a vehicle inside a closed garage. Others die or become ill at skating rinks, while attending indoor races or tractor pulls, in homes with attached garages, while stranded in their car, or while driving or riding in a vehicle with a defective exhaust system. Individual exposures to CO from vehicle exhaust caused 5 of the 14 deaths reported by Greiner.[4] The high CO concentrations produced by a cold engine when first started (over 100,000 ppm in some cases) can be trapped in the garage and move into the house.[11]

An attached garage increases the likelihood of CO problems in the home. In the typical house during winter, air flows from attached garages into the house. The amount of this flow varies from house to house. A Minnesota study found from 5 to 85% of the air leaking into the house came from the garage, carrying CO and other contaminants into the house.[13] The garage serves as a large source of CO. As the CO leaks into the house it is diluted, so CO concentrations in the house are less than those in the garage. It can take several hours for CO concentrations in the home to reach a maximum. Often CO detectors sound an alarm several hours after the vehicle left the garage. The problem is worse in the winter because cold engines produce more CO, cars are warmed up longer and more often, houses are closed, and the larger temperature differences in winter can increase airflow from the garage to the house.

Defective vehicle exhaust systems are extremely dangerous because of the high concentrations of CO in the exhaust stream. CO leaking from the exhaust system can enter the vehicle through holes in the body or open windows or doors. Exhaust systems must be gas-tight from the engine to the end of the tailpipe. The low pressure at the back of vehicles can pull combustion fumes into any holes or openings at the rear. Rear tailgates, rear windows, and trunks must be tightly closed when the engine is running. Every year people are poisoned in the back of pickup trucks. In a report on 68 Washington State pediatric patients treated for nonintentional CO poisoning, 20 cases occurred as children rode in the back of pickup trucks. In 17 of these, the children were riding under a rigid closed canopy on the rear of the truck, while three episodes occurred as children rode beneath a tarpaulin. Loss of consciousness occurred in 15 of the 20 children, 1 child died, and 1 had permanent neurological deficits.[30]

Engines tuned for maximum power and operated indoors at races and tractor pulls create large amounts of CO. Greiner[4] measured concentrations of 92 ppm at an indoor motorcycle race, even though doors were open, ventilation fans operating, and there were periods of nonracing. In Manitoba, CO concentrations had risen to 262 ppm at the end of an indoor tractor-pulling event. At the next event, additional ventilation was provided and additional time was allowed between pulls. By the end of the show the average levels had risen to 435.7 ppm. After evaluating possible solutions to correct the problem, arena management concluded the costs were prohibitive, and no more events of that nature are to be held in the arena.[31]

CO emissions from compression-ignition diesel engines are typically much lower than from gasoline engines. Diesel (compression-ignition) engines run with an excess of air, leading to nearly complete combustion of the fuel. If the diesel fuel is burned incompletely, if the engine is overloaded, or if the engine is overfueled (rich mixture), diesel engines can produce high concentrations of CO.

Contrary to popular belief, internal combustion engines fueled by LPG/propane produce CO. LPG often powers forklifts operated indoors in an attempt to reduce worker exposure to CO. The same precautions against running a gasoline engine in an enclosed space should be observed with an LPG engine. Industry sources report a properly tuned LPG engine will produce from 200 to 20,000 ppm CO, depending on design, carburetor settings, condition, and load.[32] Poorly tuned engines produce much higher concentrations. Greiner[4] measured tailpipe CO concentrations of 55,200 ppm after workers were poisoned by CO. LPG engines are sensitive to carburetor settings. Mechanics will often feel a fork truck is running well even when high CO emissions occur. One study stated[33]:

In determining the volume of CO in the exhaust emissions, it was noticed how critical the carburetion settings are. By changing the jet adjusting screw a fraction of a turn, the emissions will change from 50,000 to 50 ppm. Therefore, it was established that it is not practical to set up truck carburetion systems without using some type of (CO) gas analyzer.

LPG engines are used in ice-resurfacing machines, and have been identified as a significant source of CO in indoor skating arenas. Poor natural ventilation and nonuse of mechanical ventilation contribute to the high concentrations of CO. A study in Ontario of four skating rinks found CO levels of 5 to 110 ppm after resurfacing.[34]

Three representative case studies of CO exposure from vehicles demonstrate the dangerous and often deadly results.[4] Operating a vehicle that produces extremely high CO concentrations in a confined space as identified in Case 9510301 can result · in death.

Case 9712081

Two young men died of CO poisoning from a vehicle running in a car wash. A vehicle was found in a closed car wash bay with the ignition on, gas in the tank, a discharged battery, but not running. Investigators, in conjunction with local police and fire departments, conducted tests. The vehicle, which was poorly tuned and did not have a catalytic converter, produced approximately 90,000 ppm CO out the tailpipe. After running for 38 min, the engine stalled. When the vehicle was started in the closed car wash, CO concentrations exceeded 1200 ppm in approximately 8 min (the level considered to be immediately dangerous to life and health, IDLH). In 22 min, concentrations rose to 3500 ppm.

Operating a vehicle in an open environment can result in CO exposure if the exhaust system fails (Case 9504081) or when the enclosed vehicle environment is compromised by opening a door, window, or hatch (Case 9609291).

Case 9504081

A 20-year-old female was exposed to CO from vehicle exhaust. CO concentrations were measured in the passenger compartment of a vehicle with a cracked exhaust manifold. The loud exhaust sound emanating from the engine compartment had alerted the woman to the problem, and she had driven with the windows down. To determine the amount of CO entering the passenger compartment, the engine was started and the car driven with the windows closed. In only 2 min, CO concentrations rose to 156 ppm and were still rising. The closed window test was deemed to be too dangerous to continue. With all the car windows open, the car moving, and the heater blower operating, concentrations were 50 ppm. The car had a broken exhaust manifold and missing oxygen sensor.

Case 9609291

Three members of a family were exposed to CO from an attached garage. The family was alerted to the problem after installing a residential CO detector early in 1996. The detector repeatedly alarmed, often in the middle of the morning. A heating contractor cleaned and adjusted the furnace, and installed a new vent system for the furnace and water heater. The problem persisted. Neither the furnace nor the water heater produced CO or spilled combustion products into the house. CO was produced and entered the home when vehicles were cold-started in the tuck-under garage. Over 90,000 ppm CO came out of the tailpipe of the family's car. Even with the garage door open while the car warmed up, CO was trapped in the garage after the car was backed out. The family was advised against warming up engines in the garage.

23.2.5   GAS-FIRED UNVENTED (VENT-FREE) FIREPLACES

Some sources always produce high concentrations of CO, such as wood burning in an open fireplace, smoldering embers, and charcoal. Release of combustion products from any of these sources into enclosed areas is always extremely dangerous and must be avoided.

In contrast to CO from wood burned in an open fireplace, CO production from gas burners is highly dependent on proper design and operation. A properly operating gas flame produces little CO. Previous research has concentrated on blue flame burners. Yellow flames were generally considered undesirable, and were a precursor of soot and/or CO emission problems. For aesthetic effect, yellow flames were desired in decorative appliances. A study sponsored by the Gas Research Institute determined design parameters for a yellow flame burner with acceptable emissions and no soot deposition.[35] The Gas Appliance Manufacturers Association saw decorative yellow flame burners used in vent-free appliances as a new market. Manufacturers developed vent-free appliances, and are actively promoting sales and use.

An unvented heating appliance, also called a vent-free appliance, has no chimney vent and is located in the space being heated. Combustion products are discharged directly into the heated space rather than exhausted to the outdoors through a chimney. Because combustion products from a vent-free appliance are vented directly into the structure, the safety factor of venting to the outdoors is lost. To increase safety, oxygen depletion sensors (ODS) are now installed on unvented heaters. The ODS is designed to shut off gas flow to the burner before oxygen levels in the room drop below levels that can safely sustain human life or safe operation of the burner.

In a 1996 report, the Gas Research Institute (GRI) noted that previous studies had found, "Unacceptable concentrations of some combustion products may be found in the living space if the vent-free products are improperly sized for installation, or if they are installed or operate incorrectly."[36] GRI developed and verified a computer model to simulate the impact of vent-free gas heating products on indoor air quality (IAQ). Since there are no national indoor air quality standards for residences, GRI selected values from available guidelines. The guidelines selected were:

15 ppm for 1 h average for CO,
0.5 ppm for 1 h average for NO_2 (nitrogen dioxide),
3500 ppm for 1 h average for CO_2 (carbon dioxide),
60 to 40% relative humidity for H_2O vapor (depending on Department of
   Energy heating zone),
19.5% continuous for O_2.

In the GRI study, conditions of emission level, home structure, ventilation rate, outdoor temperature, and vent-free appliance input rate were specified.[36] The heating appliances were assumed to meet the requirements of ANSI Z21.11.2 with regard to emissions, houses were assumed to be loose, average, or tight, with ventilation rates of 1.0, 0.5, and 0.35 ACH, and six Department of Energy heating regions were considered. Under the conditions specified, the highest CO concentrations were 6 ppm. Oxygen levels remained above 19.5%. NO_2 reached the guideline level only under the most stringent conditions of outdoor temperature, H_2O vapor reached the guideline level in the coldest climate, and CO_2 reached the guideline level for all home structures in the coldest climate. To reduce pollutant concentrations in cases when the emission guideline level is reached before 4 h of operation, the recommended input rate of the appliance was adjusted downward by GRI. The report notes, "If the vent-free product is sized in accordance with Table VII (Section 3) and operated at $-10°F$ it may not be possible to create a comfortable temperature, however, all IAQ guidelines will be maintained."[37]

There are many reasons pollutants can rise to higher concentrations than assumed in the GRI study. CO emission levels from the vent-free heating appliance can rise above ANSI-specified levels. Rust or dirt on a burner, excess manifold gas pressure, flame impingement, flame disturbance, or design flaws can increase CO emission levels. An ODS protects only against a low-oxygen condition, and does not furnish protection for the preceding list of possible causes of emissions. A 2-year-old vent-free gas fireplace was producing 570 ppm CO, air-free. In a day of field testing, the manufacturer was unable to determine the cause of excess CO production, and authorized a free replacement.[4]

Pollutant concentrations rise to higher concentrations in tighter houses. Average exchange rates do not give an accurate indication of peak concentrations of pollutants under conditions of low infiltration/exfiltration (i.e., on still days). Stewart et al.[37] found natural ventilation extremes in three "very tight" houses measured over a 1-year period using tracer gas, ranging from a low of 0.02 ACH to a high of 1.04 ACH.

Control of pollutants from vent-free heating products depends on proper use, including installation of a properly sized heater used in a properly ventilated house for no more than 4 h at a time. Not all persons limit the size of heater, provide adequate ventilation, or operate the heater no more than 4 h at a time. An oversized heater, in a tighter-than-assumed house, operated for longer than 4 h will exceed the GRI selected indoor air quality guidelines.

Once combustion pollutants are in the house, outside air should be provided to dilute and remove the contaminants. The additional ventilation needed to reduce pollutants to tolerable concentrations depends on many variables, including:

Volume of the structure,
Fuel input rate,
Pollutant emission rate,
Operating time,
Tightness of the building, and
Health and tolerance levels of the occupants.

In some areas, local or state codes regulate or prohibit use of unvented appliances in living areas. Where unvented gas appliance use is permitted and desired by occupants, the following are suggested:

1. Install an Underwriters Laboratory (UL) or an Industry Applications Society (IAS) listed CO detector. Because low concentrations of CO can cause health problems, purchase a detector advertised as "sensitive" or with a digital display that will display low concentrations.

2. Use only approved gas heaters with ODS pilots.
3. Follow all operation and maintenance instructions carefully.
4. Clean the burner yearly, or more often, as required in the owner's manual.
5. Do not use an oversized heater. GRI recommends limiting the amount of pollutants by correctly sizing (limiting) heater input. A larger heater, in cold weather, will pollute the air beyond allowable guidelines.
6. Do not operate for more than 4 h at a time.
7. Use only as a decorative or supplemental heater and not as a primary heat source.
8. Do not use unvented heaters in bedrooms, bathrooms, or confined spaces.
9. Provide adequate ventilation, as required in the owner's manual. If the home has weather-stripped doors and windows, an outside air source will likely be required.
10. Provide additional ventilation, or discontinue unvented heater use, if the pollutants cause health problems.

Representative case studies (970605, 9801141, and 9711121) of fireplace CO exposures demonstrate the high level of CO production and downdrafting issues for fireplaces.[4]

Case 970605

A couple was exposed to CO from an unvented gas fireplace in the basement of their home. After installing the fireplace, the couple experienced dry throats, headaches, sleepiness, and reported the house was "stuffy." They installed a CO detector, which repeatedly alarmed. Investigators found the unvented gas heater produced 570 ppm CO, which, even with a window open, would raise CO concentrations in the room to unacceptable levels. The manufacturer's engineer checked the unit and was unable to decrease CO production rates by cleaning and adjusting the burner.

Case 9801141

A family of four was exposed repeatedly to CO from their wood-burning fireplace. The family purchased a manufactured home in 1997. They noticed a considerable amount of soot and periodic readings on their digital CO detector. The family often experienced headaches, respiratory problems, and flu-like symptoms. The home had a central air return to the furnace with no return air ducts in individual rooms. When bathroom and bedroom doors were closed, the family room and kitchen went "negative" (i.e., the pressure was less than the pressure outdoors). As the wood fire in the family room fireplace died down, the negative pressure reversed the draft, pulling outside air and wood smoke containing soot and CO into the home.

Case 9711121

Three members of family were exposed to CO from a gas insert in a wood-burning fireplace. In the evening, the husband started a gas fire in the fireplace. Just before midnight, the CO detector located at the top of the second story stairs alarmed. The fire department responded, but did not have CO instruments. The fire department aired out the house, decided the source had to be the gas insert, and allowed the family to return. The fireplace chimney was cleaned and the furnace and water heater inspected. Investigators found excessive CO production from the 2-year-old gas log insert, over 3000 ppm CO. When all exhaust appliances were operated, the negative pressure produced in the living room downdrafted the fireplace, causing all the CO to enter the living room. The family discontinued use of the gas log, and asked the builder to replace the defective unit.

23.2.6 NONSTATIONARY CARBON MONOXIDE SOURCES (SMALL ENGINES, CHARCOAL GRILLS)

Internal combustion engines have long been recognized as a source of CO. The 1938 text *Carbon Monoxide Asphyxia* stated, "Cases of acute carbon monoxide poisoning from exhaust fumes need no comment."[26] Even small engines can produce sufficient CO to cause fatalities.[39] A NIOSH study identified two Iowa cases of CO poisoning from small engines used indoors.[38] One caused death of a 33-year-old farm owner, who died while using a gasoline-powered pressure washer to clean the swine-farrowing barn. He had worked about 30 min before being overcome. In the second case a 12-year-old boy was unconscious near the door of a swine-farrowing building. He had been working alone using an 11-horsepower, gasoline-powered pressure washer for about 30 min to clean the building. NIOSH conducted several environmental measurements to estimate how quickly small engine exhaust can produce dangerous CO concentrations. In one example, they measured CO in an 8360-ft³ double-car garage. A pressure-washer powered by a 5.5-horsepower gasoline engine was started in the garage. The two double-car doors and one window were left open, and the vent was unsealed; breathing-zone concentrations of CO reached 200 ppm within 3 min and peaked at 658 ppm within 12 min. NIOSH warns, "*Do not use equipment and tools powered by gasoline engines inside buildings or other partially enclosed spaces unless the gasoline engine can be placed outdoors and away from air intakes.*"

Are LPG powered floor buffers safe to use indoors? The owner's manual for an engine used on an LPG floor buffer states, "*Exhaust gas contains poisonous CO. Avoid inhalation of exhaust gas. Never run the engine in a closed garage or confined area.*"[40]

Another supplier of engines for LPG-powered floor buffers incorporates an electronic, closed-loop LPG fuel system and three-way catalytic muffler, with diagnostic lights on an electronic control unit. Its manual warns, "*Engine exhaust is deadly! Learn the symptoms of CO poisoning in this Manual. Inspect the exhaust system every time the engine is started and after every 8 hours of operation. If exhaust noise changes, shut down the engine immediately and have it inspected. The integral exhaust system must not be modified in any way. Do not use engine-cooling air to heat a room or compartment. Make sure there is ample fresh air when operating the engine in a confined area.*" The manual further warns, "*If the engine shuts down with the light on, emissions of deadly CO gas could be excessive. Do not restart the engine. Have the engine serviced to avoid serious personal*

Injury." It further warns, *"Run the equipment only in well ventilated areas. Refer to* Industrial Ventilation — A Manual of Recommended Practice....*The user is the one responsible for complying with indoor CO regulations. Attach a CO monitor to the equipment."*

Another warning states, *"Exhaust gas is deadly! All engine exhaust (including that from an LPG engine with a catalytic muffler) contains CO — an odorless, colorless, poisonous gas that can cause unconsciousness and death."* Symptoms of CO are then listed.[40] The manufacturer's recommendations must be followed carefully. Failure to do so could lead to CO poisoning.

Operation of a floor buffer in a day-care center over a weekend sent 58 children and 11 adults to emergency centers on the following Monday morning. Sufficient CO was produced and had remained in the building to poison all occupants.[4]

Small gasoline engines on electrical generators, concrete finishers, water pumps, chain saws, and pressure power washers typically use simple carburetor systems with limited control over the air–fuel ratio. The engines run rich with high concentrations of CO, typically 30,000 ppm or more. Manufacturers stress that the engines are to be used only in well-ventilated outdoor areas, and are never to be used indoors even with ventilation. NIOSH simulated operation of a 5-horsepower gasoline engine in a 10,000 ft^3 room. With 1 ACH, CO concentrations reached over 1200 ppm (the Immediately Dangerous to Life and Health level) in less than 8 min. Even with 5 ACH, CO concentrations reached 1200 ppm in less than 12 min. NIOSH concluded it is not safe to operate gasoline engines indoors![38]

Burning charcoal produces high concentrations of CO. Two persons were found in their fifth-wheel camper dead from exposure to CO from a charcoal grill.[4] To verify that the small grill was the cause of their deaths, gas-fired appliances in the camper were first tested for proper operation. After they were found to operate correctly, approximately 1.6 lbs of charcoal was placed in the grill and burned. CO concentrations reached 750 ppm in the camper, and death from CO from the charcoal was verified.[4]

These representative case studies of CO exposure from nonstationary CO sources demonstrate how a vent-free gas heater, carpet cleaner, floor buffer, and charcoal grill are often not understood as sources of CO.[4]

Case 981016

A family of three was exposed to CO from an unvented (vent-free) gas heater. In 1996, the family purchased an unvented (vent-free) gas heater from the gas company to use as their primary heat source. Their home, previously heated with wood, was always extremely dry in winter. Now the home was wet with considerable mold growth. They began to experience numerous respiratory problems and headaches. The husband bought a CO detector, and within 2-h of heater operation, the detector alarmed. Investigators advised the family that burning fossil fuels in an unvented heater always dumps a considerable amount of water into the home and can also produce CO. The family replaced the unvented heater with a vented model, and 2 months later reported their house was dry, and their health improving.

Case 9712221

Two men poisoned by CO while their carpet was being cleaned required oxygen treatment. The carpet cleaner parked his truck in the garage, left the garage door open, and started the engine used to power the carpet cleaner. The homeowner, working in the room above the garage, suddenly became ill with a severe headache and a drunken feeling. He talked to the carpet cleaner, who was also experiencing severe headaches and then called 911. The utility company found over 1200 ppm CO in the garage. Both men were taken by ambulance for HBO treatment. After being given 100% oxygen while being transported to the hospital, the homeowner's carboxyhemoglobin (COHb) saturation was 22% (normal is from 1 to 1.5%).

Case 9610081

In this case, 69 persons (58 children, 11 adults) from a child-care center were treated for CO poisoning caused by a propane-powered floor buffer. Maintenance workers used the buffer over the weekend. On Monday morning, students and staff began exhibiting flu-like symptoms — headache and vomiting. Although the building has all-electric heat and no gas connection, the day-care director asked the utility to check for CO. At 12:20 p.m. the utility found concentrations as high as 180 ppm. The building was evacuated and the occupants taken to three area hospitals, where they were treated on 100% oxygen, with one patient kept overnight. Just a few weeks earlier, 50 children and staff at the another branch of the same day-care center required medical treatment after exposure to CO from a propane-powered floor buffer used the previous day. The fire department reported concentrations as high as 700 ppm and victims showed extreme symptoms and lost consciousness.

Case 9705311

CO from a charcoal grill in a camper caused two deaths. The couple, while camping with friends, grilled supper on a portable, outdoor charcoal grill. At 10 p.m., the couple, thinking the charcoal fire was extinguished, placed the grill in a storage area under their camper. The medical examiner determined CO from the grill penetrated the floor of the camper while the couple was sleeping. An investigator was asked to check all the gas appliances to ensure there was no other CO source. Testing revealed the LPG furnace, water heater, refrigerator, and kitchen range did not materially increase CO levels in the camper. When the grill was placed under the camper, CO concentrations in the camper increased to 750 ppm in only 5 min, a deadly concentration. There was no odor or soot in the camper.

23.3 INVESTIGATIVE TECHNIQUES FOR IDENTIFYING CARBON MONOXIDE SOURCES

Greiner,[41] in a popular press article, states, "Diagnosing indoor CO problems can be difficult because of their intermittent nature. Heating equipment in a structure operates as part of a system, and is subject to influence from depressurization caused by other devices within a building, and by weather conditions affecting the building structure."[41]

In some cases, the source of elevated indoor levels of CO might be from outdoors. Chicago passed the nation's first CO ordinance, mandating CO detectors in single-family and multiunit dwellings on October 1, 1994. On December 21–22, 1994, when an unusual thermal air inversion caused elevated ambient CO levels of 13 to 20 ppm in the Chicago area, the Chicago Fire Department responded to 3464 CO investigations over a 2-day period. The CO detectors that had alarmed had a sensitivity of 15 ppm over an 8-h period.[42] Newer CO alarms must resist alarming below 30 ppm over a 30-day period. Table 23.5 provides the CO concentrations for defined detector limits and exposures and typical levels for sources in the living environment.

The Chicago Fire Department developed the first protocol for responding to CO detector alarms. The first procedure in their protocol is issuance of a CO meter to each company designated as a meter company.[48]

CO is a colorless, odorless, tasteless, nonirritating gas that is deadly. It cannot be detected by any human senses. A meter is absolutely necessary in the performance of investigations to identify CO sources. The Building Performance Institute (BPI) has developed a draft guideline to provide a protocol for those building technicians and investigators attempting to determine whether potential exists for CO to enter a house. The draft guideline states, "Analysts shall have a digital CO analyzer and a draft gauge to assist CO investigations. Differential pressure manometer, blower doors, and duct blasters are also useful in conducting these investigations...."[49]

The atmosphere in the building must be tested for CO before entry. Action must be taken based on the levels found.

It is not possible to determine by looking at flame color if excessive amounts of CO are being produced — a blue gas flame can produce CO. The CO meter selected must be capable of measuring CO source strengths.[50] To measure source strengths, the meter must have a probe and be capable of sampling combustion products in the flue of heating equipment. Although there is no field standard for maximum allowable CO levels in combustion products, an appliance exceeding ANSI CO air-free standards should not be operated until repairs are made (i.e., 200 ppm for water heaters, 400 ppm for furnaces, and 800 ppm for gas-fired kitchen ranges). More stringent standards, as adopted by some agencies, may be appropriate (i.e., 100 ppm by BPI and Minnegasco[13]). A series of tests, measuring both CO and draft, should be conducted under specified test conditions, to ensure proper combustion under all operating conditions that will be encountered.

Exposure to CO caused by intermittent sources, spillage, or downdrafting can be difficult to detect. Spillage and downdrafting episodes can be sporadic and widely isolated. Minnesgasco, a Minnesota/St. Paul utility, responded to over 14,000 CO calls from May 1995 to April 1996. In 22% of the calls a CO source was found, while in 78% no source was found. It often had repeated calls to homes where no CO source was found, and still found no source. To determine if there were sources it was missing in its inspections, Minnegasco randomly selected 50 homes. The homes placed in the selection pool:

Had a CO alarm;
The alarm had sounded two or more times;
Minnegasco had responded two or more times and found no CO source.

TABLE 23.5
CO Concentrations for Defined Detector Limits and Exposures and Typical
Levels for Sources in the Living Environment

Conc. of CO (ppm)	Defined Detectors and Exposures Limits and Typical Levels for Sources in the Living Environment
1–2	Might be normal, from cooking stoves, spillage, outdoor traffic
>2	Raises questions about why CO is elevated; source should be identified, might be normal (i.e., traffic, kitchen range) (Gary Nelson, Personal conversation, The Energy Conservatory Minneapolis, MN)
9	The maximum allowable concentration for 8-h period in any year[43,44]
	Typical concentration after operation of unvented gas kitchen range[22]
20	Typical concentration in flue gases (chimney) of a properly operating furnace or water heater (Greiner, unpublished field studies, January, 1997)
30	UL standards for residential detectors require that they *not* alarm at 30 ppm unless exposure is continuous for 30 days
35	Maximum allowable outdoor concentration for 1-h period in any year[43,44]
50	Maximum allowable 8-h workplace exposure[45]; most fire departments require use of self-contained breathing apparatus for exposures above 50 ppm
70	UL-listed detectors must sound a full alarm within 189 min or less
150	UL listed detectors must sound a full alarm within 50 min or less
200	Maximum recommended workplace exposure 15-min maximum[46]
	Maximum flue gas concentrations for water heater[8]
400	UL listed detectors must sound a full alarm within 15 min; maximum flue gas concentrations for furnace
500	Often produced in garage when a cold car is started in an open garage and warmed-up for 2 min (Greiner, unpublished field studies, January, 1997)
800	Maximum flue gas air-free concentration from gas kitchen oven/ranges[10]
1,600	Smoldering wood fires, malfunctioning furnaces, water heaters, and kitchen ranges typically produce concentrations exceeding 1600 ppm
3,200	Concentration inside charcoal grill (Greiner, single example)
35,000	Measured tailpipe exhaust concentration from warm carbureted gasoline engines without catalytic converters (Greiner, unpublished field studies, January, 1997)
70,000	Tailpipe exhaust concentrations from cold-gasoline engine during the first minute of a cold weather start; concentrations decreased to 15 ppm after 17 min of running (Greiner, unpublished field studies, January, 1997)
100,000	Smoke from burning buildings often reaches 10%[47]

Note: 10,000 ppm (parts per million) = 1% by volume.

Minnegasco contracted with Advanced Thermography to determine if there were sources of CO in the 50 homes. Advanced Thermography identified 65 potential sources of CO in 49 of the 50 homes (several homes had more than one source). Vehicles started in attached garages were found to have caused the CO alarms to sound in 37 of the 50 homes. Six furnaces were producing excess CO, three of them in excess of 4000 ppm in the flue. One water heater was producing 3800 ppm and one was producing 4100 ppm. Other sources included kitchen ranges and heaters.[13]

In a study of 86 CO investigation, Greiner found that vent problems were implicated in 43 instances of CO exposure.[41] In addition to allowing combustion products to enter buildings, Minnegasco and Bohac found that downdrafting increased CO production in some heating appliances.[13,21] The higher CO levels produced, coupled with failure of the combustion products to draft, greatly increased the risk of CO exposure to building occupants from these appliances.

Measurement of house depressurization limits and "worst-case-testing" are two means to determine the potential for downdrafting. Excessive house depressurization causes spillage and downdrafting. Canada determined that the safe house depressurization limit (HDL) is 5 or 6 Pa for an interior B-vent.[51]

A variety of procedures have been identified to test for combustion appliance backdrafting and procedures for "worst-case-testing." The U.S. Environmental Protection Agency utilized three of the procedures to develop a backdrafting test procedure for radon mitigators. The EPA procedure explains how to establish worst-case conditions and conduct the test.[52]

23.3.1 Tests That Should Be Conducted

Interview with occupants and other persons with knowledge that might assist in determining the source of the CO;

Physical inspection of the house, including all exhaust fans and heating appliances;

Physical inspection of attached garages, including measurement of CO concentrations;

Physical inspection of all heating appliances;

Testing of all heating appliances;

Measuring gas flow rates and pressures to determine firing rate;

Inspecting and testing the heat exchanger for cracks, openings, or excessive corrosion;

Inspecting burners and observing burner operation;

Checking that the vent is intact, not blocked, and without rust or weak spots;

Inspecting chimney stack for proper operation and size;

Measuring house depressurization limits;

Conducting worst-case testing;

Determining the adequacy of combustion and makeup air;

Checking for charcoal grill operation, fireplace operation, room heaters, oven or range usage.

Other possible investigative tests include:

Blower door testing to determine structural air tightness;

Forced downdrafting testing to determine CO production under downdraft conditions;

Reenactment testing to determine operation during conditions that caused the known or suspected CO episode;

Continuous data logging.

The safety of the occupants and the tester must be protected. The air should be continuously monitored. At no time should occupants or testers be exposed to over 35 ppm CO.[53] No testing should be conducted which would endanger occupants. Appliances that are producing in excess of ANSI standards (or lower standards, if selected) should be shut down.

Conducting an extensive investigation takes a considerable amount of time. The Center for Energy and Environment reported that each test conducted in Minneapolis for the Minneapolis/St. Paul Metropolitan Airports Commission requires 3 to 5 h, even though the procedure is automated, using a data logger that is interfaced to a computer.[23] Not all fire department, utility companies, or heating contractors have the time, training, or equipment to conduct the needed investigations. Greiner,[4] in his summary of 86 cases, found that repairs had been attempted before his arrival in 65 of the cases. Repairs had successfully corrected the CO problem in only 23 (35%) of the cases. In 42 of the 65 cases (65%) repairs had not been successful, and the CO problem remained.

Recognizing the poor success rate in initially finding and correcting CO problems, clients should be encouraged to install a CO alarm. By continuously monitoring for CO, residential CO alarms are one means of reducing the number and severity of CO poisonings and protecting against CO sources that might have been missed by investigators. Yoon et al.[54] estimated that CO detectors might have prevented approximately half of the CO deaths that occurred in New Mexico. Approximately 41% of the fatalities occurred in automobiles, and 42% had a blood-alcohol content above 0.01%, which Yoon believed would have hindered their effective response to a CO alarm.[54]

23.4 PREVENTIVE STRATEGIES FOR REDUCING UNINTENTIONAL CARBON MONOXIDE POISONING

Episodes of CO can be sporadic and hard to detect, so consumers and investigators must be persistent. Just because nothing is seen, tasted, or smelled does not mean there is no problem. The CO source or sources must be found and the reason CO remained in the house determined. If an alarm sounds, there is a reason and the alarm should not be ignored. Field studies indicate there are few "false" alarms. The reason for the alarm sounding must be determined. There should always be an operating CO alarm in any house that has experienced a previous sounding of an alarm. Nonintentional deaths and injuries from CO are preventable. Individuals can reduce their risks from CO by:

1. Installing CO alarms in homes and at work. The U.S. Consumer Product Safety Commission at present recommends that all homes have at least one CO alarm near each sleeping area in the home. Only detectors meeting the UL-2034 or IAS-96 requirements should be installed. In the first 3 months after Chicago passed an ordinance requiring CO alarms in single-family and multiunit dwellings, 68 individuals were transported by Chicago paramedics because of CO detection alarming.[42]

2. Having a qualified heating contractor perform yearly maintenance on heating appliances. To perform yearly maintenance, including CO testing, adequately, a qualified contractor will have electronic equipment to measure CO in flue gases, to measure and adjust gas flow, and will have attended CO training classes. It is recommended that heating systems be inspected either annually or every-other year depending on manufacturer's recommendations and the condition of the equipment. Most persons do not inspect them that often. In an Iowa survey, 32% said they had never had their heating system professionally checked and cleaned, and another 25% said they had their system checked every 3 to 5 years.[17]
3. Becoming informed about the dangers of CO. The public does not fully realize the dangers of CO. Yearly, persons die after running vehicles in closed garages, operating charcoal grills in campers, gasoline engines in buildings, and poorly maintained heating appliances in homes. A long-term preventive measure is to provide education to the public, heating contractors, utility contractors, medical personnel, manufacturers of heating and ventilating appliances, homebuilders, home remodelers, government agencies, policy makers, architects, and engineers. Mandatory licensing, minimum competency, and continuing education requirements for professionals are also important.
4. Suspect CO when flu-like symptoms appear. It has been estimated that as many as 30% of patients with significant CO poisoning will carry an erroneous initial diagnosis. Education and screening programs can reduce the erroneous diagnosis.[42] Immediate action to protect all building occupants when an alarm sounds is a primary step, followed by correction of the problem.
5. Purchasing high-efficiency, direct-vent heating appliances. Direct-vent appliances do not require combustion air from within the house, are less affected by house depressurization, and have additional safety features.

The best way to avoid CO poisoning is prevention. Proper selection and maintenance of carbon-based fuel–burning equipment can reduce CO emissions. Adequate venting of combustion products to the outdoors can prevent buildup of CO indoors. Instruments to measure CO in combustion products, in the atmosphere, and in exhaled breath are readily available. By using these instruments, CO emissions can be monitored and reduced, adequate warning given of unacceptable CO concentrations, and erroneous medical diagnosis avoided.

REFERENCES

1. Lascaratos, J.G. and Marketos, S.G., The carbon monoxide poisoning of two Byzantine emperors, *Clin. Toxicol.*, 36, 103–107, 1998.
2. U.S. Consumer Product Safety Commission, Non-fire-related carbon monoxide death and injury incident estimates, memorandum to Elizabeth W. Leland from Kimberly Long, dated August 3, 1995.

3. Cobb, N. and Etzel, R.A., Unintentional carbon monoxide-related deaths in the United States, 1979 Through 1988, *J. Am. Med. Assoc.*, 266, 659–663, 1991.

4. Greiner, T.H., Selected case studies cases 1994–1998, February 26, 1999, unpublished paper, Iowa State University Extension, Ames, IA, 1999.

5. Wilson, A.L., Colome, S.D., and Tian, Y., *California Residential Indoor Air Quality Study*, Vol. 1, *Methodology and Descriptive Statistics*, Integrated Environmental Services, Irvine, CA, 1993.

6. Wilson, A.L., Colome, S.D., and Tian, Y., *California Residential Indoor Air Quality Study*, Vol. 3, *Ancillary and Exploratory Analysis*, Integrated Environmental Services, Irvine, CA, 1995.

7. Colome, S.D., Wilson, A.L., and Tian, Y., *California Residential Indoor Air Quality Study*, Vol. 2, *Carbon Monoxide and Air Exchange Rate: A Univariate & Multivariate Analysis*, Integrated Environmental Services, Irvine, CA, 1994.

8. American National Standard Institute, Storage Water Heaters with Input Ratings of 75,000 BTUh or Less, Standard Z21.10.1, American Gas Association, Cleveland, OH, 1998.

9. American National Standard Institute, Gas-Fired Central Furnaces, Standard Z21.47, American Gas Association, Cleveland, OH, 1993.

10. American National Standard Institute, Household Cooking Gas Appliances, Standard Z21.1, American Gas Association, Cleveland, OH, 1990.

11. Greiner, T.H. and Schwab, C.V., Carbon monoxide exposure from a vehicle in a garage, in Proceedings, Thermal Envelopes of the Exterior Envelopes of Buildings VII, U.S. Department of Energy, Clearwater Beach, FL, 1998, 209–216.

12. MidAmerican Energy] Iowa Department of Public Health, MidAmerican Energy Carbon Monoxide Calls, Report, Iowa Department of Public Health, Des Moines, IA, 1998.

13. Minnegasco, Report on Undiagnosed Carbon Monoxide Complaints, prepared by Advanced Certified Thermography, Minnegasco, Minneapolis, MN, 1997.

14. Uniform Mechanical Code, Dwelling Construction under the Uniform Mechanical Code (UMC), International Conference of Building Officials, Whittier, CA, 1994.

15. Persily, A.K., Carbon Monoxide Dispersion in Residential Buildings: Literature Review and Technical Analysis, NISTIR 5906, National Institute of Standards and Technology, Gaithersburg, MD, 1996.

16. U.S. Department of Health and Human Services, Office of Public Health and Science, Healthy People 2010 Objectives: Draft for Public Comment, Environmental Health Chapter, September 15, 1998, 5–20.

17. Lasley, P., Home heating issues, Iowa Farm and Rural Life Poll, February, 1992. Iowa State University Extension, Ames, IA, 1992.

18. Iowa Association of Professional Fire Chiefs, Fire Service Institute, and Agricultural and Biosystems Engineering Extension (Greiner, T.H.), The Iowa Model Carbon Monoxide Response Protocol (for response to residential CO detectors), March, 1996. Iowa State University Extension, Ames, IA, 1996.

19. Canadian Mortgage and Housing Corporation, Residential Combustion Venting Failure: A Systems Approach, Summary Report, July 16, 1987, Prepared by Scanada Sheltair Consortium, 1987.

20. ASHRAE, Heating, Ventilating, and Air-Conditioning Systems and Equipment, American Society of Heating, Refrigerating and Air-Conditioning Engineers, Inc., Atlanta, GA, 1996.

21. Bohac, D.L. and Brown, T.H., Results from IAQ evaluations on cold climate single family houses undergoing sound insulation, in *Conference Proceedings, Healthy Buildings: Global Issues and Regional Solutions/IAQ '97*, American Society of Heating, Refrigerating and Air-Conditioning Engineers, Inc., Atlanta, GA, 1997.

22. Tsongas, G., Field monitoring of elevated CO production from residential gas ovens, in *Proceedings: Indoor Air Quality '94*, St. Louis, MO, American Society of Heating, Refrigerating, and Air Conditioning Engineers, Inc., Atlanta, GA, 1994.

23. Reuther, J.J., Critique of ANSI Z21.1 Standard for CO Emissions from Gas-Fired Ovens/Ranges, Report GRI-96/0270, September, 1996 for Gas Research Institute by Battelle, Columbus, OH, 1996.

24. Greiner, T.H., Comments concerning carbon monoxide emissions from gas-fired ovens and ranges: with special reference to Battelle's 1996 Critique of ANSI Z21.1, prepared for Affordable Comfort Conference, Chicago, IL, available from Iowa State University Extension, Ames, IA, 1998.

25. Karg, R., personal communication, May 9, 1998.

26. Drinker, C.K., *Carbon Monoxide Asphyxia*, Oxford Medical Publications, London, 1938.

27. Ayres, S.M., Evans, R., Licht, D., Griesbach, J., Reimold, F., Ferrand, E.F., and Criscitiello, A., Health effects of exposure to high concentrations of automotive exhaust: studies in bridge and tunnel workers in New York City, *Arch. Environ. Health*, 27, 168–177, 1973.

28. Nuti, M., *Emissions from Two-Stroke Engines*, Society of Automotive Engineers, Inc., Warrendale, PA, 1998, 99–100.

29. U.S. Environmental Protection Agency, Non-Road Engine and Vehicle Emission Study, Rep. 21A-2001, November, 1991, reproduced by the U.S. Department of Commerce, National Technical Information Service, Springfield, VA, 22161, 1992, 2.

30. Hampson, N.B. and Norkool, D.M. Carbon monoxide poisoning in children riding in the back of pickup trucks, *J. Am. Med. Assoc.*, 267, 538–540, 1992.

31. Luckhurst, D.G. and Solkoski, G.R., Carbon monoxide levels in indoor tractor-pull events — Manitoba, epidemiologic report, *Can. Med. Assoc. J.*, 143, 647–648, 1990.

32. NETT Technologies, Inc., Frequently ssked questions about LPG emissions, NETT Technologies, Inc., Mississauga, Ontario, Canada, May 31, 1997.

33. Long, D., Lift truck maintenance tests show cost savings potential, *Mater. Manage. Distrib.*, Sept. 1974, as quoted in Mahoney, D.P., Carbon monoxide exposure from fork truck exhaust, Professional Safety, American Society of Safety Engineers, September, 15–17, 1990.

34. Kwok, P.W., Evaluation and control of carbon monoxide exposure in indoor skating arenas, *Can. J. Public Health*, 74, 261–265, 1983.

35. Ni, L., Roncace, E.A., Borgenson, R.A., and Thrasher, W.H., Design Recommendations for Yellow Flame Burners, Gas Appliance Technology Center for Gas Research Institute, Chicago, IL, March 20, 1996.

36. DeWerth, D.W., Borgeson, R.A., and Aronov, M.A., Development of sizing guidelines for vent-free supplemental heating products: topical report. Prepared under Contract 5095-280-3406 for the Gas Appliance Manufacturers Association and Gas Research Institute, American Gas Association Research Division, AGARD, 8501 East Pleasant Valley Road, Cleveland, OH 44131, March, 1996.

37. Stewart, M.B., Jacob, P.R., and Winston, J.G., Analysis of infiltration by tracer gas techniques, pressurization tests, and infrared scans, in *Proceedings of the ASHRAE/DOE/ORNL Conference*, December 3–5, 1979, Florida, ASHRAE SP 28, 1979, 138–148.

38. National Institute for Occupational Safety and Health, Preventing Carbon Monoxide Poisoning from Small Gasoline-Powered Engines and Tools, NIOSH Alert, DHHS (NIOSH) Publication No. 96–118, 1996.

39. Honda, Honda Engines Owner's Manual, GC135-GC160, Honda Motor Company, Ltd., 1997.

40. Onan, Onan Engine Operator's Manual, E124V, Elite Floor Care, Onan Corporation, Minneapolis, MN, Manual 965-0174, 10-97, 1997.

41. Greiner, T.H., The case of the CO leak: solving the mysteries of carbon monoxide exposures, *Home Energy*, November/December, 21–28, 1997.

42. Leikin, J.B., Carbon monoxide detectors and emergency physicians, *Am. J. Emerg. Med.*, 14, 90–94, 1996.

43. U.S. Environmental Protection Agency, National Primary and Secondary Ambient Air Quality Standards, Code of Federal Regulations, Title 40 Part 50 (40 CFR 50), 1993.

44. ASHRAE, Ventilation for Acceptable Indoor Air Quality, ASHRAE 62-1989. American Society of Heating, Refrigerating, and Air-Conditioning Engineers, Inc., Atlanta, GA, 1989.

45. OSHA, Title 29 (29 CFR 1910.1000) Occupational Safety and Health Administration, U.S. Department of Labor, Code of Federal Regulations, Washington, D.C., 1991.

46. NIOSH, Criteria for a Recommended Standard: Occupational Exposure to Carbon Monoxide, National Institute for Occupational Safety and Health, U.S. Department of Health, Education, and Welfare, Washington, D.C., 1972.

47. Ellenhorn, M.J. and Barceloux, D.G., *Medical Toxicology: Diagnosis and Treatment of Human Poisoning*, Elsevier, New York, 1988.

48. City of Chicago, Amendment of Title 13, Chapter 64 of Municipal Code of Chicago by Addition of New Sections 190 through 300 Requiring Carbon Monoxide Detectors in Various Buildings, Passed by the City Council of the City of Chicago in Regular Meeting, March 2, 1994.

49. Building Performance Institute, Carbon Monoxide Analyst Protocol, Building Performance Institute, Inc., 404 8th Ave., Suite 1801, New York, NY 10018, 1998.

50. Greiner, T.H., Checking for Complete Combustion, AEN-175, Iowa State University Extension, Department of Agricultural and Biosystems Engineering, Ames, IA, 1997.

51. Canadian Mortgage and Housing Corporation, Residential Combustion Venting Failure: A Systems Approach, Final Technical Report, Project 3: Task 1, Refinements to the Chimney Safety Tests: Determining House Depressurization Limits, Canadian Mortgage and Housing Corp., Ontario, Canada, January, 1987.

52. U.S. EPA, Radon Reduction Techniques for Existing Detached Houses: Technical Guidance, 3rd. ed., Section 11.5, Procedures for Checking Combustion Appliance Backdrafting, U.S. Environmental Protection Agency, EPA/625R-93/011, October, 1993, 1993, 265–269.

53. BPI, Carbon Monoxide Analyst Protocol — Draft, Building Performance Institute, Inc., New York, 1998.

54. Yoon, S.S., Macdonald, S.C., and Parrish, R.G., Deaths from unintentional carbon monoxide poisoning and potential for prevention with carbon monoxide detectors, *J. Am. Med. Assoc.*, 279, 685–687, 1998.

Index

A

Abdominal discomfort, 401
Abdominal pain, 405
ABO isoimmune diseases, 43
Abortion, 365
ABS, *see* Australian Bureau of Statistics
Acetic acid vapor, beneficial effect of, 5
Acetonitrile, 179
N-Acetylcysteine (NAC), 270
Acetylene gas, 244
Achievement test, Peabody individual, 455
Acid gases, 162
Acidosis, 113, 168
Acrolein, 177, 182
Acrylic fabric, fire-retarded, 183
Acrylonitrile, 179
ACT, *see* Australian Capital Territory
Adenosine phosphate (ATP), 373
Adolescent poisoning deaths, 463
Adriamycin, 481
Aggression, 412
Agitation, 470
Air
    bags, error made in designing, 509
    heat of combustion, 498
    quality monitoring sites, not meeting CO
        NAAQS, 96
AIRS, *see* EPA Aerometric Information Retrieval
        System
Alarm
    detector vs., 64
    trigger, 77
Alcohol
    consumption, 223, 225, 427
    poisoning, 2
    risk for CO exposure and, 198
Alcoholism, 247
Alkalosis, preexisting, 145
Allopurinol, 270, 279, 302
Alpha-adrenoceptors, selective blockade of, 109
Altitude, interacting effects of CO and, 135–153
    carboxyhemoglobin at altitude, 148–151
    cardiovascular effects of CO at altitude,
        143–145
    CO sources at altitude, 140–143
    compartment shifts of CO at altitude, 146–147
    effects of ascent to altitude, 138–140

effects of CO exposure, 140
endogenous production of CO, 148
measuring CO at altitude, 151–153
    conversion of units, 153
    gravimetric measurements, 152–153
    volumetric measurements, 151–152
oxygen transport, 137–138
vision effects of CO at altitude, 145–146
Alveolar ventilation, 149
American Lung Association, 242
American Medical Association, 242
Amnesia, 364
Amperometric sensor, 72
Anemia, 88, 119, 245
Anesthesia, 29
Anesthetic mask, 340
Angina, 88
    effects of CO exposure and altitude on time
        to, 144
    pectoris, 144
    time to onset of, 125
Angioparalytic neurasthenia, 236
Anxiety, 410, 412
    disorders, 243
    generalized, 444
Aortic blood acceleration, 106
Apoptosis, 7
Areflexia, 276
L-Arginine, 33
Arrhythmias, 170, 275
Arthritis, 412
Ascorbic acid, 263
Asphyxia, 173, 180
Asphyxiant
    dose, 182
    gases, 185, 189
Ataxia, 410
Atmospheric pressure, 135
ATP, *see* Adenosine phosphate
Atrial fibrillation, 364
Australian Bureau of Statistics (ABS), 208
Australian Capital Territory (ACT), 219
Automobile(s)
    air bags, 509
    CO emissions from, 136, 141
    emissions control
        milestones in, 94
        technologies, 97

543

G

H